Hydrogen Storage

SCIENCES

Energy, Field Directors –
Alain Dollet and Pascal Brault

Production, Conversion and Storage of Hydrogen,
Subject Heads – Olivier Joubert and Daniel Hissel

Hydrogen Storage

A Wide Range of Solutions

Coordinated by
Patricia de Rango
Fermin Cuevas

iSTE

WILEY

First published 2025 in Great Britain and the United States by ISTE Ltd and John Wiley & Sons, Inc.

ISTE Ltd
27-37 St George's Road
London SW19 4EU
UK

www.iste.co.uk

John Wiley & Sons, Inc.
111 River Street
Hoboken, NJ 07030
USA

www.wiley.com

Library of Congress Control Number: 2025937892

British Library Cataloguing-in-Publication Data
A CIP record for this book is available from the British Library
ISBN 978-1-78945-223-5

ERC code:
PE8 Products and Processes Engineering
PE8_6 Energy processes engineering
PE8_11 Environmental engineering, e.g. sustainable design, waste and water treatment, recycling, regeneration or recovery of compounds, carbon capture & storage

The manufacturer's authorized representative according to the EU General Product Safety Regulation is Wiley-VCH GmbH, Boschstr. 12, 69469 Weinheim, Germany, e-mail: Product_Safety@wiley.com.

Contents

Preface

Patricia DE RANGO[1] and Fermin CUEVAS[2,3]
[1]*Institut Néel, CNRS, Université Grenoble Alpes, France*
[2]*ICMPE, CNRS, UPEC, Thiais, France*
[3]*CNRS, Beijing, China*

As one of the energy vectors of the future, hydrogen is destined to play a major role in the energy transition. It is used in a wide range of fields, including transport, the decarbonization of industry and in managing the intermittency of renewable energies. As a central link in the hydrogen value chain, however, storing hydrogen downstream of its decarbonized production and upstream of its use in fuel cells or internal combustion engines is a real challenge in terms of both space and safety.

This book presents various storage solutions currently under study, with the goal of combining high mass and volume capacities, high energy efficiency, safety, social acceptability and competitive pricing. The book is intended for master's level students, engineers and researchers working in the field of hydrogen, as well as, on a more general level, scientists and decision-makers who wish to have access to fundamental knowledge and the state of the art in hydrogen storage.

A wide range of solutions are available, including hyperbaric storage, geological storage, liquid storage in the form of new hydrogen carriers such as ammonia and liquid organic hydrogen carriers (LOHCs), solid storage in metal hydrides, ionic and covalent hydrides and porous materials with a high specific surface area. Cryogenic storage at very low temperatures (20 K), the use of which is currently restricted to the aerospace sector, is not covered.

The operating conditions in terms of pressure and temperature, as well as storage capacities, differ for each of the proposed methods. The storage method ultimately chosen for a given application, whether for mobility, stationary storage or hydrogen distribution, is highly dependent on operational constraints and the actual cost of the storage system. Today, high-pressure gas storage (350 and 700 bar) is a mature technology for on-board hydrogen storage in light vehicles. For much larger quantities of hydrogen, underground geological storage of this gas could be the solution of choice for large-scale storage of hydrogen produced from renewable energies. The long-distance transportation of hydrogen in its liquid state at ambient temperature can be facilitated by the use of LOHCs or ammonia. Owing to its compactness, safety and near-ambient operating conditions, solid hydrogen storage in metal hydrides, including high-entropy alloys, is a viable solution for heavy mobility and stationary applications. Ionically or covalently bonded hydrides, which are formed from very light elements and release hydrogen at moderate temperatures and pressures, play a role in light mobility. Finally, porous materials with high specific surface areas, such as carbonaceous materials and metal–organic frameworks (MOFs), with high mass capacities and competitive costs are being considered for hydrogen compression and transport. In short, this book describes all the principles and operating conditions for each of these hydrogen storage methods.

The writing of this book was coordinated by Patricia de Rango (Institut Néel, UPR 2940) and Fermin Cuevas (ICMPE, UMR 7182), heads of the storage axis of the CNRS hydrogen research federation, FRH2. We would like to thank the authors of the various chapters for their valuable contributions to this book.

June 2025

1

Hyperbaric Storage

David CHAPELLE[1], Damien HALM[2] and Stéphane VILLALONGA[3]

[1]*Institut FEMTO-ST, CNRS, Université de Franche-Comté, Besançon, France*

[2]*Institut Pprime, CNRS, ISAE-ENSMA, Poitiers, France*

[3]*CEA Le Ripault, Monts, France*

1.1. Compressed hydrogen

Dihydrogen gas (hereafter often reduced to hydrogen), as an energy carrier or even as an energy store, is a central element in many scenarios for our future global energy mix (Espegren et al. 2021; Capurso et al. 2022; Kouckaki-Pencha et al. 2023). Key arguments include its abundance, generally in complex forms, or its high gravimetric energy content of 140 MJ/kg (but its low volumetric content of 10.8 MJ/m^3 under normal pressure and temperature conditions) associated with two possible uses: direct combustion or coupling to a fuel cell for electricity generation, or even its potential low environmental impact.

In this chapter, we will discuss the most mature way of storing hydrogen gas, namely, a solution consisting of reducing the volume occupied by dihydrogen molecules in their gaseous form by increasing pressure and confining them in a container. This solution, known as compressed hydrogen, hyperbaric hydrogen storage or high-pressure storage, is particularly interesting if the volume can be reduced sufficiently to achieve a significant gain in volumetric energy density, bearing in mind the overall energy balance. As a preamble to our considerations of these storage solutions, we offer a few general points about the element hydrogen and its gaseous form.

Hydrogen Storage,
coordinated by Patricia DE RANGO and Fermin CUEVAS.

1.1.1. *Overview of H_2*

Hydrogen is the most abundant element in the universe (75% by mass and 92% by number of atoms). On our planet, dihydrogen is not very abundant, accounting for just 0.55 ppm of atmospheric gases. The element's most common source is water, and it is also present in abundance in all organic matter (the hydrogen atom accounts for 63% of the atoms in the human body).

Table 1.1 lists the main characteristics of hydrogen.

Characteristics	Values	Units
Density of liquid hydrogen at 10 K	70.6	gH_2/L
Atomic mass	1.0079	U
Molar mass of dihydrogen	2.0159	g/mol
Hydrogen gas density at 20.3 K	1.34	gH_2/L
Hydrogen density at 273 K	0.0899	gH_2/L
Lower calorific value (LCV)	10.8	kJ/L
Gross calorific value (GCV)	12.8	kJ/L
Specific heat at constant pressure	14.3	kJ/kg.K
Specific heat at constant volume	10.3	kJ/kg.K
Autoignition temperature in air	858	K
Flame temperature in air	2,318	K
Flammability limits in air	4–75	% vol
Ignition energy	0.020	mJ
Gas thermal conductivity	0.1897	W/m.K

Table 1.1. *Main characteristics of hydrogen at atmospheric pressure and 293 K (unless otherwise stated)*

Apart from these characteristics, it is also essential to understand or remember that, for a real gas, the number of moles per liter, owing to interactions between molecules, is not a linear function of pressure. It is generally accepted that dihydrogen follows the law of perfect gases up to 100 bar. Beyond this, it is imperative to consider the deviation from this law; thus, the density of H_2 is approximately 24 g/L at 350 bar and not 30 g/L and is nearly 40 g/L at 700 bar and not 56 g/L (Figure 1.1).

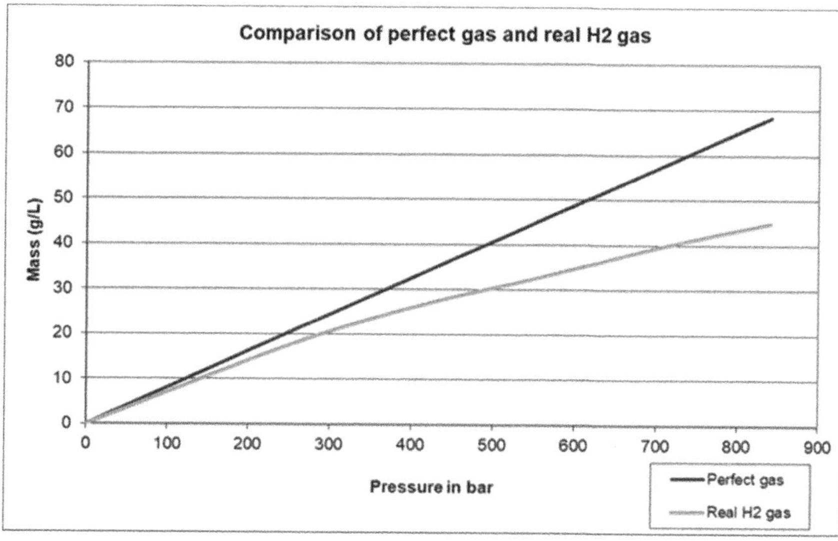

Figure 1.1. *Variation in the mass of one liter of dihydrogen as a function of pressure, compared with the variation for a perfect gas. For a color version of this figure, see www.iste.co.uk/derango-cuevas/hydrogen.zip*

Thus, the engineer in charge of developing a hydrogen tank has to find the optimum balance between the operating pressure, which increases the volumetric energy density, and the mechanical strength of the structure, which decreases the gravimetric energy density (increase in mass), bearing in mind that as pressure increases, the relative gain in energy density is reduced. There are many other factors to consider, such as economic, societal and environmental constraints, which will be discussed later in this chapter.

Table 1.2 provides the conversion factors for common hydrogen units.

	MJ (LCV)	Nm^3	GGE[1]	kWh
1 kg H_2	120	11.112	1	33.33

Table 1.2. *Conversion of common H_2 units at atmospheric pressure and 293 K*

Table 1.3 compares the energy densities of conventional fuels. The gravimetric energy density of dihydrogen is well above that of conventional fuels, but its use

1 GGE: gallon gasoline equivalent.

requires a significant reduction in the space occupied, especially if mobile applications are to be considered. It is interesting to draw a parallel with information on solid hydrogen storage. On the basis of 2% storage by mass in a material with a density of 5 kg/L, we obtain a low gravimetric energy density of 2.4 MJ/kg for this storage method and, on the other hand, the most remarkable volumetric energy density of 12 MJ/L (of solid).

Volumetric and gravimetric energy densities of common fuels.

Material	Energy per kilogram (MJ kg^{-1})	Energy per liter (MJ l^{-1})
Hydrogen (liquid)	143	10.1
Hydrogen (compressed, 700 bar)	143	5.6
Hydrogen (ambient pressure)	143	0.0107
Methane (ambient pressure)	55.6	0.0378
Natural gas (liquid)	53.6	22.2
Natural gas (compressed, 250 bar)	53.6	9
Natural gas	53.6	0.0364
LPG propane	49.6	25.3
LPG butane	49.1	27.7
Gasoline (petrol)	46.4	34.2
Biodiesel oil	42.2	33
Diesel	45.4	34.6

Table 1.3. *Gravimetric and volumetric energy densities for common fuels (Mazloomi and Gomes 2012)*

Worldwide, hydrogen production from 2000–2010 (Ewan and Allen 2005; Ball and Wietschel 2009) was close to 50 million tonnes per year, or nearly 600 billion normal cubic meters[2]. Most of this production (Figure 1.2) comes from fossil fuels, with almost half from steam reforming of natural gas and a third from refining a portion of crude oil. Other sources include coal gasification (nearly 20%) and water electrolysis (4%). Since 2021, this production has reached almost 100 million tonnes per year, but the share of steam reforming has only increased, exceeding 60% of the total contribution (IEA 2022). This dihydrogen, when it is not simply a waste product, is produced on site for internal consumption. It is used almost exclusively as a reagent in the chemical and oil industries: ammonia synthesis (50%),

2 The normal cubic meter, Nm^3, corresponds to the volume occupied by a gas under normal pressure and temperature conditions. In France, the reference (DIN1343) is 0°C for a pressure of 1 atm.

hydrocarbon refining and desulfurization (37%), and methanol synthesis (12%). Only 1% of production is exploited for its energy potential in aerospace and spacecraft propulsion.

Using hydrogen as an energy carrier requires us to consider the hydrogen energy sector in its entirety, meaning all the structures and infrastructures inherent in its operation. Earlier, we discussed hydrogen production and noted that over 95% of the hydrogen produced in 2005 came from fossil fuels. The hydrogen industry, and the economic effort it requires, should only be supported if it brings progress to our civilizations in terms of sustainable development by producing renewable energy, limiting greenhouse gas emissions and promoting access to energy for as many people as possible.

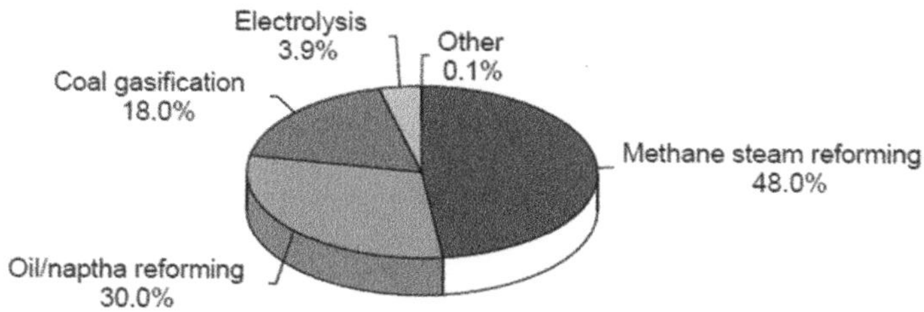

Figure 1.2. *Breakdown of primary energy sources used for hydrogen production (Ewan and Allen 2005)*

Figure 1.3 shows the impact of hydrogen production, depending on the energy source, on CO_2 production and the cost per kilogram of hydrogen produced. A number of observations can be made:

– The current means of producing hydrogen are also the means that emit the most CO_2.

– The cost of producing hydrogen from so-called renewable sources is two to three times greater than that of current solutions, which should be put into perspective by the significant drop in the cost of installing photovoltaic panels over the last 10 years.

– Carbon capture and storage can reduce emissions by a factor of 4, while the cost of hydrogen can be multiplied by 2.

In the absence of significant progress on so-called "renewable" sources, and even with higher CO_2 emissions, current means of hydrogen production still have a bright future ahead of them, especially as they will also benefit from progress in CO_2 capture and storage.

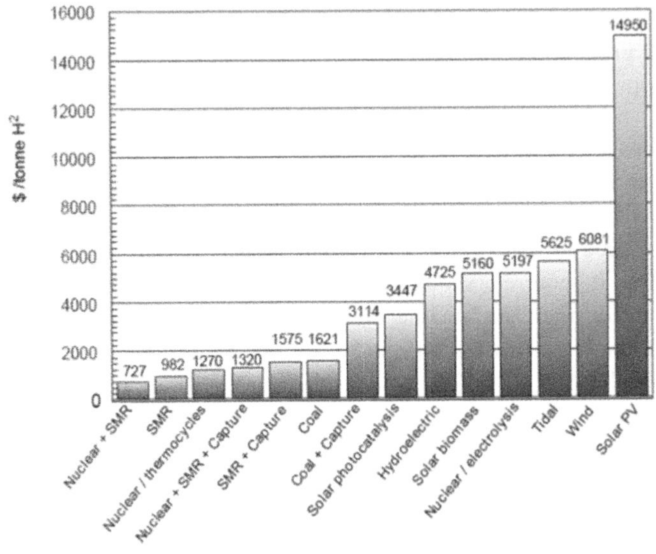

a)

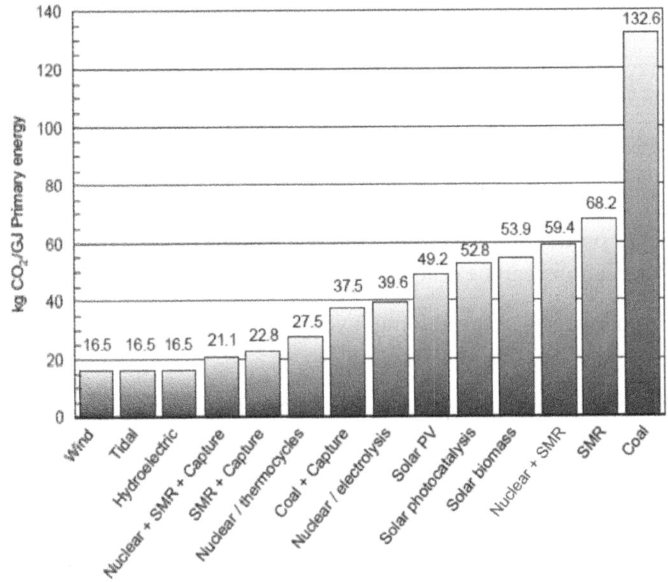

b)

Figure 1.3. *Impacts of the choice of H_2 production source (Ewan and Allen 2005) on (a) CO_2 production per conversion of 1 GJ of primary energy and (b) the cost per tonne produced*

As discussed in the regulatory and standards section below and as demonstrated by the efforts of automobile manufacturers to develop hydrogen-powered car concepts and by the efforts of manufacturers of components such as high-pressure couplings for filling stations, real-world consumer applications are gradually emerging. To illustrate this, Table 1.4 shows a comparison of costs, fuel consumption and range (which has since increased significantly for both electric and hydrogen-powered vehicles), depending on the type of vehicle propulsion.

Type of car	Fuel	Price (thousands US$)	Fuel consumption[a] (MJ per 100 km)	Fuel price (US$ per 100 km)	Driving range (km)
Conventional	Gasoline	15.3	236.8[b]	2.94	540
Hybrid	Gasoline	20.0	137.6	1.71	930
Electric	Electricity	42.0	67.2	0.901	164
Fuel cell	Hydrogen	100.0	129.5	1.69	355

[a] Fuel consumption based on 45% highway and 55% city driving.
[b] Heat content of conventional gasoline is assumed to be its lower heating value (LHV), fixed at 32 MJ l^{-1}.

Table 1.4. *Economic comparison of four automotive technologies (Granovskii et al. 2006)*

1.1.2. *Compression*

As previously specified, the volumetric energy density of hydrogen is very low at atmospheric pressure, especially compared with that of more widely used energy sources such as petroleum, for which this density reaches 34,000 MJ/m^3. A number of techniques are available to industry to increase the volumetric energy density of hydrogen: compression, liquefaction, “solid” storage, etc. Although not the cheapest, the most widely used technology is the hydrogen compression method (Amos 1999).

The two main orders of magnitude of energy costs classically evoked for increasing energy density are as follows: 15% of stored energy for compressed storage and 30% for cryogenic storage.

In the following, we list the main hydrogen compression technologies on the basis of the review of Sdanghi et al. (2019). Figure 1.4 illustrates these technologies classified into two main families: mechanical compression and non-mechanical compression.

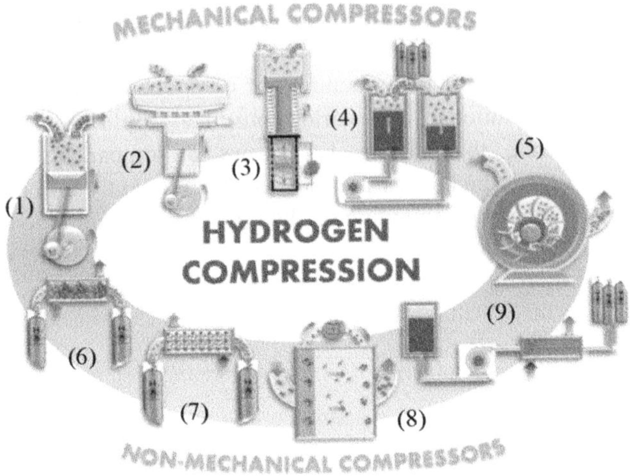

Figure 1.4. *Illustration of hydrogen compression technologies (Sdanghi et al. 2019). For a color version of this figure, see www.iste.co.uk/derango-cuevas/hydrogen.zip*

The mechanical compression solutions include the following:

– conventional piston engines (1);

– diaphragm solutions (2);

– linear compressors, in which the piston is coupled to a magnetic motor (3);

– liquid pistons (4);

– liquid-driven rotary solutions (5).

The non-mechanical compression solutions include the following:

– adsorption on high surface area carbon materials or metal organic frameworks (6);

– reversible metal hydride formation (7);

– electrochemical compression (8);

– cryogenic compression (9).

Having established this point about the various technologies that can be used to reduce the gas volume and thus increase the volumetric energy density, we will discuss the tank envelope and mechanical considerations using a simplified approach.

1.1.3. *Mechanical tank geometry*

Section 1.2 describes the four, or rather five, main families of compressed gas storage tanks. The geometry of these structures is well known: they are elongated cylinders with a cylindrical cross-section; in other words, cylinders are closed at the bases by spherical or hemispherical domes.

The structures produced in this way are designed to contain the gas, resist the useful pressure mechanically, operate over the recommended service life or number of cycles and guarantee the integrity of the envelope up to a pressure beyond which rupture is likely to occur, thus defining the safety coefficient used and specified by regulations.

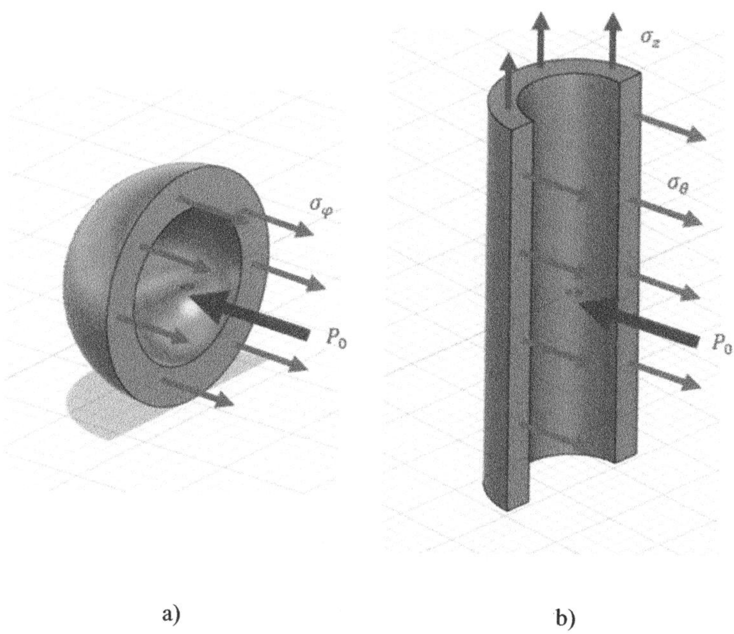

Figure 1.5. *Schematic sections of (a) a spherical tank and (b) the cylindrical part of a cylindrical tank. Representation of orthoradial ($\sigma_\varphi, \sigma_\theta$) or axial ($\sigma_z$) stresses as resultants of pressure P_0. For a color version of this figure, see www.iste.co.uk/derango-cuevas/hydrogen.zip*

The pressure is a force per unit area exerted in the direction normal to that area. Intuitively, the tank with the best mechanical strength is spherical for the same stored volume and pressure. On the inner surface of our sphere, the pressure exerts itself homogeneously and generates equal stresses at every point. This solution, although well developed, is only used on rare occasions and for rather obvious reasons: the external dimensions of this tank approach those of a cube with the diameter of the sphere as its side, and its stability and arrangement with other tanks are penalizing. Figure 1.5 schematically shows the distribution of forces and stresses on a cross-section of a sphere and a cylinder. Equations [1.1]–[1.6] establish the relationships between forces and stresses and justify that, given a fixed thickness e, inner radius R and pressure, the theoretical stress in the thin wall of a cylindrical tank is twice as high as that in the sphere and the spherical dome.

For the sphere, in a simplified way here, we can write the resultant force on the cross-sectional surface as a function of the pressure but also of the orthoradial stress if the thickness is small compared with the radius:

$$F = P_0 \times \pi \times R^2 \approx 2 \times \sigma_\varphi \times \pi \times R \times e \qquad [1.1]$$

This makes it possible to express the orthoradial stress as a function of pressure:

$$\sigma_\varphi \approx \frac{R \times P_0}{2\,e} \qquad [1.2]$$

For the cylindrical part, we obtain the following expression in the direction perpendicular to the cylinder axis of length L:

$$F = P_0 \times L \times 2 \times R \approx 2 \times \sigma_\theta \times L \times e \qquad [1.3]$$

This makes it possible to express the orthoradial stress as a function of pressure:

$$\sigma_\theta \approx \frac{R \times P_0}{e} \qquad [1.4]$$

However, for the axial component, we find:

$$F = P_0 \times \pi \times R^2 \approx 2 \times \sigma_z \times \pi \times R \times e \qquad [1.5]$$

This means:

$$\sigma_z \approx \frac{R \times P_0}{2\,e} \qquad [1.6]$$

The preferred solution for increasing the gravimetric capacity of compressed gas storage solutions is to use composite materials and, more specifically, the filament winding process (detailed in section 1.3), thereby reducing the amount of metal, the thickness of which should increase as the working pressure increases. The specificity of a long fiber is that mechanically, it works almost exclusively in the long direction and that one of the essential parameters of the process is therefore the angle at which the fiber is deposited in relation to the axis of the tank.

In the particular case of a spherical tank, or for the domes of a cylindrical tank, it is best to position the fibers so that they lie in planes passing through the origin of the sphere.

For a cylindrical tank, the preceding considerations on the state of stress enable us to establish the theoretical optimum angle α of the fiber, guaranteeing the greatest mechanical efficiency. For this purpose, we consider the volume element of the cylindrical wall, as depicted in Figure 1.6, and neglect the radial stresses.

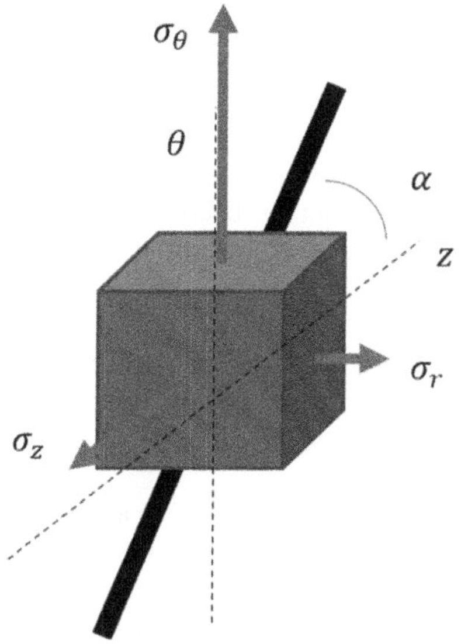

Figure 1.6. *Schematic representation of a volume element of the cylindrical part of the tank with axes z and θ, stresses σ_r, σ_θ, σ_z and the reinforcing fiber oriented at angle α to the z axis*

The stress in the fiber is denoted as σ_α. σ_z is the projection of σ_α onto the longitudinal axis z, and σ_θ is the projection of σ_α onto the longitudinal axis θ. Here, we must project the vector and the surface, which leads us to write:

$$\sigma_z = \sigma_\alpha \times \cos^2\alpha \quad [1.7]$$

$$\sigma_\theta = \sigma_\alpha \times \sin^2\alpha \quad [1.8]$$

The pressure on the closed tube generates an internal pressure with a close-end effect, and the theoretical ideal orientation of the reinforcing fiber is such that, according to [1.7] and [1.8]:

$$\tan^2\alpha = 2, \text{or } \alpha \approx 54° \qquad [1.9]$$

1.1.4. *Regulations and standards*

First, let us clarify a few essential points that will prevent the neophyte from making common mistakes. When the designer wants to confirm that his design is relevant, he refers to the recommendations established by his peers over the years, which can be found in the standards drawn up for this purpose. On the other hand, when a manufacturer wishes to market a product, it is up to him to guarantee that his product complies with the rules in force in the target market. Our directives and other regulations define the conditions for marketing and placing them on the market and specify and refer to design standards. Certification is the stage in which a certifying body checks that a product has been designed in accordance with the rules and regulations in force and issues a certificate or approval allowing it to be placed on the market.

In 2009, INERIS presented an overview of French regulations and European standardization on hydrogen energy carriers in its report DRA-09-101596-02693A (Tigreat 2009).

In France, for example, the France Hydrogène association has set up Vig'Hy, the hydrogen observatory, whose mission is to disseminate information to a wide audience. In particular, this observatory offers practical fact sheets on the regulations applicable to hydrogen in Europe. INERIS, with the support of ADEME and France Hydrogène, has published a guide for the certification of hydrogen systems (Debray and Weinberger 2021) designed to help operators in the hydrogen sector.

Readers note that the first step is to identify the intended application prior to identifying the procedure and steps involved in certifying a hydrogen tank. A distinction must be made between:

– pressure equipment devices (PEDs), for which reference should be made to Directive 97/23/EC of May 29, 1997 (2014/68/EU);

– transportable pressure equipment devices (TPEDs), for which the directive in force is no. 2010/35/EU of 16/06/10;

– equipment intended to be incorporated into an approved vehicle is covered by UN Regulation no. 134 (R134), which presents uniform provisions concerning the approval of motor vehicles and their components, particularly the safety requirements for hydrogen-powered vehicles (previous regulations: 2019-2144 and 2021/535, which repealed regulations (EC) nos. 79/2009 and 406-2010).

Note that, beginning in September 2020, Regulation 2018/858 established a general framework for type approval and placed the market of vehicles in categories M (passenger transport) and N (goods transport) within the union. Among the 80 areas of regulation, area 62 concerns hydrogen systems. R134 enables worldwide harmonization of type-approval rules for hydrogen-powered vehicles.

As far as standards are concerned, there are many standards that address pressurized gas tanks one way or another. There are even more cases in which compressed hydrogen is stored for mobility. For example, the ISO11114 series of international standards focuses on the compatibility of cylindrical and valve materials with gas contents for transportable gas cylinders (part 1: metallic materials; part 2: non-metallic materials; part 3: autoignition tests for non-metallic materials in an oxygen atmosphere, etc.).

However, we must mention here two essential standards for the use of pressure tanks:

– the French and European standard EN12245, transportable gas cylinders, fully wound cylinders, in composite materials, which aims to "provide a specification for the design, manufacture, inspection and testing of transportable and refillable fully wound composite gas cylinders". In particular, the procedures for 19 cylinder qualification tests, including a liner rupture test at ambient temperature, a hydraulic test of finished cylinders at ambient temperature, a cylinder rupture test and a repeated pressure test, are listed;

– the international standard ISO-15896.3, hydrogen gas and hydrogen gas mixtures – fuel tanks for land vehicles, "specifies the requirements for lightweight refillable fuel tanks intended for on-board storage of high-pressure compressed hydrogen gas or hydrogen mixtures" and applies to tank types 1–4 (Figure 1.7).

1.2. Design and modeling

Despite the undeniable qualities of hydrogen as an energy carrier, its transport and storage involve a number of challenges. To structure tanks to withstand high pressures (taking safety factors into account) without excessive storage system mass, high-performance composite materials are used. Figure 1.7 shows various storage structures, from Type I (all metals) to Type IV (polymer liner and composite shell) (Li et al. 2019; Rivard et al. 2019). The most recent developments are aimed at designing linerless tanks (Type V). This configuration, with no internal polymer coating separated from the hull and made entirely of composite, promises a more efficient use of volume. Moreover, manufacturing costs, operational risks and maintenance costs are reduced because of its inherently simpler construction. Solutions being considered today include the development of microcrack-resistant structural resins and/or barrier coatings (Hübner et al. 2021). This chapter will restrict itself to Type IV, which is currently the most widespread type in fuel cell vehicles (Ryu et al. 2020). The reader is referred to Figure 1.21 to identify all the components of a Type IV tank.

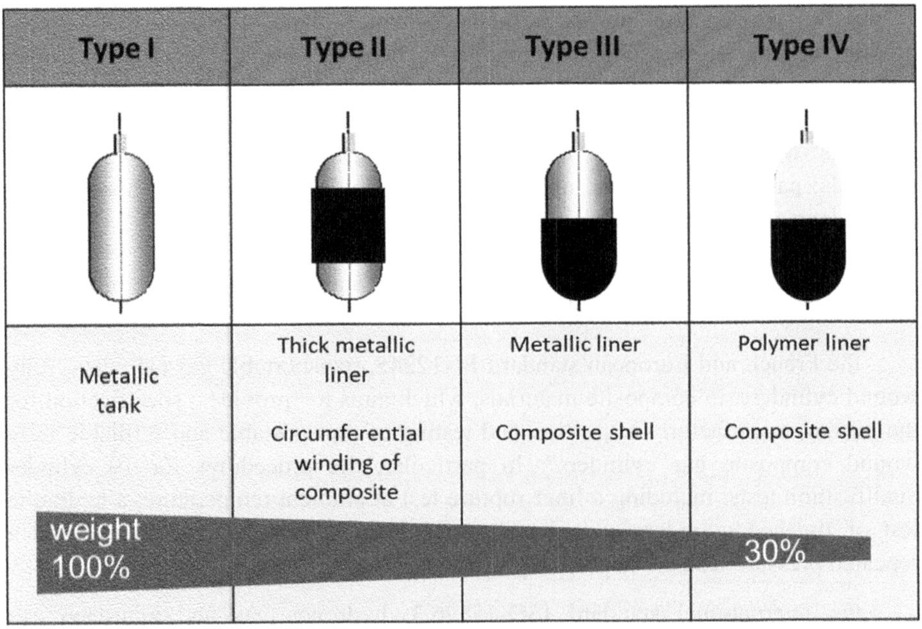

Figure 1.7. *Different types of hydrogen gas storage tanks. For a color version of this figure, see www.iste.co.uk/derango-cuevas/hydrogen.zip*

The pressures involved lead to thicker composite materials, to the detriment of the tank's price and weight. Depending on the carbon fiber used (T1000: high-performance fiber; T700: low-cost fiber), between 45% and 80% of the cost of the tank is accounted for by materials and, more specifically, by the cost of the composite shell. Tank design generally calls for the use of more or less advanced calculation codes using finite element (FE) techniques for the design of the fibrous architecture. The design and validation of this type of structure requires the use of numerical simulation tools capable of dealing with a number of difficulties. In particular, the following elements are needed:

– *An accurate representation of the tank's geometry*: the trajectory of the fibers on the tank is a non-trivial problem. While netting analysis (Peters et al. 1991; Evans and Gibson 2002; Daghia et al. 2020) can be used to optimize the orientation of the fibers in the shell (cylindrical part of the tank), their position on the dome and the thickness of composite deposited call for the resolution of more complex equations (to be coupled with the anisotropic behavior of the material), particularly in the case of non-geodesic windings (Koussios 2004; Vargas Rojas et al. 2014).

– *An adapted damage model coupled with a burst criterion*: even if, intuitively, tank failure is driven by fiber rupture, it is necessary to consider all the damage mechanisms (microcracking of the composite resin, delamination and rupture of fiber/matrix interfaces) occurring in the structure. Macroscopic criteria (Tsai and Azzi 1965; Tsai and Wu 1971) seem appropriate for large, complex structures. However, calculations based solely on elasticity, which do not distinguish between various damage modes, may prove too conservative. As countermeasures, numerous damage models for composite structures are available. For example, the most sophisticated methods are based on the microscale (Lubineau and Ladevèze 2002; Abisset et al. 2011) but are limited by their complexity and computational time cost.

– *An optimization procedure*: the value of behavior laws that are more reliable than a simple failure criterion is linked to the optimization phase, which must consider the various damage mechanisms. It is preferable to design a structure in such a way as to avoid damage rather than to observe the initiation and evolution of damage. However, an automated optimization technique integrating all laminate parameters (from geometry to damage model) poses numerical difficulties for this type of structure because of the presence of strong nonlinearities.

In addition to these numerical tools, the implementation of a storage tank design and optimization chain must include a validation phase. To this end, it is necessary to have a reference structure with a controlled geometry, which will have undergone a multi-instrument test campaign to better understand the damage mechanisms and build a database against which to compare the simulations resulting from the design phase. Following this principle, the developments presented in this chapter are based on a small-scale tank manufactured by the CEA as part of the ANR OSIRHYS IV project (Villalonga et al. 2015). It consists of a Polyamide 6 (PA6) liner. The metal bosses (end fittings that connect the tank to the distribution or supply network) are made of 316L stainless steel. The material used to manufacture the composite shell is a carbon fiber/epoxy matrix prepreg (TCR T700SC/UF3369). The capacity is 2 L. The overall length is 320 mm, with an external diameter of 114 mm (Figure 1.8).

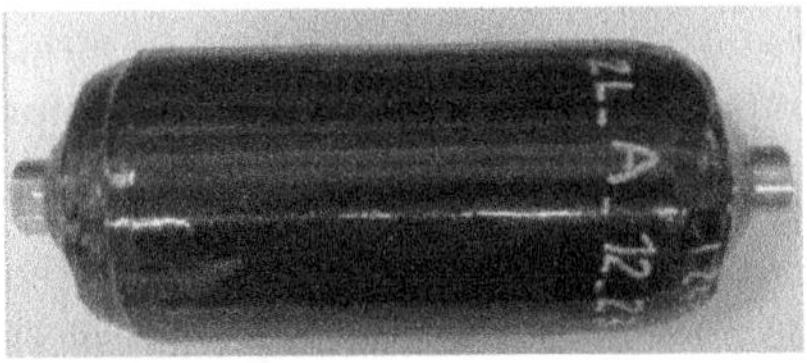

Figure 1.8. *Reference tank (CEA)*

The architecture of the composite shell consists of 24 layers (eight circumferential layers and 16 helical layers). A layer is defined as a set of adjacent plies (the basic structuring element) of composite material of identical orientation. Helical layers are deposited from the lowest to the highest angle. Circumferential layers are interposed to compact the helical layers on the cylinder.

OSIRHYS IV tank liners are manufactured via reactive rotational molding. The rotomolding machine used has two zones: a heating zone consisting of an electric oven and a polymerization and cooling zone. The mold is shaped like a 2L liner. The composite shell of the tanks is produced by the filament winding directly onto the liner. The liner is pressurized to 0.5 bar during this phase. The width of the winding strip is approximately 7 mm. A tension of 40 N is applied to the fiber during winding. After the samples were closed, the mold was rotated in the oven at 40°C for 10 min and then at 85°C for 15 min. At the end of the cycle, the mold is cooled. Second, after cooling, the mold containing the rotomolded liner was moved back into the oven at 80°C to undergo postcuring for 3 h. After cooling, the liner is removed from the mold, and the sprues are removed. The mixing quantity is set to obtain a thickness of 4 mm.

The tank is then polymerized in an oven under pressure. It is also rotated during the polymerization cycle. This cycle is chosen according to the materials used and the recommendations of the prepreg composite supplier.

The boss-liner assembly is weighed before winding. The finished tank is weighed and measured after polymerization. This enables us to calculate the mass of the composite deposited during the manufacturing of the tank shell. The average mass of the composite is 1750 g. In this chapter, the pressurization and bursting tests are detailed on the basis of the results obtained from this reference tank. Numerical tools designed to reproduce the geometry and simulate the damage sustained by the tank are also introduced. These tools are the entry points for an optimization approach, the principles of which are outlined in the final section.

1.2.1. *Tank testing*

The international hydrogen community has developed standards and regulations for on-board hyperbaric tanks (Wang et al. 2019), as well as for stationary and transportation applications. These regulations frame the technological constraints of commercialization by guaranteeing tank safety and reliability. They also provide specifications for their design, manufacture, inspection, testing and certification.

There are currently several regulations (and standards) applicable to composite hydrogen tanks. These include (see section 1.1.4) the European standard EN12245:2002 (E) and the European (EC79) and international (R134) regulations, which set out lists and/or test sequences for the qualification of hydrogen tanks and include the following:

- composite material identification;
- liner material identification;
- liner burst test at room temperature;
- hydraulic test (leak check);
- hydraulic tank testing at ambient temperature;
- tank burst tests (burst factor is reduced to 2 in 2023);
- resistance to cycling at operating pressure and ambient temperature;
- creep testing by exposure to elevated temperatures at operating pressures;
- drop test;
- high-temperature cycling;
- fire resistance test;
- high-speed impact test (bullet) (not for R134);
- permeability test;
- hydrogen cycling test.

High-temperature tests are justified by the conditions under which high-pressure tanks are filled: during this phase, which is characterized by a high flow rate to limit the supply time, the gas temperature can rise rapidly to 85°C. It is vital to ensure the integrity of the structure at this temperature. The advantage of hydraulic testing is that multiple instruments can be used to understand the mechanisms leading to failure. In this respect, they represent not only a normative obligation but also a more fundamental field of investigation for the optimization of storage structures. The results of this type of test carried out on the reference tank are detailed in the following sections. They illustrate the measurements that can be recorded in situ or postmortem to explore tank failure modes.

1.2.1.1. *Room temperature burst test*

Regulation no. 79/2009 requires a burst pressure of at least 2.25 times the working pressure. The tests described in this section are designed both to check that reference tanks meet this condition and to record as much data as possible during the test to validate the behavior models. In June 2023, the burst factor was reduced to 2.

One of the reference tanks was pressurized at an average ramp rate of 2.3 bar/s until bursting. Three linear variable differential transducers (LVDTs) were used to measure displacement: one along the axis to measure longitudinal elongation and two to measure radial swelling in the middle of the tank shell. The positioning of the transducers is shown in Figure 1.9.

The measurement results are shown in Figure 1.9. They show a nonlinearity of axial displacement at very low pressures because of the initial clearance between the composite shell and the liner, which will come into contact at the beginning of the test. The behavior of the radial LVDT sensors is linear at the start of the test.

Displacement jumps can be seen in the radial behavior, indicating the onset of significant damage at approximately 1498 bar. The final fracture occurred in the shell at a pressure of 1756 bar, which is well within the norm (burst pressure of at least 1575 bar).

The tank under study can also be instrumented with acoustic emission sensors to monitor tank damage kinetics. The curve representing the number of cumulative events (Figure 1.10) reveals that damage begins only at 300 bar (bearing in mind that the cylinder was previously calibrated at 200 bar). High-energy events, on the other hand, do not begin until the pressure reaches 500 bar. As the tank studied here is designed for a working pressure of 700 bar, we can conclude that it experiences little damage before reaching this stage.

Overall, multi-instrumentation of tanks subjected to pressure build-up not only provides a better understanding of the damage phenomena that the structure undergoes but also offers a base of experimental data that can be used to validate (or not) modeling tools. This database can be supplemented by other techniques, such as Bragg gratings (Kang et al. 2006) or digital image correlation (Lainé et al. 2021).

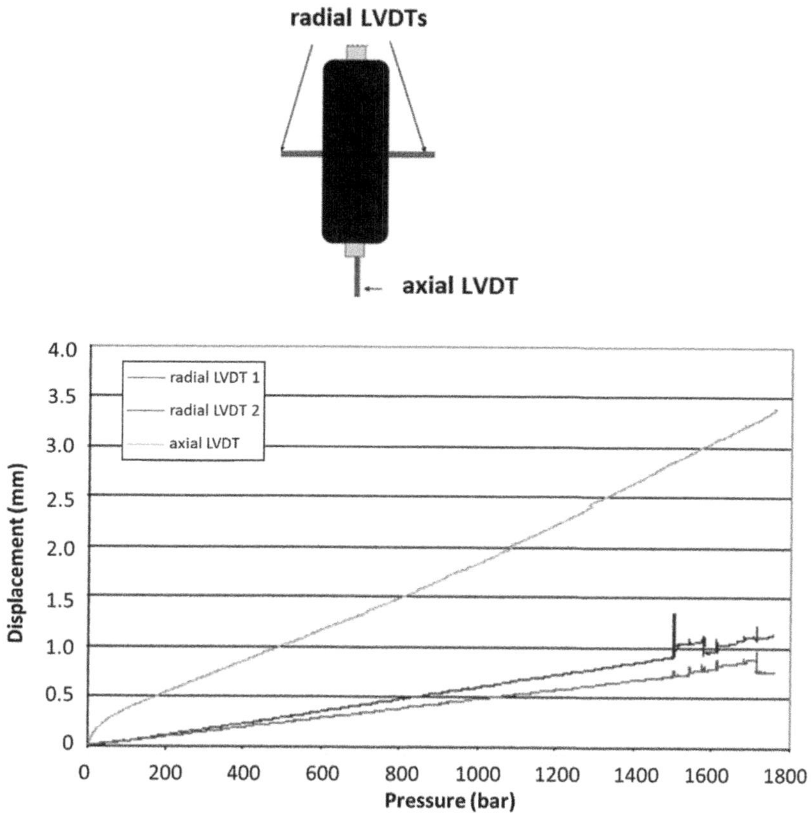

Figure 1.9. *LVDT sensor positions and displacements measured as a function of pressure during burst testing at 20°C. For a color version of this figure, see www.iste.co.uk/derango-cuevas/hydrogen.zip*

As shown in Figure 1.9, pressurization tests can be carried out until the tank bursts. European regulations require this test to be carried out at ambient temperature, with a pressure build-up rate of less than 14 bar/s. Depending on the design of the composite architecture, rupture may occur in the shell (Figure 1.11(a)) or in the domes (Figure 1.11(b)).

The latter usually results in boss projection and is due to design or manufacturing faults (resin loss in the dome during polymerization, an insufficient number of helical layers, etc.).

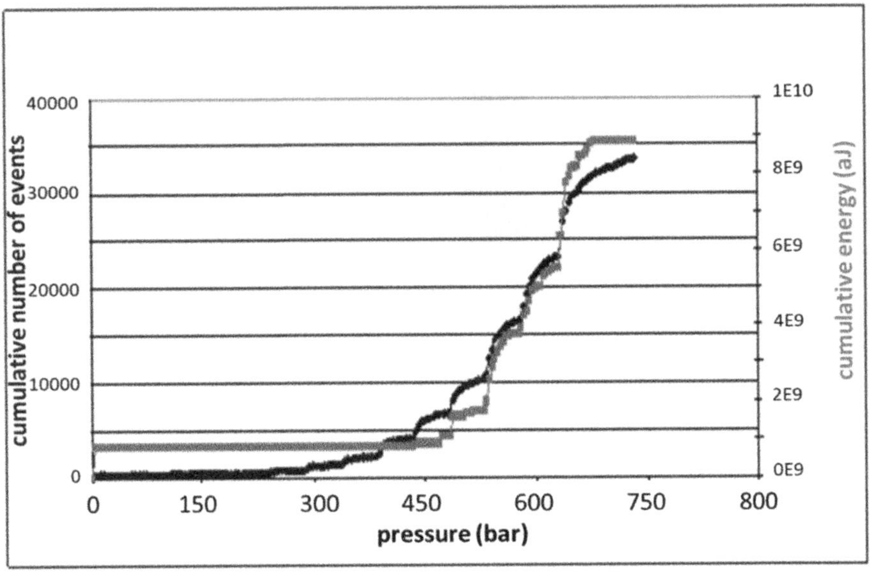

Figure 1.10. *Number of cumulative acoustic events (blue) and cumulative acoustic energy (pink) over time. For a color version of this figure, see www.iste.co.uk/derango-cuevas/hydrogen.zip*

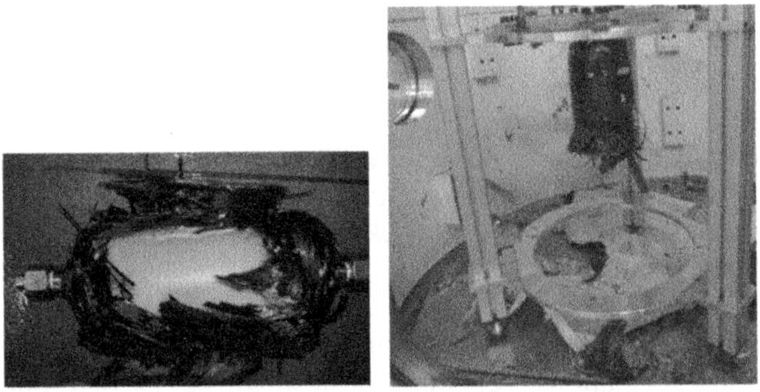

a) b)

Figure 1.11. *(a) Rupture through opening of the shell at CEA; (b) dome-shaped rupture with projection of the boss at CEA*

1.2.1.2. *Cycling test*

The burst test is just one of the tests that hydrogen gas storage tanks must pass. The R134 regulation defines other conditions that tanks must meet over their expected operating life or in extreme situations. For example, Figure 1.12 summarizes the durability test: the tank must withstand a complex sequence combining proof tests, drop tests, notch tests, exposure to chemical substances, simultaneous emptying/filling cycles, overloads, temperature maintenance, low- and high-temperature cycles, maintenance at 1.8 times the operating pressure and bursting. This test highlights the importance of the pressure cycles a tank must be able to withstand.

Specific work (Bertin et al. 2010) has identified the evolution of damage in composite samples subjected to mechanical or thermomechanical fatigue. The results show that, regardless of the loading applied, the stacking sequence of the composite significantly influences the composite, and cyclic stresses coupled with temperature have a detrimental effect on the structural strength.

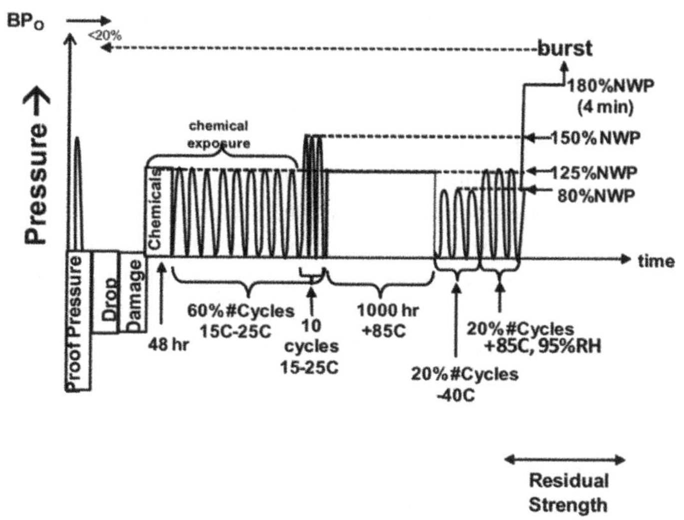

Figure 1.12. *Durability test (hydraulic sequence test) (R134) (NWP: nominal working pressure)*

1.2.1.3. *Fire resistance test*

The fire safety of hydrogen-powered vehicles is crucial for gaining public acceptance of this technology. In the event of a vehicle fire, preserving the integrity of the storage system is essential. Thermal exposure during a fire leads to degradation of the composite material and thus a reduction in mechanical strength. If there is no safety system installed to discharge the tank under pressure, cylinder rupture cannot be ruled out.

Furthermore, such a safe system must be activated by temperature, not pressure, as the pressure increase for composite cylinders in a fire is very limited due to the low thermal conductivity of the composite. In accordance with R134 regulations, storage systems must be equipped with a TPRD (thermal pressure relief device) to empty the tank in a controlled manner in the event of fire. TPRDs commonly used in composite pressure vessels are fitted with a glass bulb or fusible eutectic material.

If exposed to fire, the glass bulb shatters or the fusible material melts, and the gas vent opens, triggering a drain to prevent bursting. Even with TPRDs installed on the storage system, the risk of tank rupture during a fire is possible, for example, in the absence or delay of temperature rise detection. The latter may be caused by a localized fire affecting the tank on the side opposite the TPRD. In this case, the composite degrades without the TPRD detecting a temperature rise. To avoid such incidents, various standards attempt to unify requirements and tests for a compressed hydrogen storage system.

To address the question of safety in the event of a fire, a bonfire test must be carried out, with two different fire scenarios: a local fire and a global fire. The bonfire test is carried out on open ground at 100% nominal working pressure. The fire source used is liquefied petroleum gas, which can heat the entire length of the tank. The tank is placed horizontally approximately 100 mm above the burner. To ensure uniform heating, the experimental setup was protected by windscreens. The temperature is monitored by at least five thermocouples located along the entire length of the tank. The temperature profile for the bonfire test is shown in Figure 1.13.

During the first 10 min, local exposure to the fire is carried out. The temperature is increased to 300°C during the first minute. During the next 2 min, exposure continues at 600°C for 7 min. The temperature must not exceed 900°C. After local heating, the test includes an all-encompassing fire phase. In this case, the temperature increases for 2 min to 800°C. The temperature must remain below 1100°C during this overall heating period. The tank must not undergo an explosion under any circumstances. The particular structure of Type IV tanks (thermoplastic liner and thermoset matrix composite) gives them an intrinsic leak-before-burst property due to competition between the degradation of the outer layers of the cylinder and heat transfer through the composite shell.

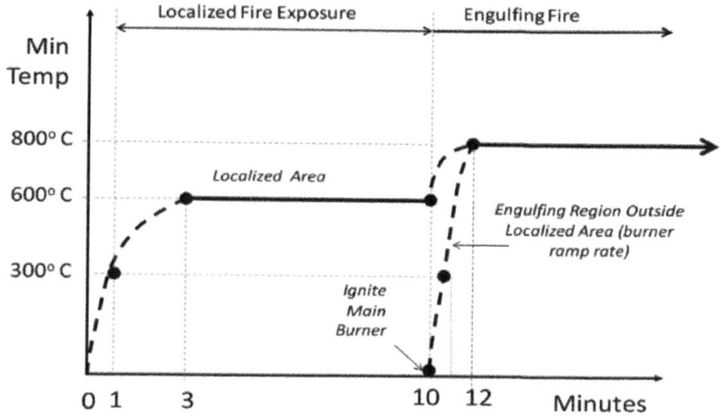

Figure 1.13. *Fire exposure test (R 134)*

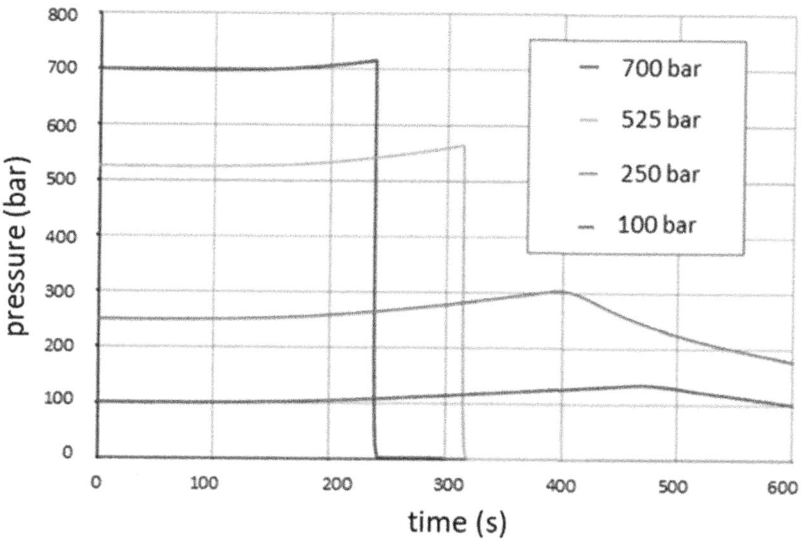

Figure 1.14. *Burst times of tanks subjected to fire as a function of pressure in a 36-L tank. For a color version of this figure, see www.iste.co.uk/derango-cuevas/hydrogen.zip*

Figure 1.14 shows the time to failure of the carbon fiber and epoxy matrix wound tanks subjected to a thermal load imposed by four hydrogen burners as a function of the pressure in the tank at the start of the test. These tests were carried out during the FireComp project (Blanc-Vannet et al. 2019).

At high pressures (700 bar, 525 bar), degradation of the tank's outer layers leads to rapid bursting. On the other hand, at lower pressures (250 bar, 100 bar), hydrogen leaks through the composite shell as a result of liner melting. This prevents an explosion, even in the absence of TPRD.

The design of hyperbaric tanks cannot be based on a trial-and-error principle. The cost of manufacturing a prototype tank and the tests required to validate it are far too high to consider multiple solutions that would be discriminated by certification tests alone. To address this difficulty, we can call on a digital tool, which must include three components:

– a tank geometry modeling module;

– a tank damage and rupture simulation module;

– an optimization module.

1.2.2. *Tank geometry modeling*

Filament winding is still the most suitable process for the production of revolutionary volumes such as composite hyperbaric tanks. A few simple predimensioning rules can be used to determine the approximate stacking sequence of the composite on the shell.

Knowing (i) the volume of the tank and the operating pressure, (ii) the material of which the boss is made and (iii) the internal diameter of the connector, it is possible to determine the external diameter of the boss, assuming that its behavior remains elastic as long as the internal pressure is lower than the test pressure.

From this outside diameter, we can estimate the minimum angle of the helical composite layers required to completely cover the boss. Net theory (which assumes that all fibers are loaded in tension and undergo no shearing) can be used to define the thickness of each layer as a first approximation. According to this approach, the thicknesses of a helical layer (t_α) and a circumferential layer (t_{90}) are given by:

$$t_\alpha = \frac{PR_C}{2\sigma_f \cos^2\alpha} \quad [1.10]$$

$$t_{90} = \frac{PR_C(2-\tan^2\alpha)}{2\sigma_f} \quad [1.11]$$

where P is the burst pressure, R_c is the tank radius, σ is the maximum permissible stress in the fiber direction and α is the angle between the tank axis and the fiber. However, the main difficulty in terms of modeling lies in the simulation of the shape of the tank dome, namely, the evolution of the angle made by the fiber with the axis of the tank and the thickness of the composite deposited. Two approaches are described here, one developed specifically for the OSIRHYS IV project and the other available in commercial FE codes.

1.2.2.1. *Multiple sequential simulations*

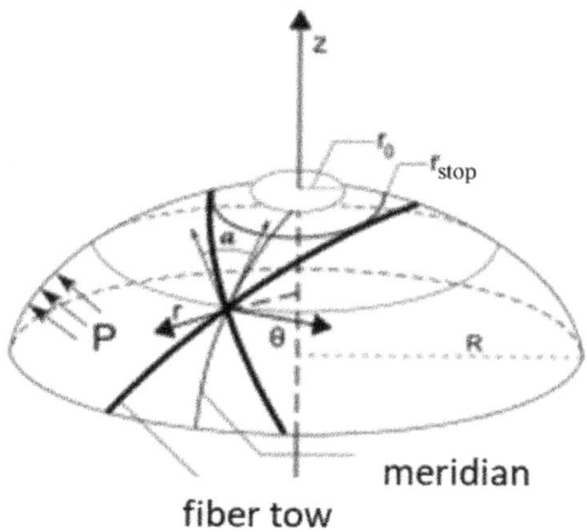

Figure 1.15. *Fiber deposited on dome (Leh 2013)*

Leh et al. (2013) proposed coupling two equations that simultaneously manage the fiber angle and the composite thickness in the dome. The general equation (geodesic or not) for the evolution of the fiber angle during deposition on the dome is written as:

$$\frac{d\alpha}{dz} = \lambda\left[\frac{\sin\alpha\tan\alpha}{r} - \frac{r''}{1+r^2}\cos\alpha\right] - \frac{r'\tan\alpha}{r} \qquad [1.12]$$

where λ is the sliding coefficient of the fiber strands (or rovings) on the laying surface, and r and z are the radial and axial coordinates of a point on the dome, as shown in Figure 1.15. By changing the value of λ, we obtain different values of the stop radius r_{stop}. Equation [1.12] can be solved iteratively. With regard to thickness determination, the work of Wang et al. (2010, 2011) decomposes its expression over two intervals:

$$t(r) = \frac{n_R m_R t_p}{\pi}\left[arccos\left(\frac{r_{stop}}{r}\right) - arccos\left(\frac{r_b}{r}\right)\right] \text{ if } r_{2b} \leq r \leq R \qquad [1.13]$$

$$t(r) = A + B.r + C.r^2 + D.r^3 \text{ if } r_{stop} \leq r \leq r_{2b} \qquad [1.14]$$

where n_R, m_R is the number of layers, t_p is the thickness of the deposited composite band (or web, both terms designating a set of strands) and b is the width. Parameters r_b and r_{2b} are the radii at one and two bandwidths, respectively, starting from the stop radius (at which the band bypasses the pole and departs in the opposite direction).

The method for determining the coefficients A, B, C and D is detailed in Wang et al. (2010, 2011). Solving the highly nonlinear system (equations [1.12]–[1.14]) thus leads to the dome shape. To address more complex shapes, more generic mathematical models based on surface theory for local curvatures are needed (Koussios 2004; Vargas Rojas et al. 2014).

1.2.2.2. *Commercial code plug-ins*

Commercial FE codes such as Abaqus also offer turnkey modules for simulating wound structures (WCM or WoundSim). These methods are based on the linear interpolation of two different geodesic trajectories. The expression for the angle is given by:

$$\alpha(r) = arcsin\left(\frac{r_{stop}}{r}\right) \pm \delta\left(\frac{r - r_{stop}}{R - r_{stop}}\right)^n \qquad [1.15]$$

The parameter δ controls the winding geodesy, and the exponent n controls the influence of non-geodesy on the angle α. The thickness is calculated directly from the radii (stop, shell), the thickness and angle of the shell layer and the bandwidth.

The advantage of this type of calculation is obvious: it provides an analytical expression for the angle and thickness without having to solve a differential equation. However, the expected thickness in the vicinity of the stop radius is overestimated and must be corrected by the operator using adjustable parameters (Leh et al. 2013). Figure 1.16 compares an X-ray of the reference tank with the dome geometry simulated by Leh's approach and the Abaqus plug-in.

Both tools provide a satisfactory response to the problem of representing the shape of the tank. However, there is a discrepancy at the end of the dome. This has no impact on the behavioral modeling, as this area is low in carbon fibers and plays a lesser role in stress absorption.

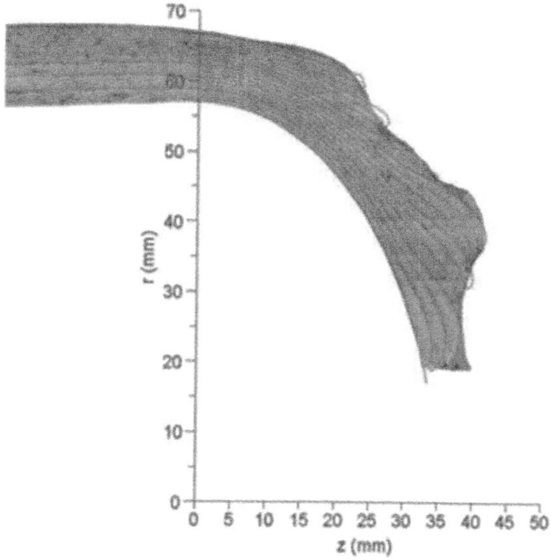

a)

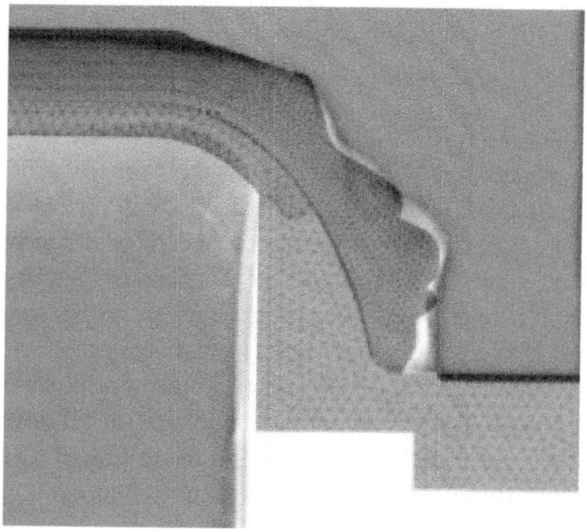

b)

Figure 1.16. *(a) Comparison between a CEA dome X-ray and the geometry (in green or red) predicted by the approach of Leh (2013) and (b) the WCM plug-in (Berro Ramirez 2013). For a color version of this figure, see www.iste.co.uk/derango-cuevas/hydrogen.zip*

1.2.3. *Behavior of the tank*

1.2.3.1. *Composite damage model*

Once the geometry is known, it is possible to carry out FE simulations of the tank's behavior and, in particular, to predict its damage and rupture under the effect of internal pressure. To achieve this, it is necessary to have a composite material behavior model that can represent the mechanical consequences of all the degradation mechanisms occurring in a tank manufactured by filament winding. There is an abundance of literature in the field of damage modeling for composite materials, but it is legitimate to ask whether all these approaches are capable of predicting tank failure in a comparable way. In particular, high composite thicknesses and potential bending phenomena in the domes may favor delamination in the composite, which is a degradation mechanism that is not systematically considered in basic versions of damage models for composite materials.

For comparison purposes, three approaches are selected here: a progressive failure analysis (which has the advantage of simplicity), a model widely used in the composite materials community (LMT Cachan meso-model without the contribution of delamination) and a model built within the framework of continuous damage mechanics integrating all degradation modes.

Progressive failure analysis: this approach is based on the use of failure criteria, whose verification makes it possible to reduce the properties of damaged layers. Damage can be abrupt (Padhi et al. 1997; Liu and Tsai 1998; Papanikos et al. 2005), considering only two states, intact or totally damaged, or it can be gradual (Reddy et al. 1995; Laurin et al. 2007; Zahari and El Zafrany 2009), considering that damage initiation and propagation occur progressively. In this simple approach, the material is degraded in such a way as to satisfy a damage criterion. With this type of model, the damage behavior of the composite is piecewise linear, which simplifies numerical analyses and saves computing time. In the following, we confine ourselves to the use of the criterion for predicting the failure of fibers or fiber/matrix interfaces (Hashin and Rotem 1973). Its application to hydrogen tanks is detailed in Leh et al. (2015b).

Mesoscopic model – Cachan model: the mesoscopic scale (i.e. the scale of the elementary layer) is the most commonly used working scale. This is probably because it combines a good level of detail in the description of behavior with sufficient flexibility to enable the calculation of large-scale structures. The best-known example in this category is the "meso-model" developed in the context of the thermodynamics of irreversible processes by P. Ladevèze at LMT CACHAN and applied to different types of composite materials (Ladevèze 1988;

Gasser et al. 1998; Lubineau and Ladevèze 2002). This model takes into account three modes of damage within the layer: fiber fracture, matrix cracking and diffuse damage. A more complete version of this model (Abisset et al. 2011) incorporates the interface (with its own damage) between layers. In the field of hydrogen storage, Magneville et al. (2015) apply the Cachan model to the dimensioning of Type IV tanks. This type of model has been successfully applied to optimize, for example, the design of Type III tanks (Chapelle and Perreux 2006).

Continuous damage model: this class includes a large number of models, also built within the framework of the thermodynamics of irreversible processes. One example is a model specifically developed by Berro Ramirez et al. (2015a, 2015b) as part of the OSIRHYS IV project. This model aims to represent the behavior of wound composites within the framework of continuous damage mechanics. The existence of preferred damage directions in composite materials (due to their architecture) enables us to rely on the concept of fixed damage directions, with which we associate a scalar variable representing the influence of the different types of damage associated with each defined system.

Other modeling approaches: the three models mentioned above are a snapshot of the simulation tools available to represent the degradation of composite materials. Other theories also exist. In particular, micromechanical models (Ladevèze et al. 2006) explicitly address damage at the scale of individual components (fibers, matrix and interfaces). While they are highly reliable at this scale, they have the disadvantage of being computationally more expensive than other types of modeling. Numerical approaches of the EF2 type (Widjaja 2020) use richer and more complex behavioral models but are cumbersome at the tank scale. For this reason, these studies focus on reducing the size of the structures modeled beforehand but generally have to do with simulations in the shell of the tank alone. In the class of continuous damage models, the work by ONERA is worth mentioning and is based on the calculation of variables at different scales of the composite material (Laurin et al. 2007). This model involves a transition between the mesoscopic scale (corresponding to the unidirectional ply) and the so-called macroscopic scale corresponding to the composite composed of several layers. This approach can be divided into five different stages: a scaling method, choice of mesoscopic behavior, definition of a mesoscopic failure criterion, property degradation and composite failure. The scaling method is based on classical laminate theory.

1.2.3.2. *Tank burst simulation*

To validate the modeling approach (geometry representation, damage simulation, stochastic fracture behavior, etc.) and thus validate the design tools, we seek to numerically reproduce the bursting of the reference tank presented in the introduction. This tank has a capacity of 2 L and is made up of two metal bosses, a polymer liner and a composite shell with a known architecture (Figure 1.17).

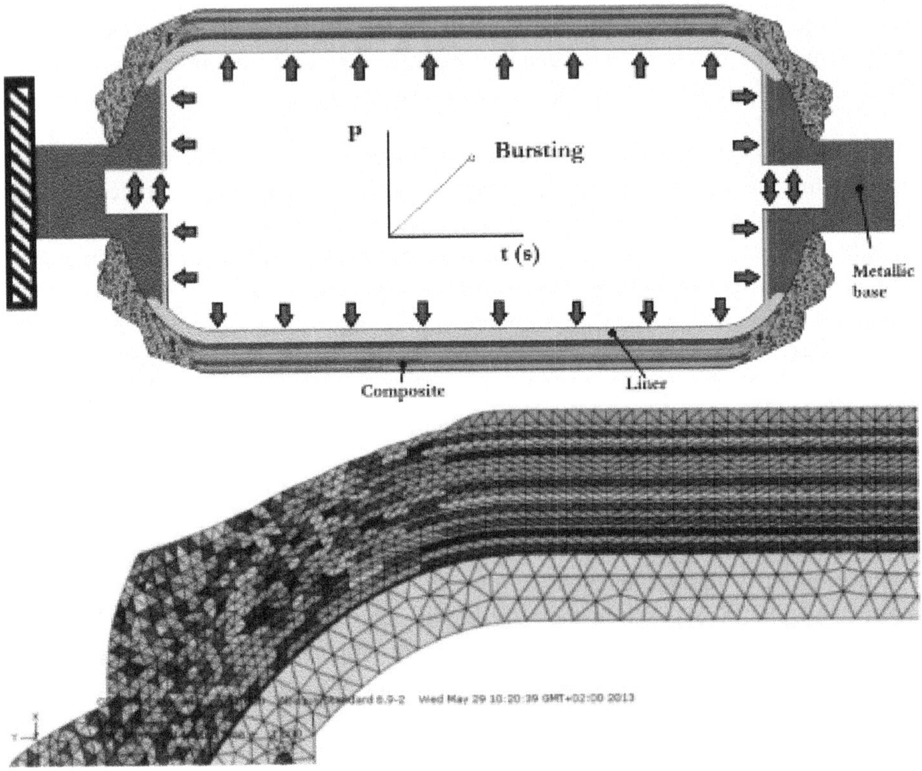

Figure 1.17. *Model of the reference wound tank – detail of the evolution of fiber angles in the dome (each color corresponds to an angle). For a color version of this figure, see www.iste.co.uk/derango-cuevas/hydrogen.zip*

Given the symmetry of the structure, it is possible to run simulations using either volumetric elements (reducing the geometry to the quarter tank and imposing kinematic conditions on two orthogonal planes of symmetry) or axisymmetric elements to reduce the calculation time. Different modeling options were selected in terms of tank symmetry, element type, damage model and dome shape modeling.

Table 1.5 summarizes these choices, the results of which are compared below.

For the sake of brevity, the detailed results of simulation 2 are presented here. The performances of the three options will be compared to determine their respective advantages and identify an effective sizing approach. The main basis of comparison for evaluating the validity of the tank burst simulation and testing the capabilities of the approach is the experimental displacement curve (axial or radial, Figure 1.9) as a function of internal pressure.

This curve clearly shows the pressure at which the tank ruptures: bursting occurs when the displacements diverge. The point at which the slope of the axial displacement becomes zero can be considered the moment of bursting.

Figure 1.18 compares the experimental (blue) and simulated (red) displacements after correction for the initial nonlinearity of the axial displacement curve, which is attributed to an initial clearance between the liner and the composite shell. Notably, the curves were initially simulated without taking into account the probabilistic aspect of the value of the stress at fiber rupture.

Simulation number	Finite element type	Number of degrees of freedom	Dome model	Damage model	EF software
1	Volumetric + symmetrical	86724	Leh	Progressive rupture	ANSYS
2	Volumetric + symmetrical	60096	WCM	Continuous damage	Abaqus
3	2D axisymmetrical	57482	Leh	LMT Cachan	SAMCEF

Table 1.5. *Modeling options*

The simulated burst point is at a pressure of 1,600 bar, whereas the experimental burst pressure is at 1,750 bar. The experimental radial displacement shows a disturbance starting at approximately 1,500 bar. This disturbance is due to the first fiber breaks in circumferential layers. These ruptures do not lead to final rupture, as the still intact layers absorb the stresses. It is only when a certain number of circumferential layers are ruptured that the tank bursts. Another piece of information can be deduced from the shape of the simulated displacement curve: the shell opens (radial swelling diverges), while at the same time, the bosses tend to retract into the tank (axial displacement decreases).

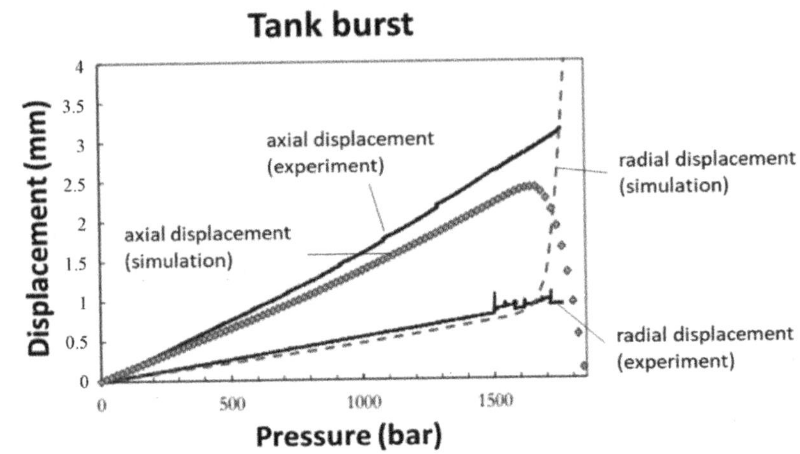

Figure 1.18. *Experimental/simulation comparison – axial and radial displacement curves (Berro Ramirez 2013). For a color version of this figure, see www.iste.co.uk/derango-cuevas/hydrogen.zip*

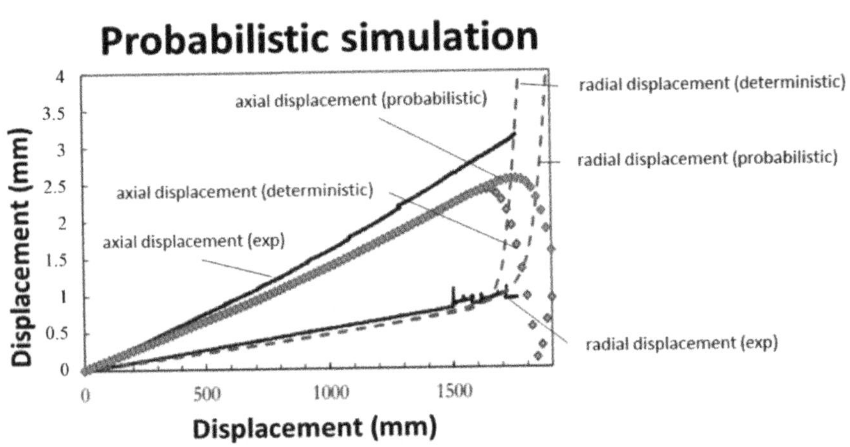

Figure 1.19. *Experimental/simulation comparison – axial and radial displacement curves and deterministic and probabilistic cases (Berro Ramirez 2013). For a color version of this figure, see www.iste.co.uk/derango-cuevas/hydrogen.zip*

The model predicts a "clean" burst. If we proceed to a probabilistic simulation incorporating the dispersion of the stress at fiber rupture (Camara et al. 2011), the prediction is improved, as shown in Figure 1.19. It can be seen that we gain approximately 70 bar in the estimation of burst pressure with the introduction of probabilistic rupture identified from the variability of fiber behavior. The explanation for this difference lies mainly in the redistribution of stresses in the tank during fiber rupture. This increase in burst pressure is attributed to the fact that, following the rupture of "weak" fibers in the tank, stresses are redistributed to "strong" zones, which take over the load from weaker elements and continue to bear the load.

However, the damage models described above offer much finer-grained analysis possibilities than simply predicting bursting. For example, Figure 1.20 shows the out-of-plane damage responsible for delamination. This damage is particularly marked at the end of the dome (dark areas). During hydraulic pressurization tests, leaks were observed precisely where the model predicted significant delamination: in the event of failure of the liner boss junction (marked by the black line), the fluid used to pressurize the tank (in this case, water) can either (a) pass through the composite at the junction, (b) diffuse into the end of the dome, or (c) follow a path along the boss/composite interface. Figure 1.20 shows an example of a leak detected during testing, which appears to correspond to path (c) (leak at the end of the dome, at the interface with the boss). This type of simulation shows the value of richer behavioral models, since only they can provide a detailed description of the state of tank damage.

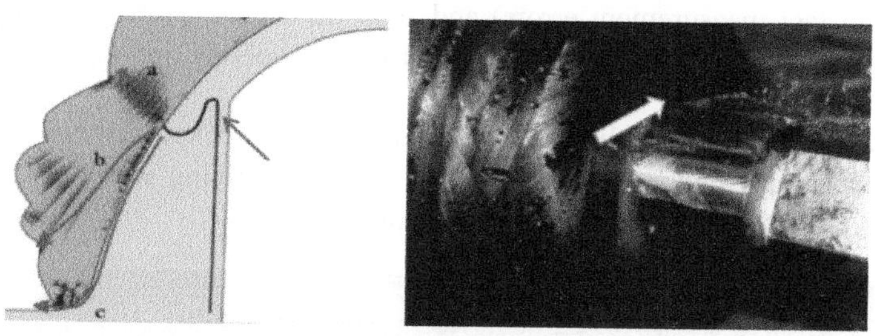

Figure 1.20. *Simulated out-of-plane damage and leakage detected during testing (Berro Ramirez 2013)*

Finally, if we consider only the burst pressure simulation objective, Table 1.6 shows that, regardless of the modeling option, the ultimate pressure values are comparable. The exception is the progressive rupture analysis approach, which predicts a value that is further below reality.

All these simulations yield common results, namely:

– the phenomenon leading to bursting is the rupture of the inner circumferential layer;

– a shell burst mode (inward projection of domes).

Simulation number	Burst pressure (bar)	Failure mode
1	1,490	Shell structure
2	1,720	Shell structure
3	1,740	Shell structure
Test	1,760	Shell structure

Table 1.6. *Burst pressure values from simulations compared with test burst pressure value*

Further postprocessing shows that matrix damage, measured by acoustic emission, is very significant at the time of bursting but that it starts very early and is not destructive. The deterministic model fails to explain the sharp increase in acoustic events after 1,500 bar (Figure 1.10). Probabilistic models explain these phenomena: the first fibers begin to break at a pressure of approximately 1,450 bar. These failures are not destructive, but we can be sure that they are "weak" fibers located in a zone of low fiber density and high stress. These ruptures give rise to displacement jumps (which are in fact measurement errors) and acoustic events. More generally, we can list the advantages and disadvantages of the different modeling options (Table 1.7).

In conclusion, each modeling approach is relevant to the objectives we wish to set (numerical implementation, reliability of results, coupling with an optimization module, etc.), and none can be excluded a priori.

Simulation number	Advantages	Disadvantages
1	Easy to implement No material parameters for damage	Non-continuous damage model Less precise burst pressure Calculation time
2	Precise burst pressure Describes tank failure mechanisms Hydrogen leakage prediction	No automation for optimization
3	Complete, simple and robust configuration for other geometries Good results Extremely fast calculation times	Simplified assumptions linked to axisymmetry (particularly for the probabilistic approach)

Table 1.7. *A comparison of approaches*

1.2.4. *Tank optimization*

In section 1.2.3, we presented several tools to help design and dimension hydrogen tanks by reliably predicting the pressure and mode of rupture. These tools can form part of an optimization chain aimed at reducing the mass (and cost) of the composite while maintaining high strength. To achieve this objective, the optimization parameters selected are as follows:

– the number of layers with different orientations;

– total number of layers;

– sequencing, that is, alternating helical and circumferential layers;

– the sliding coefficients acting on the domes;

– the winding angles on the tank shell.

The material characteristics, which are essential parameters in composite structural optimization, are static. The optimization criterion is the mass of the tank. This is a global optimization criterion that combines the following:

– the tank cost inherent in the proportion of raw materials (carbon fibers). In terms of material, it is best to select a fiber with the lowest possible cost per MPa of strength. For example, T700 has a strength of 4,900 MPa at a cost of €18–20 per kilogram of fiber;

– installation time, linked to bandwidth and number of layers;

– load transfer, rigidity.

The optimization constraints are as follows:

– location of final rupture (mode of rupture: shell/dome, a shell mode of rupture that does not eject the bosses is imperative);

– the minimum burst pressure to be met: 1,400 bar (since the new version of the R134 standard in June 2023).

Several optimization methods are available. The simplest method is manual optimization, which involves varying the number of layers and their orientation. The aim is to perform several calculations and select the best candidate.

However, given the large number of possible variables and the complexity of the structure, two optimization strategies can be considered:

– classical engineering optimization (optimization based on a linear model), which involves sizing the tank without considering material damage during pressure loading. This optimization strategy will hereinafter be referred to as linear optimization;

– nonlinear optimization, in which damage to the composite material is taken into account using the damage models presented in the previous section. This optimization strategy will hereafter be referred to as nonlinear optimization.

These two strategies differ in terms of mechanics and thus, a priori, in terms of the optimized solutions found at the end of the optimization procedure. They also differ in terms of the computation times involved in the FE resolution of each specimen (ranging from a few seconds for the first to several minutes or hours for the second, depending on the model used, and this for a single specimen in the research domain). The main difficulty is therefore to be able to carry out the optimization using the second and more realistic strategy.

The development and use of a metamodel may be one way of overcoming the high computation times associated with the integration of nonlinearity in the case of the second strategy. The construction of a metamodel consists, from FE modeling of the tank, of finding an analytical model of each response that we wish to study, in this case, mass, rupture pressure and rupture location. In this way, nonlinear optimization itself is no longer based on direct FE calculations but rather on the analytical equations of each response reflecting the tank's behavior. In this way, once the metamodel has been constructed, the computation times are very short (generally less than those for linear optimization), enabling the use of a variety of optimization methods. As a result, the same optimization methods can be used for both strategies, allowing a meaningful comparison to be made.

1.2.4.1. *Linear model optimization*

Linear optimization is a classic optimization strategy. Guided by the optimization method, it consists of performing an FE calculation representing the mechanical behavior of the tank for each specimen to be tested. FE calculations are performed under the following assumptions:

– no geometric or material nonlinearity (no composite damage);

– constant mechanical loading.

For each specimen tested, the responses are recorded: the mass and safety coefficients are derived from the failure criteria in the direction of the fibers in the shell and dome sections of the tank. The response values are linked to the input variables, that is, the number of layers and orientation, which will guide the optimization method to optimize these safety coefficients under certain constraints (e.g. pressure and location).

1.2.4.2. *Nonlinear model optimization*

The nonlinear optimization strategy is much trickier to implement, as it requires several construction and analysis stages:

– identifying the input factors or parameters that influence the responses selected: mass, rupture pressure, location, etc.;

– selecting an analytical model for each response on the basis of the factors deemed to be influential;

– integrating manufacturing constraints (increasing winding angles) into analytical models;

– selecting and developing appropriate response functions for different responses;

– checking the relevance of the metamodel and therefore of each submodel;

– integrating the metamodel into an optimization method.

It is often necessary to repeat several of these steps, as the metamodel becomes more complex and generalized. The experimental design method is chosen for the construction of the metamodel. This has the enormous advantage of gathering the maximum amount of information in a minimum number of judiciously chosen experimental trials (in this case, FE calculations). This makes it possible to study the influence of factors on responses and thus to find the most promising parameters for optimization. This, in turn, reduces the computational costs involved in finding the optimum solution.

In addition, the method makes it possible to build analytical models of the responses from the experiments carried out. This enables the value of the responses to be predicted directly and instantaneously as a function of any set of input parameter values without recourse to the FE model. However, the main difficulty of this method is the integration of manufacturing constraints, which make it impossible to preserve the orthogonality properties of conventional experimental designs.

By choosing shell FE modeling, a progressive fracture analysis composite damage model and experimental design optimization, it is possible to reduce the mass by up to 26% compared with the reference solution (Leh et al. 2015a). The solution obtained shows a decoupling between the radial and axial stresses. Indeed, between the optimized solution and the reference solution, there is a transfer from high-angle helical layers to circumferential layers. This suggests that axial forces are taken up mainly by the low-angle layers (below 35°) and that radial forces are taken up by the circumferential layers. The presence of intermediate helical layers (high angles), which are costly in terms of mass and winding time, would therefore not result in any significant advantages in terms of tank performance.

1.2.5. *Outlook*

A comparison of different damage models shows that, overall, they predict similar burst pressures, provided that they take into account the different modes of degradation in the composite.

Therefore, in regard to dimensioning and optimization, criteria such as calculation time and ease of identification should govern the selection of the behavior law. However, this conclusion is based solely on the analysis of room-temperature burst tests and could be altered by considering all the qualification tests mentioned in section 1.2.1. Beyond this spectrum of tests, the emergence of new generations of tanks is likely to modify the very approach to dimensioning by acting on the geometry or the material.

Research avenues include the following:

– Constraints on the integration of on-board tanks in vehicles have a major impact on geometry. For example, when batteries are replaced by tube networks, the storage system has to be adapted to the dimensions of the compartment.

– The advantages of linerless tanks built from thermoplastic matrix composites (future Type V) are numerous: the absence of a liner avoids collapse during filling/emptying cycles, thermoplastic materials have excellent hydrogen permeability properties, can be recycled, and so on. However, the development of Type V materials requires in-depth knowledge of the composite and, in particular, of the properties of the fiber/matrix interface, which controls not only the mechanical strength but also the tightness.

– Another major advantage of thermoplastic matrix materials is their resistance to fire. Figure 1.14 shows that the time to burst of a tank in a fire situation is quite low (approximately 5 min). Recent studies, such as the H2020 Thor project, have shown a significant increase in the failure time of a thermoplastic structure subjected to fire.

1.3. Manufacturing processes

This chapter describes the main stages in the process used to manufacture polymer-lined composite tanks for the pressurized storage of gaseous dihydrogen. For the record, this storage technology is referred to as a Type IV tank. For hydrogen mobility applications at 700 bar, it is currently the most mature technology (U.S. DOE 2023, CEA and DGEC 2018). We will deal with the general case, namely, a bottle-shaped tank with a connector at both bosses. These components have axisymmetric shapes for which revolution processes are suitable.

A Type IV tank consists of three main components:

– metal bosses for connecting the tank to the storage system;

– a polymer liner for hydrogen tightness;

– a composite shell that mechanically structures the tank.

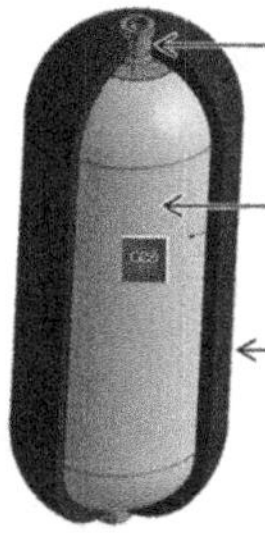

Figure 1.21. *CEA CAD of a Type IV tank with metal bosses, a polymer liner and a composite shell. For a color version of this figure, see www.iste.co.uk/derango-cuevas/hydrogen.zip*

Type IV tanks are manufactured using the following processes:

– machining to obtain metal bosses;

– rotational molding, extrusion blow-molding or injection-welding to produce the polymer liner;

– filament winding for composite manufacturing.

The interfaces between the various components are also very important. Appropriate processes can be used to deal with them, such as bonding if this solution is chosen.

1.3.1. *Boss manufacturing*

Metal bosses are usually manufactured using a machining process. Mechanical turning is used to remove material from a blank. This blank is usually obtained by casting or forging. In general, the boss consists of a plate and a neck. The neck can be fitted with a number of flats to prevent rotation of the boss when the tank is tightened to connect it to the storage system. The materials used are generally compatible with the constraints of hydrogen applications, with aluminum 6061 T6 being one of the most widely used.

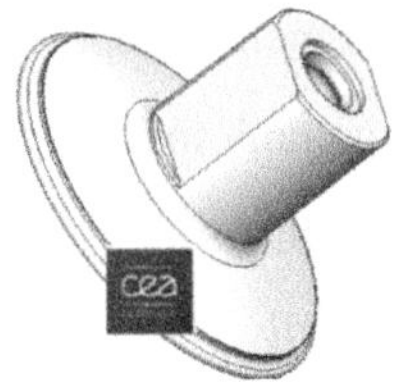

Figure 1.22. *CEA CAD of a metal boss with two flats (antirotation function) running the full height of the boss neck*

1.3.2. *Polymer liner manufacturing*

Polymer liners are manufactured using plastic processing techniques. Currently, there are at least three ways to manufacture a polymer liner:

– rotational molding process;

– extrusion blow-molding or injection blow-molding process;

– injection welding process.

For each of these processes, the behavior of the plastic material is a key parameter that the manufacturer must master to obtain quality parts. The materials used are generally polyethylene or polyamide.

1.3.2.1. *Rotational-molded liner*

Rotational molding is a plastic processing technique used to manufacture polymer parts in a rotating mold.

Figure 1.23. *CEA high-density polyethylene liner obtained by rotational molding (one-piece part)*

In the melt rotational molding process, micronized thermoplastic polymer powder is introduced into a mold, which is heated and rotated to produce a liner. The mold's two-axis rotation program enables the powder to flow, gradually starting to melt and set on the mold walls or on the polymer that has already melted. Once the material has set, the mold is cooled at a controlled speed to manage any crystallization of crystalline or semicrystalline polymers.

Once cooling is complete, the mold is opened, and the liner is removed from the mold. Liner production takes approximately 40 min. In most cases, the bosses are placed inside the mold at the ends before production begins, and rotational molding is used to overmold the bosses with a polymer, which more or less partially covers the boss. In this case, the resulting liner is a one-piece part fitted with its bosses. Programming of mold movements enables axisymmetric control of the liner thickness required in the desired zone.

The process can be optimized through material formulation, heat transfer (time–temperature diagram), polymer grain coalescence and densification, cooling, solidification and crystallization. Simulation of this process enables faster optimization.

Figure 1.24. *Rotational molding machine at the CEA in Tours, France, for manufacturing a polymer liner. For a color version of this figure, see www.iste.co.uk/derango-cuevas/hydrogen.zip*

There are several variants to the melt rotational molding process. The reactive rotational molding process can be used to manufacture liners with thermoplastic or thermoset end materials, depending on the nature of the precursors injected into the mold. In general, this variant of rotational molding results in materials with better intrinsic properties (compared with the same materials used by the molten route, which have undergone several transformations), with reduced energy costs depending on the exothermicity of the chemical reaction. However, this process is more difficult to develop because, in addition to the rheology of the material, there is the constraint of the chemical kinetics of the polymerization reaction (rheokinetic constraint).

1.3.2.2. *Liner manufactured by extrusion blow-molding and by injection blow-molding*

Extrusion or injection blow molding processes can produce polymer liners in just a few minutes (2–3 min). These two processes are widely used in the high-volume plastics industry.

In the case of extrusion blow-molding, the extruded tube is clamped in a mold and a gas is blown through it to press the polymer against the mold walls to form a liner.

In injection blow-molding, the preform is injected into the mold and then blown into its final shape against the mold walls.

These processes have the advantage of short cycle times (a few minutes). However, molds are very expensive, as they require very high closing pressures. In addition, the constrained flow of material tends to orientate polymer molecular chains. This orientation can vary depending on the location of the component and causes anisotropies in the final mechanical behavior of the component.

Once the liner has been manufactured, the bosses need to be attached to the liner before the composite is applied. Adding the bosses requires additional operations, which tend to slow down the production cycle of a liner fitted with its bosses.

1.3.2.3. *Liner manufactured by injection-welding*

The injection-welding manufacturing process is a two-stage process. In the first stage, the polymer is injected into a mold to match the geometry of the liner base. This mold may or may not be fitted with a boss.

When the mold is not equipped with a boss, the boss is then attached to each liner base, with easy access to the inner side of the base (which is not possible with other processes). The bosses are then welded together to complete the liner. Welding can be thermal (e.g. laser or ultrasonic heating) or thermomechanical (rotation-friction).

1.3.3. *Composite manufacturing*

Filament winding is currently the most suitable process for manufacturing Type IV composite tanks. It is used to manufacture a multilayer, axisymmetric composite shell around the liner with its bosses. Some variants are not yet ready for industrial production of 700-bar tanks for mobility applications.

Figure 1.25. *CEA's 700 bar H_2 Type IV tanks*

First, the nominal manufacturing process is described. This involves wet filament winding with the impregnation of continuous carbon fibers with a thermosetting matrix.

The nominal fiber best suited to the constraints of mass-produced hydrogen mobility is 24K carbon fiber, with an ultimate axial strength of approximately 5000 MPa, a modulus of 230 GPa and a price of €20–€25/kg. The most suitable matrix for structural applications is epoxy.

Like all composite processes, this process comprises three manufacturing phases:

– fiber impregnation;

– layer compaction;

– polymerization (curing) of the composite matrix.

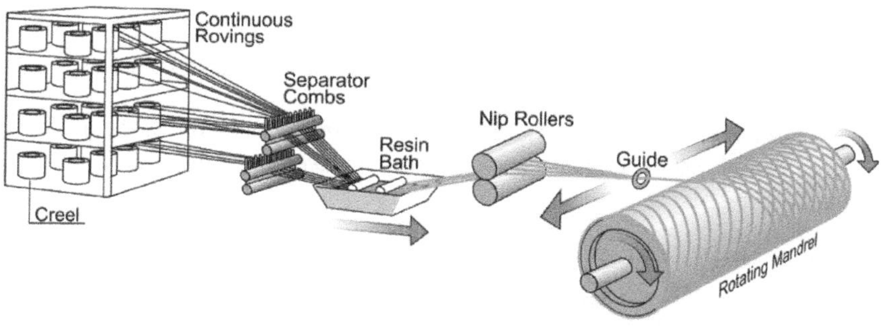

Figure 1.26. *Schematic diagram of the filament winding process with fiber impregnation followed by deposition around the rotating liner (Peters 2011)*

Figure 1.27. *Robotic filament winding machine (in towpreg configuration) at the CEA in Tours, France, for manufacturing composite around a liner. For a color version of this figure, see www.iste.co.uk/derango-cuevas/hydrogen.zip*

A coil is made up of a wick (also known as a strand) containing several thousand carbon monofilaments (12,000 filaments = 12K, 24K, 36K, 48K) that are a few micrometers in diameter (7 μm). The use of several wicks in parallel forms a web. The web is laid around the rotating liner in a helical pattern (several turns around the liner before joining the two bosses), in a polar pattern (less than one turn to join the two bosses) or in a circumferential pattern (only on the cylindrical part of the liner). These different fiber orientations enable the composite walls to be structured according to the mechanical stresses the tank will experience during pressurization (Figures 1.28 and 1.29).

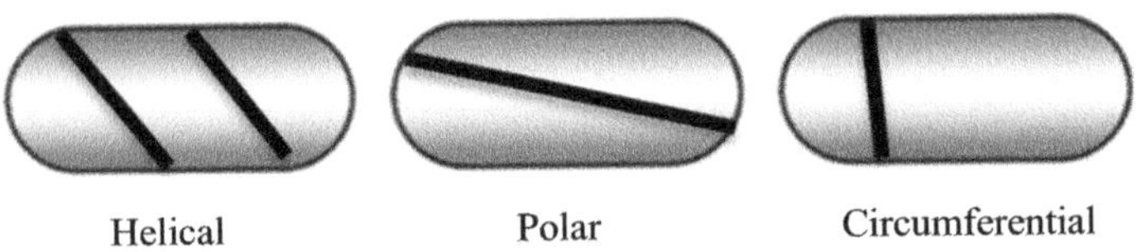

Figure 1.28. *Schematic illustration of the possibilities for winding a web around a liner using a filament winding machine (CEA)*

– *Carbon fiber impregnation*: Today, the majority of composite tank manufacturers use the wet filament winding process with open impregnation tanks (Figure 1.30). This technique is available in two variants: contact impregnation and dip impregnation. These tanks have numerous mechanical parts that need to be methodically cleaned after each winding. The adjustments required to obtain the desired resin ratio are difficult. These adjustments depend on the viscosity of the resin and its evolution during the winding phase. In both cases, whether contact or dipping, the resin is constantly agitated and sheared, resulting in the inclusion of air bubbles, which are detrimental to the mechanical strength of the laminate (high porosities in the composite of 4%–8%), as well as heat, which reduces the useful life (also known as potlife) of the resin. In some cases, resin scrapers can damage fibers. First, the fiber strands on each spool (spools arranged on a reel) are conveyed under mechanical tension to the impregnation tank (tension of several tens of Newtons per strand).

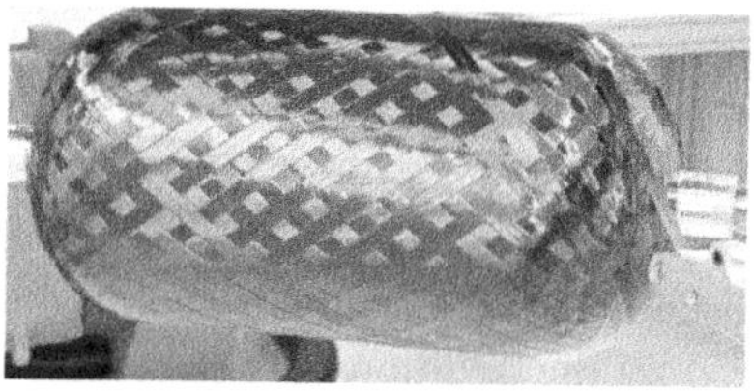

Figure 1.29. *Example of helical winding on CEA's robotized filament winding machine*

Figure 1.30. *Two threads of 24K carbon are impregnated in the CEA contact impregnation tank (the surface of the roller is loaded with resin by soaking in the bottom of the tank, and the fiber is loaded with resin by contact with the roller on the upper part of the roller)*

Contact impregnation tanks are most frequently used to manufacture tank composites. The webs fed from the reel come into contact with the surface of the rotating roller. The roller surface is coated with a thermosetting resin. The contact time with the roller and the pressure of the web on the roller allow the resin to be transferred to the web and then into the web through the thousands of filaments. The rotating roller is then refilled with resin by dipping into the bottom of the impregnation tank. Depending on the roller's rotation speed, the viscosity of the resin and its evolution over time, and the tank's singularities, air bubbles may be created in the resin, which the operator must minimize to avoid porosity in the final composite.

– *Compacting carbon–epoxy layers*: Once impregnated with resin, the fiber webs are conveyed to a moving deposition head, where they are applied to the rotating mandrel (liner) to form a composite layer. To build a layer, the machine must make a certain number of to-and-fro movements along the axis of the liner to completely cover the desired area (Figures 1.31 and 1.32). Each wire is tensioned by mechanical systems at the reel. The tension is generally a few tens of Newtons per wire. The compaction of the deposited layer is caused by the tension applied to each fiber coil. This tension is transferred to the surface of the mandrel, compacting the composite. Compaction is also ensured by the addition of a circumferential layer (or a layer at an angle >45°) above the layer in question. The number of layers wound around the mandrel depends on the desired mechanical performance. Each layer can have a different winding angle. The angle is measured as a function of the main axis of the liner with the axis of the layer deposit.

Figure 1.31. *Production of the first composite layer around the liner by a wet filament winding on the CEA's robotized installation*

– *Polymerization of the composite matrix*: Once all the composite layers have been wound around the liner, the tank is moved from the filament winding machine to an oven. In most cases, the tank is rotated horizontally during its stay in the oven. This rotation prevents the resin from flowing by gravity. The resin is usually a thermosetting epoxy resin, which takes a long time to polymerize (several half-dozen hours), especially as the maximum polymerization temperature (EC79, R134) must not exceed the liner softening temperature minus 10°C. In the case of a polyethylene liner with a softening temperature of approximately 115°C, the maximum temperature allowed for curing the composite is 105°C. In addition to a very long curing time, a low curing temperature leads to matrix glass transition temperatures that are at most equal to the maximum curing temperature in the material.

Figure 1.32. *CEA's deposition head during wet filament winding composite production for depositing epoxy-impregnated carbon fiber filaments around a liner (using two filaments in this example)*

1.3.4. *Developments and prospects (materials, thermoplastics, etc.)*

To conclude this chapter, we will take a look at the latest developments in hyperbaric storage and the prospects for further improvements.

– *Tank cost*: in 2023, the cost of manufacturing Type IV 700 bar composite tanks for on-board storage of gaseous hydrogen was close to €700 per kilogram of hydrogen stored. The European Union's objective is to reach a cost of approximately €300/kg H_2. Carbon fiber is a major contributor, accounting for 50% of the cost of the entire storage system (European FCH-JU Copernic 325330 project). Better composite architectures could help reduce costs. This involves optimizing the architecture as well as alternative manufacturing processes to filament winding, where the fiber is better positioned. This remains to be developed. Another solution is to find manufacturers capable of producing carbon fiber at a lower cost. At present, the reference carbon fiber is Toray T700 at a cost of between €20 and €25/kg.

– *Tank manufacturing time*: it takes approximately 1 h to manufacture a liner with bosses and a composite around the liner. For the composite part, the filament winding time for 700 bar tanks is generally between 1 and 2 h, depending on the quantity of composite to be deposited. The curing time is approximately 15 h, depending on the type of epoxy resin used. Today, new rapid-cure systems reduce the composite curing time to approximately 7 h.

For example, during the ANR Fastcure 2 project (Projet-ANR-19-CE05-0018), the CEA, INSA Lyon and Vitech Composites patented an epoxy resin that cures in

6 hours 45 minutes. Another example is the use of a robotized installation (Figure 1.33) to double production rates while reducing the risk of web overturning. The outlook is for further reductions in tank manufacturing times, with the aim of moving toward automotive production rates. The use of a towpreg (a semifinished product consisting of a carbon roving impregnated with a polymer matrix) enables us to reduce the winding time with a virtually constant fiber volume rate, regardless of the zone of the composite under consideration.

– *Porosity of the composite material*: composite porosity is a subject not often raised by composite tank manufacturers, unlike in the aeronautical sector, where porosity must be less than 1%. The presence of porosity greatly reduces certain mechanical properties, such as the interlaminar shear of composites. A 1% increase in porosity reduces the interlaminar shear strength by an average of 6%. The wet filament winding process is conducive to the creation of porosity in the composite by air entrainment during the impregnation phase. The current average porosity of industrial composite tanks is approximately 4–8%. The towpreg winding process (the impregnation phase is no longer necessary, as the roving is already impregnated with resin) results in porosities of less than 4%. This variant is rarely used by manufacturers, as it is too costly. Today, the European EC79 or international R134 regulations do not require any level of porosity on composite high-pressure tanks for on-board hydrogen storage.

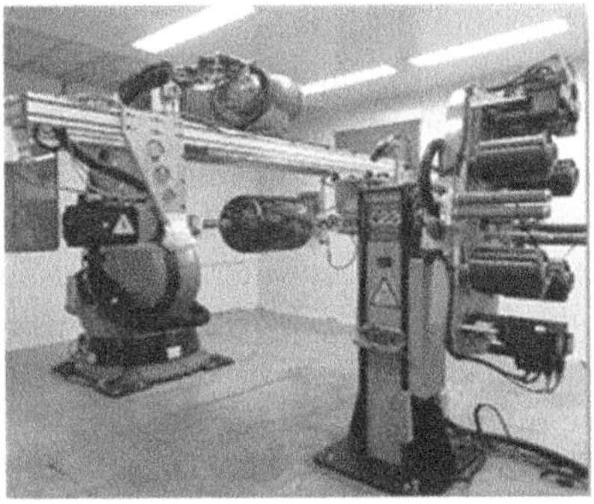

Figure 1.33. *CEA robotized installation with a new rotary winder (Magnier et al. 2022) to avoid web reversals during winding and increase the winding speed. For a color version of this figure, see www.iste.co.uk/derango-cuevas/hydrogen.zip*

– *Permeation level of liners*: hydrogen permeation is a very important issue. The function of the liner is to ensure hydrogen-tight storage with the lowest possible permeation levels, below regulatory thresholds (6 $Ncm^3/h/L$ of tank according to

R134). The liner material must therefore have intrinsic hydrogen barrier properties. Hydrogen permeability through a dense organic material is explained by the organization of molecular chains and the presence of free volumes between macromolecular chains. Diffusion is the limiting factor in the permeation process. There are two categories of polymers: amorphous polymers, where the chains are disordered, and semicrystalline polymers, where ordered crystalline zones are present. Crystalline zones are impermeable to gas, which explains why semicrystalline polymers are less permeable than amorphous polymers. High-density polyethylene (PE) is an amorphous polymer used for tank liners with permeation levels meeting the EC79 and R134 regulatory requirements. Polyamides (PA: polyamide 6, polyamide 11 and polyamide 12) are semicrystalline polymers with better hydrogen permeation performance than polyethylene. They are also used in high-pressure tank technologies. The outlook is for a further reduction in permeation levels, with the ideal aim of achieving a tightness level close to that of metallic materials. Another solution would be to transfer the hydrogen-tightness function to the composite shell (Type V tank) with a polymer matrix that also has a low permeation level.

– *New composite architecture concepts for bottle shapes*: today, with the filament winding process, fiber loading across the entire structure is highly heterogeneous. When the tank reaches its limit of resistance to internal pressure increase, very few fibers are loaded to 100% of their tensile strength capacity. For bottle-like shapes, the composite yield is estimated at approximately 20% (Villalonga 2023). This is largely because too much composite is deposited close to the boss, and a significant proportion of the fiber is not oriented along the main stress axes. This constraint of excess thickness on the boss is intrinsic to the filament winding process, in which the quantity of composite deposited for a layer is the same, regardless of the diameter of the mandrel on which the composite is deposited. A smaller diameter corresponds to greater composite thickness on the bosses. Furthermore, it is not possible to deposit circumferential layers on the domes when circumferential stresses are present. The boss is therefore the part of the tank on which the composite is least effective. The composite can be more effective when the bosses have less impact, as in the case of tubes. A notable avenue for improvement, with the goal of reducing the mass and cost of the structure, lies in the use of new methodologies such as hybrid solutions exploiting filament winding and fiber placement.

– *SHM*: structural health monitoring (SHM) is used to monitor or track the behavior of a structure by installing sensors on, inside or around the tank. The introduction of SHM will enhance the safety of storage systems, particularly when the nominal pressure is 700 bar. This point needs to be developed.

Figure 1.34. *CEA's prototype 36 L 700 bar Type IV H_2 "conformable" tank*

– *Compliance*: the trend in 2023 for light vehicles was to replace cylinder shapes (two or three tanks in the vehicle) with a network of tubes under the vehicle floor. This type of design is currently under development (Figure 1.34). The volume capacity of the storage system can be improved. On the other hand, the number of high-pressure connections is greater, and the temperature rise during the filling phase is faster (which can become a limiting factor for the filling time).

– *Filling time*: today, the filling time for 700 bar tanks in a light vehicle is approximately 5 min when the outside temperature is not high and when hydrogen is dispensed at –40°C. This filling time can be reduced if the current maximum operating temperature of 85°C is increased. This could be the case if the thermomechanical properties of the polymer liners were improved. This would also reduce the selling price of hydrogen at the pump, as it would no longer be necessary to cool the hydrogen to –40°C.

– *Blistering*: in certain situations and for certain polymers, notably elastomers, rapid depressurization under hydrogen can create "internal explosions" in materials, as the dissolved hydrogen has not had time to diffuse outwards quickly enough. This leads to degradation of the polymer and its functions, reducing a liner's hydrogen barrier properties as well as its mechanical and thermomechanical properties. Configurations and the probability of rapid depressurization must therefore be taken into account when considering this phenomenon.

– *Collapse*: today, bonding the liner to the composite structure of a 700-bar tank is very difficult. In most cases, when the composite is manufactured, there is a gap between the liner and the composite. The liner is not integral with the composite. When the tank is rapidly depressurized, internal collapse of the liner can occur. The aim is to find solutions to avoid this phenomenon.

– *Recyclability and thermoplastic matrix composites*: most composite pressure vessels for hydrogen storage are made from a thermosetting epoxy matrix. The cross-linked structure of this polymer makes it very difficult to recycle. The use of a thermoplastic matrix (a non-cross-linked polymer that can melt when heated) makes composite tank structures more easily recyclable. This is crucial in the development of hyperbaric storage solutions, not only from an economic point of view but also above all from an ecological perspective. Our goal is to provide solutions whose environmental impact is known and controlled and to provide levers for reducing this impact and managing the end-of-life of these products. This work is one of the major expectations of HYperStock, the seventh target project of the Priority Research Program and Equipment, Decarbonated Hydrogen (Chapelle 2022).

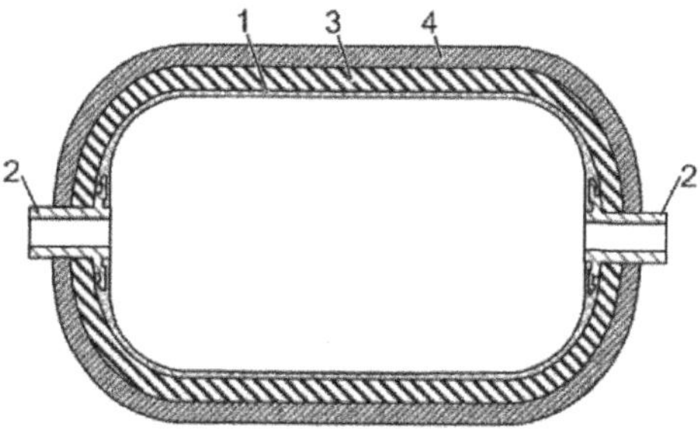

Figure 1.35. *Intrinsically hot-melt tank concept (Molkov et al. 2018)*

– *Fire resistance*: the safety of 700-bar tanks in the event of fire relies on the opening of hydrogen discharge valves via thermofuses (TPRD). The intrinsic resistance of a tank during a fire is between 3 and 5 min (ANR Hype project). The tank bursts not because of an increase in pressure (pressure increases by only a few tens of bar) but because of degradation of the composite structure. The intrinsic fire resistance of a tank can be improved with more suitable materials. Tests carried out during the ANR Hype project revealed that the fire resistance of thermoplastic matrix composites is improved. There are also other material solutions, such as coatings (Villalonga et al. 2013) or fillers, in the composite matrix to improve fire resistance. Molkov et al. (2018) present an architecture (Figure 1.35) that provides intrinsic tank safety against fire. The tank consists of several layers of material. The first layer comprises an inner liner normally impermeable to gas, which becomes permeable to gas at a predetermined temperature (liner). A second intermediate composite layer is designed to withstand the internal pressure exerted by the compressed gas to be stored in the container. A third outer thermal protection layer is formed from a material with a lower thermal conductivity than the intermediate layer, whereby the outer thermal protection layer has a higher temperature gradient throughout its thickness than the intermediate layer when the container is exposed to fire. This concept is a good defense against localized fires where the thermal fuses known as TPRDs are not activated or in the event of TPRD failure.

1.4. Acknowledgments

All the elements addressed and discussed in this chapter, and a very large proportion of the illustrations used, come from collaborative work carried out over the last 10 years. The authors would like to thank the ANR and the FCH-JU for their financial support (ANR OSIRHYS IV, FCH-JU FireComp 325329, FCH-JU Copernic 325330, Project-ANR-07-PANH-0006, Project-ANR-19-CE05-0018, ANR-22-PEHY-0004 2022).

1.5. References

Abisset, E., Daghia, F., Ladevèze, P. (2011). On the validation of a damage mesomodel for laminated composites by means of open-hole tensile tests on quasi-isotropic laminates. *Composites Part A*, 42, 1515–1524.

Amos, W.A. (1999). Costs of storing and transporting hydrogen. Report, National Renewable Energy Lab., Golden, NREL/TP-570-25106 [Online]. Available at: https://www.osti.gov/scitech/biblio/6574 [Accessed 27 February 2017].

Ball, M. and Wietschel, M. (2009). The future of hydrogen – Opportunities and challenges. *Int. J. Hydrog. Energy*, 34, 615–627.

Berro Ramirez, J.P. (2013). Caractérisation et modélisation de l'endommagement des composites bobinés. Application à la prédiction de l'éclatement des réservoirs bobinés hyperbares. PhD Thesis, ISAE-ENSMA, Chasseneuil-du-Poitou.

Berro Ramirez, J.P., Halm, D., Grandidier, J.C., Villalonga, S. (2015a). A fixed directions damage model for composite materials dedicated to hyperbaric type IV hydrogen storage vessel. Part I: Model formulation and identification. *Int. J. Hydrog. Energy*, 40, 13165–13173.

Berro Ramirez, J.P., Halm, D., Grandidier, J.C., Villalonga, S. (2015b). A fixed directions damage model for composite materials dedicated to hyperbaric type IV hydrogen storage vessel. Part II: Validation on notched structures. *Int. J. Hydrog. Energy*, 40, 13174–13182.

Bertin, M., Touchard, F., Lafarie-Frenot, M.C. (2010). Experimental study of the stacking sequence effect on polymer/composite multi-layers submitted to thermomechanical cyclic loadings. *Int. J. Hydrog. Energy*, 35, 11397–11404.

Blanc-Vannet, P., Jallais, S., Fuster, B., Fouillen, F., Halm, D., van Eekelen, T., Welch, S., Breuer, P., Hawksworth, S. (2019). Fire tests carried out in FCH JU Firecomp project, recommendations and application to safety of gas storage systems. *Int. J. Hydrog. Energy*, 44, 9100–9109.

Camara, S., Bunsell, A.R., Thionnet, A., Allen, D.H. (2011). Determination of lifetime probabilities of carbon fibre composite plates and pressure vessels for hydrogen storage. *International Journal of Hydrogen Energy*, 36(10), 6031–6038.

Capurso, T., Stefanizzi, M., Torresi, M., Camporeale, S.M. (2022). Perspective of the role of hydrogen in the 21st century energy transition. *Energy Conversion and Management*, 251, 114898. doi: 10.1016/j.enconman.2021.114898.

CEA and DGEC (2018). Plan de déploiement de l'hydrogène pour la transition énergétique [Online]. Available at: https://www.ecologie.gouv.fr/sites/default/files/Rapport%20H220 MTES%20CEA%200106.pdf [Accessed 16 August 2023].

Chapelle, D. (2022). Stockage hyperbare de l'hydrogène : référentiel et méthodologies matériaux. HYperStock Project ANR-22-PEHY-0004, PEPR-H2 [Online]. Available at: https://www.pepr-hydrogene.fr/projets/hyperstock/ [Accessed 20 July 2023].

Chapelle, D. and Perreux, D. (2006). Optimal design of a Type 3 hydrogen vessel. Part I: Analytic modeling of the cylindrical section. *Int. J. Hydrog. Energy*, 31, 627–638.

Daghia, F., Baranger, E., Tran, D.T., Pichon, P. (2020). A hierarchy of models for the design of composite pressure vessels. *Compos. Struct.*, 235, 111809.

Debray, B., Weinberger, B., INERIS (2021). Guide pour l'évaluation de la conformité et la certification des systèmes à hydrogène [Online]. Available at: https://s3.production. france-hydrogene.org/uploads/sites/5/2022/08/Guide_20certification_20H2_20charte_ 20ADEME_20_28ID_202714844_29_2.pdf [Accessed 14 April 2023].

Espegren, K., Damman, S., Pisciella, P., Graabak, I., Asgeir Tomasgard, A. (2021). The role of hydrogen in the transition from a petroleum economy to a low-carbon society. *Int. J. Hydrog. Energy*, 36, 23125–23138. doi: 10.1016/j.ijhydene.2021.04.143.

Evans, J.T. and Gibson, A.G. (2002). Composite angle ply laminates and netting analysis. *Proc. R. Soc. Lond. A*, 458, 3079–3088.

Ewan, B.C.R. and Allen, R.W.K. (2005). A figure of merit assessment of the routes to hydrogen. *International Journal of Hydrogen Energy*, 30, 809–819.

Gasser, A., Ladevèze, P., Peres, P. (1998). Damage modelling for a laminated ceramic composite. *Mater. Sci. Eng. A*, 250, 249–255.

Granovskii, M., Dincer, I., Rosen, M.A. (2006). Economic and environmental comparison of conventional, hybrid, electric and hydrogen fuel cell vehicles. *J. Power Sources*, 159, 1186–1193.

Hashin, Z. and Rotem, A. (1973). A fatigue failure criterion for fiber reinforced materials. *J. Comp. Mater.*, 7, 448–464.

Hübner, F., Brückner, A., Dickhut, T., Altstädt, V., Rios de Anda, A., Ruckdäschel, H. (2021). Low temperature fatigue crack propagation in toughened epoxy resins aimed for filament winding of type V composite pressure vessels. *Polym. Test.*, 102, 107323.

IEA (2022). Global hydrogen review 2022 [Online]. Available at: https://www.iea.org/reports/global-hydrogen-review-2022 [Accessed 15 January 2024].

Kang, D.H., Kim, C.U., Kim, C.G. (2006). The embedment of fiber Bragg grating sensors into filament wound pressure tanks considering multiplexing. *NDT and E International*, 39, 109–116.

Kouchaki-Penchah, H., Bahn, O., Bashiri, H., Bedard, S., Bernier, E., Elliot, T., Hammache, A., Vaillancourt, K., Levasseur, A. (2023). The role of hydrogen in a net-zero emission economy under alternative policy scenarios. *Int. J. Hydrog. Energy*. doi: 10.1016/j.ijhydene.2023.07.196.

Koussios, S. (2004). Filament winding: A unified approach. PhD Thesis, University of Technology, Delft.

Ladevèze, P. (1988). Sur la modélisation et le calcul des structures composites. *Journées Nationales sur les Composites – JNC 6*, Paris, 11–13 October, 679–712.

Ladevèze, P., Lubineau, G., Violeau, D. (2006). A computational damage micromodel of laminated composites. *Int. J. Fracture*, 137, 139–150.

Lainé, E., Dupré, J.C., Grandidier, J.C., Cruz, M. (2021). Instrumented tests on composite pressure vessels (type IV) under internal water pressure. *Int. J. Hydrog. Energy*, 46, 1334–1346.

Laurin, F., Carrère, N., Maire, J.F. (2007). A multiscale progressive failure approach for composite laminates based on thermodynamical viscoelastic and damage models. *Composites Part A*, 38, 198–209.

Leh, D. (2013). Optimisation du dimensionnement d'un réservoir composite type IV pour stockage très haute pression d'hydrogène. PhD Thesis, Université de Grenoble, Grenoble.

Leh, D., Saffré, P., Francescato, P., Arrieux, R. (2013). Multi-sequence dome lay-up simulations for hydrogen hyper-bar composite pressure vessels. *Composites Part A*, 52, 106–117.

Leh, D., Magneville, B., Saffré, P., Francescato, P., Arrieux, R., Villalonga, S. (2015a). Optimisation of 700 bar type IV hydrogen pressure vessel considering composite damage and dome multi-sequencing. *Int. J. Hydrog. Energy*, 40, 13215–13230.

Leh, D., Saffré, P., Francescato, P., Arrieux, R., Villalonga, S. (2015b). A progressive failure analysis of a 700-bar type IV hydrogen composite pressure vessel. *Int. J. Hydrog. Energy*, 40, 13206–13214.

Li, M., Bai, Y., Zhang, C., Song, Y., Jiang, S., Grouset, D., Zhang, M. (2019). Review on the research of hydrogen storage system fast refueling in fuel cell vehicle. *Int. J. Hydrog. Energy*, 44, 10677–10693.

Liu, K.S. and Tsai, S.W. (1998). A progressive quadratic failure criterion for a laminate. *Comp. Sci. Tech.*, 58, 1023–1032.

Lubineau, G. and Ladevèze, P. (2002). An enhanced mesomodel for laminates based on micromechanics. *Comp. Sci. Tech.*, 62, 533–541.

Magneville, B., Gentilleau, B., Villalonga, S., Nony, F., Galiano, H. (2015). Modeling, parameters identification and experimental validation of composite materials behavior law used in 700 bar type IV hydrogen high pressure storage vessel. *Int. J. Hydrog. Energy*, 40, 13193–13205.

Magnier, C., Villalonga, S., Louvat, E. (2022). Device for unwinding a reel of carbon fibers. Patent, WO2022229554A1.

Mazloomi, K. and Gomes, C. (2012). Hydrogen as an energy carrier: Prospects and challenges. *Renewable and Sustainable Energy Reviews*, 16, 3024–3033.

Molkov, V., Makarov, D., Sergii Kashkarov, S. (2018). Composite pressure vessel for hydrogen storage. Patent, WO2018149772A1.

Padhi, G.S., Shenoi, R.A., Moy, S.S.J., Hawkins, G.L. (1997). Progressive failure and ultimate collapse of laminated composite plates in bending. *Composite Structures*, 40, 277–291.

Papanikos, P., Tserpes, K.I., Labeas, G., Pantelakis, S. (2005). Progressive damage modelling of bonded composite repairs. *Theoretical and Applied Fracture Mechanics*, 43, 189–198.

Peters, S.T. (2011). *Composite Filament Winding*. ASM International, Materilas Park, OH.

Peters, S.T., Humphrey, W.D., Foral, R.F. (1991). *Filament Winding Composite Structure Fabrication*. Sampe Publications, Diamond Bar, CA.

Reddy, Y.S.N., Dakshina Moorthy, C.M., Reddy, J.N. (1995). Non-linear progressive failure analysis of laminated composite plates. *Int. J. Non-Linear Mech.*, 30, 629–649.

Rivard, E., Trudeau, M., Zaghib, K. (2019). Hydrogen storage for mobility: A review. *Materials*, 12, 1973.

Ryu, C., Jung, C., Bae, J. (2020). A study on the analysis of the technology of hydrogen fuel cell vehicle parts – With focus on the patent analysis of co-assignees. *Trans. Korean Soc. Automot. Eng.*, 28, 227–237.

Sdanghi, G., Maranzanab, G., Celzarda, A., Fierro, V. (2019). Review of the current technologies and performances of hydrogen compression for stationary and automotive applications. *Renewable and Sustainable Energy Reviews*, 102, 150–170.

Tigreat, D. (2009). Points réglementaire et normatif sur l'hydrogène en 2009 [Online]. Available at: https://primarisk.ineris.fr/node/3489 [Accessed 14 April 2023].

Tsai, S.W. and Azzi, V.D. (1965). Anisotropic strength of composites. *Exp. Mech.*, 5, 283–288.

Tsai, S.W. and Wu, E.M. (1971). A general theory of strength of anisotropic materials. *J. Composite Mater.*, 5, 55–80.

U.S. DOE (2023). U.S. national clean hydrogen strategy and roadmap. [Online]. Available at: https://www.hydrogen.energy.gov/pdfs/us-national-clean-hydrogen-strategy-roadmap.pdf [Accessed 16 August 2023].

Vargas Rojas, E., Chapelle, D., Perreux, D., Delobelle, B., Thiébaud, F. (2014). Unified approach of filament winding applied to complex shape mandrels. *Composite Structures*, 116, 805–813.

Villalonga, S. (2023). A new dimensionless number for type IV composite pressure vessel designer to increase efficiency and reduce cost. In *International Conference on Hydrogen Safety*. HySafe and the RQEI, Quebec.

Villalonga, S., Nony, F., Yvernes, J.L., Garonne, F. (2013). Fire protection material, high-pressure storage tank coated with said material, methods for preparing same, and uses thereof. Patent, WO2013127902A1.

Villalonga, S., Gentilleau, B., Halm, D. (2015). Preface to the special section on "OSIRHYS IV project: Type IV hydrogen high pressure storage vessel simulation and optimization". *Int. J. Hydrog. Energy*, 40(38), 13146–13147.

Wang, R., Jiao, W., Liu, W., Yang, F. (2010). A new method for predicting dome thickness of composite pressure vessels. *J. Reinf. Plast. Compos.*, 29, 3345–3352.

Wang, R., Jiao, W., Liu, W., Yang, F. (2011). Dome thickness prediction of composite pressure vessels by a cubic spline function and finite element analysis. *Polym. Compos.*, 19, 227–234.

Wang, D., Liao, B., Zheng, J., Huang, G., Hua, Z., Gu, C., Xu, P. (2019). Development of regulations, codes and standards on composite tanks for on-board gaseous hydrogen storage. *Int. J. Hydrog. Energy*, 44, 22643–22653.

Widjaja, M.P. (2020). Towards a full-scale type IV pressure vessel burst simulation using a multiscale model and the ergodic functions theory. PhD Thesis, Mines ParisTech, Paris.

Zahari, R. and El-Zafrany, A. (2009). Progressive failure analysis of composite laminated stiffened plates using the finite strip method. *Composite Structures*, 87, 63–70.

2

Geological Storage

Laurent TRUCHE and Frédéric-Victor DONZÉ
ISTerre, CNRS, Université Grenoble Alpes, France

2.1. Introduction

Hydrogen is the decarbonized fuel of choice, since its combustion, or use in fuel cells, emits only water. At present, storage is the limiting factor for mass deployment of hydrogen energy technologies. Hydrogen gas has a low energy density: 1 kg of H_2 occupies 11 m^3 at 25°C and 1 bar, but in terms of energy by mass, it is equivalent to approximately 3.7 l of gasoline (142 MJ/kgH_2). This means that we need to consider either very large-volume storage (underground geological storage in salt caverns or aquifers, as for natural gas), pressurized or liquefied H_2 tanks, or storage in the form of solid hydrides. In any case, only underground tanks in geological formations are capable of storing very large quantities of energy (from 10 MWh to 10 GWh) for high-power installations (from 10 to 1,000 MW), as shown in Figure 2.1.

The development of renewable energies such as wind and solar power, which are by their very nature intermittent, has produced increasing quantities of electricity. Increasing the share of these renewable energies in the primary energy supply therefore requires the implementation of intermediate energy storage systems to manage production intermittencies and ensure a match between production system supply and consumer demand. Among these technical solutions, water electrolysis can be used to produce hydrogen from carbon-free sources. In such a system, storing hydrogen is equivalent to storing electricity (Figure 2.2). Hydrogen is then considered an energy carrier. Other hydrogen applications, notably in the petrochemical industry or for fertilizer production, may also require storage

Hydrogen Storage,
coordinated by Patricia DE RANGO and Fermin CUEVAS.

solutions to overcome production and consumption contingencies. Storage would also increase the energy efficiency of production and limit its environmental impact.

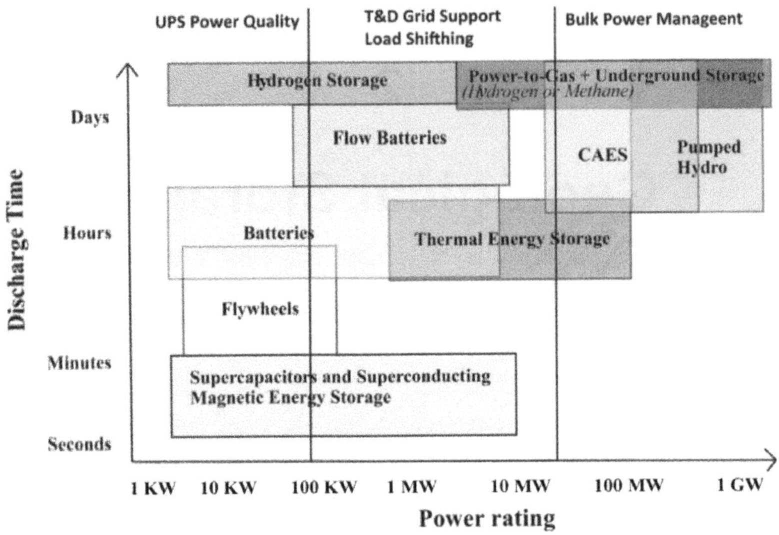

Figure 2.1. *Comparison of different energy storage technologies in terms of power and charge/discharge cycle time. Shown in blue are technologies based on storage in geological reservoirs (compressed air (CAES), heat and geological gas reservoirs: CH_4 and/or H_2). Adapted from Carneiro et al. (2019)*

Geological storage is probably the most promising technology for storing large quantities of hydrogen at low cost, which would make it economically viable in the medium term (Crotogino et al. 2010). Underground storage also offers safety guarantees, as there is no contact with atmospheric oxygen. As shown in Figure 2.3, there are various geological hydrogen storage technologies with different properties in terms of injection/withdrawal cycle time, stored energy quantity and power: porous rock (aquifers or depleted hydrocarbon reservoirs) and artificial cavities (salt caverns or underground mining galleries).

Hydrogen storage in salt caverns is already an effective solution in the United States and Great Britain. Storage in porous media (depleted hydrocarbon fields or aquifers) could be an alternative, especially given that the recovery rate is economically viable and regulations are appropriate. Storage in lined cavities (polymers or steels) could also be an option in geographical areas where other reservoir-type structures are not present but where rocks exist that allow large underground excavations to be made.

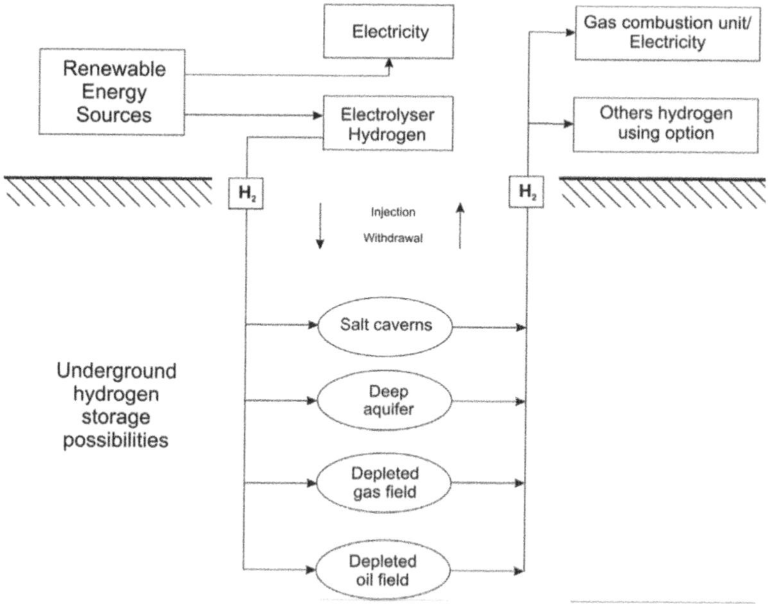

Figure 2.2. *Schematic diagram of geological hydrogen storage. Adapted from Tarkowski (2019)*

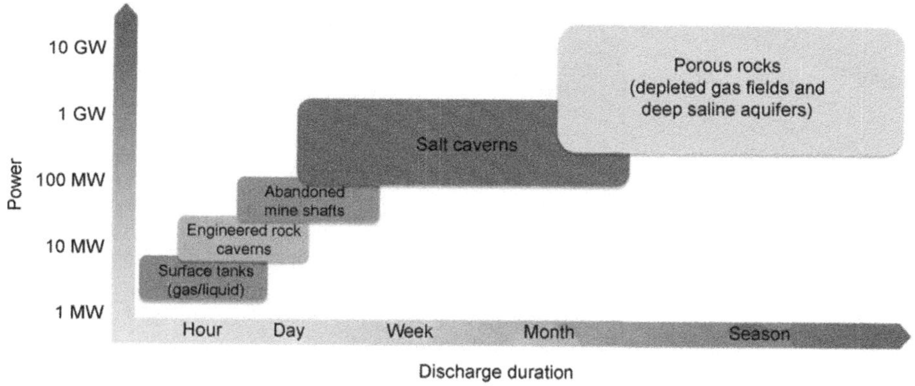

Figure 2.3. *Comparison of different geological hydrogen storage technologies (mined caverns, abandoned mines, salt caverns, aquifers and depleted hydrocarbon tanks). Adapted from Miocic et al. (2023). For a color version of this figure, see www.iste.co.uk/derango-cuevas/hydrogen.zip*

For all of these storage solutions, we need to better understand and quantify the phenomena that could affect the quantity and quality of the resource (Figure 2.4). These include leakage due to the high mobility of hydrogen molecules, diffusive phenomena through the surrounding rock (salt, marl, clay) and hydrogen bioconsumption processes that could produce methane or H_2S. For storage in porous media, numerous parameters need to be taken into account when assessing storage efficiency, such as the geometry of the structure, the petrophysical properties of the reservoir and the cap rock, pressure and temperature differences during reservoir charge/discharge cycles and possible interactions with fluids (water, brines, hydrocarbons), as well as the mineralogical components of the rocks making up the reservoir.

Numerous modeling tools already exist and can be adapted for the evaluation and sizing of hydrogen storage facilities. However, these tools need to be adapted to take into account the specific characteristics of the hydrogen molecule. In addition, many of the parameters specific to the physical and biogeochemical behavior of H_2 needed to drive these models have yet to be acquired.

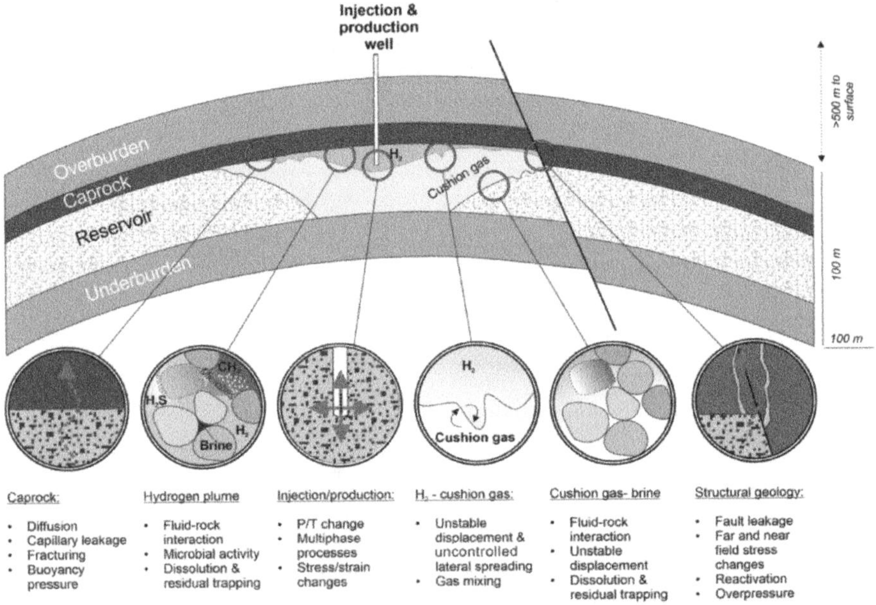

Figure 2.4. *Inventory of the various phenomena that can compromise the effectiveness of H_2 geological storage. Adapted from Heinemann et al. (2021)*

2.2. Underground gas storage

2.2.1. *General concepts*

Gas storage in underground geological formations is not a new concept, with the first installations dating back to 1916 for natural gas (mainly CH_4) in Ontario, Canada (Taylor et al. 1986). An underground gas storage facility is an artificially created accumulation of gas in the natural environment at a significant depth underground, typically between 500 and 2,000 m. It is made up of working gas and buffer gas, known as cushion gas. The working gas is the quantity of gas that is made available to customers, meaning that it can be withdrawn without compromising the properties and reversibility of the storage. Cushion gas is the volume of unextracted gas that acts as a buffer to manage and maintain pressure during injection and withdrawal cycles. Cushion gas does not necessarily have the same composition as working gas. Indeed, in the case of hydrogen storage, some studies suggest choosing nitrogen or CO_2 as the cushion gas because of their wettability properties and low cost (Pfeiffer et al. 2017).

For example, the use of nitrogen facilitates the displacement of water (due to its higher density and viscosity when compared to H_2) to achieve optimal storage space. Feldmann et al. (2016) also suggested that the difference in density between the working gas and the cushion gas is a criterion for cushion gas selection. However, the use of H_2 itself, as a cushion gas, remains a possibility to avoid gravitational instabilities (digitation or fingering) and therefore gas mixtures in the event of too great a density contrast between the cushion gas and the working gas (Lysyy et al. 2021). The purity of the hydrogen stored and, above all, withdrawn is discussed in section 2.2.6.

Underground gas storage is useful because it ensures (i) the safety of the stored gas (less susceptible to fire and damage); (ii) a gain in space; (iii) an economic advantage; and (iv) strategic independence owing to the availability of suitable geological formations around the world (Tarkowski 2019). In addition, it is the best option for massively storing excess electricity generated in the form of H_2 for future use when consumption needs exceed production, as discussed above.

As a result, several initiatives have been funded around the world to study the feasibility of producing, transporting, storing and using H_2 to stimulate the hydrogen economy. Among the projects launched are Roads 2HyCOM (Anderson et al. 2009), HyStorPor (Edlmann et al. 2021), HyUnder (Simon et al. 2015), H2STORE (Henkel et al. 2013), ANGUS+ (Kabuth et al. 2017), SUN.STORAGE (Rag 2020), Hychico (Pérez et al. 2016) and ADMIRE (Hashemi et al. 2021b).

The different characteristics of conventional H_2 underground storage facilities – salt caverns, depleted hydrocarbon aquifers and reservoirs, abandoned coal mines, lined hard rock caverns and refrigerated mined caverns – are detailed below.

2.2.2. *Salt cavern storage*

Salt cavities or salt caverns (mainly halite and gypsum or anhydrite) are artificial chambers created within salt domes or stratified salt deposits. They have a cylindrical shape formed by controlled injection of fresh water into the zone of interest and dissolution of the salt. Figure 2.5 shows a typical representation of the structure of a salt cavern and the effects of different dissolution techniques. A typical salt cavern can reach a volume of 1,000,000 m^3, a height of 300–500 m, and a diameter of 50–100 m, all at a depth of up to 2,000 m (Michalski et al. 2017). These characteristics can enable massive storage quantities at pressures ranging from 30% to 80% of the lithostatic pressure or between 100 and 500 bar depending on the storage depth (Stolten and Grube 2010). Notably, approximately 22%–33% of a cavern's total volumetric capacity is used for cushion gas (Wallace et al. 2021).

Salt caverns built on stratified salt formations are thinner and less stable than those in salt domes (diapirs) because of their high degree of heterogeneity (Bruno and Dusseault 2002; Han et al. 2007). Salt creep processes are essential phenomena to understand as a function of the stress state of the medium. Accurate prediction of the resulting stress evolution and deformation over time is therefore essential to ensure the safety of the storage facility. Creep deformation mechanisms in salt caverns are controlled at the scale of the salt grains (millimeter size): laboratory tests indicate that dislocation creep and solution precipitation creep are the two main deformation mechanisms (Kumar and Hajibeygi 2021).

Salt caverns are often considered the best option for H_2 storage because of their excellent impermeability and ability to redistribute stress via viscoplastic deformation (Lemieux et al. 2019). They are ideally suited for medium- and short-term storage (Sainz-Garcia et al. 2017). Furthermore, owing to their high salinity, caverns may not be affected by autochthonous or allochthonous microbiological activities that consume H_2 and alter the quality of the resource (Jaakkola et al. 2016; Gregory et al. 2019; Dopffel et al. 2021).

In general, the construction of salt caverns faces environmental problems due to the need for large quantities of water for leaching and management of the collected brine. However, they remain the most attractive alternative for the storage of so-called peak H_2 due to their simplicity of control during injection and withdrawal (Tarkowski 2019; Muhammed et al. 2022).

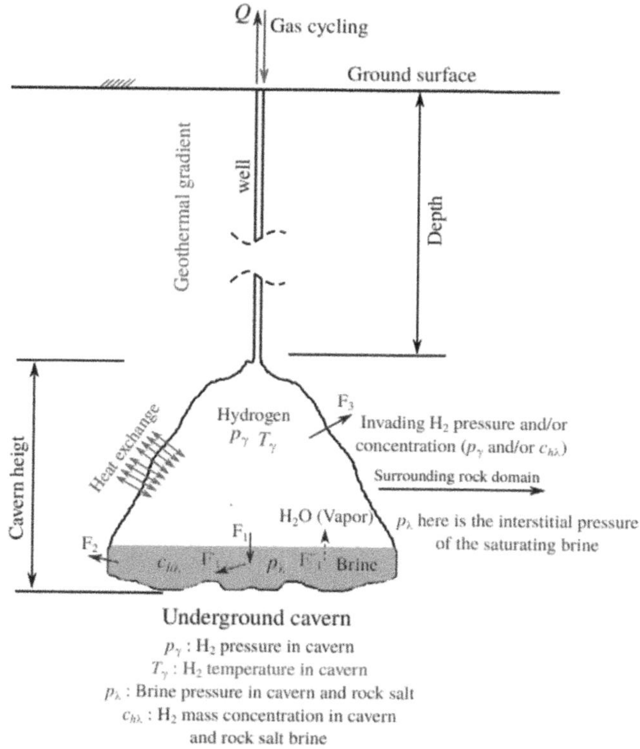

Figure 2.5. *Schematic diagram of H_2 storage in a salt cavern, showing exchanges between the brine phase, the gaseous phase (H_2 and water vapor, depending on temperature) and the surrounding salt formation. Adapted from Wallace et al. (2021). For a color version of this figure, see www.iste.co.uk/derango-cuevas/hydrogen.zip*

2.2.3. *Aquifer storage*

An aquifer is an underground layer of permeable, porous rock filled with fresh or salt water. In the absence of salt formations or depleted hydrocarbon reservoirs, they offer interesting properties for H_2 storage. The mechanisms for storage in aquifers or depleted hydrocarbon reservoirs are similar since both are defined as porous and permeable. The host rock for storage must have good reservoir properties (permeability, porosity), whereas the cap rock must be impermeable to prevent leakage of stored gas.

Figure 2.6 shows the simplified geometry of an aquifer trap within a folded geological formation (anticline type), preventing lateral and vertical gas migration.

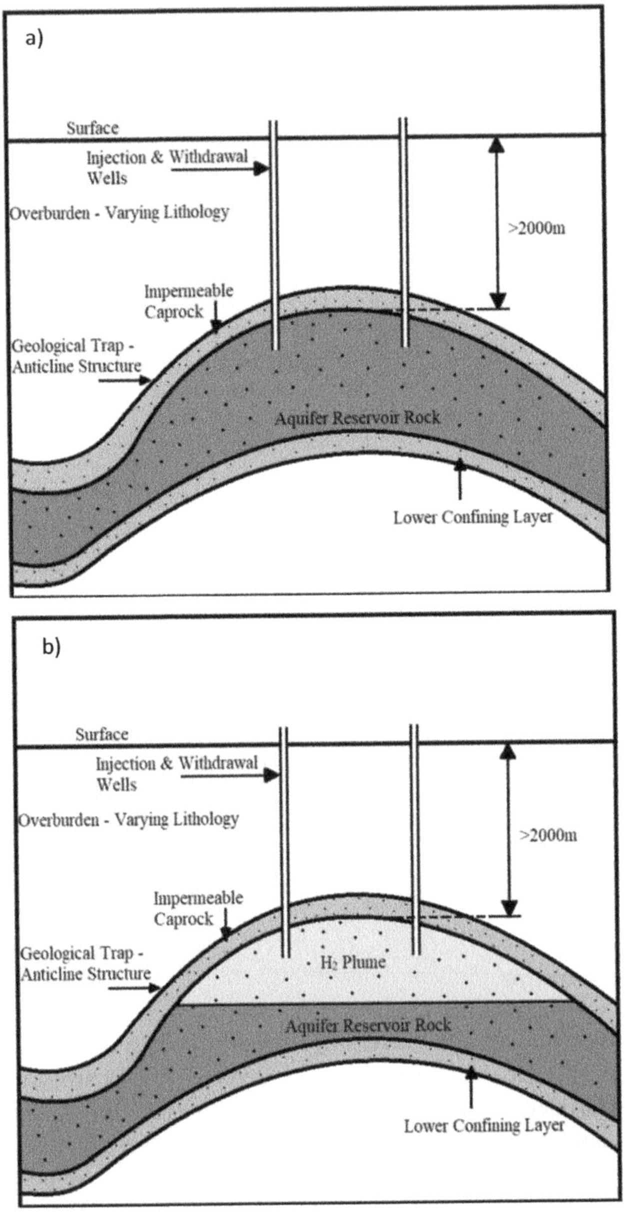

Figure 2.6. *Schematic of aquifer gas storage (a) before injection and (b) after H_2 injection. Adapted from Wallace et al. (2021)*

When H_2 is injected under pressure into a water-filled aquifer, a change in density occurs between the injected gas and the liquid. This causes the liquid to move downward or sideways to create storage space. A liquid–gas interface is then formed within the aquifer. This interface can become a drawback during the withdrawal process if it leads to the simultaneous production of water and H_2 (Pfeiffer and Bauer 2015; Zivar et al. 2021). Importantly, the volume of gas that can be stored is essentially determined by the size and porosity of the reservoir, as well as the temperature and pressure of the stored gas. Cushion gas requirements for aquifers are generally higher than those for depleted natural gas reservoirs because, in the latter, some of the preexisting gas still remains in place after exploitation. As a result, the amount of cushion gas in aquifers can be as high as 80% of the total reservoir capacity. The amount of working gas actually available for exploitation is only 20% (Lord et al. 2014).

Several factors would have a negative impact on the operation of H_2 storage in aquifers. These include leakage along faults, biological consumption processes and the abiotic reactivity of H_2 as a reducing agent with the oxidized phases (mineralized or in solution) present (Truche et al. 2010, 2013a). For these reasons, aquifer storage could be more costly than salt cavern storage (Lord et al. 2014; Tarkowski 2019; Zivar et al. 2021; Muhammed et al. 2022).

2.2.4. *Storage in depleted oil or gas reservoirs*

Depleted reservoirs can be considered as a part of an aquifer where only residual quantities of water exist in the pore space – the latter being mainly occupied by oil or natural gas (Wallace et al. 2021). Generally, they are easy to develop, operate and maintain because of the existing infrastructure.

Furthermore, the integrity of reservoirs has already been proven since they have previously trapped hydrocarbons that have migrated naturally from underlying source rocks to the trap formed by their accumulation at depth. However, this last point needs to be put into perspective in the event of a significant drop in reservoir pressure. If this is the case, the effective stress may reach the mechanical instability range and reactivate the fracture networks present in the cap rock, or worse, reactivate any throughgoing faults.

On the other hand, the gas remaining in the reservoir (mainly CH_4 but also H_2S or CO) can affect the purity of the injected H_2. The trapping system often combines the geometric or structural properties of geological layers (e.g. anticlines) and petrophysical properties (e.g. impermeable clay layers). Between 50% and 60% cushion gas is needed, whereas 80% cushion gas is needed for the aquifer

(Lord et al. 2014; Wallace et al. 2021). In the case of a depleted oil reservoir that has already been exploited for liquid or gaseous hydrocarbons, more in-depth knowledge is needed, as various reactions with oils, or more broadly with organic matter, could compromise the quality of the injected H_2 (Zivar et al. 2021).

2.2.5. *Storage in former coal mines*

Three coal mines have been used to store natural gas worldwide. Two of these are in Belgium (the Péronnes and Anderlues mines) and one is in the United States (the Leyden mine). Although technically anecdotal, gas storage in abandoned coal mines is still being studied as an alternative option for storing large volumes of gas due to the flexibility of use (Ridge 1998).

The geological formation of the Leyden mine in Colorado consists of an underlying aquifer sealed with 20-m-thick impermeable clay. The mine's access shafts were sealed with concrete. In the 1950s, this storage of over 750 million Nm^3 (m^3 normalized to 0°C and 1 bar) of natural gas was able to operate for more than 100 days a year, supplying gas to the Denver area during periods of high demand (Ridge 1998). Similarly, the Anderlues and Péronnes mines enabled the seasonal storage of natural gas. More specifically, the Anderlues operation lasted from 1857–1969 and was capable of storing approximately 180 million Nm^3 of methane.

Although only two cases of H_2 storage in coal seams have been documented to date (Iglauer et al. 2021; Keshavarz et al. 2021), this type of reservoir is considered highly attractive for H_2 storage, particularly in areas where salt caverns, aquifers and depleted oil and gas reservoirs are unavailable. One of the main scientific questions surrounding this type of storage concerns the adsorption of gases (CH_4 or H_2) on the coal surface and in microporous volumes.

2.2.6. *Storage in lined hard rock caverns or refrigerated cavities*

Here, we discuss galleries or artificial chambers excavated in different types of sedimentary, metamorphic or igneous rock. These excavated spaces can then be lined with an impermeable polymer membrane (Lord et al. 2014; Lemieux et al. 2019). A layer of concrete is then constructed during the design phase to transfer the pressure load from the cavern to the surrounding rock and provide a smooth bearing surface for the impermeable lining. Stainless steel or polypropylene plastic can be used to manufacture these linings (Lord 2009; Lord et al. 2014). It is also possible to use an artificial water curtain around the cavern as an impermeable layer. In the case of a refrigerated cavern, this involves freezing water around an artificial cavity to create an impermeable envelope. In addition, the cushion gas injected to maintain pressure inside these storage systems will be minimal because hard rock caverns are structurally stable. A groundwater drainage system is installed around the cavern to relieve hydrostatic pressure against the liner. This storage system has, for example, been used to store natural gas in hard rock in Sweden (Skallen project) (Lemieux et al. 2019). This type of installation could therefore also be suitable for seasonal H_2 storage.

2.2.7. *Feedback from the past*

Underground storage of hydrogen or manufactured gas (a CH_4-H_2-CO_2-CO mixture obtained mainly by pyrolysis or thermal cracking of coal) is not a new concept. A major study of underground hydrogen gas storage was carried out in 1979 by the Gas Technology Institute (GTI) in the United States (Foh et al. 1979). Over the last few decades, there have been several examples of underground hydrogen storage, which are briefly presented below, and in Table 2.1 (Panfilov 2010):

– In France, at Beynes (Ile-de-France), Gaz de France stored manufactured gas containing 45–50% H_2 between 1956 and 1974 in an aquifer with a capacity of 330 million Nm^3 (m^3 normalized to 0°C and 1 bar). No safety problems or losses were reported during the 18 years of operation.

– In Kiel, Germany, gas manufactured with 62% H_2 was stored at 80–100 bar in a 32,000 m^3 salt cavern.

– In Lobodice, Czechoslovakia, manufactured gas containing 50% H_2 was stored in an aquifer.

– In England, at Teesside in Yorkshire, the British company Sabic has stored 1 million Nm^3 of pure H_2 (95% H_2 and 3–4% CO_2) in three salt caverns at a depth of approximately 400 m.

– For several years, Praxair has been operating an underground H_2 salt cavern storage facility to smooth out its hydrogen production. This facility, located in Texas, is connected to Praxair's hydrogen pipeline network, which serves Texas and Louisiana for petrochemical needs.

– In Spindeltop (Texas, USA), Air Liquide recently commissioned the world's largest underground hydrogen salt cavern storage facility. The cavity is located at a depth of 1,500 m, with a diameter of approximately 70 m and a volume of 580,000 Nm^3.

– In Russia, pure hydrogen is stored in a lined cavity at 90 bar for the aerospace industry.

Location	Storage type	Gas composition	Storage volume (m^3)	Mean depth (m)	Status	Year
Teesside, UK	Salt cavern (bedded salt)	95% H_2, 3–4% CO_2	3 × 70 000	350	Active	1972
Clemens, USA	Salt cavern (domal salt)	95% H_2	580 000	1000	Active	1983
Moss Bluff, USA	Salt cavern (domal salt)	H_2	566 000	1200	Active	2007
Spindletop, USA	Salt cavern (domal salt)	95% H_2	906 000	1340	Active	2017
STOPIL-H_2, Etzel, France	Salt cavern		570 000		Under development	
Kiel, Germany	Salt caverns	Town gas	7.8×10^7	1330	Repurposed as natural gas storage	
Bad Lauchstädt, Germany	Salt cavern and porous reservoir (depleted gas field)	Town gas	6.7×10^8	800	Repurposed as natural gas storage	
Underground Sun Storage, Austria	Porous reservoir (depleted field)	10% H_2	115.000	1200	Under development	2017
Yakshunovskoe Field, Russia	Porous reservoir (depleted field)				Active	2010
Hychico, Argentina	Porous reservoir (depleted field)	10% H_2	750 000	815	Under development	2010–18
Kirchheiligen, Germany	Porous reservoir (depleted field)	Town gas	$2.4 * 10^8$	900	Repurposed as natural gas storage	
Hähnlein, Germany	Porous reservoir (aquifer)	Town gas	$1.6 * 10^8$	500	Repurposed as natural gas storage	
Eschenfelden, Germany	Porous reservoir (aquifer)	Town gas	$1.68 * 10^8$	600	Repurposed as natural gas storage	
Engelborstel, Germany	Porous reservoir (aquifer)	Town gas	?	?	Decommissioned	1955–98
Ketzin, Germany	Porous reservoir (aquifer)	Town gas	1.30×10^8	250–400	Decommissioned	1964–2000
Lobodice, -Czech Republic	Porous reservoir (aquifer)	Town gas	$1 * 10^8$	400–500	Repurposed as natural gas storage	1965–95
Beynes, France	Porous reservoir (aquifer)	Town gas	3.3×10^8	430	Repurposed as natural gas storage	1956–72
HyBRIT, Sweden	Rock cavern	100% hydrogen	100	30	Under development	2016

Table 2.1. *Main characteristics of different types of H_2 geological storage facilities that have existed in the past or are still in operation today (adapted from Miocic et al. (2023))*

A few historical articles have been written on pure hydrogen storage (Carden and Paterson 1979; Paterson 1983; Lindblom 1985). They tend to conclude that these storages do not present any major particularities compared with natural gas storage. Nevertheless, the very low density of hydrogen seems to induce very specific gas behavior, including, among other things, a significant spreading of the gas bubble, a source of potential leaks and fingering phenomena during injection.

Variations in gas composition have also been observed in manufactured gas storage (Buzek et al. 1994; Panfilov 2010). Such variations were studied at Lobodice in the former Czechoslovakia, where after several months of storage, the composition of the gas withdrawn showed higher methane levels than those of the initial gas. Moreover, the quantity of acid gas (CO_2 + CO) decreased significantly, and the observed pressure was systematically lower than the pressure calculated on the basis of the material balance. Isotopic analysis of the extracted methane revealed a different isotopic signature to that of the injected methane. These observations were interpreted as indicative of in situ biogenic methane production.

Storage methods are therefore highly dependent on the objectives being pursued (e.g. duration, volume, gas purity and nature of the subsoil). If the aim is to store pure H_2 for later use in fuel cells, for example, any chemical modification of H_2 during storage must be avoided. In contrast, if H_2 is to be used in gas turbines or injected into gas pipelines, it will probably be possible to add methane to H_2.

In the case of underground storage of pure H_2 (for example, use in electrolyzers), salt caverns or mined cavities will certainly be the most practical storage sites. Aquifers or depleted hydrocarbon reservoirs could be reserved for H_2 storage containing other gases, such as methane. The latter could be coinjected (as was the case with city gas storage mentioned above) or produced in situ by the activity of methanogenic bacteria (Panfilov 2010).

2.3. Hydrodynamic properties of hydrogen in a storage context

2.3.1. *Fluid–H_2 interactions*

2.3.1.1. *Hydrogen density*

The H_2 density as a function of P and T can be calculated according to equation [2.1] (Lemmon et al. 2018):

$$\rho_{H_2}(P,T) = \frac{P_{H2}}{RT}\left(1 + \sum_{i=1}^{9} a_i \left(\frac{100}{T}\right)^{b_i} \left(P_{H_2}\right)^{c_i}\right)^{-1} \quad [2.1]$$

where P_{H_2} is the partial pressure of H_2 gas expressed in MPa, T is the temperature in Kelvin, R is the perfect gas constant (8.314 J/(mol·K)) and a_i, b_i and c_i are coefficients given in Table 2.2. This equation is valid over a pressure and temperature range from 1 to 2,500 bar and from 0 to 250°C, respectively. The combined effect of temperature and pressure on H_2 density is shown in Figure 2.7.

The H_2 density increases significantly with pressure (P) and marginally with temperature (T), as shown in Figure 2.7. For example, when the pressure increases from 5 to 120 MPa, at 50°C, the H_2 density increases from approximately 5 to 50 g/L, whereas at 20 MPa, when the temperature increases from 25°C to 150°C, the H_2 density decreases from 15 to 10 g/L. Compared with methane or CO_2, the variation in H_2 density with temperature and pressure is much smaller. Notably, the density of H_2 is much lower than that of water, CO_2 and methane at a given pressure and temperature.

This large difference in density between water and hydrogen is responsible for the gravitational instabilities (digitations or fingering) of such a mixture.

i	a_i	b_i	c_i
1	0.05888460	1.325	1.0
2	−0.06136111	1.87	1.0
3	−0.002650473	2.5	2.0
4	0.002731125	2.8	2.0
5	0.001802374	2.938	2.42
6	−0.001150707	3.14	2.63
7	0.9588528×10^{-4}	3.37	3.0
8	$-0.1109040 \times 10^{-6}$	3.75	4.0
9	0.1264403×10^{-9}	4.0	5.0

Table 2.2. *Parameter values used in equation [2.1]*

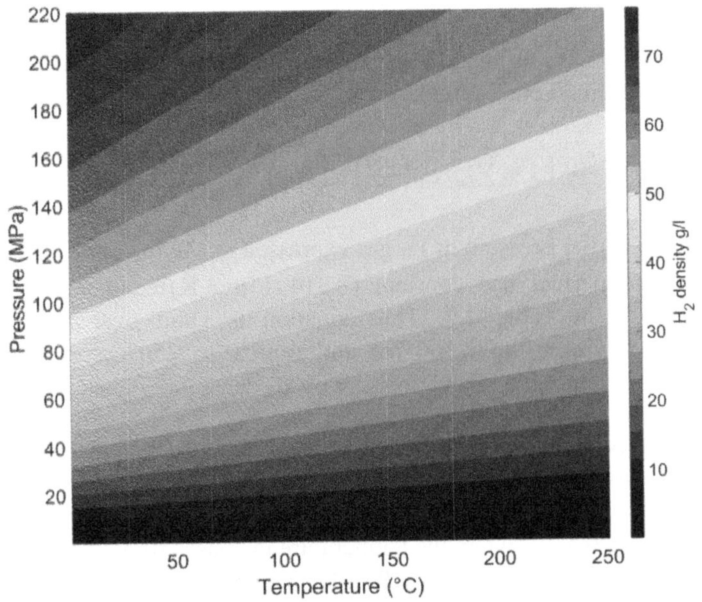

Figure 2.7. *Effect of temperature and pressure on H_2 gas density calculated according to equation [2.1]. For a color version of this figure, see www.iste.co.uk/derango-cuevas/hydrogen.zip*

2.3.1.2. *Hydrogen viscosity*

Hydrogen viscosity is an essential parameter during the injection and withdrawal stages since it measures the fluid's ability to flow through the porous medium. When fluid viscosity is combined with relative permeability, it can be used to determine fluid–fluid interfacial stability, indicating the propensity to generate digitation- or fingering-type gravity instabilities (Wang et al. 2013). The viscosity of H_2 can be calculated via the formalism proposed by (Muzni et al. 2013) from triple-point P/T conditions (13.7 K and 7.2 MPa) up to 1000 K and 200 MPa. The viscosity of H_2 is affected by pressure and temperature, as shown in Figure 2.8.

For example, when the pressure increases from 5 to 130 MPa, at 50°C, the H_2 viscosity increases by approximately 10–15 μPa. s, whereas at 20 MPa, when the temperature increases from 25°C to 150°C, the H_2 viscosity increases by only 9 to 11 μPa. s. While the viscosities of CH_4 or CO_2 are relatively similar to those of H_2 at low pressure (<10 MPa), they are relatively higher at high pressure. At identical temperature and pressure, the viscosity of water is one to two orders of magnitude greater than that of H_2. This large difference in viscosity between water and H_2 contributes to gravity instabilities during H_2 injection, as do differences in density.

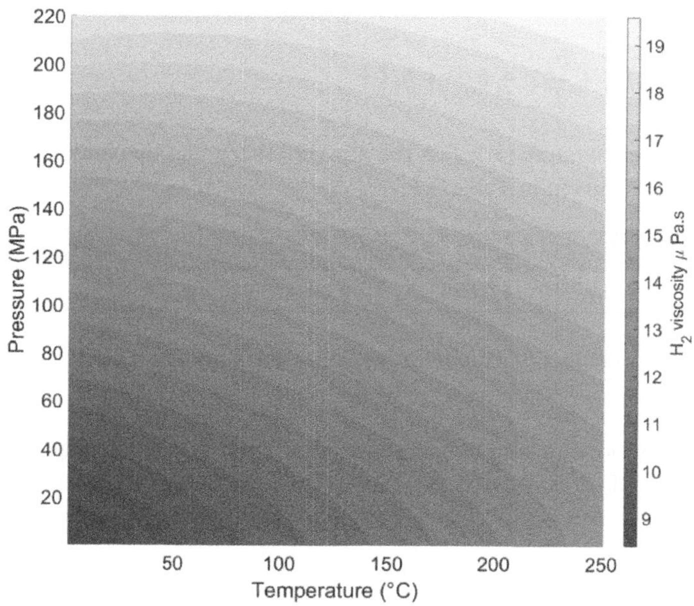

Figure 2.8. *Effect of temperature and pressure on the viscosity of H_2 calculated according to the formalism of Muzni et al. (2013). For a color version of this figure, see www.iste.co.uk/derango-cuevas/hydrogen.zip*

2.3.1.3. *Hydrogen solubility in water*

Hydrogen is a gas with very low solubility in water. Under normal temperature and pressure conditions (1 bar and 25°C), its solubility is 0.784 mol/m^3 in water (Ortiz et al. 2002). On the basis of experimental measurements, a simple formulation for fast, accurate calculations of H_2 solubility in water has been proposed by Chabab et al. (2020). Using the number of moles of H_2 (n_{H_2}) and H_2O (n_{H_2O}), the solubility of H_2 in water is determined by:

$$\chi_{H_2}^0 = \frac{n_{\mathrm{H}_2}}{n_{\mathrm{H}_2} + n_{\mathrm{H}_2\mathrm{O}}} \qquad [2.2]$$

The solubility, expressed as a mole fraction χ_{H_2}, can then be converted into a molality m_{H_2} (in mol/kgw) according to the following equation:

$$m_{\mathrm{H}_2} = \frac{100\ \chi_{\mathrm{H}_2}^0}{M_{\mathrm{H}_2\mathrm{O}}\left(1 - \chi_{\mathrm{H}_2}^0\right)} \qquad [2.3]$$

The solubility of H_2 in water can be approximated using the following empirical equation (Chabab et al. 2020):

$$\chi_{\mathrm{H}_2}^0 = b_1 PT + \frac{b_2}{T} T + b_3 P + b_4 P^2 \qquad [2.4]$$

where T is the temperature (K); P is the pressure (bar); and the values of parameters b_1, b_2 b_3 and b_4 are given in Table 2.3.

b_1	b_2	b_3	b_4
3.338844×10^{-7}	0.0363161	–0.00020734	$-2.1301815 \times 10^{-9}$

Table 2.3. *Parameter values used in equation [2.4]*

The range of pressure and temperature values used to establish equation [2.4] for the solubility of H_2 in pure water is $273 < T(K) < 373$ and $1 < P$ (bar) < 203. Indeed, the values of parameters b_1 to b_4 were established on the basis of experimental data up to 200 bar.

However, it is possible to reasonably extend the scope of equation [2.4] to higher pressures (up to 1 kbar) and temperatures (up to 250°C), as it appears that the H_2 solubility data provided by Seward and Franck (1981) can be well predicted while retaining the same coefficient values (Figure 2.9(a)).

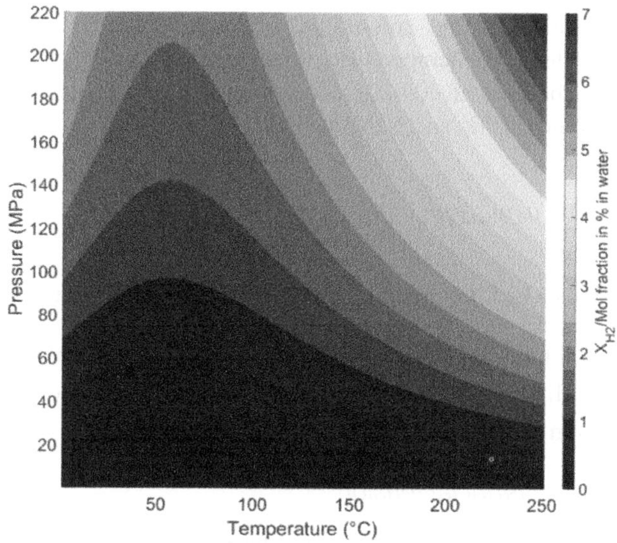

a)

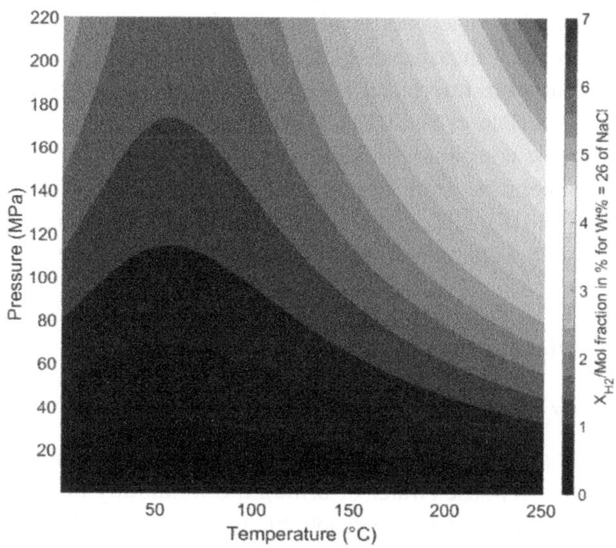

b)

Figure 2.9. *Effect of temperature and pressure on H_2 solubility: (a) Pure water; (b) 26 wt% NaCl solution. For a color version of this figure, see www.iste.co.uk/derango-cuevas/hydrogen.zip*

Note that to calculate the solubility of H_2 in brine, a simple correlation considering the effects of temperature, pressure and NaCl molality has been developed by Chabab et al. (2020). The proposed correlation is based on a Setschenow-type relationship and is denoted by:

$$\ln\left(\frac{\chi_{\mathrm{H_2}}^{\mathrm{brine}}}{\chi_{\mathrm{H_2}}^{0}}\right) = a_1 m_{\mathrm{NaCl}}^2 + a_2 m_{\mathrm{NaCl}} \qquad [2.5]$$

with a value of 0.018519 for parameter a_1 and – 0.30185103 for parameter a_2.

Knowing the solubility $\chi_{H_2}^0$ of H_2 in pure water at system temperature and pressure, the solubility $\chi_{H_2}^{brine}$ of H_2 in brine at a given mNaCl molality is therefore obtained by equation [2.5] and is presented in Figure 2.9(b).

2.3.1.4. *Fluid–fluid interfacial tension*

The separation between two immiscible phases is determined by the types of molecules on either side of the meniscus and by the intermolecular interaction between these molecules. The interfacial tension between the phases present is a critical characteristic, as it measures the amount of energy required to produce more surface area for separation. Interfacial tension is a function of pressure, temperature and fluid composition (Pan et al. 2020). In the case of water and H_2, the interfacial tension remains almost constant with increasing pressure but decreases significantly with increasing temperature from approximately 70–40 mN/m between 25°C and 175°C (Chow et al. 2018). CH_4 and CO_2 behave differently with water, since their interfacial tension with water decreases significantly with pressure.

2.3.2. *Solid–H_2 interactions*

2.3.2.1. *Wettability*

In a three-phase system comprising the rock matrix, the non-aqueous phase (oil or gas) and the aqueous phase (water or brine), the wettability of the rock surface is generally defined in terms of the hydrophobicity or hydrophilicity of the aqueous phase and the ability of the phase to spread or adhere to the rock surface (Figure 2.10). The wetting phase may be oil, water or gas, whereas the solid phase is reservoir rock. This phenomenon is governed by the interaction effects of intermolecular forces (Yekeen et al. 2020). In reservoir rock previously saturated with water, the rock prefers water to oil or gas so that the oil/gas occupies the center of the pores (Deng et al. 2020). Wettability is determined at the pore scale by the local contact angle (the angle formed by the interface between two fluid phases with the solid, usually measured through the denser phase) (Iglauer 2017).

There are several techniques for characterizing wettability, but here, we focus on the contact angle method since it is the most suitable when pure fluids and rocks or minerals are present (Yekeen et al. 2020). However, the wettability of a surface can be greatly influenced by local changes in porosity, surface heterogeneities and the chemistry of the aqueous phase.

Owing to the variety of mineral compositions, rock surface chemistry and pore geometry, rock wettability is heterogeneously distributed in the storage rock pore network (Singh et al. 2016; Scanziani et al. 2017).

However, if the wettability (contact angle) for a given rock type is known, then this quantity can be used to perform pore-scale modeling to estimate the upscaled capillary pressure and relative permeability curves (Hashemi et al. 2021a).

These latter parameters can then be used as input parameters for reservoir-scale simulations (Blunt 2017; Kunz et al. 2018). Generally speaking, the wettability (contact angle) increases with pressure for a given surface area and temperature.

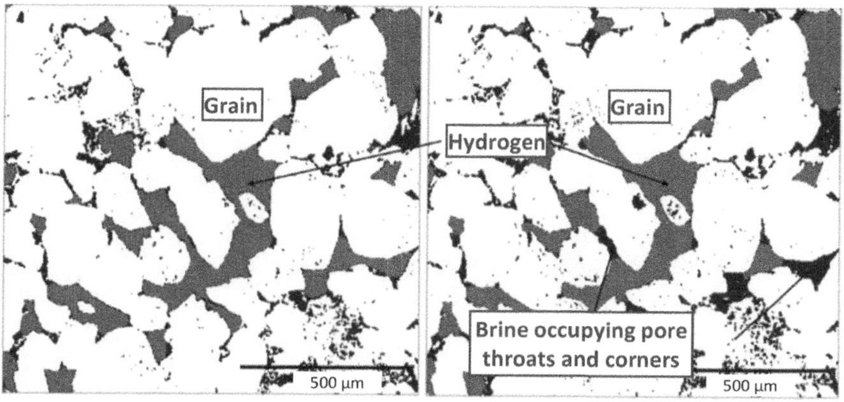

a) b)

Figure 2.10. *Microtomography image of the pore network in a sandstone rock (adapted from Jha et al. (2021)). Brine is shown in dark blue, whereas hydrogen is represented in light blue: (a) H_2-saturated rock and (b) residual H_2 saturation after gas withdrawal. For a color version of this figure, see www.iste.co.uk/derango-cuevas/hydrogen.zip*

2.3.2.2. *Capillary pressure*

The pressure between two immiscible fluids in the pore space results from the interaction of forces between the fluids and the solid walls. This pressure is known as capillary pressure (Pc). This is an essential aspect in the study of multiphase fluid flow in porous media (Iglauer 2017):

$$Pc = \left(\frac{2\gamma\cos\theta}{r}\right) \quad [2.6]$$

where Pc, γ, θ and r denote the capillary pressure, interfacial tension, contact angle and radius of the largest pore, respectively.

Capillary pressure is usually combined with relative permeability curves (both as a function of water or brine saturation) to assess fluid movement and different fluid configurations at the pore scale. When H_2 is injected into porous media, the liquid phase (water or brine) initially present in the aquifer formation is replaced by H_2. As a result, H_2 moves toward the top of the reservoir and remains at the level of the caprock due to gravity segregation. The pressure required to displace this brine is equal to the capillary pressure. It is therefore clear that during storage operation, it will be necessary to maintain a low water saturation; or in other words, a high hydrogen saturation, hence the importance of the cushion gas.

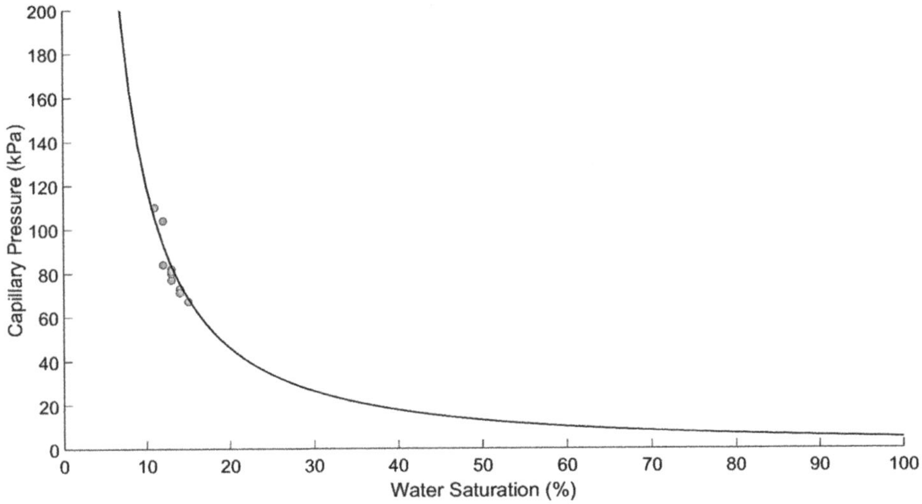

Figure 2.11. *Capillary H_2 pressure as a function of water saturation was calculated according to equation [2.7] (black curve). The red dots correspond to experimental measurements by Yekta et al. (2018). For a color version of this figure, see www.iste.co.uk/derango-cuevas/hydrogen.zip*

Yekta et al. (2018) carried out sandstone core experiments to measure capillary pressure, which is defined as the pressure difference across the interface between two phases represented by H_2 and water in porous sandstone. A capillary pressure dataset was obtained from measurements of semidynamic capillary pressure and mercury injection capillary pressure. Combining these two types of data allows us to determine the capillary pressure evolution for the H2–water system over the entire water saturation range:

$$P_c(S_l) = P_d(S_w)^{-1/\lambda} \quad [2.7]$$

where $\lambda = 0.725$ is the Brooks–Corey parameter and Pd = 5.0 kPa is the inlet pressure. The variation in capillary pressure as a function of water saturation, calculated according to equation [2.7], is shown in Figure 2.11.

2.3.2.3. *Relative permeability*

Several factors can affect the permeability of hydrogen gas in the porous medium. In the case of dry flow, meaning in the absence of water, the permeability corresponds to the absolute permeability of the medium (for sandstone, see Yekta et al. (2018)). A gas can be subject to two types of interactions within a porous medium: molecule–molecule interactions and molecule–pore wall interactions. The smaller the pores are, the greater the molecule–pore wall interactions and therefore the faster the molecules will move because of interactions with the walls. Gas transport in low-permeability rocks can therefore be affected by slip (Klinkenberg effect), leading to deviations from Darcy's law. Slip occurs when the mean free path of gas molecules approaches the mean diameter of the pore opening, resulting in an acceleration (slip) of the velocity of individual gas molecules as they come into contact with the pore walls. Under slip conditions, the measured (apparent) gas permeability $k_{app\_H_2}$ is a function of the mean pore pressure such that:

$$k_{app\_H_2} = k_\infty\left(1 + \frac{b_k}{P_f}\right) \quad [2.8]$$

where (k_∞) is the value of the intrinsic permeability as a function of the mean H_2 pressure (P_f) and the coefficient b_k, known as the slip factor or Klinkenberg coefficient. On the basis of the Klinkenberg phenomenon, the apparent gas permeability coefficients $k_{app\_H_2}$ approach a limiting value at an infinite mean pore pressure. This limiting permeability value is commonly referred to as the Klinkenberg corrected permeability. Increasing the gas pressure during testing reduces the effect of gas slip but also decreases the effective stress, which in turn influences permeability. The coupled effect of gas slip and effective stress on rock permeability is still not well defined (Yang et al. 2021), and as a first approximation, we neglect the contribution of gas slip, that is:

$$k_{H_2} = k_{app\_H_2} \quad [2.9]$$

Water saturation (Sw) is another factor influencing H_2 gas flow in porous media (Pan et al. 2021). Water saturation in geological reservoirs generally varies between 15% and 35%. The apparent permeability can be modified using the relative permeability ratio $k_{r\_H_2}$ such that $K_{H_2} = k_{H_2} \times k_{r\_H_2}$. It is well established that $k_{r\_H_2}$ is a function of Sw and rock wettability. However, data are very scarce for H_2 and water: only one dataset was found by the authors, where $k_{r\_H_2}$ and the associated relative water permeability ratio k_{r_w} were measured under in situ reservoir conditions (Yekta et al. 2018) (Figure 2.12). These results indicate a very low capacity of H_2 to flow in a saturated medium. The dataset (Yekta et al. 2018) was fitted with an exponential function, for water saturation values between 13% and 90%, as follows:

$$k_{r\_H_2} = e^{-0.12(\mathrm{Sw}-10)} \quad [2.10]$$

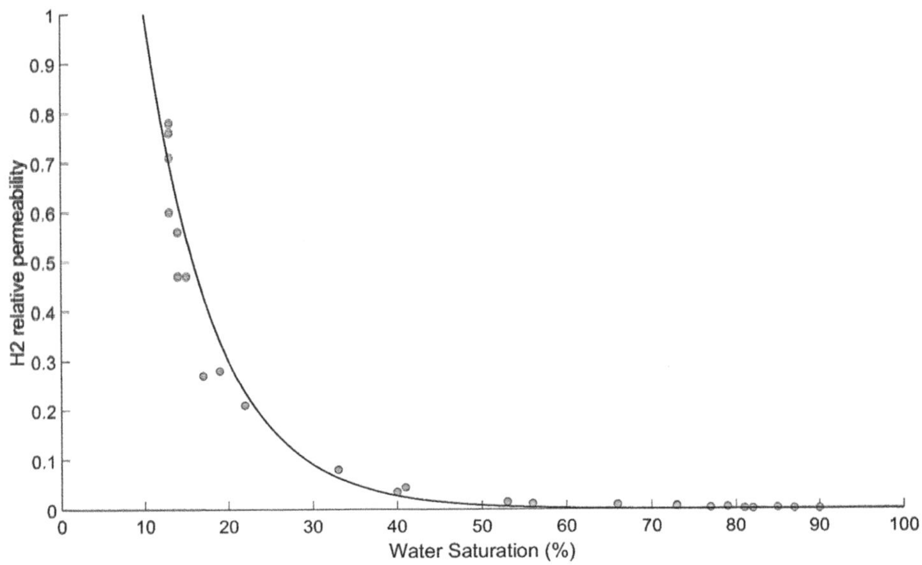

Figure 2.12. *Relative H_2 permeability as a function of water saturation calculated according to equation [2.9] (curve in black). The red dots correspond to experimental measurements from Yekta et al. (2018). For a color version of this figure, see www.iste.co.uk/derango-cuevas/hydrogen.zip*

Among the influencing properties (Pc, k_r, θ and γ) used to quantify fluid–fluid and rock–fluid interactions in underground storage, the effects of k_r and Pc are the two main quantities that define storage efficiency (Zivar et al. 2021). Notably, the relative permeability is a function of the reservoir's state of saturation and therefore of the storage history. Each injection and withdrawal phase implies different trapping conditions for H_2. The same observation applies to the inlet capillary pressure, as shown in Figure 2.13.

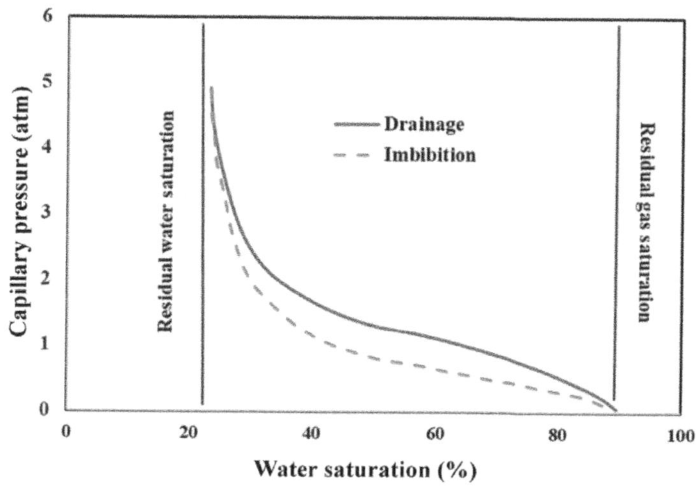

Figure 2.13. *Illustration of the hysteresis phenomenon affecting H_2 inlet capillary pressure as a function of water saturation during an injection (= "imbibition")/ withdrawal (= "drainage") cycle. Adapted from Carden and Paterson (1979). For a color version of this figure, see www.iste.co.uk/derango-cuevas/hydrogen.zip*

2.3.2.4. *Adsorption–desorption*

Adsorption, or physisorption of H_2, has been the subject of very few studies in geology. Indeed, the vast majority of work on the subject focuses on synthetic materials (e.g. metal organic frameworks (MOFs), zeolites and nanocarbons) and cryogenic temperatures (<77 K), so their conclusions are difficult to extrapolate to the conditions of geological hydrogen storage. However, we can anticipate that clay minerals with high specific surface area and microporosity will certainly be of major importance to the adsorption phenomenon.

Indeed, H_2 intercalation in the interfoliar space, on the basal and lateral surfaces and in the micropores between the stacks of clay particles, is favored by (i) the presence of a very strong electrostatic field inside the interfoliar cavities due to the presence of exchangeable cations (this electrostatic field induces a dipole moment to the H_2 molecule – normally apolar – and thus favors physisorption); (ii) the geometry of the interfoliar space, which has no bottleneck likely to prevent H_2 diffusion; and (iii) the size of the micropore network, which is close to the kinetic diameter of H_2 (2.89 Å).

Interactions between hydrogen gas and clay rock have been studied mainly in the context of radioactive waste disposal. Hydrogen is produced by the anoxic corrosion of steel waste containers and, to a lesser extent, by the radiolysis of water. A series of pioneering studies on hydrogen sorption on Callovo–Oxfordian argillites (a mixture of illite/smectite, quartz and carbonates) and montmorillonite have been carried out in this context (Didier et al. 2012; Bardelli et al. 2014; Mondelli et al. 2015). The results of these studies show that montmorillonite can adsorb up to 0.2% by weight (or 5 kg_{H2}/m^3 density = 2.4) at an H_2 pressure of 30 bar and at 90°C. However, these results are open to question. First, the quantity of H_2 adsorbed appears to be little affected by temperature between 25°C and 90°C, and second, the measured adsorption rates are very similar to those reported for the best synthetic nanomaterials (MOF type) under the same temperature and pressure conditions. The specific surface area is a major parameter controlling the quantities of gases adsorbed. The specific surface area of natural clays does not exceed a few tens of square meters per gram, whereas that of synthetic nanoadsorbents is often in excess of 1,000 m^2/g.

Edge et al. (2014) reported adsorbed H_2 quantities of no more than 0.1% by weight at 20°C and 100 bar on laponites: synthetic hectorite-type clays (smectite) with specific surface areas in excess of 100 m^2/g. Recently, Ziemiański and Derkowski (2022) studied H_2 adsorption on montmorillonite and illite at temperatures ranging from 25°C to 70°C and pressures up to 145 bar. Under the conditions tested, the maximum H_2 adsorption never exceeded 0.04% by weight.

It is important to note, however, that even if the quantities of H_2 adsorbed by clays in the 25–90°C temperature range remain relatively modest (certainly around 0.04% at 20°C and 100 bar H_2), this phenomenon can take considerable proportions when the volume or mass of rock impacted by geological storage in a porous medium is considered.

Finally, even if the adsorption phenomenon seems reversible on the scale of a powder sample weighing a few grams in the laboratory (as used for the adsorption measurements reported above), it will be important to consider any hysteresis between the adsorption and desorption phases under storage conditions.

2.4. Hydrogen reactivity in underground environments

The reactivity of hydrogen as a reducing agent in the natural environment is still poorly understood, as it does not correspond to common geological reactions. In recent years, however, interest within the scientific community in the question of hydrogen-induced redox processes has increased. This has focused on three main areas of research: (i) the production of natural hydrogen in different geological contexts and its role in the development of the deep biosphere and as a primary carbon-free energy resource; (ii) the reactivity and transport of hydrogen produced by corrosion in geological storage of radioactive waste; and (iii) the massive storage of hydrogen in certain geological formations.

Depending on the field of investigation, the scientific motivations are different: (i) transport of metals, origin of life, assessment of energy resources and material flows; (ii) destabilization of the geological environment and modification of prevailing geochemical conditions; and (iii) loss or alteration of the resource through possible leakage or biotic or abiotic interactions with the host formation. All of these studies focus on the abiotic or microbial reactivity of hydrogen toward oxidized species, whether dissolved or mineralized. In section 2.4.1, we propose an analysis of the situation, distinguishing between abiotic and microbial reactivity, bearing in mind that these two processes can act in tandem in underground H_2 storage.

2.4.1. *Abiotic reactivity and surface catalysis*

2.4.1.1. *Sulfate reduction*

The reduction of sulfates by organic species is well documented in the oil industry. A distinction is usually made between low-temperature redox reactions catalyzed by microorganisms and reactions occurring spontaneously at relatively high temperatures (TSR: thermochemical sulfate reduction; see equation [2.10]). The activation energy of the TSR reaction is very high, implying a very strong temperature dependence.

On the basis of measurements of sulfate reduction by organic compounds, half-lives for sulfate are estimated at between 100,000 and 200,000 years for temperatures of approximately 100°C (Kiyosu and Krouse 1993; Goldhaber and Orr 1995; Cross et al. 2004; Zhang et al. 2008a, 2008b). The parameters controlling the kinetics are still poorly understood, but it appears that the presence of H_2S and an acidic pH are required to initiate the reaction:

$$SO_4^{2-} + 4H_2 + 2H^+ = H_2S + 4H_2O \qquad [2.11]$$

Truche et al. (2009) quantified the kinetics of the TSR reaction using H_2 as the electron donor. Measurements were carried out in a homogeneous aqueous medium between 250°C and 300°C and then extrapolated to the temperatures of interest for storage. Experiments were carried out with aqueous sulfate concentrations representative of the water of formation (a few millimoles per liter), under hydrogen pressure ranging from 0 to 16 bar, and in a pH range from 1.5 to 7. The main conclusion of interest for storage is that this reaction is unlikely to occur at temperatures below 90°C for less than 100,000 years. Indeed, the high activation energy of the reaction (131 ± 10 kJ/mol) and the half-life of sulfate of 210,000 years at 90°C rule out the possibility of such a reaction in the absence of a catalyst.

2.4.1.2. *Pyrite reduction*

The reduction of pyrite to pyrrhotite (see [2.11]) has been the subject of thermogravimetric research at temperatures ranging from 350 to 700°C in the vapor phase (Wiltowski et al. 1987; Lambert et al. 1998). Under hydrated conditions and at lower temperatures, published studies have focused on the reverse reaction (pyrrhotite + H_2S => pyrite) in the context of prebiotic chemistry (Rickard and Luther 1997). Truche et al. (2010, 2013a) studied the kinetics of the pyrite–hydrogen reaction in aqueous solution under pressure and temperature conditions of interest for storage. The experiments were carried out in a batch autoclave and in a three-phase environment, with solid pyrite powder in contact with liquid water, itself in equilibrium with a sky of gas:

$$FeS_2 + H_{2(aq)} = FeS + H_2S_{(aq)} \qquad [2.12]$$

The mechanism and kinetics of the reaction were described by combining a textural study and solution analysis for different temperatures (90–180°C) and H_2 partial pressure (0–30 bar). When the pH of the solution is controlled by calcite or Callovo–Oxfordian clay ($6.5 < pH < 9$), pyrite is partially reduced to pyrrhotite in the presence of hydrogen (Figure 2.14). The reaction mechanism results from a coupling between the dissolution–reduction of pyrite in the presence of H_2 and the precipitation of pyrrhotite in pyrite dissolution pits. The pyrrhotite precipitated at the reaction front takes the form of small, micrometer-sized crystals.

The reaction rate increases with increasing hydrogen partial pressure and decreases with increasing sulfide concentration. These observations are consistent with a dissolution–precipitation mechanism. The composition of the fluid at the dissolution front and the role of the pyrite surface as a substrate for pyrrhotite precipitation are key parameters controlling the progress of the reaction. The activation energy of the reaction (53 kJ/mol) suggests a kinetic regime controlled by both diffusive mechanisms and surface reactions.

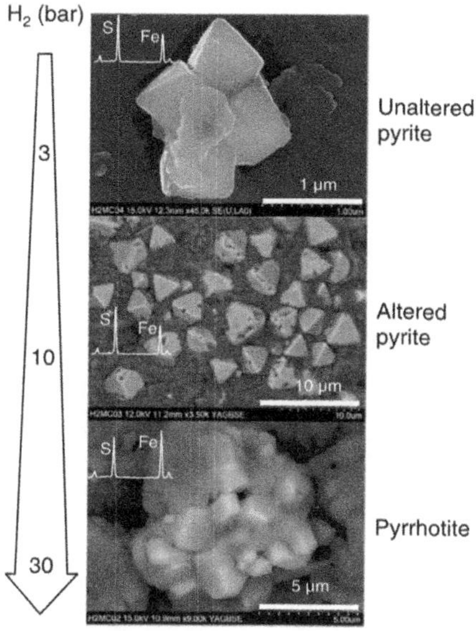

Figure 2.14. *Scanning electron microscopy images taken in backscattered electron mode showing the progressive alteration of framboidal pyrites from Callovo–Oxfordian argillite to pyrrhotite after three months of water–rock interaction at 90°C and three different H_2 partial pressures: 3, 10 and 30 bar. Energy dispersive spectra (EDS) of pyrite and pyrrhotite appear in the upper left-hand corner of the images. Adapted from Truche et al. (2013a)*

2.4.1.3. *Hematite reduction*

The reduction of hematite to magnetite, and even to native iron in the presence of H_2, is well documented at temperatures above 400°C (Pineau et al. 2006; Zielinski et al. 2010) (see [2.13] and [2.14]):

$$3Fe_2O_3 + H_2 = 2Fe_3O_4 + H_2O \quad [2.13]$$

$$Fe_3O_4 + 4H_2 = 3Fe + 4H_2O \quad [2.14]$$

Recently, Thüns et al. (2019) have shown that the reaction also occurs at 200°C dry (with no water initially in the system) and ultimately leads to the production of native iron after a few days at H_2 partial pressures on the order of 10 bar. The kinetics of the reaction at lower temperatures and the effect of the presence of water still need to be assessed before these results can be extrapolated to geological storage conditions. However, it would appear that the reduction of hematite to magnetite is still possible at 70°C under 100 bar H_2 partial pressure (Alpermann and Ostertag-Henning 2020).

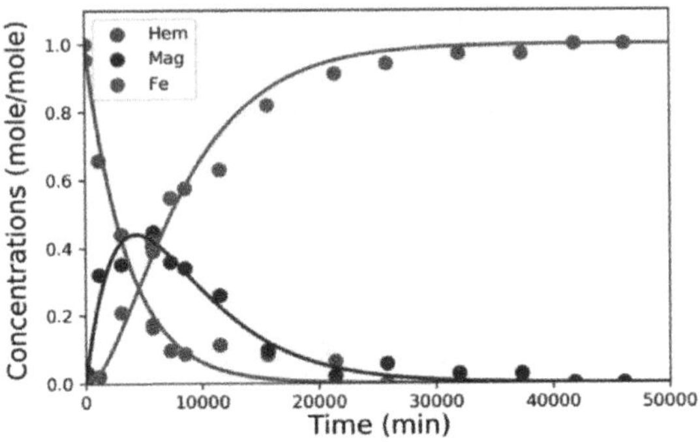

a)

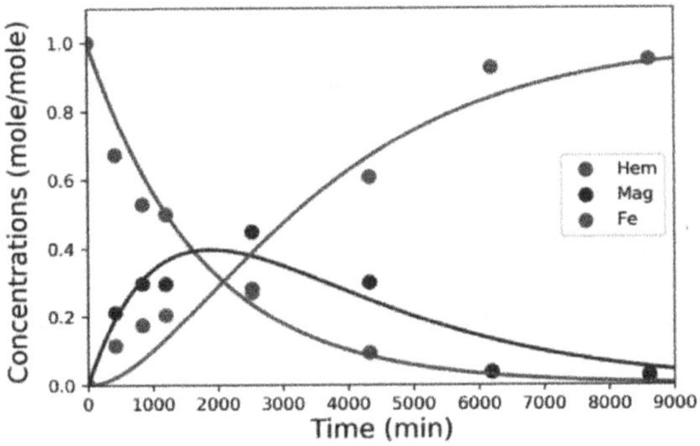

b)

Figure 2.15. *Reduction of hematite to magnetite and native iron at 200°C in the presence of (a) 30 and (b) 60 bar H_2 without water in the system initially. Adapted from Thüns et al. (2019). For a color version of this figure, see www.iste.co.uk/derango-cuevas/hydrogen.zip*

2.4.1.4. *Nitrate reduction*

From a thermodynamic point of view, the presence of nitrates and H_2 makes a redox reaction possible, which ultimately leads to the formation of more reduced species, nitrite (NO_2^-), nitrogen (N_2), ammonium (NH_4^+) and an increase in pH:

$$NO_3^- + 4H_2 + 2H^+ = NH_4^+ + 3H_2O \qquad [2.15]$$

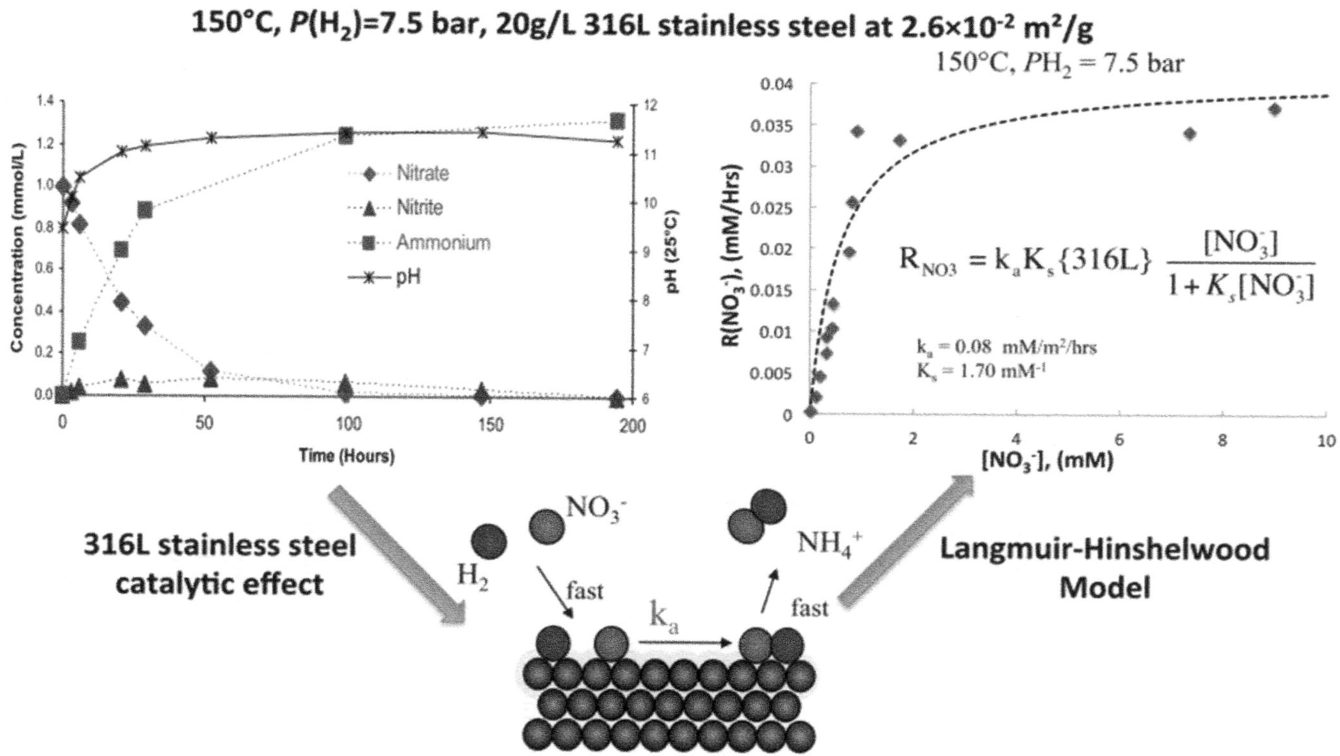

Figure 2.16. *Kinetic model of nitrate reduction in the presence of stainless steel and H_2 in a neutral environment (pH < 9). For a color version of this figure, see www.iste.co.uk/derango-cuevas/hydrogen.zip*

Truche et al. (2013b, 2013c) have shown that nitrate reduction does not occur in the absence of a catalyst or iron metal (native iron or carbon steel), even at 200°C. In contrast, Hastelloy C276 alloys and 316L stainless steel are formidable catalysts for the nitrate reduction reaction in the presence of H_2, since this occurs in a few hundred hours at 150°C (Figure 2.16) and remains measurable at 25°C over a few months.

These two Ni- and Cr-rich steels do not corrode under the experimental conditions, nor do they act as electron donors (catalysts in the strict sense): nitrate reduction requires hydrogen and produces ammonium directly – nitrite and N_2 being produced in negligible and transient quantities. The measured rates are independent of pH within the usual range found in nature (4–10). However, the reaction is inhibited in the presence of phosphate. The reaction is of order 1 with steel 316L and of order 0 with HC276, but it is suggested to test an Eley–Rideal formalism (reaction between a sorbed species and a free species).

Carbon steel behaves radically differently: no inhibiting effect of bentonite, a strong pH effect, the alteration of steel into oxides and effective nitrate reduction, even without H_2 for low nitrate concentrations. The order of the reaction is more difficult to establish because of the protective effect of secondary phases (magnetite in particular).

The abiotic reduction of nitrates by H_2 is therefore a reaction of secondary interest in the context of storage. This is due to the low availability of nitrates in the deep geological environment and the need for metallic materials for its catalysis, which limits its range of prevalence in contact with boreholes.

2.4.2. *Microbial reactivity*

Hydrogen is a highly efficient energy source for deep-sea microorganisms. Therefore, serious consideration should be given to hydrogen-trophic microbial activity in the context of hydrogen storage. It is generally accepted that the temperature limit for the development of life is approximately 120°C, but microbial activity above 90°C is already severely limited. It is therefore understandable that temperature will be a key parameter to consider in anticipating possible H_2 consumption by microbial catalysis. While abiotic reactions are favored at temperatures above 100°C, microbial reactions are most important at lower temperatures (Jacops et al. 2015; Bagnoud et al. 2016). Most microorganisms in nature grow in biofilms attached to surfaces in the form of aggregated microbial cells embedded in a secreted matrix of extracellular polymeric substances (Yin et al. 2019; Dopffel et al. 2021; Thaysen et al. 2021). This biofilm can reduce pore size and increase the tortuosity of the flow path, leading to a significant reduction in permeability (Coombs et al. 2010).

The underground environment is a complex incubator of biochemical reactions that can influence the chemical and structural integrity of underground gas storage sites. Microbial activity in the Earth's crust can be limited by high salt concentrations, high temperature and pressure, reduced pore space, limited nutrient availability and strongly reducing conditions.

Artificially increasing the concentration of H_2 in the geological environment can therefore stimulate undesirable microbial activity, leading to reduced permeability, loss of hydrogen, mechanical weakening of the reservoir, activation of rock faults and corrosion of metallic infrastructure. The extent of these interactions varies according to the type of geological storage method chosen.

Depleted hydrocarbon reservoirs and aquifers are more vulnerable to these interactions than salt caverns. The importance of hydrogen-trophic microbial activity has already been the subject of numerous studies, but a quantitative assessment of the impact of high H_2 concentrations on subsurface microbial activity is currently lacking. Studies have reported changes in gas composition, clogging near injection wells and biofilm growth. The four main microbial processes with implications for underground hydrogen storage are shown in Figure 2.17 and briefly discussed below. Metabolic reactions differ according to the electron acceptor considered.

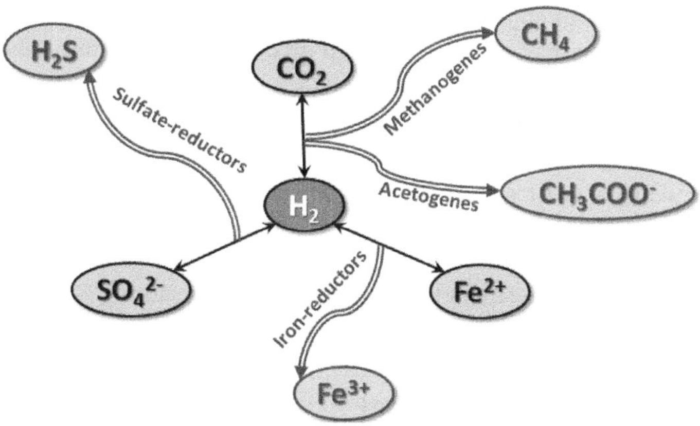

Figure 2.17. *Expected microbial reactivity at an H_2 storage site. Adapted from Ebigbo et al. (2013)*

These processes (methanogenesis, acetogenesis, sulfate reduction and iron(III) reduction) can contribute to H_2 consumption, whether in salt cavities or porous media (aquifers).

2.4.2.1. *Ferric iron reduction*

$$3Fe_2O_3 + H_2 = 2Fe_3O_4 + H_2O \tag{2.16}$$

Under subsurface conditions, iron-reducing bacteria fall into two categories: heterotrophic bacteria (which consume organic matter for food) and autotrophic bacteria (which synthesize their own organic matter). In addition to consuming H_2, the main negative impact of this reaction is that it can alter the porosity and permeability of the reservoir, affecting the overall safety of the containment (Porsch et al. 2009). These microorganisms also lead to water production, which reduces storage capacity (Porsch et al. 2009; Pichler 2013; Gregory et al. 2019). Some common iron-reducing bacteria are *Geobacter metallireducens* and *Shewanella putrefaciens* (Panfilov 2016).

2.4.2.2. *Methanogenesis*

$$CO_2 + 4H_2 = CH_4 + 2H_2O \tag{2.17}$$

Methanogenesis is the biological production of methane. It occurs in the presence of hydrogen-trophic methanogenic microorganisms (archaea) that use carbonates as electron acceptors (Panfilov 2016). This type of reaction occurs optimally at pressures of approximately 10 MPa and temperatures between 30°C and 40°C. Methanogenesis has already been reported to operate at the Lobodice town gas storage site.

2.4.2.3. *Acetogenesis*

$$2CO_2 + 4H_2 = CH_3COOH + 2H_2O \tag{2.18}$$

Acetate production occurs slowly in acidic aqueous aquifers and depleted hydrocarbon reservoirs. The main limiting factor is water salinity.

2.4.2.4. *Sulfate reduction*

$$SO_4^{2-} + 4H_2 = H_2S + 2H_2O + 2OH^- \tag{2.19}$$

The micro-organism, also known as an archaea, then produces H_2S after consuming the H_2 in the presence of sulfate ions. The production of H_2S in the subsurface of a geological H_2 storage site represents a major threat, as this reaction would not only reduce the quantity of H_2 available but also greatly alter its quality, as H_2S is a highly toxic acid gas.

2.5. Scientific perspectives and challenges

Among all hydrogen storage techniques, geological storage has recently received a great deal of attention as part of the energy transition and massive deployment of renewable energies. It is the only technology capable of storing quantities of energy equivalent to several hundred MWh. While salt cavern storage already seems to be well mastered, as demonstrated by the few sites already in operation, the same cannot be said for storage in porous media (aquifers or depleted hydrocarbon reservoirs).

In addition to the purely technical difficulties (geological site, management of leakage, flow in porous media and biogeochemical reactions), it is also important to consider purely legal, economic and above all societal issues, which have received very little attention (Kabuth et al. 2017).

While some of these factors (financial, political, governmental, legal and social) are "adjustable", the same is obviously not true of everything to do with the geological environment itself. It is therefore essential to ensure that all of the criteria mentioned below are properly taken into account to ensure safe and effective storage.

The geological setting must be assessed first, as it determines the choice of storage site. In this context, caverns are considered to be the best option, as they are a proven solution. However, their volume remains limited compared to reservoirs in porous media: deep aquifers and depleted hydrocarbon reservoirs.

The latter, nevertheless, present a number of scientific and technical challenges. They are currently the subject of laboratory work and pilot projects. Reservoir properties (e.g. volume, porosity and permeability), as well as reservoir depth, are crucial for assessing storage performance and capacity.

H_2 density and viscosity change with temperature and depth. The same applies to biogeochemical reactions that can alter the quantity and quality of the resource. Understanding fluid–fluid and solid–fluid interactions during storage operation is also essential. Although not mentioned in this summary, it will also be important to take into account the seismic risk induced by storage operation (injection and withdrawal cycle). This risk is linked to the loss of integrity of storage structures (e.g. fracturing of mother rock or caverns), with not only possible H_2 leakage but also a direct impact on surface dwellings or infrastructures.

Methods for monitoring leaks both at depth and on the surface must be developed to guarantee the integrity of the storage facility. Finally, as a new technology impacting the subsoil as a common good, geological hydrogen storage must be the subject of public debate to ensure a democratic and informed decision-making process.

2.6. References

Alpermann, T. and Ostertag-Henning, C. (2020). Kinetics of the abiotic oxidation of hydrogen (H2) by hematite at subsurface storage conditions. In *Proceedings of the Goldschmidt Conference*, 21–26.

Anderson, P.D., Norgaard, P.H., Olesen, M.H., Tanner, A.N. (2009). Final report of the roads2hycom project: Fuel cells and hydrogen in a sustainable energy economy. In *Contra 2009* [Online]. Available at: https://orbit.dtu.dk/en/publications/final-report-of-the-roads2hycom-project-fuel-cells-and-hydrogen-i.

Andrew, M., Bijeljic, B., Blunt, M.J. (2014). Pore-scale contact angle measurements at reservoir conditions using X-ray microtomography. *Adv. Water Resour.*, 68, 24–31. doi: 10.1016/j.advwatres.2014.02.014.

Arif, M., Al-Yaseri, A.Z., Barifcani, A., Lebedev, M., Iglauer, S. (2016). Impact of pressure and temperature on CO_2-brine-mica contact angles and CO_2-brine interfacial tension: Implications for carbon geo-sequestration. *J. Colloid Interface Sci.*, 462, 208–215. doi: 10.1016/j.jcis.2015.09.076.

Arif, M., Abu-khamsin, S.A., Iglauer, S. (2019). Wettability of rock/CO_2/brine and rock/oil/CO_2-enriched-brine systems: Critical parametric analysis and future outlook. *Adv. Colloid Interface Sci.*, 268, 91–113. doi: 10.1016/j.cis.2019.03.009.

Bagnoud, A., Leupin, O., Schwyn, B., Bernier-Latmani, R. (2016). Rates of microbial hydrogen oxidation and sulfate reduction in Opalinus Clay rock. *Appl. Geochem.*, 72, 42–50. doi: 10.1016/j.apgeochem.2016.06.011.

Bardelli, F., Mondelli, C., Didier, M., Vitillo, J.G., Cavicchia, D.R., Robinet, J.-C., Leone, L., Charlet, L. (2014). Hydrogen uptake and diffusion in Callovo-Oxfordian clay rock for nuclear waste disposal technology. *Appl. Geochem.*, 49, 168–177. doi: 10.1016/j.apgeochem.2014.06.019.

Blunt, M.J. (2017). *Multiphase Flow in Permeable Media*. Cambridge University Press, Cambridge. doi: 10.1017/9781316145098.

Bruno, M.S. and Dusseault, M.B. (2002). Geomechanical analysis of pressure limits for thin bedded salt caverns. In *Spring 2002 Tech Meet Banff*. Solution Mining Research Institute, Alberta, 29–30 April.

Buzek, F., Onderka, V., Vančura, P., Wolf, I. (1994). Carbon isotope study of methane production in a town gas storage reservoir. *Fuel*, 73(5), 747–752. doi: 10.1016/0016-2361(94)90019-1.

Carden, P.O. and Paterson, L. (1979). Physical, chemical and energy aspects of underground hydrogen storage. *Int. J. Hydrogen Energy*, 4, 559–569. doi: 10.1016/0360-3199(79)90083-1.

Carneiro, J.F., Matos, C.R., Van Gessel, S. (2019). Opportunities for large-scale energy storage in geological formations in mainland Portugal. *Renewable and Sustainable Energy Reviews*, 99, 201–211. doi: 10.1016/j.rser.2018.09.036.

Chabab, S., Théveneau, P., Coquelet, C., Corvisier, J., Paricaud, P. (2020). Measurements and predictive models of high-pressure H_2 solubility in brine (H_2O+NaCl) for underground hydrogen storage application. *Int. J. Hydrogen Energy*, 45, 32206–32220. doi: 10.1016/j.ijhydene.2020.08.192.

Chow, Y.T.F., Maitl, G.C., Trusler, J.P.M. (2018). Interfacial tensions of (H_2O + H_2) and (H_2O + CO_2 + H_2) systems at temperatures of (298–448) K and pressures up to 45 MPa. *Fluid Phase Equilib.*, 475, 37–44. doi: 10.1016/j.fluid.2018.07.022.

Coombs, P., Wagner, D., Bateman, K., Harrison, H., Milodowski, A.E., Noy, D., West, J.M. (2010). The role of biofilms in subsurface transport processes. *Q. J. Eng. Geol. Hydrogeol.*, 43, 131–139. doi: 10.1144/1470-9236/08-029.

Cross, M.M., Manning, D.A.C., Bottrell, S.H., Worden, R.H. (2004). Thermochemical sulphate reduction (TSR): Experimental determination of reaction kinetics and implications of the observed reaction rates for petroleum reservoirs. *Org. Geochem.*, 35, 393–404. doi: 10.1016/j.orggeochem.2004.01.005.

Crotogino, F., Schneider, G.S., Evans, D.J. (2018). Renewable energy storage in geological formations. *Proc. Inst. Mech. Eng. A J. Power Energy*, 232, 100–114. doi: 10.1177/0957650917731181.

Deng, X., Kamal, M.S., Patil, S., Muhammad, S., Hussain, S.M.S., Zhou X. (2020). A review on wettability alteration in carbonate rocks: Wettability modifiers. *Energy & Fuels*, 34(1), 31–54.

Didier, M., Leone, L., Greneche, J.M., Giffaut, E., Charlet, L. (2012). Adsorption of hydrogen gas and redox processes in clays. *Environ. Sci. Technol.*, 46, 3574–3579. doi: 10.1021/es204583h.

Dopffel, N., Jansen, S., Gerritse, J. (2021). Microbial side effects of underground hydrogen storage – Knowledge gaps, risks and opportunities for successful implementation. *Int. J. Hydrogen Energy*, 46, 8594–8606. doi: 10.1016/j.ijhydene.2020.12.058.

Ebigbo, A., Golfier, F., Quintard, M. (2013). A coupled pore-scale model for methanogenic microbial activity in underground hydrogen storage. *Adv. Water Resour.*, 61, 74–85. doi: 10.1016/j.advwatres.2013.09.004.

Edge, J.S., Skipper, N.T., Fernandez-Alonso, F., Lovell, A., Srinivas, G., Bennington, S.M., Garcia Sakai, V., Youngs, T.G. (2014). Structure and dynamics of molecular hydrogen in the interlayer pores of a swelling 2:1 clay by neutron scattering. *The Journal of Physical Chemistry C*, 118(44), 25740–25747. doi: 10.1021/jp5082356.

Edlmann, K., Haszeldine, S., Heinemann, N., Hassanpouryouzb, A., Thaysen, E. (2021). Hystorpor: Unlocking the UK's hydrogen storage. In *HyStorPor Brief*, Eltom, H.A., Alqubalee, A.M., Sultan, A.S., Barri, A.A., Abdelbasit, K. (eds). University of Edinburgh, Edinburgh [Online]. Available at: https://blogs.ed.ac.uk/hystorpor/wp-content/uploads/sites/951/2021/06/HyStorPor__storage_briefing_FINAL.pdf [Accessed 17 December 2024].

Feldmann, F., Hagemann, B., Ganzer, L., Panfilov, M. (2016). Numerical simulation of hydrodynamic and gas mixing processes in underground hydrogen storages. *Environmental Earth Sciences*, 75, 115. doi: 10.1007/s12665-016-5948-z.

Foh, S., Novil, M., Rockar, E., Randolph, P. (1979). Underground hydrogen storage. Salt caverns, excavated caverns, aquifers and depleted fields. Report, Brookhaven National Lab. doi: 10.2172/6536941.

Goldhaber, M.B. and Orr, W.L. (1995). Kinetic controls on thermochemical sulfate reduction as a source of sedimentary H_2S. In *Geochemical Transformations of Sedimentary Sulfur*, Vairavamurthy, M.A. and Schoonen, M.A.A. (eds). American Chemical Society, Washington, D.C.

Gregory, S.P., Barnett, M.J., Field, L.P., Milodowski, A.E. (2019). Subsurface microbial hydrogen cycling: Natural occurrence and implications for industry. *Microorganisms*, 7(2), 53. doi: 10.3390/microorganisms7020053.

Han, G., Bruno, M., Lao, K., Young, J., Dorfmann, L. (2007). Gas storage and operations in single bedded salt caverns: Stability analyses. In *SPE Production & Operations*, 22(3), 368–376. doi: 10.2118/99520-ms.

Hashemi, L., Blunt, M., Hajibeygi, H. (2021a). Pore-scale modelling and sensitivity analyses of hydrogen-brine multiphase flow in geological porous media. *Sci. Rep.*, 11, 1–13. doi: 10.1038/s41598-021-87490-7.

Hashemi, L., Glerum, W., Farajzadeh, R., Hajibeygi, H. (2021b). Advances in water resources contact angle measurement for hydrogen/brine/sandstone system using captive-bubble method relevant for underground hydrogen storage. *Adv. Water Resour.*, 154, 103964. doi: 10.1016/j.advwatres.2021.103964.

Heinemann, N., Alcalde, J., Miocic, J.M., Hangx, S.J.T., Kallmeyer, J., Ostertag-Henning, C., Hassanpouryouzband, A., Thaysen, E.M., Strobel, G.J., Schmidt-Hattenberger, C. et al. (2021). Enabling large-scale hydrogen storage in porous media – The scientific challenges. *Energy Environ. Sci.*, 14, 853–864. doi: 10.1039/d0ee03536j.

Henkel, S., Pudlo, D., Gaupp, R. (2013). Research sites of the H2STORE project and the relevance of lithological variations for hydrogen storage at depths. *Energy Procedia*, 40, 25–33. doi: 10.1016/j.egypro.2013.08.004.

Iglauer, S. (2017). CO_2-water-rock wettability: Variability, influencing factors, and implications for CO_2 geostorage. *Acc. Chem. Res.*, 50, 1134–1142. doi: 10.1021/acs.accounts.6b00602.

Iglauer, S., Abid, H., Al-Yaseri, A., Keshavarz, A. (2021). Hydrogen adsorption on subbituminous coal: Implications for hydrogen geo-storage. *Geophys. Res. Lett.*, 48, 2–5. doi: 10.1029/2021GL092976.

Jaakkola, S.T., Ravantti, J.J., Oksanen, H.M., Bamford, D.H. (2016). Buried alive: Microbes from ancient halite. *Trends Microbiol.*, 24, 148–160. doi: 10.1016/j.tim.2015.12.002.

Jacops, E., Wouters, K., Volckaert, G., Moors, H., Maes, N., Bruggeman, C., Swennen, R., Likte, R. (2015). Measuring the effective diffusion coefficient of dissolved hydrogen in saturated Boom Clay. *Appl. Geochem.*, 61, 175–184. doi: 10.1016/j.apgeochem.2015.05.022.

Jha, N.K., Al-Yaseri, A., Ghasemi, M., Al-Bayati, D., Lebedev, M., Sarmadivaleh, M. (2021). Pore scale investigation of hydrogen injection in sandstone via X-ray microtomography. *Int. J. Hydrogen Energy*, 46, 34822–34829. doi: 10.1016/j.ijhydene.2021.08.042.

Kabuth, A., Dahmke, A., Beyer, C., Bilke, L., Dethlefsen, F., Dietrich, P., Duttmann, R., Ebert, M., Feeser, V., Görke, U.-J. et al. (2017). Energy storage in the geological subsurface: Dimensioning, risk analysis and spatial planning: The ANGUS+ project. *Environ. Earth Sci.*, 76. doi: 10.1007/s12665-016-6319-5.

Keshavarz, A., Abid, H., Ali, M., Iglauer, S. (2021). Hydrogen diffusion in coal: Implications for hydrogen geo-storage. *J. Colloid Interface Sci.* doi: 10.1016/j.jcis.2021.10.050.

Kiyosu, Y. and Krouse, H.R. (1993). Thermochemical reduction and sulfur isotopic behavior of sulfate by acetic acid in the presence of native sulfur. *Geochem. J.*, 27, 49. doi: 10.2343/geochemj.27.49.

Kumar, K.R. and Hajibeygi, H. (2021). Multiscale simulation of inelastic creep deformation for geological rocks. *J. Comput. Phys.*, 440, 110439. doi: 10.1016/j.jcp.2021.110439.

Kunz, P., Hassanizadeh, S.M., Nieken, U. (2018). A two-phase SPH model for dynamic contact angles including fluid–solid interactions at the contact line. *Transp. Porous Media*, 122, 253–277. doi: 10.1007/s11242-018-1002-9.

Lambert, J.M., Simkovich, G., Wlaker, P.L. (1998). The kinetics and mechanism of the pyrite-to-pyrrhotite transformation. *Metall. Mater. Trans. B*, 29B, 385–396. doi: 10.1007/s11663-998-0115-x.

Lemieux, A., Sharp, K., Shkarupin, A. (2019). Preliminary assessment of underground hydrogen storage sites in Ontario, Canada. *Int. J. Hydrogen Energy*, 44, 15193–15204. doi: 10.1016/j.ijhydene.2019.04.113.

Lemmon, E.W., Bell, I.H., Huber, M.L., McLinden, M.O. (2018). NIST standard reference database 23: Reference fluid thermodynamic and transport properties-REFPROP, Version 10.0. Standard, National Institute of Standards and Technology. Standard Reference Data Program, Gaithersburg.

Lindblom, U.E. (1985). A conceptual design for compressed hydrogen storage in mined caverns. *Int. J. Hydrogen Energy*, 10, 667–675. doi: 10.1016/0360- 3199(85)90006-0.

Lord, A.S. (2009). Overview of geologic storage of natural gas with an emphasis on assessing the feasibility of storing hydrogen. Report, Sandia. doi: 10.2172/975258.

Lord, A.S., Kobos, P.H., Borns, D.J. (2014). Geologic storage of hydrogen: Scaling up to meet city transportation demands. *Int. J. Hydrogen Energy*, 39, 15570–15582. doi: 10.1016/j.ijhydene.2014.07.121.

Lysyy, M., Fernø, M., Ersl, G. (2021). Seasonal hydrogen storage in a depleted oil and gas field. *Int. J. Hydrogen Energy*, 46, 25160–25174. doi: 10.1016/j.ijhydene.2021.05.030.

Michalski, J., Bünger, U., Crotogino, F., Donadei, S., Schneider, G.S., Pregger, T., Cao, K.-K., Heide, D. (2017). Hydrogen generation by electrolysis and storage in salt caverns: Potentials, economics and systems aspects with regard to the German energy transition. *Int. J. Hydrogen Energy*, 42, 13427–13443. doi: 10.1016/j.ijhydene.2017.02.102.

Miocic, J., Heinemann, N., Edlmann, K., Scafidi, J., Molaei, F., Alcalde, J. (2023). Underground hydrogen storage: A review. *Geological Society, London*, 528(1), SP528–2022.

Mondelli, C., Bardelli, F., Vitillo, J.G., Didier, M., Brendle, J., Cavicchia, D.R., Robinet, J.C., Charlet, L. (2015). Hydrogen adsorption and diffusion in synthetic Na-montmorillonites at high pressures and temperature. *International Journal of Hydrogen Energy*, 40(6), 2698–2709. doi: 10.1016/j.ijhydene.2014.12.038.

Muhammed, N.S., Haq, B., Al Shehri, D., Al-Ahmed, A., Rahman, M.M., Zaman, E. (2022). A review on underground hydrogen storage: Insight into geological sites, influencing factors and future outlook. *Energy Reports*, 8, 461–499. doi: 10.1016/j.egyr.2021.12.002.

Muzny, C.D., Huber, M.L., Kazakov, A.F. (2013). Correlation for the viscosity of normal hydrogen obtained from symbolic regression. *Journal of Chemical & Engineering Data*, 58(4), 969–979. doi: 10.1021/je301273j.

Ortiz, L., Volckaert, G., Mallants, D. (2002). Gas generation and migration in Boom Clay, a potential host rock formation for nuclear waste storage. *Engineering Geology*, 64(2/3), 287–296. doi: 10.1016/s0013-7952(01)00107-7.

Pan, B., Gong, C., Wang, X., Li, Y., Iglauer, S. (2020). The interfacial properties of clay-coated quartz at reservoir conditions. *Fuel*, 262, 116461. doi: 10.1016/j.fuel.2019. 116461.

Pan, B., Yin, X., Iglauer, S. (2021). Rock-fluid interfacial tension at subsurface conditions: Implications for H_2, CO_2 and natural gas geo-storage. *Int. J. Hydrogen Energy*, 46, 25578–25585. doi: 10.1016/j.ijhydene.2021.05.067.

Panfilov, M. (2010). Underground storage of hydrogen: In situ self-organisation and methane generation. *Transp. Porous Media*, 85, 841–865. doi: 10.1007/s11242-010-9595-7.

Panfilov, M. (ed.) (2016). Underground and pipeline hydrogen storage. In *Com. Pend. Hydrog. Energy*. Elsevier, Amsterdam. doi: 10.1016/B978-1-78242-362-1.00004-3.

Paterson, L. (1983). The implications of fingering in underground hydrogen storage. *Int. J. Hydrogen Energy*, 8, 53–59. doi: 10.1016/0360- 3199(83)90035-6.

Pérez, A., Pérez, E., Dupraz, S., Bolcich, J. (2016). Patagonia wind-hydrogen project: Underground storage and methanation. In *Proceedings of the WHEC 2016-21st World Hydrogen Energy Conference 2016*, Zaragoza.

Pfeiffer, W.T. and Bauer, S. (2015). Subsurface porous media hydrogen storage. Scenario development and simulation. *Energy Procedia*, 76, 565–572. doi: 10.1016/j.egypro.2015.07.872.

Pfeiffer, W.T., Beyer, C., Bauer, S. (2017). Hydrogen storage in a heterogeneous sandstone formation: Dimensioning and induced hydraulic effects. *Pet Geosci.*, 23, 315–326. doi: 10.1144/petgeo2016-050.

Pichler, M.P. (2013). Assessment of hydrogen rock interaction during geological storage of CH_4-H_2 mixtures. In *Second EAGE Sustainable Earth Sciences (SES) Conference and Exhibition.* European Association of Geoscientists & Engineers.

Piessens, K. and Dusar, M. (2004). Feasibility of CO_2 sequestration in abandoned coal mines in Belgium. *Geol. Belgica*, 7, 165–180.

Pineau, A., Kanari, N., Gaballah, I. (2006). Kinetics of reduction of iron oxides by H_2 Part I: Low temperature reduction of hematite. *Thermochim Acta*, 447, 89–100. doi: 10.1016/j.tca.2005.10.004.

Porsch, K., Meier, J., Kleinsteuber, S., Wendt-Potthoff, K. (2009). Importance of different physiological groups of iron reducing microorganisms in an acidic mining lake remediation experiment. *Microb. Ecol.*, 57, 701–717. doi: 10.1007/s00248-009-9505-0.

Rag, R. (2020). Austria AG-Underground Sun Storage. Report, public13 [Online]. Available at: https://www.underground-sun-storage.at/en/.

Rickard, D. and Luther, G.W. (1997). Kinetics of pyrite formation by the H_2S oxidation of iron (II) monosulfide in aqueous solution between 25 and 125°C: The mechanism. *Geochim. Cosmochim. Acta*, 61, 135–147. doi: 10.1016/s0016-7037(96)00322-5.

Ridge, R. (1998). Gas storage at the abandoned leyden coal mine near denver, colorado. Report, US Environmental Protection Agency, 14.

Sainz-Garcia, A., Abarca, E., Rubi, V., Grandia, F. (2017). Assessment of feasible strategies for seasonal underground hydrogen storage in a saline aquifer. *Int. J. Hydrogen Energy*, 42, 16657–16666. doi: 10.1016/j.ijhydene.2017.05.076.

Scanziani, A., Singh, K., Blunt, M.J., Guadagnini, A. (2017). Automatic method for estimation of in situ effective contact angle from X-ray micro tomography images of two-phase flow in porous media. *J. Colloid Interface Sci.*, 496, 51–59. doi: 10.1016/j.jcis.2017.02.005.

Seward, T.M. and Franck, E.U. (1981). The system hydrogen-water up to 440 C and 2500 bar pressure. *Berichte der Bunsengesellschaft für physikalische Chemie*, 85(1), 2–7. doi: 10.1002/bbpc.19810850103.

Simon, J., Ferriz, A.M., Correas, L.C. (2015). HyUnder hydrogen underground storage at large scale: Case study Spain. *Energy Procedia*, 73, 136–144. doi: 10.1016/j.egypro.2015.07.661.

Singh, K., Bijeljic, B., Blunt, M.J. (2016). Imaging of oil layers, curvature and contact angle in a mixed-wet and a water-wet carbonate rock. *Water Resour. Res.*, 52, 1716–1728. doi: 10.1002/2015WR018072.

Šmigáň, P., Greksák, M., Kozánková, J., Buzek, F., Onderka, V., Wolf, I. (1990). Methanogenic bacteria as a key factor involved in changes of town gas stored in an underground reservoir. *FEMS Microbiol. Lett.*, 73, 221–224. doi: 10.1016/0378-1097(90)90733-7.

Stolten, D. and Grube, T. (2010). Storage systems/policy perspectives. Initiatives and cooperations. In *18th World Hydrogen Energy Conference 2010.*

Tarkowski, R. (2019). Underground hydrogen storage: Characteristics and prospects. *Renew. Sustain. Energy Rev.*, 105, 86–94. doi: 10.1016/j.rser.2019.01.051.

Taylor, J.B., Alderson, J.E.A., Kalyanam, K.M., Lyle, A.B., Phillips, L.A. (1986). Technical and economic assessment of methods for the storage of large quantities of hydrogen. *Int. J. Hydrogen Energy*, 11, 5–22. doi: 10.1016/0360-3199(86)90104-7.

Thaysen, E.M., McMahon, S., Strobel, G.J., Butler, I.B., Ngwenya, B.T., Heinemann, N., Wilkinson, M., Hassanpouryouzband, A., McDermott, C., Edlmann, K. (2021). Estimating microbial growth and hydrogen consumption in hydrogen storage in porous media. *Renew. Sustain. Energy Rev.*, 151, 111481. doi: 10.1016/j.rser.2021.111481.

Thüns, N., Krooss, B.M., Zhang, Q., Stanjek, H. (2019). The effect of H_2 pressure on the reduction kinetics of hematite at low temperatures. *International Journal of Hydrogen Energy*, 44(50), 27615–27625. doi: 10.1016/j.ijhydene.2019.08.178.

Truche, L., Berger, G., Destrigneville, C., Pages, A., Guillaume, D., Giffaut, E., Jacquot, E. (2009). Experimental reduction of aqueous sulphate by hydrogen under hydrothermal conditions: Implication for the nuclear waste storage. *Geochimica et Cosmo-chimica Acta*, 73(16), 4824–4835. doi: 10.1016/j.gca.2009.05.043.

Truche, L., Berger, G., Destrigneville, C., Guillaume, D., Giffaut, E. (2010). Kinetics of pyrite to pyrrhotite reduction by hydrogen in calcite buffered solutions between 90 and 180°C: Implications for nuclear waste disposal. *Geochimica et Cosmochimica Acta*, 74(10), 2894–2914. doi: 10.1016/j.gca.2010.02.027.

Truche, L., Jodin-Caumon, M.C., Lerouge, C., Berger, G., Mosser-Ruck, R., Giffaut, E., Michau, N. (2013a). Sulphide mineral reactions in clay-rich rock induced by high hydrogen pressure. Application to disturbed or natural settings up to 250°C and 30 bar. *Chem. Geol.*, 351, 217–228. doi: 10.1016/j.chemgeo. 2013.05.025.

Truche, L., Berger, G., Albrecht, A., Domergue, L. (2013b). Abiotic nitrate reduction induced by carbon steel and hydrogen: Implications for environmental processes in waste repositories. *Applied Geochemistry*, 28, 155–163. doi: 10.1016/j.apgeochem.2012.10.010.

Truche, L., Berger, G., Albrecht, A., Domergue, L. (2013c). Engineered materials as potential geocatalysts in deep geological nuclear waste repositories: A case study of the stainless steel catalytic effect on nitrate reduction by hydrogen. *Applied Geochemistry*, 35, 279–288. doi: 10.1016/j.apgeochem.2013.05.001.

Tudek, J., Crandall, D., Fuchs, S., Werth, C.J., Valocchi, A.J., Chen, Y., Goodman, A. (2017). In situ contact angle measurements of liquid CO_2, brine, and Mount Simon sandstone core using micro X-ray CT imaging, sessile drop, and Lattice Boltzmann modeling. *J. Pet Sci. Eng.*, 155, 3–10. doi: 10.1016/j.petrol.2017.01.047.

Ugarte, E.R. and Salehi, S. (2021). A review on well integrity issues for underground hydrogen storage (UHS). *J. Energy Resour. Technol.*, 144, 1–27. doi: 10.1115/1.4052626.

U.S. Energy Information Administration (2017). International Energy Outlook 2017. Report.

Vivian, J.E. and King, C.J. (1964). Diffusivities of slightly soluble gases in water. *AIChE J.*, 10, 220–221. doi: 10.1002/aic.690100217.

Wallace, R.L., Cai, Z., Zhang, H., Zhang, K., Guo, C. (2021). Utility-scale subsurface hydrogen storage: UK perspectives and technology. *Int. J. Hydrogen Energy*, 46, 25137–25159. doi: 10.1016/j.ijhydene.2021.05.034.

Wang, Y., Zhang, C., Wei, N., Oostrom, M., Wietsma, T.W., Li, X., Bonneville, A. (2013). Experimental study of crossover from capillary to viscous fingering for supercritical CO_2-water displacement in a homogeneous pore network. *Environ. Sci. Technol.*, 47, 212–218. doi: 10.1021/es3014503.

Wiltowski, T., Hinckley, C.C., Smith, G.V., Nishizawa, T., Saporoschenko, M., Shirley, R.H., Webster, J.R. (1987). Kinetics and mechanisms of iron sulphide reduction in hydrogen and carbon monoxide. *J. Solid State Chem.*, 71, 95–102. doi.org: 10.1016/0022-4596(87)90146-0.

Yang, Z., Yuan, B., Clarkson, C.R., Ghanizadeh, A. (2021). Evaluation of surface diffusion in microporous/mesoporous media using a numerical model applied to rate-of-adsorption data: Implications for improved gas permeability estimation in shales/tight rocks using drill cuttings. *Fuel*, 285, 118974. doi: 10.1016/j.fuel.2020.118974.

Yekeen, N., Padmanabhan, E., Sevoo, T.A.L., Kanesen, K.A.L., Adenike, O. (2020). Wettability of rock/CO_2/brine systems: A critical review of influencing parameters and recent advances. *Journal of Industrial and Engineering Chemistry*, 88, 1–28. doi: 10.1016/j.jiec.2020.03.021.

Yekta, A.E., Manceau, J.C., Gaboreau, S., Pichavant, M., Audigane, P. (2018). Determination of hydrogen–water relative permeability and capillary pressure in sandstone: Application to underground hydrogen injection in sedimentary formations. *Transp. Porous Media*, 122, 333–356. doi: 10.1007/s11242-018-1004-7.

Yin, W., Wang, Y., Liu, L., He, J. (2019). Biofilms: The microbial protective clothing in extreme environments. *Int. J. Mol. Sci.*, 20. doi: 10. 3390/ijms20143423.

Zhang, T., Amrani, A., Ellis, G.S., Ma, Q., Tang, Y. (2008a). Experimental investigation on thermochemical sulfate reduction by H_2S initiation. *Geochim. Cosmochim. Acta*, 72(14), 3518– 3530. doi: 10.1016/j.gca.2008.04.036.

Zhang, T., Ellis, G.S., Walters, C.C., Kelemen, S.R., Wang, K.S., Tang, Y. (2008b). Geochemical signatures of thermochemical sulfate reduction in controlled hydrous pyrolysis experiments. *Org. Geochem.*, 39, 308–328. doi: 10.1016/j.orggeochem.2007.12.007.

Zhang, C., Oostrom, M., Grate, J.W., Wietsma, T.W., Warner, M.G. (2011). Liquid CO_2 displacement of water in a dual-permeability pore network micromodel. *Environ. Sci. Technol.*, 45, 7581–7588. doi: 10.1021/es201858r.

Zielinski, J., Zglinicka, I., Znak, L., Kaszkur, Z. (2010). Reduction of Fe_2O_3 with hydrogen. *Appl. Catal. Gen.*, 381, 191e6. doi: 10.1016/j.apcata.2010.04.003.

Ziemiański, P.P. and Derkowski, A. (2022). Structural and textural control of high-pressure hydrogen adsorption on expandable and non-expandable clay minerals in geologic conditions. *International Journal of Hydrogen Energy*, 47(67), 28794–28805. doi: 10.1016/j.ijhydene.2022.06.204.

Zivar, D., Kumar, S., Foroozesh, J. (2021). Underground hydrogen storage: A comprehensive review. *Int. J. Hydrogen Energy*, 46, 23436–23462. doi: 10.1016/j.ijhydene.2020.08.138.

3

Liquid Storage: LOHC

Xiaolong JI, Valérie MEILLE and Catherine PINEL
IRCELYON, CNRS, Université Claude Bernard Lyon 1, Villeurbanne, France

3.1. Introduction

Among many hydrogen storage techniques, the hydrogenation of some organic liquids, called "liquid organic hydrogen carriers (LOHCs)", is a promising alternative (Pawelczyk et al. 2022). The storage of hydrogen is carried out by a catalytic hydrogenation reaction, allowing a hydrogen-rich compound to be obtained, which is able to return its hydrogen atoms in the form of molecular H_2 via a catalytic dehydrogenation reaction (Figure 3.1(a)). The advantage of this method is the simplicity of transporting hydrogen-rich compounds as a standard liquid, in general non-volatile, non-toxic and stable, over months or years.

Note that this chapter does not consider what is called "circular hydrogen carriers" (Figure 3.1(b)), including formic acid, methanol, ammonia, etc., as LOHC compounds (Oner and Khalilpour 2022). The reversibility of the reaction is not considered the same as that for LOHCs, even if the gaseous byproducts of the carriers (CO_2, N_2) can be recycled.

A comparison of LOHC with other H_2 storing methods has been proposed by Andersson and Grönkvist (2019), Nazir et al. (2020) and Faye et al. (2022). In the latter work, the main advantages of the LOHC over the other methods are its safety, price and high gravimetric and volumetric storage, whereas the main drawbacks are the high pressure required for storing (hydrogenation) and the high temperature for H_2 recovery (dehydrogenation). To consider LOHC candidates relevant for H_2 storage, a list of necessary conditions is presented, including a high storage capacity

Hydrogen Storage,
coordinated by Patricia DE RANGO and Fermin CUEVAS. © ISTE Ltd 2025.

(> 56 kg/m³, > 6 wt-%), a low melting point (< –30°C), a high boiling point (> 300°C), low toxicity, desorption heat and production costs.

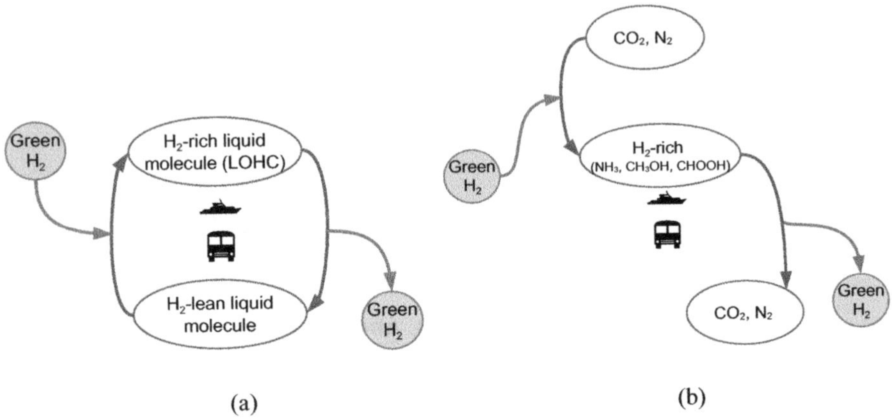

Figure 3.1. *Principle of the LOHC (a) compared to circular storage (b). For a color version of this figure, see www.iste.co.uk/derango-cuevas/hydrogen.zip*

Many bibliographic reviews exist on the most studied LOHC systems (Aakko-Saksa et al. 2018; Makaryan et al. 2020; Rao and Yoon 2020); the following paragraphs summarize the state of knowledge on the different possible LOHC pairs. The catalytic reactions must be very selective in both directions (hydrogenation/dehydrogenation). No byproduct can be tolerated to allow many reversible cycles. Particular attention is given to these aspects while analyzing the state-of-the-art.

3.2. Aromatic hydrocarbons

3.2.1. *Introduction*

Aromatic hydrocarbons are one of the groups of molecules that can be used as LOHCs; these are unsaturated molecules with π bonds in resonance comprising solely C and H atoms. Despite their simple atomic composition, aromatic hydrocarbons have a large variety of molecular sizes and configurations, ranging from benzene to multiring compounds with a significant number of isomers (Speight 2017).

In general, the storage and release of hydrogen with aromatic hydrocarbons follow the reaction pathway illustrated in Figure 3.2.

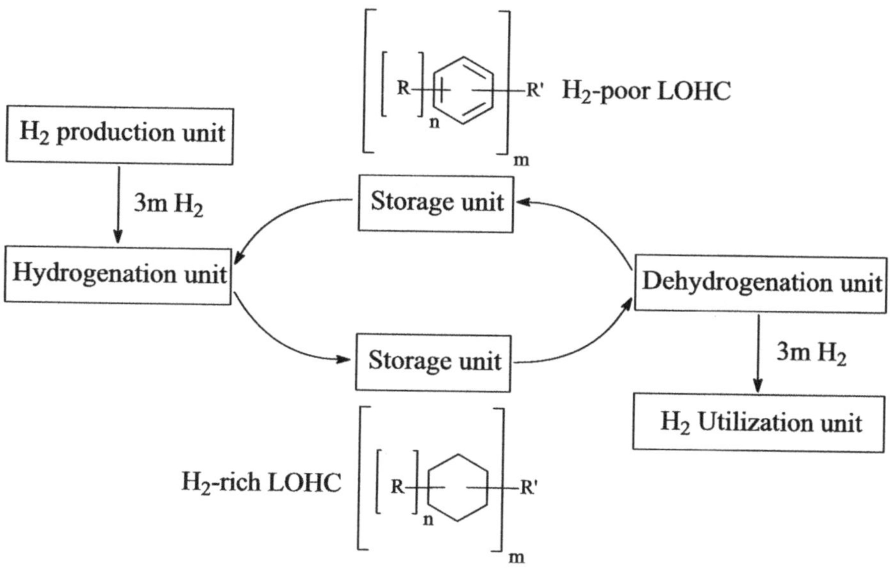

Figure 3.2. *Storage, transport and release of H_2 using aromatic hydrocarbons, where n represents the length of the side chains and m represents the number of monomers*

The storage of hydrogen into aromatic hydrocarbons is typically achieved by three-phase (gas–liquid–solid) or two-phase (gas–solid) heterogeneous catalytic hydrogenation of the aromatic rings. Depending on the molecules and processes, the reaction temperature varies from 80°C to 250°C, and the gas-phase hydrogen pressure varies from 10 to 50 bar (Rao and Yoon 2020). Raney-Ni and supported catalysts based on noble metals have been proven to be the most effective (Sisáková et al. 2021). Mass transfer phenomena should also be considered, and the optimization of operating conditions such as the stirring speed in a batch reactor or of the LHSV/GHSV in a continuous reactor could influence the overall efficiency of the process (Lee et al. 2021a). The catalytic hydrogenation of aromatic hydrocarbons is a very exothermic reaction with an enthalpy of approximately – 65 kJ/mol H_2. Therefore, the management of temperature by a cooling system installed in reactors is necessary. Alternatively, the heat generated from the reaction could be harnessed and utilized elsewhere (Modisha et al. 2019a). Hydrogen stored inside the molecules exists in the form of C–H bonds, and aromatic hydrocarbons are therefore hydrogenated into their saturated aliphatic form. Many types of reactors can be used for this process, such as batch reactors (Pez et al. 2008) and fixed bed reactors (Okada et al. 2006).

A simplified mechanism for aromatic hydrocarbon hydrogenation is illustrated in Figure 3.3. The aromatic ring adsorbs on a metal active site as an L_3-type ligand with π-bonding, whereas the hydrogen molecule adsorbs on adjacent sites as X ligands through dissociative coordination (Chou and Vannice 1987). The hydrogenation of the benzene ring is composed of six consecutive steps of hydrogen adatom addition to unsaturated bonds, followed by desorption of the saturated product molecule (Saeys et al. 2005).

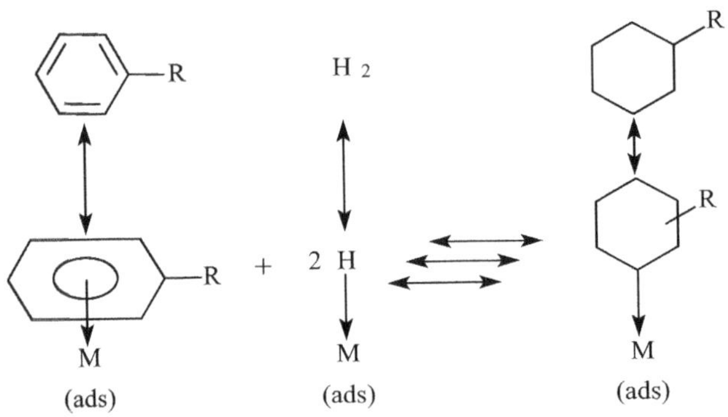

Figure 3.3. *General aromatic hydrocarbon hydrogenation mechanism*

Few side reactions take place during the hydrogenation process because of the relatively low operating temperature (Van Der Steen and Scholten 1990). However, small amounts of cyclohexene or its derivatives, which are intermediate molecules in the hydrogenation process, can be desorbed from the catalytic surface before going through the final hydrogenation step (Schoenmaker-Stolk et al. 1987).

High-purity hydrogen is conventionally used for hydrogenation. However, new studies have explored the possibility of using hydrogen gas mixtures without a purification step. Compared with pure hydrogen, the mixed gas hydrogenation (MGH) process has potentially greater energy and economic efficiency (Jorschick et al. 2021).

The hydrogenation of aromatic hydrocarbons and the dehydrogenation of saturated hydrocarbons are one pair of reversible reactions in which the chemical equilibrium depends on the temperature and concentration of different components. The release of hydrogen is realized through the catalytic dehydrogenation of aliphatic hydrocarbons. Dehydrogenation regenerates H_2, which could be further utilized in fuel cells, and aromatic hydrocarbons, which could be recycled to serve again as hydrogen carrier. The hydrogen released from polycyclic compounds is reported to have very high purity (Preuster et al. 2017). However, some side reactions can still take place, such as dealkylation (Amende et al. 2014). Compared with hydrogenation, this reaction normally takes place at higher temperatures from 200°C to 400°C and with lower hydrogen pressures from 1 to 5 bar (Rao and Yoon 2020). The most effective catalysts reported in studies are Pt- and Pd-based catalysts (Ali et al. 2020, 2021). As the reverse reaction of hydrogenation, this process is very endothermic (Geburtig 2019) and requires a constant external energy input, which hinders the global energy efficiency of the LOHC process when aromatic hydrocarbons are used. It is estimated that the energy cost for hydrogen release for H_2-rich LOHC accounts for approximately 26% of the energy provided by the released hydrogen.

The most studied molecules are benzene, toluene, naphthalene and their derivatives. The majority of these compounds remain in the liquid phase under reaction conditions for both hydrogenation and dehydrogenation (Aakko-Saksa et al. 2018). However, gas-phase reaction for smaller molecules such as benzene and toluene has also been explored (Lin and Vannice 1993). The molecules produced in their hydrogen-rich form could be stored and transported under mild conditions by existing infrastructures for petroleum and are stable over a relatively wide range of temperature and pressure (Niermann et al. 2019).

Compound	H_2-lean form	H_2-rich form	Gravimetric storage capacity (wt%)
Benzene/ cyclohexane	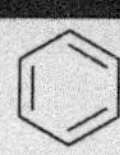	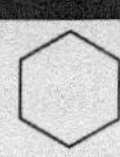	7.1%
Toluene/ methylcyclohexane	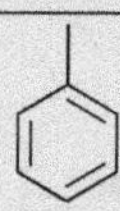	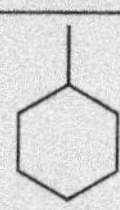	6.1%
Terphenyl/ tercyclohexane	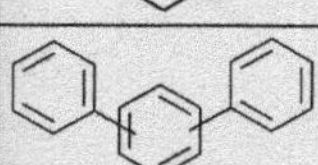	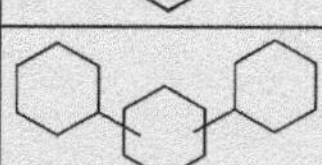	7.2%
Benzyltoluene/ perhydro-benzyltoluene	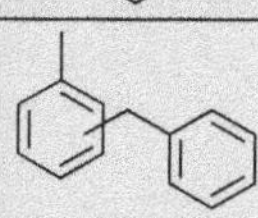	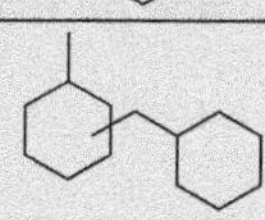	6.2%
Dibenzyltoluene/ perhydro-dibenzyltoluene	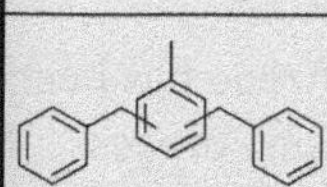	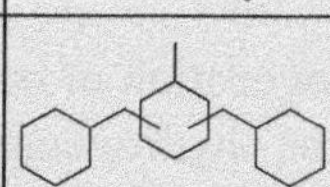	6.2%
Naphthalene/ decalin	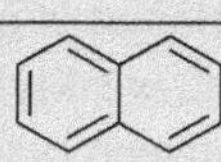	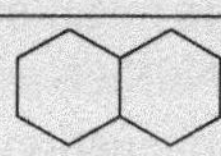	7.2%

Table 3.1. *List of hydrocarbon LOHC molecules*

3.2.2. *Aromatic hydrocarbon LOHC couples*

The most studied aromatic hydrocarbons are listed in Table 3.1. These molecules meet the target requirements set by the U.S. Department of Energy (DOE) by 2020 and the recommendations for chemical hydrides (H)75 made by the European Association for Storage of Energy (EASE) and the European Energy Research Alliance (EERA) for the European energy storage technology development roadmap toward 2030, as shown in Table 3.2.

Target	Volumetric energy density	Gravimetric energy density
U.S. DOE	1 kWh/L (system) 30 kg H_2/m^3	1.5 kWh/kg (system) 4.5 wt.% H_2
EASE/EERA	30 kg H_2/m^3	3–5 wt.% H_2 (system)

Table 3.2. *DOE and EASE/EERA H_2 storage targets (EASE-EERA 2017)*

3.2.2.1. *Benzene/cyclohexane*

Benzene is produced worldwide on a large scale each year via multiple methods, including extraction from crude oil or coal, dehydrogenation of cycloparaffins and transalkylation of toluene (Garcia Villaluenga and Tabe-Mohammadi 2000). Purified benzene could be a candidate for hydrogen storage with a gravimetric storage capacity of 7.1 wt-%. One benzene molecule could store three equivalents of H_2, as shown in Figure 3.4.

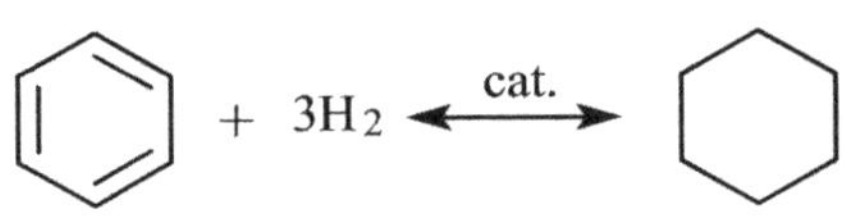

Figure 3.4. *Storage and release of H_2 with benzene*

3.2.2.1.1. Hydrogenation

The partial and total hydrogenation of benzene has been profoundly studied since the last century. During the hydrogenation process, benzene first undergoes π-complex adsorption on a single metal atom (Basset et al. 1975), and H_2 undergoes dissociative adsorption. On a Ni/SiO_2 catalyst, weakly adsorbed hydrogen is the main form involved in the hydrogenation of benzene (van Meerten et al. 1977). For benzene, the molecules adsorbed on hollow sites over a Pt (111) catalyst are the reactive species (Saeys et al. 2005).

The desired product for benzene hydrogenation is cyclohexane for maximum H_2 storage capacity. Cyclohexene is a potential side product of this process, but such a reaction is thermodynamically not favored, as cyclohexane is at least 75 kJ/mol more stable than cyclohexene in terms of ΔG (Foppa and Dupont 2015). The enthalpy of total hydrogenation of benzene is reported to be approximately 69 kJ/mol H_2 (Carey and Sundberg 2007). Supported Ni, Pt, Pd and Ru are effective catalysts for this reaction (Qi et al. 2013). Co-Pt bimetallic (Chen et al. 2010) and MOF (Chen et al. 2017) based catalysts have been found to be effective at low temperatures. Fixed-bed reactors and trickle-bed reactors are applicable for the hydrogenation process (Metaxas and Papayannakos 2006).

Although benzene has the advantages of higher H_2 storage capacity and lower reaction temperature than larger LOHC molecules containing saturated carbons, several major disadvantages still exist. The difficult separation of the benzene/cyclohexane mixture hinders the economic and energy efficiency of the process. Conventional distillation is not applicable for benzene/cyclohexane separation, as their binary mixture liquid–vapor equilibrium is nearly a close-boiling system. The azeotropic and extractive distillation methods used by the industry (Rautenbach and Albrech 1989) require high-cost and complex systems. Studies on membrane (Garcia Villaluenga and Tabe-Mohammadi 2000) and MOF (Sapianik et al. 2020) have shown promising results, with high separation rate and lower energy toll. The toxicity of benzene and its high vapor pressure may also cause harm to the human body during operation, and the use of benzene is therefore restricted under REACH in the EU.

3.2.2.1.2. Dehydrogenation

The dehydrogenation of cycloalkane is thermodynamically favored at a high temperature contrary to hydrogenation of aromatics. Lowering the concentration of hydrogen will also push the equilibrium toward the formation of aromatic compounds. Therefore, the release of hydrogen from cyclohexane is typically carried out at temperatures higher than 275°C and pressures lower than 1 atm, where the cyclohexane is in its vapor phase. DFT calculations revealed seven elementary reactions for the dehydrogenation process, and the initial transformation of C_6H_{12}* to C_6H_{11}* was identified as the rate-determining step (Li et al. 2021).

Pd- and Pt-based catalysts showed better reactivity for the dehydrogenation reaction. The addition of Au to Pt(111) (Sachtler and Somorjai 1984) or Cu to Ni/SiO_2 (Xia et al. 2017) greatly increased the production of benzene. The use of carbon nanotubes was also found to provide enhanced reactivity (Ombaka et al. 2013).

As cyclohexane does not possess any methyl group, its dehydrogenation did not result in any side reaction and the hydrogen released was highly pure (Bourane et al. 2016). However, the heating of cyclohexane to its gas phase and the recondensation of products complexify the process.

3.2.2.2. *Toluene (TOL)/methylcyclohexane (MCH)*

Toluene is produced from catalytic reformate from refineries and petroleum-derived pyrolysis gasoline (Fishbein 1985), and is isolated by distillation. Owing to the presence of the methyl group, its gravimetric H_2 storage capacity is 6.1 wt-%, which is lower than that of benzene. The wider liquid state temperature range of MCH and toluene than that of CH and benzene means that hydrogen storage and transport via the TOL/MCH couple could be more convenient than the benzene/CH couple. In addition, the TOL/MCH couple has lower toxicity toward human health in general (Fishbein 1985), but the poor biodegradability of MCH might lead to its persistence in the environment (Markiewicz et al. 2019). One toluene molecule could store three equivalents of H_2, as shown in Figure 3.5.

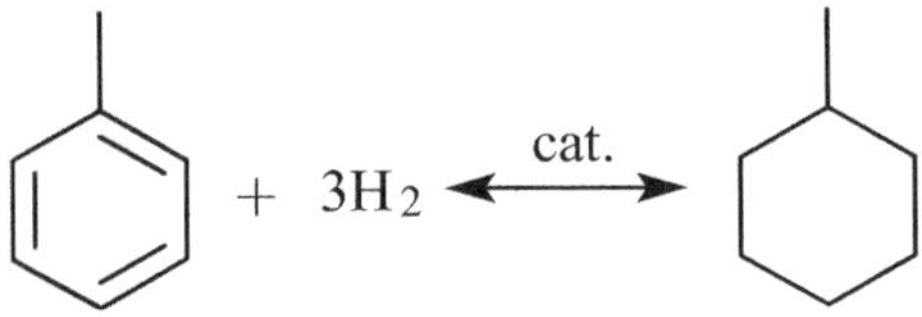

Figure 3.5. *Storage and release of H_2 with toluene*

3.2.2.2.1. Hydrogenation

Like benzene, the adsorption of toluene is achieved via the interaction of π-electrons of the aromatic ring with d-orbitals of the catalytic metal sites. Both gas- and liquid-phase reactions have been studied. Rh has been found to have better activity than Pt at temperatures lower than 170°C but not at higher temperatures (Ali et al. 1998). Toluene hydrogenation activity is lower than that of benzene under the same conditions because of its greater strength of adsorption or steric effects caused by the introduction of a methyl group to the aromatic ring (Völter et al. 1968). The decoration of the classic Pt/Al_2O_3 catalyst by Rh, Ir and Re could also enhance its activity (Ali et al. 1998). The hydrogenation enthalpy is 68.3 kJ/mol H_2 (Müller et al. 2013). Taube et al. (1983) explored the feasibility of a hydrogen-powered vehicle system using toluene as a hydrogen carrier, which they named the MTH system. Toluene hydrogenation kinetics have been found to differ across various studies of different catalysts under different conditions. Hydrogenation over Ni/SiO_2 was found to be a zero-order reaction (Klvana et al. 1988), and that over Ni/Al_2O_3

was found to be positive for hydrogen and negative for toluene (Lindfors et al. 1993).

The utilization of low-quality hydrogen containing CO and H_2S was also studied by Gonda et al. (2014), who reported that Pd catalysts supported on acidic oxides were resistant to CO inhibition and that Pt catalysts were resistant to H_2S.

3.2.2.2.2. Dehydrogenation

Owing to the presence of a methyl side group, the dehydrogenation of MCH is easier to achieve than that of CH (Okada et al. 2006) and therefore requires a lower temperature for the reaction to take place. The dehydrogenation of MCH is usually carried out at temperatures higher than its boiling point (110°C). Therefore, the system is composed of a gas–solid two-phase reaction using a batch or flow system at temperatures between 210°C and 350°C. Spray pulse injection of MCH in a flow system has been reported to increase system activity (Biniwale et al. 2008). These widely used reactors have a relatively low cost and are easier to establish owing to their technical maturity and simplicity but are limited by thermodynamic equilibrium. Ali et al. (1994) explored the possibility of exceeding the equilibrium conversion for MCH dehydrogenation using a catalytic membrane reactor containing a Pd–Ag membrane to separate the generated H_2 in situ. A conversion that was four times greater than the equilibrium value was achieved after 300 h of reaction. A catalytic membrane reactor could also reduce the amount of heat to be provided for the reaction and make the process more compact by eliminating the purification unit after hydrogen release. Byun et al. (2021) performed cost and economic analysis for such a process and reported that catalytic membrane reactors may have an even lower cost than a packed bed reactor. Acharya et al. (2021) summarized suitable membranes for this reaction in their review. A heatpipe reformer was also studied by Badakhsh et al. (2022) and integrated with a hydrogen burner with stainless steel as the reformer wall material and benzyltoluene as the heat transfer oil, and this system is proven to be feasible.

One of the difficulties for this system is the dealkylation of toluene, which tends to occur simultaneously with the dehydrogenation of MCH. Another difficulty is the deactivation of catalysts. No catalyst has stable performance for MCH dehydrogenation due to severe coking, which causes catalyst deactivation. Pt-based catalysts have higher activity for this process compared to other metals. A detailed review covering different Pt and Pt-M bimetallic catalysts for MCH dehydrogenation was provided by Sekine and Higo (2021). Yolcular and Olgun (2008) reported that high-concentration nickel-based catalysts (20 wt.%) have activities similar to those of a commercial R-50 (UOP) (UK) catalyst consisting of 0.25 wt.% Pt at temperatures between 653 and 713 K with no production of benzene. Moreover, the increase in the calcination temperature during catalyst preparation hinders its activity toward toluene production, which is probably due to the transformation of NiO toward nickel aluminates. Coughlin et al. (1984) found that presulfurization of Pt and PtRe catalysts enhanced the yield of toluene and that coke deposition could be retarded by sulfur. A successful demonstration of the toluene/MCH LOHC system on an industrial scale is the SPERA Hydrogen™ system implemented by Chiyoda Corporation. A novel catalyst consisting of partially sulfided Pt nanoparticles dispersed on alumina was developed. This catalyst had a long lifecycle of over 10,000 h and had a high selectivity toward toluene of more than 99% (Okada et al. 2018).

3.2.2.3. *Benzyltoluene/perhydro-benzyltoluene*

Even though benzene/CH and TOL/MCH couples could be ideal candidates for hydrogen storage, their operating conditions and storage temperatures are rather limited because of their relatively low boiling points. Their toxicity and high vapor pressure also make them hazardous for handling.

Therefore, other couples with higher boiling points and lower toxicity have been the focus of recent studies. The addition of more aromatic rings to the molecule could meet the requirements without hindering its gravimetric storage capacity. Benzyltoluene has been widely used as a heat transfer oil in industry and is named Marlotherm LH®, but its potential as an LOHC was first reported by Brückner et al. (2014). The gravimetric storage capacity of benzyltoluene is similar to that of toluene at 6.2 wt.%, and its volumetric storage capacity is 56 kg H_2/m^3. Some properties of benzyltoluene (H0-BT) and its H_2-rich form (H12-BT) are listed in Table 3.3. Commercially available BT is a mixture of three isomers with the benzyl group in the ortho, meta or para position of the methyl group. One BT molecule could store six equivalents of H_2, as shown in Figure 3.6. The first addition of three equivalents of H_2 could take place on either aromatic ring.

	Benzyltoluene	Perhydro-benzyltoluene
Molar mass (g/mol)	182.3	194.3
Melting point (K)	243.2 (Brückner et al. 2014)	–
Boiling point (K)	553.2 (Brückner et al. 2014)	–
Density (kg/m³)	996 (20°C) (Müller et al. 2015)	876 (20°C) (Müller et al. 2015)

Table 3.3. *Properties of benzyltoluene and perhydrobenzyltoluene*

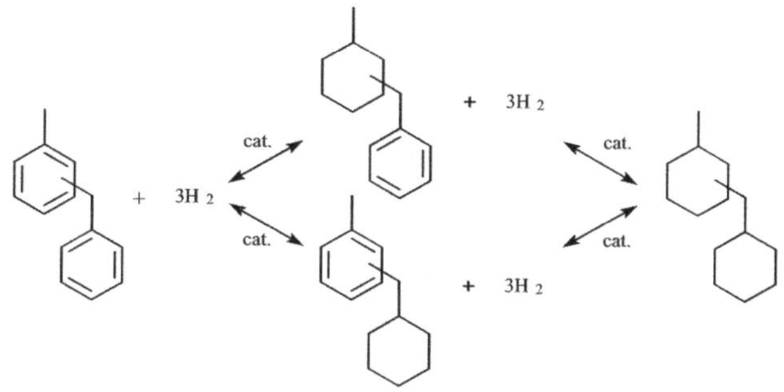

Figure 3.6. *Storage and release of H_2 with benzyltoluene*

3.2.2.3.1. Hydrogenation

The benzene hydrogenation mechanism and kinetics were studied by Leinweber and Müller (2018) on a 0.5 wt.% Ru/Al_2O_3 catalyst. H6-BT was the main intermediate, whereas H4 and H10-BT could only be found in small amounts during the reaction, which is in accordance with the reaction thermodynamics. Both reactions shown in Figure 3.6 were observed, but side ring (without the methyl group) hydrogenation was favored over main ring (with the methyl group) hydrogenation. Therefore, the intermediate H6-BT with one hydrogenated ring has nine isomers, and the fully hydrogenated H12-BT has six isomers. In terms of reaction kinetics, the para isomers have the highest reaction rate, which is in agreement with the results of tests performed by Kim et al. (2023), and the commercial mixture has a lower reaction rate. The reaction is a non-zero-order reaction for BT reactants, contrary to what has been observed for monocyclic hydrocarbons such as benzene (Leinweber and Müller 2018).

Rüde et al. (2022) studied the hydrogenation and dehydrogenation cycles of BT with a 0.3 wt.% Pt/Al_2O_3 catalyst under semicontinuous conditions via the hot pressure swing method (P_{H2} = 30 bar for hydrogenation and P_{H2} = 1 bar for dehydrogenation) at 290°C and reported that both the catalyst and the LOHC molecules were stable for more than 200 h on stream after seven cycles, with the formation of traces of fluorene.

Oh et al. (2022) synthesized mesoporous 3 wt.% Pt/Al_2O_3 (M3PtA) via one-pot solvent-deficient precipitation, which was used for BT (de)hydrogenation. This catalyst showed better activity than that prepared by the impregnation method in the temperature range between 230°C and 270°C for dehydrogenation at 1 bar and 110°C and 170°C for hydrogenation at 30 bar. Reversible H_2 charge and release cycles also indicated the better stability of M3PtA.

Kim et al. (2018) studied BT hydrogenation in 2018 over a 5 wt.% Ru/ZrO_2 catalyst prepared from $Ru_3(CO)_{12}$ and precipitated $ZrO_x(OH)_{4-2x}$ subjected to different aging periods and reported that BT hydrogenation activity exhibited a volcano-shaped dependency on aging time, as it is affected both by the surface hydroxyl content and surface area of the catalyst. Kim et al. (2021) further developed a 0.5 wt.% Ru/MgO catalyst that showed higher activity for BT hydrogenation than any other support because of the greater amount and strength of BT and H_2 adsorption. Both homolytic and heterolytic adsorption of H_2 were identified on 0.5Ru/MgO and confirmed by DFT calculations. Kim et al. (2023) synthesized a BT mixture with a 72% fraction of the para isomer H0-BT-*p*72 and a H-Beta zeolite (SiO_2/Al_2O_3 = 25), which showed better (de)hydrogenation activity than Marlotherm LH.

3.2.2.3.2. Dehydrogenation

Conventional dehydrogenation of perhydro-benzyltoluene (H12-BT) is carried out at elevated temperature (> 250°C) with an absolute hydrogen partial pressure of 1 bar. Liquid-phase dehydrogenation at atmospheric pressure is limited by its boiling point (270°C). As a higher global energy efficiency of the system is required, a lower operation temperature is needed for this process. Geisselbrecht et al. (2020) developed a reactive distillation column under reduced pressure, which allows the reaction temperature to be reduced to 200°C. Rüde et al. (2023), from the same group, further developed this concept, and their experimental results showed a hydrogen productivity of 0.35 gH_2/g_{Pt}/min at 267°C, which is four times greater than that of a conventional liquid-phase reaction.

The formation of methylfluorene during dehydrogenation with a Pt-based catalyst was detrimental for subsequent hydrogenation over a Ru-based catalyst. This issue was resolved by the use of a $Pt\text{-}S/Al_2O_3$ catalyst (Kim et al. 2023), similar to the method developed by Chiyoda Corporation for MCH dehydrogenation. Similarly, Jo et al. (2022) decorated M3PtA synthesized with the same method as Oh et al. (2022) with a different amount of sulfur and found that its activity for H12-BT dehydrogenation showed a volcano-shaped dependency on the sulfur content and reached a maximum value at 0.21 wt.% S. The addition of a small amount of sulfur increased the reaction rate, whereas an amount greater to 0.36 wt.% S hindered the reaction rate. Such a catalyst also showed great robustness and less selectivity toward side reaction products such as methylfluorene. Other modifications of Pt-based catalysts have also been explored. Kim et al. (2022) synthesized different $W_y\text{-}Pt_x/Al_2O_3$ catalysts with different impregnation sequences and reported an increase in dehydrogenation activity compared to Pt/Al_2O_3 catalysts with a W ratio (W/W+Pt) smaller than 0.42.

Compared to perhydro-dibenzyltoluene (H18-DBT), the lower boiling point of H12-BT limits the liquid phase dehydrogenation temperature. However, its lower viscosity makes it easier to handle during reaction and transport than DBT. Its significantly lower toxicity and vapor pressure and wider liquid state range than benzene and toluene imply that using the H0-BT/H12-BT couple would be a safer process (Zhang et al. 2020).

3.2.2.4. *Dibenzyltoluene/perhydro-dibenzyltoluene*

Like benzyltoluene, dibenzyltoluene (H0-DBT) has risen to the attention of the scientific and industrial community for its potential as a LOHC candidate. This mixture, which contains more than 10 isomers, has also been used as heat transfer oil via the name Marlotherm SH®, manufactured by Sasol, or Jarytherm®, manufactured by Arkema. Mixtures are becoming the focus of researchers because of their high thermal stability, wide liquid state range and low cost. Some properties of dibenzyltoluene (H0-DBT) and its H_2-rich form (H18-DBT) are listed in Table 3.4. Having three aromatic rings, the molecule has the same gravimetric H_2 storage capacity as toluene and benzyltoluene at 6.2 wt.%. The loading of hydrogen involves three consecutive reactions, as shown in Figure 3.7. The addition of H_2 to side rings is favored, as reported by Do et al. (2016).

	Dibenzyltoluene	**Perhydro-dibenzyltoluene**
Molar mass (g/mol)	272.4	290.4
Melting point (K)	239.2 (Jorschick et al. 2021)	215.2 (Jorschick et al. 2021)
Boiling point (K)	659.2 (Jorschick et al. 2021)	624.2 (Jorschick et al. 2021)
Density (kg/m³)	1,044 (20°C) (Müller et al. 2015)	913 (20°C) (Müller et al. 2015)
Viscosity (mPa.s)	3.79 (20°C) (Müller et al. 2015)	6.09 (20°C) (Müller et al. 2015)

Table 3.4. *Properties of dibenzyltoluene and perhydro-dibenzyltoluene*

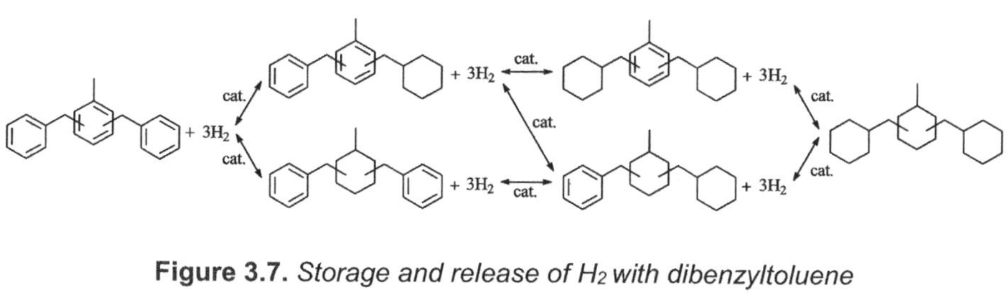

Figure 3.7. *Storage and release of H_2 with dibenzyltoluene*

3.2.2.4.1. Hydrogenation

Owing to the presence of three aromatic rings, the hydrogenation pathway of H0-DBT is much more complicated than that of benzyltoluene. The reaction could follow four paths: side–side–middle (SSM), side–middle–side (SMS), middle–side–side (MSS) or simultaneous hydrogenation. Do et al. (2016) used NMR spectroscopy to determine the possible reaction pathways and found that the SSM pathway is highly favored at temperatures between 120°C and 200°C under 50 bar of H_2 with a Ru/Al_2O_3 catalyst, which was further proven via high-performance liquid chromatography (HPLC) analysis. It was also found that the transformation from H12-DBT to H18-DBT could be the rate-determining step for this process. As mentioned in the previous section, the three isomers of benzyltoluene (BT) were shown to have different activities toward hydrogenation. Since the H0-DBT mixture contains six dibenzyltoluene isomers and nine benzyl(benzyltoluene) (methyl group on the side aromatic cycle) isomers (Ji et al. 2023), it is also expected that those isomers would have different activities. However, this has not yet been explored in the literature, as the separation and synthesis of these isomers are very challenging. Modisha et al. (2018) developed two-dimensional gas chromatography coupled with time-of-flight mass spectrometry (2D-GC-tof-MS) and single quadrupole-mass spectrometry (GC-SQ-MS), which enabled some isomers to be analyzed and the degree of hydrogenation of a reaction mixture to be determined. Important physicochemical properties, such as hydrogen solubility as a function of temperature and pressure, were measured by Aslam et al. (2016). Other properties, such as density, viscosity, vapor pressure and heat capacity, as a function of temperature were studied by Müller et al. (2015). The thermal conductivities of BT and DBT were studied by Berger Bioucas et al. (2020). The higher boiling point of the DBT mixtures allows a wider operation range for H_2 loading and release. Over a 0.25 mol% Ru/Al_2O_3 catalyst, the hydrogenation rate of DBT is more than 50% slower than that of BT under the same reaction conditions (Brückner et al. 2014). The high viscosity of DBT poses challenges for the establishment of LOHC systems. The addition of BT to DBT mixtures was potentially improved by Jorschick et al. (2020). A hot pressure swing reactor was used by Jorschick et al. (2017) to study the hydrogenation of H0-DBT in the temperature range between 160°C and 310°C. The initial reaction was found to be zero-order in terms of the DBT species. Ru/Al_2O_3 was found to be more active than Pt/Al_2O_3 at temperatures below 200°C, but a significant increase in activity was observed for Pt from 7 g_{H2}/g_{Pt}/min at 230°C to 25 g_{H2}/g_{Pt}/min at 300°C. Modisha and Bessarabov (2020) reported that a NiSat 310 catalyst with 50 wt% Ni could yield a 98% degree of hydrogenation (DoH) with only 15 bars of H_2 pressure. Ali et al. (2021) also compared the activity of different catalysts at low pressure (9 bar) and temperature below 200°C and reported that Raney-Ni could be more active than 5% Pd/Al_2O_3 and 5% Ru/Al_2O_3 in a batch reactor. Efforts have also been made by Dürr et al.

(2021) to experimentally determine the hydrogenation/dehydrogenation equilibrium. An empirical function was successfully established for the equilibrium in the pressure range between 1 and 9.9 bar (absolute) and the temperature range between 490 and 635 K.

3.2.2.4.2. Dehydrogenation

The dehydrogenation rate of DBT is also reported by Brückner et al. (2014) to be slower than that of BT using 0.5 wt.% Pt/Al_2O_3 as a catalyst. Pt/C was found to be more active for dehydrogenation at 270°C than Pt/SiO_2 and Pt/Al_2O_3. Pt was found to be more active than Pd supported on alumina at 320°C but could produce more methane as a by-product (Modisha et al. 2019b).

Chen et al. (2021) synthesized Pt/TiO_2 (0.3 wt-%) and S-Pt/TiO_2 (0.3 wt-% Pt, 0.1 wt-% S) catalysts via wet impregnation, which showed better selectivity toward H0-DBT during dehydrogenation than S-Pt/Al_2O_3. The catalysts also exhibited high dehydrogenation activity. It could reach 98% dehydrogenation at 310°C. These results are in agreement with those obtained by Auer et al. (2019). CeO_2 was also studied as a Pt support. Pt/CeO_2 synthesized by the glycine nitrate process showed much better performance than Pt/Al_2O_3, with 80.5% conversion in 2.5 h compared with 3.5% (Lee et al. 2021b).

Pushing the thermodynamic equilibrium toward the release of hydrogen could also be achieved by the extraction of produced hydrogen via process optimization. Mrusek et al. (2021) demonstrated the coupling of a dehydrogenation unit and electrochemical hydrogen compression (EHC), which accelerates the hydrogen release process. Furthermore, EHC has the benefit of producing high-pressure hydrogen with high purity. The dehydrogenation temperature could be reduced to 240°C with a platinum-based catalyst, the production of side products such as methane decreased to 0.09 ppm from 120 ppm, and no CO or CO_2 production was detected. Process intensification integrating PdAg membranes into a radial flow reactor was also proven to be applicable by Wunsch et al. (2018, 2020). DFT calculations carried out by Huynh et al. (2021) showed that by modifying H18-DBT with perhydro-lithium dibenzyl phenolate (H18-DBT-OLi), the dehydrogenation enthalpy could be reduced from 52.6 to 48.1 kJ/mol. The calculations also indicated that the rate-determining step for dehydrogenation is the release of the first hydrogen molecule from H12-DBT to form H10-DBT.

A kinetic model of H18-DBT dehydrogenation was established between 260°C and 310°C by Peters et al. (2019) to couple the LOHC dehydrogenation unit with a solid oxide fuel cell. The feasibility of such a system was successfully demonstrated, and the maximum H_2-electricity efficiency was found to be 45%.

3.2.2.5. *Terphenyl/tercyclohexane*

Commercial terphenyl product contains three isomers: *o*-terphenyl, *m*-terphenyl and *p*-terphenyl, which have very different boiling and melting points (m.p. at 58°C, 86°C and 212°C, respectively, and b.p. at 337°C, 379°C, and 389°C, respectively). This mixture is widely used as a heat transfer and storing material. As a conjugated aromatic compound without an alkane group, this type of molecule is more compact and therefore potentially has fewer steric problems entering and exiting catalyst pores than toluene derivatives with the same number of benzene rings, such as dibenzyltoluene (Sung et al. 2008). However, considering that terphenyl and perhydroterphenyl are solid under ambient conditions, they cannot be transported using pipelines. The enthalpy of fusion of terphenyl isomers (17.2 kJ/mol for *o*-terphenyl, 31.0 kJ/mol for *m*-terphenyl and 35.6 kJ/mol for *p*-terphenyl) could also hinder the energy efficiency of the process using these molecules as LOHC because energy is required in each usage to transform the solid into a liquid to carry out hydrogenation and dehydrogenation reactions (Murthy et al. 1995; Verevkin 1997). The hydrogen storage and release reactions are presented in Figure 3.8.

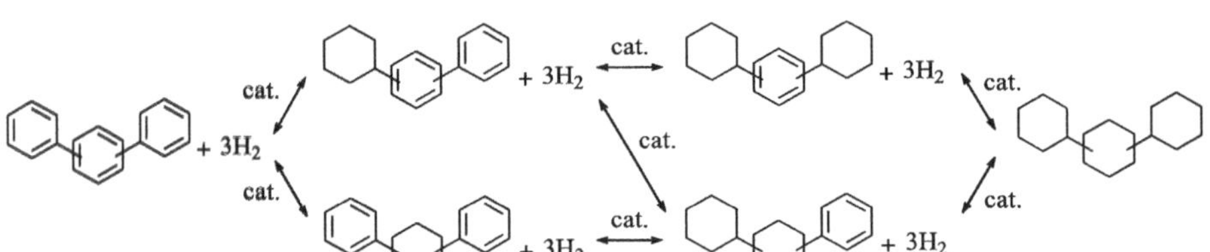

Figure 3.8. *Storage and release of H_2 with terphenyl*

3.2.2.5.1. Hydrogenation

The hydrogenation mechanism of terphenyl reactions depends largely on the isomer and catalyst used. According to the exploration carried out by Scola et al. (2002) at temperatures between 200°C and 275°C and pressures between 103 and 172 bar, center-ring attack by hydrogen was found to favor *o*-terphenyl and *m*-terphenyl over copper chromite, whereas a random addition pattern was observed for *p*-terphenyl, with only a slight preference toward side rings. However, with a nickel catalyst, a side-ring preference was found for *m*-terphenyl and *p*-terphenyl groups. The hydrogenation of the single isomer *o*-terphenyl could reach 99% toward perhydro-*o*-terphenyl with a commercial 3% Pt/C catalyst at 180°C and 50 bar of hydrogen (Kalenchuk et al. 2018). A lower temperature hydrogenation at 160°C was studied where 50 atm of hydrogen was applied for the reaction over 10% Pt/C, 5% Pt/SiO_2, 5% Pt/Al_2O_3, Pt/Sibunit and Pd/Sibunit, which were prepared via impregnation by Sung et al. (2008). It was found that 5% Pt/SiO_2 showed better activity (70.1% conversion after 2 h) than 10% Pt/C when the reaction mixture was not stirred despite its lower metal content, which could provide the possibility of using non-stirred reactors for hydrogenation process that could potentially save energy. Pd-based catalyst was found to be more active than Pt-based one at the same loading of metal and has higher stability over hydrogenation/dehydrogenation cycles.

3.2.2.5.2. Dehydrogenation

While the hydrogenation of terphenyl is usually carried out in stirred reactors, both stirred and flow reactors are used for the dehydrogenation of perhydroterphenyl. Sung et al. (2008) also studied the dehydrogenation of perhydro-terphenyl using the same method used for the hydrogenation of terphenyl at 320°C with 1 bar of H_2. Compared with 5% Pt/SiO_2, 10% Pt/C was found to be more active. An initial rate of 82 $L_{H2}/g_{Pt}/h$ was obtained for the Pt/C catalyst. Pt-based catalysts were found to be more stable than Pd-based catalysts, and deactivation of Pt-based catalysts was observed over hydrogenation/dehydrogenation cycles. Studies on a 3% Pt/C catalyst at temperatures between 280°C and 340°C in a flow reactor have shown that dehydrogenation is preferred for terminal rings on perhydro-*m*-terphenyl and perhydro-*p*-terphenyl and that total dehydrogenation could be achieved for these two isomers without side reactions, whereas total dehydrogenation cannot be realized for perhydro-o-terphenyl, and products such as triphenylene from isomerization are important for this isomer (Kalenchuk et al. 2016, 2018). Therefore, considering the high melting point of p-terphenyl, m-terphenyl could be a better candidate for hydrogen storage and release. Different catalysts have been tested for m-terphenyl dehydrogenation, and their activity order is as follows: Pt/C > Pd/C > Pt/Al_2O_3 > Pt/SiO_2. The doping of Pt-based catalysts with Ni and Cr was found to enhance the selectivity toward m-terphenyl. Ni-, Cr- or Au-based catalysts all exhibited lower activity than do Pt- and Pd-based catalysts (Kalenchuk et al. 2015).

3.2.2.6. *Naphthalene/Decalin*

Naphthalene is the simplest polycyclic aromatic hydrocarbon, which is largely used as a precursor to a variety of chemicals, such as phthalic anhydride. Its melting point is 80.2°C, and its boiling point is 217.9°C. It is therefore in the solid phase under ambient conditions. Naphthalene is produced mainly from coal tar and heavy petroleum fractions. The hydrogenation of naphthalene first gives tetralin and then *cis*- and *trans*-decalin, as illustrated in Figure 3.9 (Weitkamp 1968). Tetralin is the partially hydrogenated intermediate of naphthalene. It is a colorless liquid at room temperature with a melting point of – 35.8°C and a boiling point of 207°C. The final product, decalin, is also in its liquid state at ambient temperature. Its melting point and boiling point are different for *cis* and *trans* molecules. The melting points are – 42.9°C and – 30.4°C, and the boiling points are 196°C and 187°C for the *cis* and *trans* isomers, respectively (Haynes 2016). The decalin mixture is widely used as an industrial solvent for organic compounds. The naphthalene/decalin couple has a high storage capacity (7.3 wt-%), similar to the terphenyl/tercyclohexyl couple, and considering that, unlike the terphenyl/tercyclohexyl couple, its hydrogenated products are liquids under ambient conditions, it could pose less energetic and technological challenges for industrialization. However, owing to the high hydrogenation enthalpy of naphthalene (66.3 kJ/mol H_2 compared with 63.5 kJ/mol H_2 for benzyltoluene) and its enthalpy of fusion (19.1 kJ/mol) (Sharma et al. 2008), the energy efficiency of the hydrogenation process could be disadvantageous compared with that of other linear aromatic hydrocarbon LOHC systems.

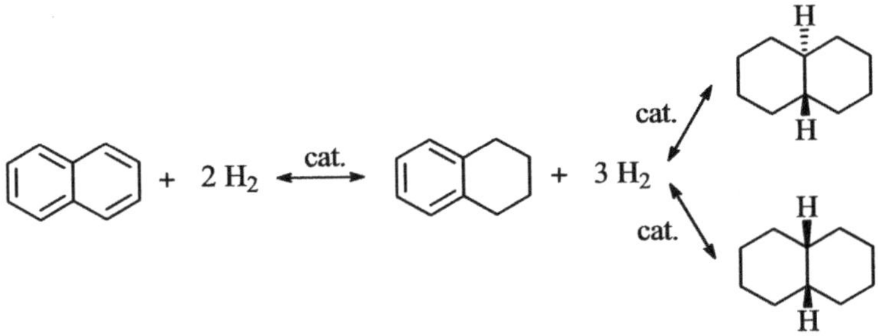

Figure 3.9. *Storage and release of H_2 with naphthalene*

3.2.2.6.1. Hydrogenation

The hydrogenation selectivity toward tetralin and decalin depends largely on the catalyst and reaction temperature. Naphthalene is hydrogenated first to tetralin through stepwise dissociated hydrogen *cis*-addition. Tetralin then transforms into octahydronaphthalene, which could be hydrogenated to *cis*-decalin and *trans*-decalin (Rautanen et al. 2002). Kirumakki et al. (2006) studied the hydrogenation of naphthalene on NiO-SiO_2-Al_2O_3 catalysts in a fixed-bed reactor at 200°C and established a two-site Langmuir–Hinshelwood model for this reaction, which demonstrated that naphthalene adsorbed twice as strongly as tetralin on the catalyst surface. This indicated that the weak adsorption of tetralin could be the reason for its low yield of decalin. Despite the low conversion of naphthalene to decalin, total conversion of naphthalene to tetralin was achieved without side product. Similar phenomena were observed for naphthalene hydrogenation at 275°C with 50 bar of hydrogen over the 1Pd-4Ni/SA (1 wt-% Pd, 4 wt-% Ni) catalyst, but the main product was found to be decalin (more than 90%) over 1Pd/SA and 1Pd-8Ni/SA, with *trans*-decalin being the favored isomer. 1Pd-8Ni/SA was also found to have better resistance toward deactivation (Barrio et al. 2003). The deactivation of Ni-based catalysts was possibly due to the σ-adsorption of tetralin, which led to coke formation (Rautanen et al. 2002). Using supercritical CO_2 as a solvent instead of organic solvents, the catalyst activity could be enhanced because of the higher H_2 solubility in supercritical CO_2 and lower reaction temperature (Hiyoshi et al. 2005). The activity of charcoal-supported metal catalysts for naphthalene hydrogenation in supercritical CO_2 was found to be in the following order: Rh > Pd > Pt > Ru (Shirai et al. 2006; Hiyoshi et al. 2007).

Methylnaphthalene/methyldecalin (6.6 wt-% H_2 storage capacity) could also be used for hydrogenation/dehydrogenation cycles. As methylnaphthalene is in its liquid state, it does not have the disadvantages that naphthalene poses. In a trickle-bed reactor on a Ni/Al_2O_3 catalyst, the selectivity toward *cis*-methyldecalin was greater at lower temperatures with higher Ni contents. The impregnation of the catalyst with Ru favored the formation of *cis*-methyldecalin, whereas the use of Pt drastically increased the selectivity toward *trans*-methyldecalin (Kline et al. 2022).

3.2.2.6.2. Dehydrogenation

Dehydrogenation of decalin takes place at higher temperatures than hydrogenation does, thus posing more risk for catalyst coking and by-product formation. Operating at lower temperatures with suitable catalysts is therefore crucial for the process. Kalenchuk et al. (2020) studied the reaction kinetics of decalin dehydrogenation over a 3% Pt/C catalyst in flow reactors at temperatures between 260°C and 340°C with a volume hourly space velocity of 1 h^{-1} and reported that the dehydrogenation rate of *trans*-decalin was much lower than that of *cis*-decalin. DFT calculations have demonstrated that Pd(111) is more active for the dehydrogenation of tetralin to naphthalene, whereas Pt(111) is more active for the transformation of decalin to tetralin, which has been further proved by experimental dehydrogenation of decalin at 270°C over 5 wt-% Pt/C and Pd/C catalysts synthesized by wet impregnation (Kim et al. 2017). $Pt/MgAl_2O_4$ prepared by incipient wet impregnation was reported to have good activity for decalin dehydrogenation in the temperature range between 210°C and 250°C, as the strong electronic interaction between Pt and the support created partially positively charged Pt, which weakened the bond between Pt and the dehydrogenation product naphthalene, thus favoring its desorption (Tuo et al. 2021). Note that 4 wt-% Pt/C prepared by wet impregnation exhibited better activity at 260°C than 5 wt-% Pt/C owing to its better Pt dispersion, which led to 60% decalin conversion, which was limited by reaction thermodynamics (Sebastián et al. 2008). A study applying superheated liquid-film conditions with Pt-Re/granular active carbon led to an increase in the conversion of decalin to naphthalene to nearly 100% in the temperature range between 210°C and 280°C (Hodoshima et al. 2005), which could be a solution to thermodynamic limitations.

3.2.2.7. *Eutectic aromatic hydrocarbon mixtures*

As the crystallization of certain LOHC couples under ambient conditions could hinder their application for hydrogen storage and release, the utilization of eutectic mixtures containing these molecules could solve this problem. Some examples of eutectic mixtures include *o*-MBN-*p*-MBN/*o*-H16-MBN-*p*-H16-MBN (MBN: methylbenzyl naphthalene) (Rao et al. 2023) and biphenyl-(diphenyl ether)/bicyclohexyl-(dicyclohexyl ether) (Jang et al. 2020). These mixtures have similar H_2 storage capacities to the pure substances that compose them while remaining in the liquid phase at room temperature.

3.2.2.7.1. Hydrogenation

As eutectic mixtures are composed of different LOHC molecules, their hydrogenation and dehydrogenation mechanisms could be considered similar to those of their pure components. The MBN mixture could be completely hydrogenated by a Ru-based catalyst at 150°C, and the resulting perhydro-MBN mixture could release hydrogen on a Pt/CeO_2 catalyst at 330°C (Rao et al. 2023). The low hydrogenation temperature of the MBN mixture makes it a potential LOHC candidate.

3.2.2.7.2. Dehydrogenation

Pt/KIT-6 (ordered mesoporous silica) (Cao et al. 2016) prepared by wet impregnation was reported to be a good catalyst for the dehydrogenation of bicyclohexyl-dicyclohexylmethane eutectic mixture at 340°C, with higher total dehydrogenation selectivity than commercial Pt/SiO_2 and Pt/Al_2O_3, albeit with slightly higher cracking and isomerization rate (Ahn et al. 2022). An in-depth study of Pt/Al_2O_3 on a catalyst support demonstrated that a minimum BET surface area is needed (above 35 m^2/g) for the catalyst to be active in the eutectic mixture and that a higher Pt dispersion could enhance its activity (Kwak et al. 2021). A system evaluation of a biphenyl-diphenylmethane mixture using a temperature-cascade process was carried out to compare it with other LOHC systems by Brigljević et al. (2020a, 2020b) and was found to be more sensitive to variations in dehydrogenation parameters. DFT calculations over biphenyl-(diphenyl ether)/bicyclohexyl-(dicyclohexyl ether) have demonstrated that LOHCs containing oxygen atoms have more favorable dehydrogenation thermodynamics than do their similar hydrocarbon molecules (Jang et al. 2020).

3.3. N-containing molecules

3.3.1. *Introduction*

With a much lower enthalpy of dehydrogenation than cycloalkanes due to weaker bond strengths, N-containing molecules are presented as energy-saving LOHC candidates (Crabtree 2008). The presence of two N atoms in the cycle decreases the dehydrogenation temperature by 150°C because of a much weaker dehydrogenation enthalpy (Cooper et al. 2006). The system showing the highest H_2 capacity is perhydro-4,7-phenanthroline (7.2 wt-%), but the most studied and mature system is dodecahydro-N-ethylcarbazole (Zhou et al. 2020), which was first proposed by Air Products in 2006 (Pez et al. 2006). The main problems of this system today are the high raw material price and the (eco)toxicity and non-biodegradability of the molecule (Markiewicz et al. 2019). Other N-heterocycles have been proposed within the pyridine and pyrrole families (including polycycles).

The most detailed reviews of these systems and the catalysts used for both hydrogenation and dehydrogenation are those of Crabtree (2017a) and, not specifically dedicated to N-containing molecules, those of Rao and Yoon (2020) and Aakko-Saksa et al. (2018). Except for carbazole families, which are sometimes studied at large scale, other systems are less mature and are sometimes studied in a solvent (quinolines, nitriles, etc.).

3.3.2. N-heterocycles

3.3.2.1. *Introduction*

The N-heterocycle families that are considered LOHC candidates are presented in Table 3.5. Substituted forms of these molecules (generally on the N atom) should also be considered.

	H_2-lean form	H_2-rich form	Gravimetric storage capacity (wt%)
Carbazole/perhydrocarbazole	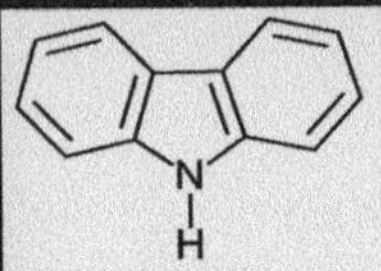	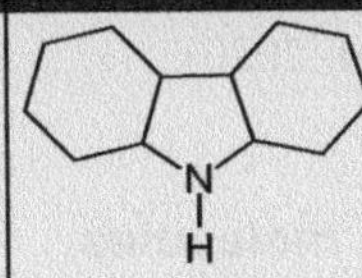	5.8%
Pyrrole/pyrrolidine		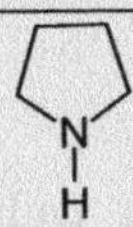	5.6%
Indole/octahydroindole	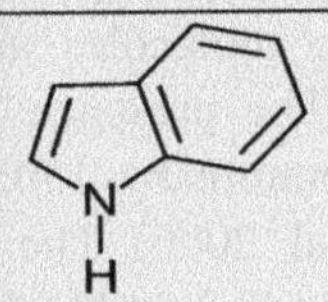	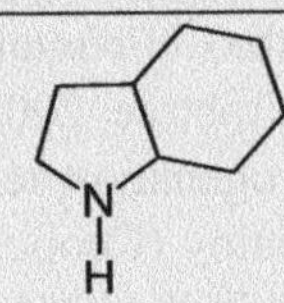	6.4%
Pyridine/piperidine	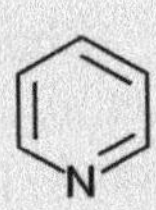	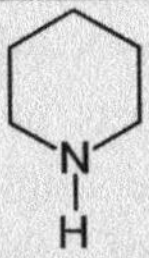	7.1%
Quinoline/perhydroquinoline	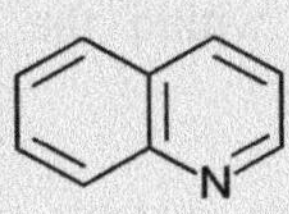	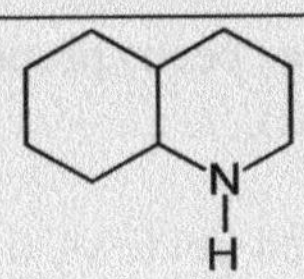	7.2%

Table 3.5. *Families of N-heterocyclic LOHCs*

3.3.2.2. *N-ethylcarbazole (NEC)/perhydro-N-ethylcarbazole*

3.3.2.2.1. Thermodynamic considerations

With a standard dehydrogenation enthalpy of 50.5 kJ/molH_2, compared with ca. 65 kJ/molH_2 for cycloalkanes, perhydro-N-ethylcarbazole has appeared to be an attractive alternative (Emel'yanenko et al. 2019), especially for mobility (Niermann et al. 2019).

Considering the whole process, including hydrogenation of the NEC at 433 K, dehydrogenation of the perhydro compound at 473 K, and the use of pumps, heaters and coolers, the overall exergy efficiency of the storage and release process was estimated to be greater than 86% for this molecule (Choi et al. 2016).

3.3.2.2.2. Hydrogenation

The hydrogenation of NEC (H_2 storage) is typically performed in batch reactors on heterogeneous Ru or Ni catalysts (Zhou et al. 2020) at temperatures lower than 180°C and pressures of 8–100 bars. Bimetallic RuNi may also be used (Wang and Bai 2023). Some kinetic studies can be found for both types of catalysts (Ye et al. 2011; Wan et al. 2013). Among various nickel catalysts, the 70 wt-% Ni/SiO_2-Al_2O_3 catalyst developed by Ding et al. (2021) is said to be as active as a 5 wt-% Ru/Al_2O_3 catalyst at 150°C, 70 bars H_2 and thus appears to be a more economical solution.

3.3.2.2.3. Dehydrogenation

The majority of perhydro-N-ethylcarbazole dehydrogenation experiments use heterogeneous catalysts (Zhou et al. 2020). In a recent review, Sekine and Higo (2021) reported that Pd is the most active at temperatures near 200°C and is sometimes associated with other metals, such as Ru and Au, as well as with Ni and Cu, which reduce the amount of precious metals and improve the stability of the catalyst. Various supports have been reported, with a preference for carbon and alumina. The standard dehydrogenation conditions were 160°C–200°C and atmospheric pressure. Recently, a new Pd-Ni supported on an ordered mesoporous silica, KIT-6, showed a dehydrogenation performance that was 1.7 times greater than that of Pd/KIT-6 at 180°C, with less leaching and agglomeration (Feng and Bai 2022).

As the dehydrogenation temperature is far below the boiling point of the hydrogen-rich compound, the reaction occurs in the liquid phase. Most of the studies have been performed in batch reactors of 10–100 mL. Kiermaier et al. (2021) studied dehydrogenation under reduced pressure to mimic the sucking effect of the fuel cell during operation. A few groups have studied continuous reactions. Li et al. (2022) performed dehydrogenation in a continuous flow fixed bed reactor involving 0.4 g of different catalysts but using a solvent (xylene). The Pd/$MgAl_2O_4$ catalyst showed the best performance, with a conversion of 98.3% at 180°C. However, working without any solvent, continuous operation must manage the large volume expansion due to hydrogen: for 1 mL of perhydro-N-ethylcarbazole, 600 mL of H_2 is recovered after cooling at RT (Peters et al. 2015a). These authors used a tube reactor filled with spheres coated with a thin layer of Pt/Al_2O_3 catalyst and performed dehydrogenation between 220°C and 260°C at atmospheric pressure and at a flow rate of 0.5 to 4 mL/min; the highest turnover frequency (TOF) observed (lowest catalyst thickness to reduce mass transfer limitations) was 10.9 g_{H2}/min/g_{Pt}. This productivity is among the highest found for all LOHC systems. The same group also tried to use a coated cellular structured reactor (Peters et al. 2015b) but with a lower productivity. At a larger scale, Yang et al. (2022) modeled H_2 release in a 3-m long reactor tube (3 cm diameter) filled with Pd/Al_2O_3 and fed with an NEC flow rate of 24 mL/min. This resulted in a conversion of 90% at 250°C.

Different studies have focused on dehydrogenation kinetics, but as mass transfer is not limited (Peters et al. 2015a), most of the activation energies and kinetic constants have been underestimated. Yang et al. (2022) proposed a two-dimensional model for a dehydrogenation tube reactor for NEC dehydrogenation. To reach high conversion, the temperature at the exit of the tube, where the dehydrogenation of 4H-NEC occurs, must be high because of the slow kinetics of this rate-determining step. Note that 235°C was used for this last step. To reach high conversion and fast kinetics, 247°C was recommended for the subsequent dehydrogenation steps (the optimized value to have high yield and low energy consumption).

Even if it is a widely studied system, the produced N-ethylcarbazole presents the major drawback of being solid at RT (Table 3.6). Thus, it may be necessary to limit the conversion (e.g. 90%) to avoid crystallization (Peters et al. 2015b; Stark et al. 2016).

Consequently, as perhydro-N-ethylcarbazole only stores 5.8 wt% H_2, the actual capacity would therefore be reduced to less than 5.5 wt%, which is below the US DoE requirements. The use of other systems should also be considered.

	N-Ethylcarbazole	Perhydro-N-ethylcarbazole
Molar mass (g/mol)	195.26	207.26
Melting point (K)	342.4	188.6 (T_G)
Boiling point (K)	> 513	–
Density (kg/m^3)	1,059 (80°C)	940.3 (20°C)

Table 3.6. *Properties of N-ethylcarbazole and perhydro-N-ethylcarbazole*

3.3.2.3. *Other N-substituted carbazoles*

Other N-substituted carbazoles have been proposed as alternatives to N-ethylcarbazole; some are only studied from a thermodynamic point of view, without any experiments, for example, N-phenylcarbazole, which has a higher H_2 capacity (6.9 wt-% compared with 5.8 for NEC) but still presents a melting point that is too high (90°C). The dehydrogenation enthalpy is much higher than that of NEC (10 kJ/molH_2 higher) but still lower than that of the aromatic/cycloalkane couples (Emel'yanenko et al. 2019). This system was further studied by Brigljević et al. (2020a) in terms of costs and in combination with SOFC (Brigljević et al. 2020b).

N-Acetylcarbazole was also proposed for computational approaches, but no experiments have been performed (Choi et al. 2016; Shin et al. 2018).

Many other publications mention the combination of N-propylcarbazole/perhydro-N-propylcarbazole. These studies concern both hydrogenation and dehydrogenation reactions. Similar conditions and catalysts to those described for NEC were used: Ru for hydrogenation (Yang et al. 2013) and Pd for dehydrogenation (Dong et al. 2018). However, the melting point of N-propylcarbazole, which is lower than that of the N-ethyl compound, is still above RT (47.3°C). Other options could include binary or ternary mixtures of N-alkylcarbazoles, which offer a eutectic composition (N-ethylcarbazole + N-propylcarbazole is a liquid above 24°C, N-ethylcarbazole + N-propylcarbazole + N-butylcarbazole above 12.5°C) (Stark et al. 2016).

3.3.2.4. *Pyrroles, indoles*

The use of couples based on pyrrole or indole is much less mature than the use of alkyl carbazoles, and the studies are often far from direct use as a LOHC system since they are sometimes performed in a solvent, thus leading to a very low effective hydrogen storage capacity. Some studies have nevertheless been performed with pure perhydro 2-methylindole and octahydroindole with different Pd catalysts. The yield of 2-methylindole reached 99% at 250°C after 4 h of reaction with a metal-to-reactantology ratio of 0.6 mol% (Oh et al. 2019).

Cui et al. (2008) tested a variety of catalysts for both hydrogenation and dehydrogenation. They reported that RuO_2 was a suitable catalyst for the hydrogenation of 1-methylindole. Rh/C was also well adapted to the hydrogenation of indole and 1-methylindole. With respect to the dehydrogenation reactions, the Pd catalysts had the best performance, especially Pd/SiO_2, and a dehydrogenation temperature of 165–200°C was sufficient to dehydrogenate pure 1-methyloctahydroindole. The major product was 1-methylindole, but some intermediates (dehydrogenation of only the pyrrole ring) were still present, especially in closed vessels. The 1-methylindole yield reached 80% at 170°C in a N_2-flushed vessel.

Other alkylindoles have been studied as LOHCs, and hydrogenation and dehydrogenation have been performed under relatively mild conditions in batch reactors but with the use of a solvent. A 5 wt% Ru/Al_2O_3 catalyst can be used for the hydrogenation of N-ethylindole (Dong et al. 2015) and 2-methylindole (Li et al. 2016) at temperatures less than 200°C and 70–90 bar. The dehydrogenation of the products was quantitative using a 5 wt% Pd/Al_2O_3 catalyst at the same temperature and atmospheric pressure. The reaction can also be catalyzed by platinum. Owing to theoretical calculations, Ouma et al. (2021) proposed doping Pt with chalcogens to stabilize the surface adsorption of the products. S and Se seemed to be the best boosters for the catalytic dehydrogenation of octahydroindole to indole. The Papp group studied the surface reactions of the LOHC system indole/indoline/octahydroindole on Pt(111) (Bachmann et al. 2018) and on Ni(111) (Steinhauer et al. 2020) using XPS assisted by temperature-programmed desorption (TPD) to elucidate the dehydrogenation mechanism. Even if all of the molecules are studied for their ability to be dehydrogenated, only octahydroindole may be of interest as a hydrogen-rich molecule owing to its storage capacity of 6.4 wt-%. Platinum allowed the selective formation of indolides (deprotonation of the NH group), whereas nickel also produced some ring-opening by-products. In both cases, indole seemed to be only a short-lived intermediate, whereas the other publications mention it as the final product.

Damodara et al. (2014) used a $Cu(0)/Al_2O_3$ catalyst for the dehydrogenation of various indolines and carbazoles and quinolines in dimethylformamide (DMF) solutions under a N_2 atmosphere at 120°C in sealed pressure tubes. The dehydrogenated counterparts were produced in good yields. Even if the turnover number (TON) and TOF were moderate compared with those of Pd catalysts, the use of copper as the catalyst was very encouraging.

With respect to molecular catalysts, Brayton and Jensen (2015) used an Ir pincer complex for the dehydrogenation of butyl pyrrolidine and methyl perhydroindole between 140°C and 200°C. The aromatic ring without a heteroelement was not dehydrogenated; only the pyrrolidine part was dehydrogenated, making the actual hydrogen storage greater for butyl pyrrolidine (3.1 wt-%) than for methyl perhydroindole (2.9 wt-%). Zubar et al. (2020) reported a manganese complex as a catalyst for the hydrogenation of indole-based compounds and the dehydrogenation of corresponding indolines. The system did not lead to octahydroindole but stopped at the hydrogenation of the pyrrole ring. Reactions were carried out in a solvent with the use of a base at 100°C and 50 bar for hydrogenation and at 120–160°C for dehydrogenation. The same complex was evaluated for its ability to hydrogenate a variety of N-containing heterocycles (quinoxalines and benzoxazine). The hydrogenation of the N-ring was quantified after 24 h at 120°C with 50 bar of H_2 using 1 mol% Mn complex, tert-potassium butoxide (tBuOK) as the base and toluene as the solvent. Note that the ring without any heteroelements was not hydrogenated.

Very recently, simple or multiple alkyl substitutions of indoles were shown to lower the enthalpy of the reaction for dehydrogenation, thus increasing the overall energy efficiency of the process with dehydrogenation at lower temperatures

(Safronov et al. 2023). In conclusion, alkylindoles can be interesting LOHC candidates (Stepanenko et al. 2022).

3.3.2.5. *Pyridines and quinolines*

Similar to what they did for indoles, an important work was performed by Cui et al. (2008) to determine which structural features of substituted piperidines could lower the enthalpy of dehydrogenation, both using theoretical Gaussian calculations and experiments. Attaching electron-donating or electron-conjugated substituents to the piperidine ring was demonstrated to significantly increase the rate of catalytic dehydrogenation. Thus, 4-aminopiperidine and piperidine-4-carboxamide were calculated to be the most promising structures as reversible hydrogen storage compounds. The authors tested a variety of catalysts for both hydrogenation and dehydrogenation. They reported that RuO_2 was a suitable catalyst for the hydrogenation of 4-dimethylaminopyridine, whereas Rh/C was well adapted for the hydrogenation of piperidine-4-carboxamide. The hydrogenation of 4-aminopyridine was not possible because of the disproportionation of the product into dipiperidylamine. With respect to the dehydrogenation reactions, the Pd catalysts had the best performance, especially Pd/SiO_2, and dehydrogenation temperatures lower than 200°C were sufficient for most of the molecules studied.

2-[(n-Methylcyclohexyl)methyl]piperidine and 2-(n-methylbenzyl)pyridine are liquids at a wide range of temperatures (lower than – 20°C to more than 290°C) and appear to constitute an interesting LOHC system, despite having a higher dehydrogenation enthalpy (67.3 $kJ/molH_2$) than other N-heterocyclic LOHCs. The dehydrogenation of 2[(n-methylcyclohexyl)methyl]piperidine was thus studied using different Pd catalysts (Bathula et al. 2019; Oh et al. 2019). A dehydrogenation efficiency of 99% was obtained at 270°C in 4 h using a one-pot solvent-deficient precipitation Pd/Al_2O_3 catalyst with a metal-to-reactant ratio of 0.1 mol%.

Many studies have been performed using a solvent, either for hydrogenation or dehydrogenation. The hydrogenation of various quinoline derivatives was performed in various solvents at 30°C, 20 bars using nitrogen-doped carbon-supported Ru nanoparticles (Nie et al. 2022). Ethanol reached 99.6% yield of tetrahydroquinoline, but perhydroquinoline was not obtained.

A solid Co/graphene oxide catalyst prepared from $Co(acac)_3$ and further pyrolyzed was evaluated for the dehydrogenation of 1,2,3,4-tetrahydroquinoline. The reaction requires t-BuOK as a base to accelerate substrate binding to the catalyst surface by activating the N–H bond (Jaiswal et al. 2019). The reaction was performed at 150°C in n-decane as a solvent. The catalyst was evaluated for a broad scope of tetrahydroquinolines. The reverse reaction was performed at 120°C in toluene with the same catalyst.

Quinaldine (2-methylquinoline), a liquid present between – 2 and 248°C, can store 6.6 wt% hydrogen. Decahydroquinaldine can be dehydrogenated at temperatures as low as 138°C with iridium complexes in solution under p-xylene reflux (Yamaguchi et al. 2009). The system was also studied by Ryabchuk et al. (2019) with an intermetallic nickel silicide catalyst on a fumed silica for both hydrogenation and dehydrogenation in a continuous setup, with triglyme as the solvent. Hydrogenation was performed at 120°C and 30 bars, and dehydrogenation was performed at 200°C under 1 bar of Ar. They compared their results with those of Ir, Pt, Pd, and Rh complexes that were used in the literature, demonstrating the advantages of their nickel silicate system. Decahydroquinaldine appears to be a good LOHC candidate and, furthermore, presents low to moderate (eco)toxicity, comparable to that of diesel oil, but a weaker biodegradability (Markiewicz et al. 2019).

The dehydrogenation enthalpy is not as low as that of perhydroalkylcarbazoles but is still advantageous compared with non-N-substituted molecules (Naseem et al. 2021). Nonetheless, to the best of our knowledge, no experiments involving pure compounds have been reported.

Acridine may also be cited in this section, as it consists of a quinoline with one more benzenic cycle (Figure 3.10). The hydrogenation of the molecule was studied using a 5% Ru/Al_2O_3 catalyst in the temperature range of 140–190°C in mesitylene in an autoclave at 70 bars of hydrogen (Yang et al. 2020). The perhydrogenated product was obtained, representing a gravimetric hydrogen storage of 7.25 wt-%. To our knowledge, dehydrogenation has not been studied. Notably, owing to the high melting point of acridine (111°C), this molecule should be used in eutectic mixtures to obtain a convenient LOHC system.

Among the quinoline derivatives, trisphaeridine (Figure 3.10), which was proposed because of its natural origin (amaryllidacae alkaloids), is also interesting to consider (Tang et al. 2020).

The dehydrogenation thermodynamics was studied, and the enthalpy for perhydro-trisphaeridine dehydrogenation was similar to that of perhydroNEC, making this natural molecule a promising candidate, with a storage capacity of 5.9 wt-%. Nevertheless, other properties, such as the melting point, toxicity and kinetics, still need to be explored.

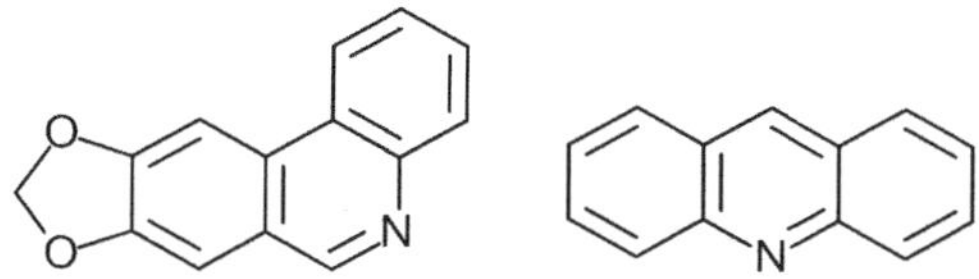

Figure 3.10. *Trisphaeridine and acridine*

3.3.2.6. *Other N-heterocycles*

New ideas of LOHC systems appear in the literature and cannot be categorized in the previous paragraphs.

Phenazine (Figure 3.11) was proposed as a molecule to store hydrogen since it can be synthesized from ammonia and cyclohexane-1,2-diol, the latter being itself easily obtained from lignin (Forberg et al. 2016). A $Pd_2Ru/SiCN$ catalyst was used to hydrogenate phenazine at 115°C and 50 bar in dioxane and to dehydrogenate perhydrophenazine at 190°C in diglyme.

Many Ir complexes were reviewed by Cho et al. (2021) for the hydrogenation and dehydrogenation of N-heterocyclic compounds. One of the salient results with this family of catalysts is the dehydrogenation of 2,5-dimethylpiperazine to yield 2,5-dimethylpyrazine (Figure 3.11, H_2 capacity of 5.3 wt-%) used under neat conditions and quantitatively dehydrogenated under reflux (163°C) (Fujita et al. 2017). The reverse reaction was also performed without a solvent or the same Ir catalyst. The same authors also studied the release and uptake of five molecules of H2 from the couple 2,6-dimethyldecahydro-1,5-naphthyridine/2,6-dimethyl-1,5-naphthyridine (Fujita et al. 2014).

Owing to its very low dehydrogenation enthalpy (35.9 kJ/molH_2), 1,2-dihydro-1,2-azaborine, a heterocycle bearing N and B (Figure 3.11) and its hydrogenated form storing 7.1 wt% H_2 (Campbell et al. 2010; Müller et al. 2012), is worth mentioning. However, obtaining these compounds is not straightforward, and the literature lacks data on the stability and toxicity of the compounds.

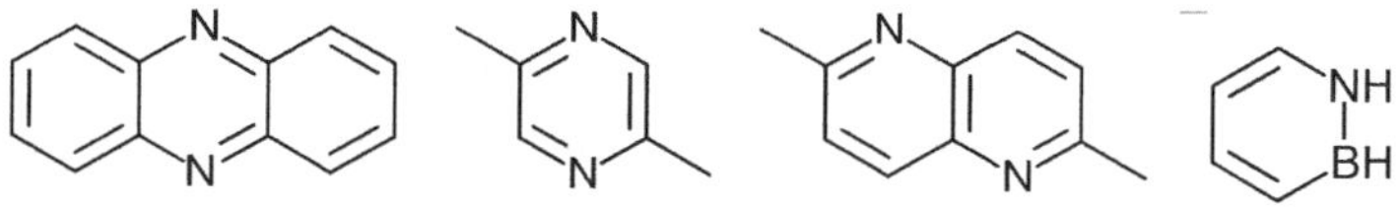

Figure 3.11. *Phenazine, 2,5-dimethylpyrazine, 2,6-dimethyl-1,5-naphthyridine and 1,2-dihydro-1,2-azaborine*

3.3.2.7. *Conclusion*

Many systems containing an N-heterocycle have been explored for their potential for use in hydrogen storage. The knowledge gathered is uneven due to the nature of the LOHC molecules and the catalysts involved. Using molecular catalysts or non-PGM catalysts, the reaction is usually limited to the transformation of the N-heterocycle, and the adjacent rings without heteroelements are not modified. N-ethylcarbazole is, by far, the most studied system. However, emerging options, including biomass-derived components, deserve interest but are currently only studied in very dilute systems and not as neat compounds.

3.3.3. *Nitriles/amines*

The amine/nitrile pairs could be potential LOHC systems because one H_2 molecule can be recovered from each CH_2NH_2 of the compound (13.3 wt% hydrogen per CH_2NH_2) (Choi et al. 2016). Most of these studies use a homogeneous Ru catalyst (Tseng et al. 2013; Grellier and Sabo-Etienne 2014; Ventura-Espinosa et al. 2016). The hydrogenation of nitriles is carried out at RT, whereas the dehydrogenation is performed at 120°C, both of which are in solvents. The immobilization of the complex by grafting on graphene oxide was nevertheless proposed to heterogenize the catalyst (Ventura-Espinosa et al. 2016). The hybrid material was used in the dehydrogenation of benzylamine in toluene (Figure 3.12), with 2 mol% Ru used for successive runs.

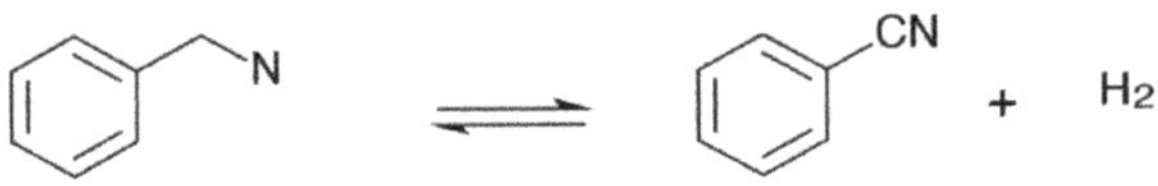

Figure 3.12. *Reversible dehydrogenation of benzylamine to benzene nitrile*

A few examples of heterogeneous catalysts can be found in the literature. Among several Cu catalysts, $Cu(0)/Al_2O_3$ was found to allow the selective production of nitriles (Damodara et al. 2014) at 120°C in DMF, starting from various substituted benzylamines. Aliphatic amines also lead to the production of nitriles, but owing to their slower kinetics. The hydrogen storage capacity is low, reaching only 2–3 wt-%, but the study did not focus on LOHC applications. Nevertheless, the use of a non-PGM catalyst for this reversible reaction is worth mentioning. However, a recent work focused on the possible use of amine/nitrile couples without the use of inert gas (Müller 2022) and concluded that thermodynamics prevents reaching reasonable conversions during the dehydrogenation of amines at 110 °C without dilution. The use of electrochemical compression to selectively withdraw hydrogen was presented as a potential option to reach higher conversions. In fact, the acceptorless dehydrogenation of amines to nitriles presents similar thermodynamic restrictions to those of other LOHCs. The reaction at approximately 100°C is only possible under reduced partial pressure.

3.4. O-containing molecules

3.4.1. *Introduction*

The dehydrogenation of primary or secondary alcohols to carbonyl or carboxyl groups is a preferred alternative to aromatic substrates for intermediate hydrogen storage. As in the case of amines, homogeneous or heterogeneous catalysts have been described to perform these transformations. The reverse hydrogenation reaction is well known, especially for carbonyl hydrogenation to alcohols, which is carried out under mild conditions. Harder conditions are needed to perform hydrogenation of the acid or ester to the corresponding alcohol derivative.

The hydrogenation/dehydrogenation of oxygenated compounds has been recently reviewed both using noble metal homogeneous complexes (Os, Ru, Ir, Rh) and non-precious metals (Co, Mn, Fe) (Bonitatibus et al. 2015; Pandey et al. 2016; Crabtree 2017b; Filonenko et al. 2018; Yadav et al. 2021); however, far fewer studies have been reported in comparison to (hetero)aromatic/(hetero)cyclic substrate couples. Most reports address the thermal catalytic dehydrogenation of alcohols, but catalytic electrooxidation or photocatalytic dehydrogenation are also possible (Shibata et al. 2017; Trincado et al. 2021).

	H_2-rich form	H_2-lean form	
Primary alcohol/aldehyde – acid or ester	$R{-}CH_2OH$	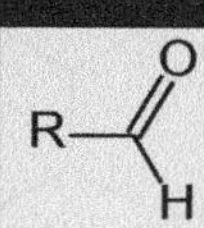	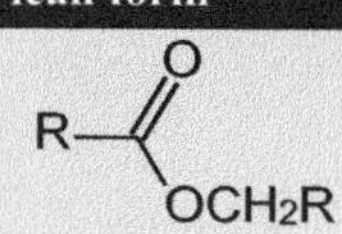
Secondary alcohol/ketone	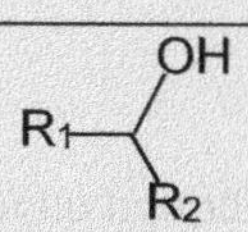	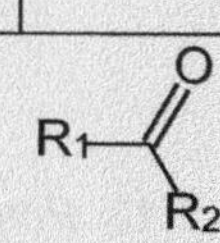	
1,4 -Butanediol/γ-butyrolactone	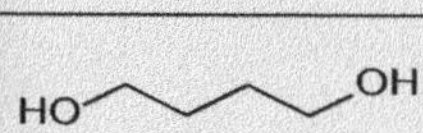	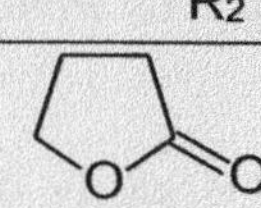	

Table 3.7. *Alcohols and their dehydrogenated counterparts*

3.4.2. *Monoalcohols*

A large variety of monoalcohols have been studied. With respect to the storage capacity, the shorter the carbon chain is, the higher the storage capacity, starting from 12.2 wt% for methanol. Hydrogen production via steam reforming is described using supported heterogeneous catalysts under medium pressure (25–50 bar) and at high temperature (Ranjekar and Yadav 2021) or even some homogeneous complexes. Only gaseous products, including CO and CO_2, are obtained via this approach, so the system is not considered an LOHC.

Heavier monoalcohols were chosen as LOHC donors. Two families must be considered: secondary alcohols, which yield ketones as the only product, and primary alcohols, whose selectivity to aldehydes or esters must be considered. While few studies have focused on the use of these alcohols for LOHC applications, their dehydrogenation has been largely studied. Indeed, in the 1960s, several bulk Cu catalysts were described for the dehydrogenation of iPrOH to acetone at 300°C (Kawamoto 1961).

3.4.2.1. *Primary alcohols to aldehydes or esters*

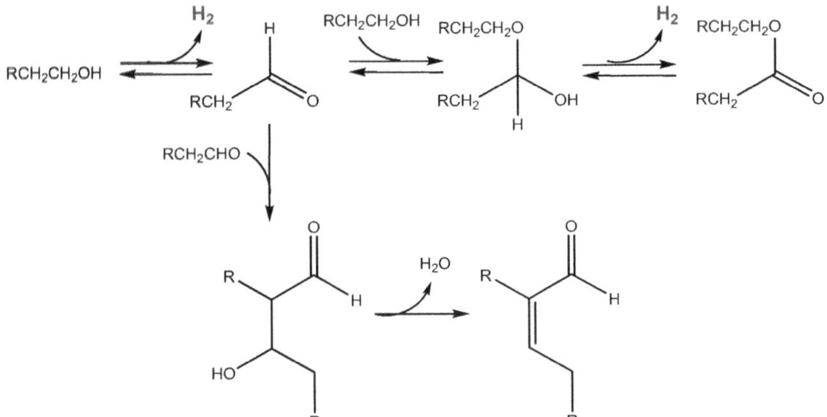

Figure 3.13. *Potential products formed under dehydrogenation conditions*

Ethanol is easily dehydrogenated to acetaldehyde (Guan and Hensen 2009). Many supported metallic catalysts, as well as some homogeneous or supported complexes, have been described for this transformation. An overview of the different types of homogeneous catalysts was recently published (Campos 2018). Notably, most of the reports focused on organic product selectivity but not on hydrogen production. The activity and selectivity are dependent mainly on the pretreatment of the catalysts and on the reaction conditions. High selectivity for acetaldehyde is the main issue of this reaction. Indeed, depending on the properties of the catalyst and the reaction conditions, either a dehydration product is formed in parallel (ethylene), or a C4-product is produced from the Guerbet reaction (Figure 3.13) (Quesada et al. 2018). These parallel reactions are detrimental considering the hydrogen yield, and water is also a coproduct that could be deleterious for the stability of the catalysts. The formation of these undesired products can be avoided by proper design of the catalysts since they involve both dehydrogenation characteristics (metallic properties) and dehydration or coupling properties at acid-basic sites.

Copper-based heterogeneous materials are the main catalysts described for acceptorless dehydrogenation transformations of alcohols. The reaction was carried out in continuous flow reactors, and the alcohol was diluted in inert gas (He or Ar) at temperatures ranging from 250°C to 350°C. Recently, the relationships among the catalytic activity and textural and structural properties of Cu-modified alumina–zirconia catalysts have been studied. The most efficient catalyst was obtained when the appropriate Cu^+/Cu^0 reduction state ratio was stabilized. Up to 97% acetaldehyde selectivity was achieved at 86% conversion (Zhukova et al. 2021). The tuning of the Cu oxidation state was shown to be essential to optimize performances of supported catalysts. After N_2O pretreatment, the restructuring of the surface of the copper nanoparticles was established via transmission electron microscopy (TEM) and X-ray absorption spectrometry (XAS) analyses, which revealed a fourfold increase in the reaction rate (He et al. 2019). STM analysis of the Cu(111) and Cu(110) model surfaces was performed and it was shown that dehydrogenation occurred at surface defects in the form of step edges and kinks. It was also shown that the Cu model surfaces exhibited no deactivation after several temperature-programmed reactions, with a stable 100% selectivity for alcohol dehydrogenation to aldehydes and hydrogen (Wang et al. 2017). Cu/SiO_2 catalysts were also evaluated for the dehydrogenation of longer primary alcohols. Starting from 1-decanol, high selectivity (98%) for decenal was reported, with the observed by-products being decyldecanoate and, to a minor extent, decene (Sun et al. 2019). Cu/La-MgAlO catalysts were also used for transfer dehydrogenation of 1-decanol to decanal in the liquid phase, but in that case, styrene was also present to trap the formed hydrogen (Zhang et al. 2016). Cr-modified Cu/Al_2O_3-La_2O_3 was used in n-pentanol dehydrogenation via a continuous flow reactor, resulting in up to 97% selectivity in n-pentanal at 70% conversion (Liu et al. 2014).

The presence of a second metal (Ag or Ni) affected the performance of the catalytic system. For example, it has been shown that AgCu catalysts are less active than the corresponding Cu catalysts because of the blockage of Cu surface active sites by silver atoms (Conesa et al. 2019). On the other hand, Ni_xCu alloys ($0.001 < x < 0.01$) were evaluated in the reaction, and the authors have shown that catalytic activity as well as long-term stability increased (Shan et al. 2017). This was attributed to the stabilization of the Cu particles against sintering. Some cobalt-based catalysts were also evaluated for the dehydrogenation of ethanol, but ethylene was obtained simultaneously with acetaldehyde (Chernov et al. 2021). In the presence of Pd-based catalysts, high selectivity for acetaldehyde can be achieved only at low conversion of ethanol (Ouyang et al. 2020). The gas-phase dehydrogenation of n-butanol to n-butanal can be carried out with high selectivity over Au deposited on hydrotalcite. Under similar reaction conditions, Pd supported on hydrotalcite favored aldehyde decarbonylation (Shylesh et al. 2015). Stable conversion was observed over a 10-h reaction time.

The dehydrogenation of activated alcohols such as aromatic or allylic primary alcohols was also studied. The storage capacity of these alcohols is much lower (1.6 wt-%) since their molecular weight is higher; however, since they are more reactive, milder reaction conditions can be applied. Noble metal-based catalysts are reported to be active for both acceptorless dehydrogenation of activated alcohols as well as the reverse reaction (Cui et al. 2020). Iron- or cobalt-based single-atom catalysts are efficient in dehydrogenation of benzyl alcohol at 140°C under an Ar atmosphere. A synergistic effect was observed by using mixed catalysts prepared from iron and cobalt (named double-atom catalysts). The activation energy decreased with the use of FeCo catalysts, and computational studies suggested that the reaction route was modified. This system was very efficient for a large range of aromatic derivatives; however, moderate activity was achieved for allylic alcohol dehydrogenation (Liu et al. 2022). MoN_2 nanobelts were prepared via nitridation of MoO_3, and Mo_2C nanobelts were prepared via carburization of MoO_3. Both catalysts were efficient for the selective dehydrogenation of benzyl alcohol under moderate conditions (Li et al. 2014a, 2014b). The mixed oxide ZnO-CeO_2 was reported to dehydrogenate benzyl alcohol in a fixed bed reactor under a N_2 atmosphere. High selectivity for benzaldehyde was obtained together with a small amount of toluene; however, moderate conversion was observed (Enjamuri et al. 2016). Furthermore, after recycling, the selectivity for benzaldehyde decreased in favor of toluene. This was attributed to the reduction of Ce^{4+} to Ce^{3+} during the different tests.

Silver nanoparticles supported on MgAl oxides are efficient at the dehydrogenation of benzyl alcohols, even at 100°C (Liu et al. 2018). Small deactivation was observed over five recycle tests. Considering these results, it would be difficult to use these activated alcohols as LOHC.

	Achievements	Ref
Cu/(Zr + Ce)O_2/Al_2O_3	EtOH in He flow (25 mL/min), 220–360°C. Up to CH_3CHO 97% selectivity	(Zhukova et al. 2021)
Cu/SiO_2	30 mL/min, 5 kPa EtOH in N_2, 200°C TOF increased after N_2O pretreatment	(He et al. 2019)
Cu/SiO_2	1-Decanol under N_2 flow. 275°C. 77% alcohol conversion to form 98% decanal selectivity	(Sun et al. 2019)
Cu-Cr/γ-Al_2O_3-La_2O_3.	PenOH/N_2, 280°C. 100% Selectivity up to 70% conv	(Liu et al. 2014)
Ag-Cu/C	5 µL/min EtOH/25 mL/min He 100% selective to CH_3CHO	(Conesa et al. 2019)
Ni0.01Cu/SiO_2	5.6% EtOH in He, 300°C, H_2 formation rate 0.55 mmol/Cu m^2s	(Shan et al. 2017)
Co@N-doped Carbon	5% EtOH in He flow, 200–400°C. Up to 66% conversion and 84% selectivity to CH_3CHO	(Chernov et al. 2021)
Pd/ZnO	140–170°C; 2% EtOH/He (15 mL/min) 100% CH_3CHO selectivity at 20% conv.	(Ouyang et al. 2020)
Au/HT	1-BuOH in He, 200°C, TOF = 40 h^{-1}	(Shylesh et al. 2015)
FeCo	2.5 wt-% cat, benzyl alcohol in xylene, 140°C, 24 h, benzaldehyde sel =98%; up to 862 × 10^{-3} h^{-1} TOF.	(Liu et al. 2022)
MoN_2 nanobelts	benzyl alcohol in DMSO, N_2, 150°C, full conversion	(Li et al. 2014b)
ZnO-CeO_2	Benzyl alcohol (3 mL/h) in N_2 flow (30 mL/min), 400°C, ZnO/CeO_2 = 30/70); benzaldehyde sel = 95% (44% conv.)	(Enjamuri et al. 2016)
Ag/MgAlO	Benzyl alcohol in xylene, 100°C, Ar, 96% conv, benzaldehyde sel = 90%	(Liu et al. 2018)
BaCuCrO	EtOH, 220–240°C, 20 bars, >98% selectivity to AcOEt at 60–65% conv.	(Santacesaria et al. 2012)
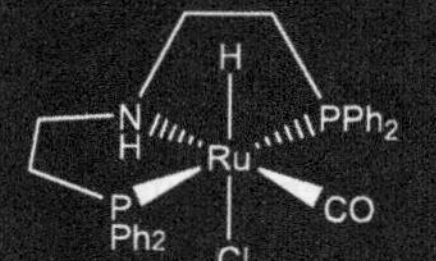	25 ppm complex, 0.6 mol% NaOEt, pure EtOH, 90°C, 46 h, TOF = 15,400 AcOEt production	(Nielsen et al. 2012)

Table 3.8. *Selected achievements in the literature for primary alcohols*

Alternatively, ethanol can be dehydrogenated to a greater degree than the corresponding ethyl acetate. For this reaction, the hydrogen capacity is not increased since condensation of two ethanol molecules was involved. This reaction involved heterogeneous catalysts (Santacesaria et al. 2012) but also some homogeneous complexes (Nielsen et al. 2012; Milstein et al. 2017). In this latter case, the authors have shown that ruthenium complexes exhibited very high activity in acceptorless dehydrogenation of ethanol to ethyl acetate but in the presence of a catalytic amount of base at low temperature (90°C). Recent calculation has shown that with this type of catalytic system, up to 85 MWh of energy can be achieved by storing 5 MT of hydrogen (Tran et al. 2021). Fe- and Mn-based complexes have also been evaluated for their ability to dehydrogenate primary alcohols to carboxylic acid under alkaline conditions (Andérez-Fernández et al. 2017; Nguyen et al. 2017). Some copper-based catalysts are also efficient in this reaction at high temperatures (240°C). At 65% conversion (continuous reaction), pure hydrogen was obtained together with > 98% selectivity for ethyl acetate. A global analysis of the system has been recently published by Mevawala et al. (2023a). They reported that a significant efficiency of 88% can be achieved when the EtOH/AcOEt coupling system is considered.

3.4.2.2. *Secondary alcohols to ketones*

Considering the dehydrogenation of secondary alcohols, the highest hydrogen storage reached only 3.3 wt%, starting from isopropyl alcohol, iPrOH. However, owing to the easy availability of this substrate and the mild reaction conditions of both hydrogenation and dehydrogenation reactions, these alcohols may be considered potential LOHC. iPrOH is produced on a large scale by propene hydration, but acetone can also be obtained from the ABE fermentation process. Endothermic dehydrogenation of iPrOH produced acetone, which can be easily removed from the reaction mixture because of its lower boiling point ($\Delta T = 26°C$). Catalysts based on Ni, Ru, Pt or Pd are known to be efficient for both acetone hydrogenation and iPrOH dehydrogenation under mild conditions. At 82°C, in the presence of a Raney nickel catalyst, an initial reaction rate of up to 1.3 $mmol_{H2}/min/g_{Ni}$ was reported for pure iPrOH dehydrogenation (Kim et al. 1992). Raney nickel is also efficient in acetone hydrogenation under atmospheric pressure of hydrogen with H_2/acetone = 4. High productivity was reported recently in the presence of a Pt/C catalyst and via microwave irradiation (Ichikawa et al. 2019). Copper-based catalysts were also described to be efficient for this dehydrogenation reaction. Indeed, a Cu/SiO_2 catalyst pretreated with N_2O was evaluated for the dehydrogenation of iPrOH to yield acetone. Under similar reaction conditions, the reactivity of secondary alcohols is greater than that of primary alcohols ($TOF_{iPrOH}/TOF_{nPrOH} = 3$) (He et al. 2019). The same catalyst efficiently reduced acetone to iPrOH at 200°C under atmospheric pressure. As previously reported with primary alcohols, some homogeneous ruthenium complexes are efficient in acceptorless dehydrogenation of isopropanol under neutral and mild conditions (82°C, atmospheric pressure) (Nielsen et al. 2011). To date, the best performance involving the iPrOH/acetone couple was reported in the presence of an iridium complex supported on silica but at high temperature (350°C) (Polukeev et al. 2022). With this supported complex, almost 100,000 h^{-1} TOF was initially obtained, but a regular decrease was observed (approximately 20% within 20 h), together with the formation of iridium nanoparticles detected by TEM analysis.

Dehydrogenation of isoamyl alcohol was performed in the presence of a zinc-modified copper catalyst. The best performance was obtained with the addition of 5 wt% zinc, which was also the catalyst with the highest stability (Shiau et al. 2000). Unsupported rhenium nanoparticles selectively dehydrogenated secondary alcohols at 180°C (Yi et al. 2014). Dehydrogenation of aliphatic secondary alcohols was investigated for complete conversion and high selectivity for the corresponding ketone. However, this system is not efficient for the dehydrogenation of primary alcohols. The authors demonstrated that the mechanism of dehydrogenation of alcohols in the presence of this catalyst is quite specific and involves γ C–H activation. Cobalt catalysts supported on C, Al_2O_3, ZnO, ZrO_2 and TiO_2 were efficient in dehydrogenation of 2-octanol (Kaźmierczak et al. 2020). Cobalt supported on commercial TiO_2 P25 was the most active, and a 69% yield of 2-octanone with > 90% selectivity was obtained when decane was used as the solvent.

The acceptorless dehydrogenation of phenylethanol was evaluated with a series of heterogeneous and homogeneous catalysts. Alcohol was solubilized in xylene or toluene, and the reaction was performed under inert gas (N_2 or Ar) at temperatures ranging from 110°C to 130°C. Using noble metal-based catalysts, a quantitative yield of ketone was achieved (Kim et al. 2006). High stability was observed when tests were performed over 48 h on stream in the presence of a Pd/C catalyst with up to 0.683 $g_{(product)}$/mL/h (Nicolau et al. 2019). However, during the hydrogenation of acetophenone with this catalyst (2 bar H_2, 130°C), not only phenylethanol was obtained but also ethylbenzene to a significant extent (20%). Supported silver catalysts have also been described for the dehydrogenation of phenylethanol (Mitsudome et al. 2008; Kurhe et al. 2015). Silver catalysts supported on hydrotalcite were very efficient for the dehydrogenation of secondary alcohols with TON higher than 20,000. No Ag leaching was detected in solution, and under the reaction conditions, the catalyst was reused for five cycles with full conversion (Mitsudome et al. 2008). This catalyst was also active for the dehydrogenation of a series of secondary aliphatic alcohols, whereas primary aliphatic alcohol such as 1-octanol was not dehydrogenated efficiently. Few reports have described the use of silver catalysts for the hydrogenation of ketones (Claus and Hofmeister 1999).

Homogeneous complexes prepared in situ from $Fe(acac)_3$ were efficient for acceptorless dehydrogenation of secondary benzylic alcohols but in the presence of a base (Song et al. 2014). However, this catalytic system was not efficient for primary and aliphatic alcohols. Well-defined homogeneous iron or cobalt complexes bearing diphosphorus/nitrogen ligands were described for efficient reversible acceptorless dehydrogenation-hydrogenation of alcohols under mild conditions (Zhang et al. 2013; Bonitatibus et al. 2015). The use of earth-abundant metal catalysts is essential as an alternative to noble metal catalysts for such applications.

	Achievements	Ref
Ni Raney	IPrOH, 82°C, 1.3 $mmol_{H2}/min/g_{cata}$	(Kim et al. 1992)
Cu/SiO_2	30 mL/min, 3 kPa iPrOH in N_2, 200°C iPrOH/nPrOH activity = 3	(He et al. 2019)
Pt/C	iPrOH, 2 MPa 10 W Microwave irradiation >95% H_2 purity at 81% conv	(Ichikawa et al. 2019)
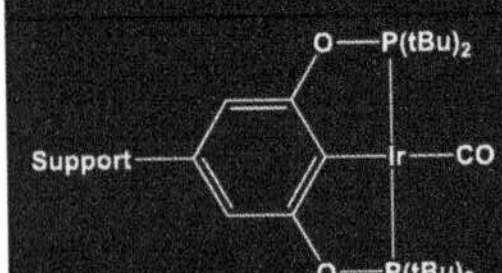	IPrOH/350°C TON > 1,500,000 17.4 g_{H2}/g_{Ir}/min	(Polukeev et al. 2022)
$CuZn/SiO_2$	Isoamylalcohol, 210–290°C,	(Shiau et al. 2000)
NH_4ReO_4	3-Octanol neat, 180°C, 2wt-% cat., 10 h, 100% conv	(Yi et al. 2014)
Supported Co	2-Octanol in decane, 145°C, 100% selectivity	(Kaźmierczak et al. 2020)
Pd/C	Phenylethanol in xylene, 130°C, N_2, selectivity > 98%	(Nicolau et al. 2019)
Ru nanoparticles	Phenylethanol in toluene, 110°C, 3wt-% cat, quant. yield	(Kim et al. 2006)
Ag/HT	Phenylethanol in xylene, 130°C, Ar, TON = 22,000	(Mitsudome et al. 2008)
$Fe(acac)_3$/phenantroline	Phenylethanol in toluene, 110°C, K_2CO_3, Ar, quantitative	(Song et al. 2014)
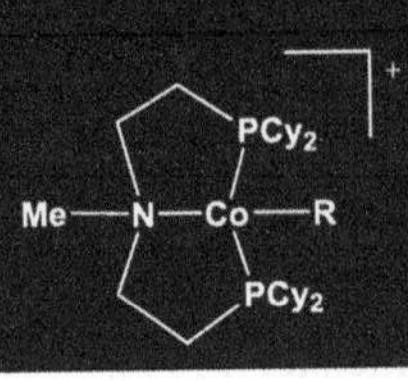	Phenylethanol in toluene, 120°C, 95% acetophenone yield	(Zhang et al. 2013)

Table 3.9. *Selected achievements in the literature for secondary alcohols*

Recently, a two-step dehydrogenation of 1-cyclohexylethanol to phenylacetophenone was patented (D'Ambra et al. 2023). In that case, up to 6.2% H_2 weight storage was expected. In such an approach, the first dehydrogenation was performed in the presence of a copper-based catalyst to yield acetylcyclohexane, which was further dehydrogenated to acetophenone in the presence of a platinum-based catalyst. This two-step reaction was preferable to a single one because the latter yielded a significant amount of undesired by-products. Reverse hydrogenation of acetophenone to 1-cyclohexylethanol was also claimed in this patent.

3.4.3. *Polyols*

Considering the availability of biosourced oxygenated compounds, their use as liquid organic hydrogen storage could contribute to the lower carbon footprint of this approach. The use of LOHC couples in which at least one of the compounds was obtained from biosourced derivatives was claimed in a recent patent (Chappaz and Bengaouer 2020). Several reports have described the dehydrogenation of glycerol to lactic acid, but this reaction is not reversible (Li et al. 2015). On the other hand, the dehydrogenation of primary diol to lactone corresponds to hydrogen storage up to 4.5 wt% (starting from 1,4-butanediol (1,4-BDO)). Several reports have described the synthesis of lactones by dehydrogenation followed by intramolecular cyclization of diols in the presence of a molecular iron-based catalyst with a modest TON (Bonitatibus et al. 2015; Peña-López et al. 2015). Heterogeneous catalysts, mainly copper-based catalysts, are also efficient at dehydrogenating diol into lactone in the gas phase (Ichikawa et al. 2004; Bhanushali et al. 2019). It was shown that 1,4-BDO was initially dehydrogenated to 4-hydroxybutanal, which was then hemiacetalized to 2-hydroxytetrahydrofuran, followed by a second dehydrogenation to gamma-butyrolactone (GBL) (Ichikawa et al. 2004). The influence of pretreatment of the catalyst as well as the reaction conditions significantly influences the selectivity of the reaction to avoid the production of by-products such as THF. Dehydrogenation of 1,4-BDO combines with the hydrogenation step of benzaldehyde or furfural to combine endo- and exothermic reactions (Javaid and Bildea 2014; Bhanushali et al. 2020). A full thermodynamic analysis of hydrogen storage through a 1,4-BDO/GBL couple was performed, and the efficiency of the global process under gas-phase or liquid-phase conditions was similar (Mevawala et al. 2023b).

Dehydrogenation of diols containing both primary and secondary alcohols was also reported. In the presence of a homogeneous iron complex, the secondary alcohol was selectively dehydrogenated to the corresponding ketone (Bonitatibus et al. 2015). In the presence of Co/TiO_2, 1,2-octanediol was selectively dehydrogenated to 1-hydroxy-2-octanone (90% selectivity) together with H_2 production but at low conversion (33%) (Kaźmierczak et al. 2020).

2,3-Butanediol is also a potential LOHC (Alagy and Trambouze 2020). It was dehydrogenated to acetoin in the presence of a copper heterogeneous catalyst. The dehydrogenation of the second alcohol group could increase the hydrogen storage of this molecule. However, only low selectivity for dione was observed when the reaction was performed at relatively high temperatures (Al-Auda et al. 2022).

These diols are potential because very high selectivity was achieved.

	Reaction conditions	Ref
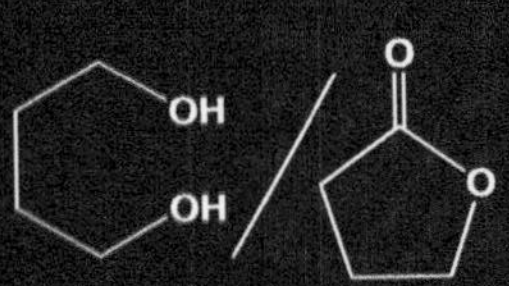 1,4-BDO / GBL	0.5% [Fe], 10% base Batch reactor BDO in *t*-amylalcohol, 150°C NMR yield = 80%	(Peña-López et al. 2015)
	$6CuO/2ZnO/2ZrO_2/2Al_2O_3$ Fixed-bed reactor BDO in He, 240°C Up to 84% conv; 97% sel	(Ichikawa et al. 2004)
	20% Cu/CeO_2 Fixed-bed reactor BDO in N_2, 240°C Up to 86% conv; 100% sel	(Bhanushali et al. 2019)
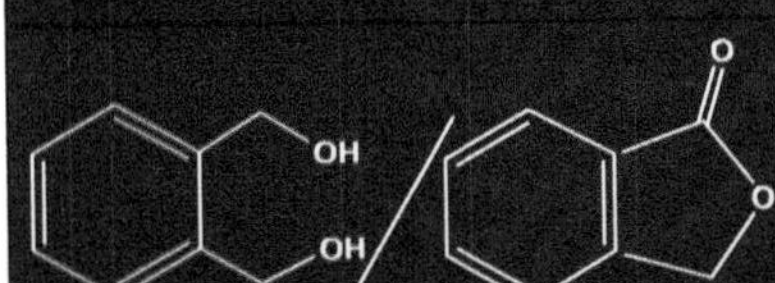	0.5% [Fe], 10% base Batch reactor diol in toluene, 120°C yield = 80%	(Bonitatibus et al. 2015)
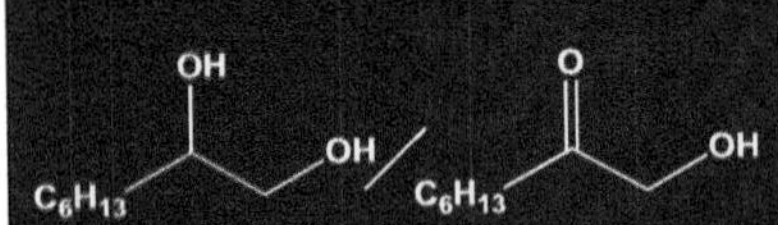	5% Co/TiO_2 Diol in dioxane, 145°C Yield ketone = 30% Yield H_2 = 28%	(Kaźmierczak et al. 2020)
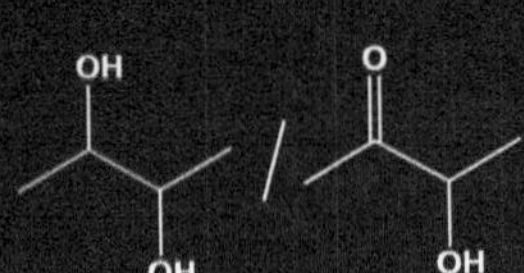	15% Cu/Al_2O_3 Fixed-bed reactor Diol in N_2, 170°C 63% conv, 96% sel	(Al-Auda et al. 2022)

Table 3.10. *Selected achievements in the literature for polyols*

3.5. Amines/amides

More exotic systems have also been proposed that still consider reversible reactions between hydrogen-rich and hydrogen-lean compounds but putting in relation amine and alcohol functions, either on the same molecule or not, to allow the formation of amides, peptides or oligopeptides using ruthenium complexes (Hu et al. 2015; Kothandaraman et al. 2017; Zou et al. 2020) or manganese complexes (Shao et al. 2020). An example with neat molecules (without solvent) has been published, but in the presence of K_3PO_4 as a base additive, it enhances the catalytic activity (Kothandaraman et al. 2017). All of these systems, which allow the storage of 5.3–6.6 wt% H_2, were reviewed by Cho et al. (2021).

3.6. Further comparison and conclusion

Many LOHCs have been proposed to store hydrogen over long periods and long distances, with more or less mature systems. Several studies have compared some of these systems. They focus on the most mature couples: toluene/methylcyclohexane, DBT/perhydro-DBT and NEC/perhydroNEC.

Considering the hydrogenation unit only, DBT is far ahead of the others, with the lowest unit cost and the maximum hydrogen storing rate. The NEC system is better for only one criterion, which is the lowest CO_2 emission of the unit (Byun et al. 2022).

Considering dehydrogenation, it is clear that the required heat must be supplied externally to make the LOHC an interesting option. Asif et al. (2021) proposed the use of the heat of dehydrogenation products to preheat the feed of a dehydrogenation reactor. DBT is again said to be the most interesting option for long-distance transportation. NEC cannot be used for transport purposes due to its cost (Niermann et al. 2021). Considering a ranking based on gravimetric and volumetric hydrogen storage, TRL, reversibility, material handling, toxicity, safety, environmental impact and cost, DBT is presented as the best solution before TOL and NEC. It is presented as the best H_2 storage option for international energy transport (Oner and Khalilpour 2022).

3.7. List of abbreviations

ABE: Acetone–butanol–ethanol

BT: Benzyltoluene

BDO:	Butanediol
CH:	Cyclohexane
DBT:	Dibenzyltoluene
DFT:	Density-functional theory
DMF:	Dimethylformamide
DoH:	Degree of hydrogenation
GBL:	Gamma-butyrolactone
GC:	Gas chromatography
GHSV:	Hourly space velocity of gas
HPLC:	High-pressure liquid chromatography
iPrOH:	Isopropanol
LHSV:	Liquid-hourly space velocity
LOHC:	Liquid organic hydrogen carrier
MCH:	Methylcyclohexane
MOF:	Metal Organic Framework
MS:	Mass spectrometry
NEC:	N-ethylcarbazole
PGM:	Platinum Group Metal
STM:	Scanning tunnelling microscopy
t-BuOK:	Potassium tert-butoxide
TEM:	Transmission electron microscopy
TPD:	Temperature-programmed desorption

TOF: Turnover frequency (syn. activity)

TON: Turnover number

TOL: Toluene

TRL: Technology Readiness Level

XAS: X-ray absorption spectroscopy

3.8. References

Aakko-Saksa, P.T., Cook, C., Kiviaho, J., Repo, T. (2018). Liquid organic hydrogen carriers for transportation and storing of renewable energy review and discussion. *J. Power Sources*, 396, 803–823.

Acharya, D., Ng, D., Xie, Z. (2021). Recent advances in catalysts and membranes for MCH dehydrogenation: A mini review. *Membranes*, 11(12), 955.

Ahn, C.-I., Kwak, Y., Kim, A.-R., Jang, M., Badakhsh, A., Cha, J., Kim, Y., Jo, Y.S., Jeong, H., Choi, S.H. et al. (2022). Dehydrogenation of homocyclic liquid organic hydrogen carriers (LOHCs) over Pt supported on an ordered pore structure of 3-D cubic mesoporous KIT-6 silica. *Applied Catalysis B: Environmental*, 307, 121169.

Al-Auda, Z., Li, X., Hohn, K.L. (2022). Dehydrogenation of 2,3-butanediol to acetoin using copper catalysts. *Ind. Eng. Chem. Res.*, 61(10), 3530–3538.

Alagy, J. and Trambouze, P. (2020). Système et procédé de stockage et de restitution de dihydrogène. Report, WO2020141282.

Ali, J.K., Newson, E.J., Rippin, D.W.T. (1994). Exceeding equilibrium conversion with a catalytic membrane reactor for the dehydrogenation of methylcyclohexane. *Chemical Engineering Science*, 49(13), 2129–2134.

Ali, A., Ali, L.I., Aboul-Fotouh, S.M., Aboul-Gheit, A.K. (1998). Hydrogenation of aromatics on modified platinum-alumina catalysts. *Applied Catalysis A: General*, 170(2), 285–296.

Ali, A., Udaya Kumar, G., Lee, H.J. (2020). Parametric study of the hydrogenation of dibenzyltoluene and its dehydrogenation performance as a liquid organic hydrogen carrier. *Journal of Mechanical Science and Technology*, 34(7), 3069–3077.

Ali, A., Kumar, G.U., Lee, H.J. (2021). Investigation of hydrogenation of dibenzyltoluene as liquid organic hydrogen carrier. *Materials Today: Proceedings*, 45, 1123–1127.

Amende, M., Gleichweit, C., Werner, K., Schernich, S., Zhao, W., Lorenz, M.P.A., Höfert, O., Papp, C., Koch, M., Wasserscheid, P. et al. (2014). Model catalytic studies of liquid organic hydrogen carriers: Dehydrogenation and decomposition mechanisms of dodecahydro-N-ethylcarbazole on Pt(111). *ACS Catal.*, 4(2), 657–665.

Andérez-Fernández, M., Vogt, L.K., Fischer, S., Zhou, W., Jiao, H., Garbe, M., Elangovan, S., Junge, K., Junge, H., Ludwig, R. et al. (2017). A stable manganese pincer catalyst for the selective dehydrogenation of methanol. *Angewandte Chemie International Edition*, 56(2), 559–562.

Andersson, J. and Grönkvist, S. (2019). Large-scale storage of hydrogen. *International Journal of Hydrogen Energy*, 44(23), 11901–11919.

Asif, F., Hamayun, M.H., Hussain, M., Hussain, A., Maafa, I.M., Park, Y.-K. (2021). Performance analysis of the perhydro-dibenzyl-toluene dehydrogenation system. A simulation study. *Sustainability*, 13(11), 6490.

Aslam, R., Minceva, M., Müller, K., Arlt, W. (2016). Development of a liquid chromatographic method for the separation of a liquid organic hydrogen carrier mixture. *Separation and Purification Technology*, 163, 140–144.

Auer, F., Blaumeiser, D., Bauer, T., Bösmann, A., Szesni, N., Libuda, J., Wasserscheid, P. (2019). Boosting the activity of hydrogen release from liquid organic hydrogen carrier systems by sulfur-additives to Pt on alumina catalysts. *Catalysis Science & Technology*, 9(13), 3537–3547.

Bachmann, P., Schwarz, M., Steinhauer, J., Späth, F., Düll, F., Bauer, U., Nascimento Silva, T., Mohr, S., Hohner, C., Scheuermeyer, M. et al. (2018). Dehydrogenation of the liquid organic hydrogen carrier system indole/indoline/octahydroindole on Pt(111). *The Journal of Physical Chemistry C*, 122(8), 4470–4479.

Badakhsh, A., Song, D., Moon, S., Jeong, H., Sohn, H., Woo Nam, S., Soon Kim, P., Hui Seo, J., Kim, Y., Lee, J. et al. (2022). COX-free LOHC dehydrogenation in a heatpipe reformer highly integrated with a hydrogen burner. *Chemical Engineering Journal*, 449, 137679.

Barrio, V.L., Arias, P.L., Cambra, J.F., Güemez, M.B., Pawelec, B., Fierro, J.L.G. (2003). Aromatics hydrogenation on silica-alumina supported palladium-nickel catalysts. *Applied Catalysis A: General*, 242(1), 17–30.

Basset, J.M., Dalmai-Imelik, G., Primet, M., Mutin, R. (1975). A study of benzene hydrogenation and identification of the adsorbed species with $PtAl_2O_3$ catalysts. *Journal of Catalysis*, 37(1), 22–36.

Bathula, H.B., Oh, J., Jo, Y., Suh, Y.-W. (2019). Dehydrogenation of 2-[(n-methylcyclohexyl)methyl]piperidine over mesoporous Pd-Al_2O_3 catalysts prepared by solvent deficient precipitation: Influence of calcination conditions. *Catalysts*, 9(9), 719.

Berger Bioucas, F.E., Piszko, M., Kerscher, M., Preuster, P., Rausch, M.H., Koller, T.M., Wasserscheid, P., Fröba, A.P. (2020). Thermal conductivity of hydrocarbon liquid organic hydrogen carrier systems: Measurement and prediction. *Journal of Chemical & Engineering Data*, 65(10), 5003–5017.

Bhanushali, J.T., Prasad, D., Patil, K.N., Babu, G.V.R., Kainthla, I., Rao, K.S.R., Jadhav, A.H., Nagaraja, B.M. (2019). The selectively regulated vapour phase dehydrogenation of 1,4-butanediol to γ-butyrolactone employing a copper-based ceria catalyst. *New Journal of Chemistry*, 43(30), 11968–11983.

Bhanushali, J.T., Prasad, D., Patil, K.N., Reddy, K.S., Rama Rao, K.S., Jadhav, A.H., Nagaraja, B.M. (2020). Simultaneous dehydrogenation of 1,4-butanediol to γ-bu-tyrolactone and hydrogenation of benzaldehyde to benzyl alcohol mediated over competent $CeO_2Al_2O_3$ supported Cu as catalyst. *International Journal of Hydrogen Energy*, 45(23), 12874–12888.

Biniwale, R.B., Rayalu, S., Devotta, S., Ichikawa, M. (2008). Chemical hydrides: A solution to high capacity hydrogen storage and supply. *International Journal of Hydrogen Energy*, 33(1), 360–365.

Bonitatibus, P.J., Chakraborty, S., Doherty, M.D., Siclovan, O., Jones, W.D., Soloveichik, G.L. (2015). Reversible catalytic dehydrogenation of alcohols for energy storage. *Proceedings of the National Academy of Sciences*, 112(6), 1687–1692.

Bourane, A., Elanany, M., Pham, T.V., Katikaneni, S.P. (2016). An overview of organic liquid phase hydrogen carriers. *International Journal of Hydrogen Energy*, 41(48), 23075–23091.

Brayton, D.F. and Jensen, C.M. (2015). Dehydrogenation of pyrrolidine based liquid organic hydrogen carriers by an iridium pincer catalyst, an isothermal kinetic study. *International Journal of Hydrogen Energy*, 40(46), 16266–16270.

Brigljević, B., Byun, M., Lim, H. (2020a). Design, economic evaluation, and market uncertainty analysis of LOHC-based, CO_2 free, hydrogen delivery systems. *Applied Energy*, 274, 115314.

Brigljević, B., Lee, B., Dickson, R., Kang, S., Liu, J.J., Lim, H. (2020b). Concept for temperature-cascade hydrogen release from organic liquid carriers coupled with SOFC power generation. *Cell Reports Physical Science*, 1(3), 100032.

Brückner, N., Obesser, K., Bösmann, A., Teichmann, D., Arlt, W., Dungs, J., Wasserscheid, P. (2014). Evaluation of industrially applied heat-transfer fluids as liquid organic hydrogen carrier systems. *ChemSusChem*, 7(1), 229–235.

Byun, M., Kim, H., Choe, C., Lim, H. (2021). Conceptual feasibility studies for cost-efficient and bifunctional methylcyclohexane dehydrogenation in a membrane reactor for H_2 storage and production. *Energy Conversion and Management*, 227, 113576.

Byun, M., Lee, A., Cheon, S., Kim, H., Lim, H. (2022). Preliminary feasibility study for hydrogen storage using several promising liquid organic hydrogen carriers: Technical, economic, and environmental perspectives. *Energy Conversion and Management*, 268, 116001.

Campbell, P.G., Zakharov, L.N., Grant, D.J., Dixon, D.A., Liu, S.-Y. (2010). Hydrogen storage by boron-nitrogen heterocycles: A simple route for spent fuel regeneration. *Journal of the American Chemical Society*, 132(10), 3289–3291.

Campos, J. (2018). Dehydrogenation of alcohols and polyols from a hydrogen production perspective. *Physical Sciences Reviews*, 3(6), 20170017.

Cao, Y., Ran, R., Chen, Y., Wu, X., Weng, D. (2016). Nanostructured platinum in ordered mesoporous silica as novel efficient catalyst for propane total oxidation. *RSC Advances*, 6(36), 30170–30175.

Carey, F.A. and Sundberg, R.J. (2007). *Advanced Organic Chemistry. Part A: Structure and Mechanisms*. Springer, New York.

Chappaz, A. and Bengaouer, A. (2020). Use of hydrogenated organic liquids, in particular in devices for converting energy. Report, EP3725738A1.

Chen, J., Qi, S.-T., Humbert, M., Menning, C., Zhu, Y.-X. (2010). Rational design of low-temperature hydrogenation catalysts: Theoretical predictions and experimental verification. *Acta Physico-Chimica Sinica*, 26(4), 869–876.

Chen, Z., Chen, J., Li, Y. (2017). Metal-organic-framework-based catalysts for hydrogenation reactions. *Chinese Journal of Catalysis*, 38(7), 1108–1126.

Chen, X., Gierlich, C.H., Schötz, S., Blaumeiser, D., Bauer, T., Libuda, J., Palkovits, R. (2021). Hydrogen production based on liquid organic hydrogen carriers through sulfur doped platinum catalysts supported on TiO_2. *ACS Sustainable Chemistry & Engineering*, 9(19), 6561–6573.

Chernov, A.N., Astrakova, T.V., Koltunov, K.Y., Sobolev, V.I. (2021). Ethanol dehydrogenation to acetaldehyde over Co@N-doped carbon. *Catalysts*, 11(11), 1411.

Cho, J.-Y., Kim, H., Oh, J.-E., Park, B.Y. (2021). Recent advances in homogeneous/heterogeneous catalytic hydrogenation and dehydrogenation for potential liquid organic hydrogen carrier (LOHC) systems. *Catalysts*, 11(12), 1497.

Choi, I.Y., Shin, B.S., Kwak, S.K., Kang, K.S., Yoon, C.W., Kang, J.W. (2016). Thermodynamic efficiencies of hydrogen storage processes using carbazole-based compounds. *International Journal of Hydrogen Energy*, 41(22), 9367–9373.

Chou, P. and Vannice, M.A. (1987). Benzene hydrogenation over supported and unsupported palladium: II. Reaction model. *Journal of Catalysis*, 107(1), 140–153.

Claus, P. and Hofmeister, H. (1999). Electron microscopy and catalytic study of silver catalysts: Structure sensitivity of the hydrogenation of crotonaldehyde. *The Journal of Physical Chemistry B*, 103(14), 2766–2775.

Conesa, J.M., Morales, M.V., López-Olmos, C., Rodríguez-Ramos, I., Guerrero-Ruiz, A. (2019). Comparative study of Cu, Ag and Ag-Cu catalysts over graphite in the ethanol dehydrogenation reaction: Catalytic activity, deactivation and regeneration. *Applied Catalysis A: General*, 576, 54–64.

Cooper, A.C., Campbell, K.M., Pez, G.P. (2006). An integrated hydrogen storage and delivery approach using organic liquid-phase carriers. In *Proceedings of WHEC 16*, Lyon.

Coughlin, R.W., Hasan, A., Kawakami, K. (1984). Activity, yield patterns, and coking behavior of Pt and PtRe catalysts during dehydrogenation of methylcyclohexane: II. Influence of sulfur. *Journal of Catalysis*, 88(1), 163–176.

Crabtree, R.H. (2008). Hydrogen storage in liquid organic heterocycles. *Energy & Environmental Science*, 1(1), 134–138.

Crabtree, R.H. (2017a). Nitrogen-containing liquid organic hydrogen carriers: Progress and prospects. *ACS Sustainable Chemistry & Engineering*, 5(6), 4491–4498.

Crabtree, R.H. (2017b). Homogeneous transition metal catalysis of acceptorless dehydrogenative alcohol oxidation: Applications in hydrogen storage and to heterocycle synthesis. *Chemical Reviews*, 117(13), 9228–9246.

Cui, Y., Kwok, S., Bucholtz, A., Davis, B., Whitney, R.A., Jessop, P.G. (2008). The effect of substitution on the utility of piperidines and octahydroindoles for reversible hydrogen storage. *New J. Chem.*, 32(6), 1027–1037.

Cui, X., Huang, Z., van Muyden, A.P., Fei, Z., Wang, T., Dyson, P.J. (2020). Acceptorless dehydrogenation and hydrogenation of N- and O-containing compounds on $Pd_3Au_1(111)$ facets. *Science Advances*, 6(27), eabb3831.

D'Ambra, F., Faucheux, V., Nicolas, E., Silva Costa, R. (2023). New liquid organic hydrogen carriers, uses thereof for transporting and storing hydrogen, and methods for generating hydrogen using same. Report, WO2023006956.

Damodara, D., Arundhathi, R., Likhar, P.R. (2014). Copper nanoparticles from copper aluminum hydrotalcite: An efficient catalyst for acceptor- and oxidant-free dehydrogenation of amines and alcohols. *Advanced Synthesis & Catalysis*, 356(1), 189–198.

Ding, Y., Dong, Y., Zhang, H., Zhao, Y., Yang, M., Cheng, H. (2021). A highly adaptable Ni catalyst for liquid organic hydrogen carriers hydrogenation. *International Journal of Hydrogen Energy*, 46(53), 27026–27036.

Do, G., Preuster, P., Aslam, R., Bösmann, A., Müller, K., Arlt, W., Wasserscheid, P. (2016). Hydrogenation of the liquid organic hydrogen carrier compound dibenzyltoluene reaction pathway determination by ^1H NMR spectroscopy. *Reaction Chemistry & Engineering*, 1(3), 313–320.

Dong, Y., Yang, M., Yang, Z., Ke, H., Cheng, H. (2015). Catalytic hydrogenation and dehydrogenation of N-ethylindole as a new heteroaromatic liquid organic hydrogen carrier. *International Journal of Hydrogen Energy*, 40(34), 10918–10922.

Dong, Y., Yang, M., Zhu, T., Chen, X., Cheng, G., Ke, H., Cheng, H. (2018). Fast dehydrogenation kinetics of perhydro-N-propylcarbazole over a supported Pd catalyst. *ACS Applied Energy Materials*, 1(8), 4285–4292.

Dürr, S., Zilm, S., Geißelbrecht, M., Müller, K., Preuster, P., Bösmann, A., Wasserscheid, P. (2021). Experimental determination of the hydrogenation/dehydrogenation. Equilibrium of the LOHC system H0/H18-dibenzyltoluene. *International Journal of Hydrogen Energy*, 46(64), 32583–32594.

EASE-EERA (2017). Joint EASE/EERA recommendations for a European energy storage development roadmap. Report [Online]. Available at: https://ease-storage.eu/publication/ease-eera-energy-storage-technology-development-roadmap-2017/.

Emel'yanenko, V.N., Zaitsau, D.H., Pimerzin, A.A., Verevkin, S.P. (2019). N-phenyl-carbazole as a potential liquid organic hydrogen carrier: Thermochemical and computational study. *The Journal of Chemical Thermodynamics*, 132, 122–128.

Enjamuri, N., Hassan, S., Auroux, A., Pandey, J.K., Chowdhury, B. (2016). Nobel metal free, oxidant free, solvent free catalytic transformation of alcohol to aldehyde over ZnO-CeO_2 mixed oxide catalyst. *Applied Catalysis A: General*, 523, 21–30.

Faye, O., Szpunar, J., Eduok, U. (2022). A critical review on the current technologies for the generation, storage, and transportation of hydrogen. *International Journal of Hydrogen Energy*, 47(29), 13771–13802.

Feng, Z. and Bai, X. (2022). Enhanced activity of bimetallic Pd-Ni nanoparticles on KIT-6 for production of hydrogen from dodecahydro-N-ethylcarbazole. *Fuel*, 329, 125473.

Filonenko, G.A., Putten, R.V., Hensen, E.J.M., Pidko, E.A. (2018). Catalytic (de)hydrogenation promoted by non-precious metals – Co, Fe and Mn: Recent advances in an emerging field. *Chemical Society Reviews*, 47(4), 1459–1483.

Fishbein, L. (1985). An overview of environmental and toxicological aspects of aromatic hydrocarbons II. Toluene. *Science of The Total Environment*, 42(3), 267–288.

Foppa, L. and Dupont, J. (2015). Benzene partial hydrogenation: Advances and perspectives. *Chemical Society Reviews*, 44(7), 1886–1897.

Forberg, D., Schwob, T., Zaheer, M., Friedrich, M., Miyajima, N., Kempe, R. (2016). Single-catalyst high-weight% hydrogen storage in an N-heterocycle synthesized from lignin hydrogenolysis products and ammonia. *Nature Communications*, 7(1), 13201.

Fujita, K.-I., Tanaka, Y., Kobayashi, M., Yamaguchi, R. (2014). Homogeneous perdehydrogenation and perhydrogenation of fused bicyclic N-heterocycles catalyzed by iridium complexes bearing a functional bipyridonate ligand. *Journal of the American Chemical Society*, 136(13), 4829–4832.

Fujita, K.-I., Wada, T., Shiraishi, T. (2017). Reversible interconversion between 2,5-dimethylpyrazine and 2,5-dimethylpiperazine by iridium-catalyzed hydrogenation/dehydrogenation for efficient hydrogen storage. *Angewandte Chemie International Edition*, 56(36), 10886–10889.

Garcia Villaluenga, J.P. and Tabe-Mohammadi, A. (2000). A review on the separation of benzene/cyclohexane mixtures by pervaporation processes. *Journal of Membrane Science*, 169(2), 159–174.

Geburtig, D. (2019). Transfer hydrogenation using liquid organic hydrogen carrier systems as hydrogen source. PhD Thesis, Friedrich-Alexander University, Erlangen-Nürnberg.

Geisselbrecht, M., Mrusek, S., Müller, K., Preuster, P., Bösmann, A., Wasserscheid, P. (2020). Highly efficient, low-temperature hydrogen release from perhydro-benzyltoluene using reactive distillation. *Energy & Environmental Science*, 13(9), 3119–3128.

Gonda, M., Ohshima, M., Kurokawa, H., Miura, H. (2014). Toluene hydrogenation over Pd and Pt catalysts as a model hydrogen storage process using low grade hydrogen containing catalyst inhibitors. *International Journal of Hydrogen Energy*, 39(29), 16339–16346.

Grellier, M. and Sabo-Etienne, S. (2014). New perspectives in hydrogen storage based on RCH_2NH_2/RCN couples. *Dalton Trans.*, 43(17), 6283–6286.

Guan, Y. and Hensen, E.J.M. (2009). Ethanol dehydrogenation by gold catalysts: The effect of the gold particle size and the presence of oxygen. *Applied Catalysis A: General*, 361(1), 49–56.

Haynes, W.M. (ed.) (2016). *CRC Handbook of Chemistry and Physics*, 97th edition. CRC Press, Boca Raton. doi: 10.1201/9781315380476.

He, X., Wang, Y., Zhang, X., Dong, M., Wang, G., Zhang, B., Niu, Y., Yao, S., He, X., Liu, H. (2019). Controllable in situ surface restructuring of Cu catalysts and remarkable enhancement of their catalytic activity. *ACS Catalysis*, 9(3), 2213–2221.

Hiyoshi, N., Miura, R., Rode, C.V., Sato, O., Shirai, M. (2005). Enhanced selectivity to decalin in naphthalene hydrogenation under supercritical carbon dioxide. *Chemistry Letters*, 34(3), 424–425.

Hiyoshi, N., Osada, M., Rode, C.V., Sato, O., Shirai, M. (2007). Hydrogenation of benzothiophene-free naphthalene over charcoal-supported metal catalysts in supercritical carbon dioxide solvent. *Applied Catalysis A: General*, 331, 1–7.

Hodoshima, S., Takaiwa, S., Shono, A., Satoh, K., Saito, Y. (2005). Hydrogen storage by decalin/naphthalene pair and hydrogen supply to fuel cells by use of superheated liquid-film-type catalysis. *Appl. Catal. A*, 283, 235–242.

Hu, P., Fogler, E., Diskin-Posner, Y., Iron, M.A., Milstein, D. (2015). A novel liquid organic hydrogen carrier system based on catalytic peptide formation and hydrogenation. *Nature Communications*, 6(1), 1–7.

Huynh, N.-D., Hur, S.H., Kang, S.G. (2021). Tuning the dehydrogenation performance of dibenzyl toluene as liquid organic hydrogen carriers. *International Journal of Hydrogen Energy*, 46(70), 34788–34796.

Ichikawa, N., Sato, S., Takahashi, R., Sodesawa, T., Inui, K. (2004). Dehydrogenative cyclization of 1,4-butanediol over copper-based catalyst. *Journal of Molecular Catalysis A: Chemical*, 212(1), 197–203.

Ichikawa, T., Matsuo, T., Tachikawa, T., Teranishi, W., Yamada, T., Sawama, Y., Monguchi, Y., Sajiki, H. (2019). Microwave-mediated continuous hydrogen abstraction reaction from 2-PrOH catalyzed by platinum on carbon bead. *Catalysts*, 9(8), 655.

Jaiswal, G., Subaramanian, M., Sahoo, M.K., Balaraman, E. (2019). A reusable cobalt catalyst for reversible acceptorless dehydrogenation and hydrogenation of N-He-terocycles. *ChemCatChem*, 11(10), 2449–2457.

Jang, M., Shin, B.S., Jo, Y.S., Kang, J.W., Kwak, S.K., Yoon, C.W., Jeong, H. (2020). A study on hydrogen uptake and release of a eutectic mixture of biphenyl and diphenyl ether. *Journal of Energy Chemistry*, 42, 11–16.

Javaid, A. and Bildea, C.S. (2014). Design and control of an integrated 1,4-butanediol dehydrogenation and furfural hydrogenation plant. *Chemical Engineering & Technology*, 37(9), 1515–1524.

Ji, X., Louarn, E., Fache, F., Vanoye, L., Bonhommé, A., Pitault, I., Meille, V. (2023). Analysis of dibenzyltoluene mixtures: From fast analysis to in-depth characterization of the compounds. *Molecules*, 28(9), 3751.

Jo, Y., Wan Kim, T., Oh, J., Kim, D., Suh, Y.-W. (2022). Mesoporous sulfur-decorated Pt-Al_2O_3 for dehydrogenation of perhydro benzyltoluenes: Activity-favorable adsorption of reaction species onto electron-deficient Pt atoms. *Journal of Catalysis*, 413, 127–137.

Jorschick, H., Preuster, P., Dürr, S., Seidel, A., Müller, K., Bösmann, A., Wasserscheid, P. (2017). Hydrogen storage using a hot pressure swing reactor. *Energy & Environmental Science*, 10(7), 1652–1659.

Jorschick, H., Geißelbrecht, M., Eßl, M., Preuster, P., Bösmann, A., Wasserscheid, P. (2020). Benzyltoluene/dibenzyltoluene-based mixtures as suitable liquid organic hydrogen carrier systems for low temperature applications. *International Journal of Hydrogen Energy*, 45(29), 14897–14906.

Jorschick, H., Preuster, P., Bösmann, A., Wasserscheid, P. (2021). Hydrogenation of aromatic and heteroaromatic compounds a key process for future logistics of green hydrogen using liquid organic hydrogen carrier systems. *Sustainable Energy & Fuels*, 5(5), 1311–1346.

Kalenchuk, A., Bogdan, V., Kustov, L.M. (2015). Comparing the activities of catalysts in perhydro-m-terphenyl dehydrogenation. *Catalysis in Industry*, 7(1), 60–63.

Kalenchuk, A., Bogdan, V., Bogorodskii, S.E., Kustov, L. (2016). Effect of the structure of the ortho, meta, and para isomers of perhydroterphenyl on their reactivity in heterogeneous catalytic dehydrogenation. *Kinetics and Catalysis*, 57(2), 219–223.

Kalenchuk, A., Bogdan, V., Dunaev, S., Kustov, L. (2018). Effect of isomerization on the reversible reaction of hydrogenation-dehydrogenation of ortho-terphenyl on a Pt/C catalyst. *Chemical Engineering & Technology*, 41(9), 1842–1846.

Kalenchuk, A., Bogdan, V., Dunaev, S., Kustov, L. (2020). Influence of steric factors on reversible reactions of hydrogenation-dehydrogenation of polycyclic aromatic hydrocarbons on a Pt/C catalyst in hydrogen storage systems. *Fuel*, 280, 118625.

Kawamoto, K. (1961). Dehydrogenation of secondary alcohols with reduced copper. V. Catalytic dehydrogenation of sec-butyl alcohol. *BCSJ*, 34(6), 799–803.

Kaźmierczak, K., Pinel, C., Loridant, S., Besson, M., Michel, C., Perret, N. (2020). Supported cobalt catalysts for acceptorless alcohol dehydrogenation. *ChemPlusChem*, 85(6), 1315–1324.

Kiermaier, S., Lehmann, D., Bösmann, A., Wasserscheid, P. (2021). Dehydrogenation of perhydro-N-ethylcarbazole under reduced total pressure. *International Journal of Hydrogen Energy*, 46(29), 15660–15670.

Kim, T.G., Yeo, Y.K., Song, H.K. (1992). Chemical heat pump based on dehydrogenation and hydrogenation of i-propanol and acetone. *International Journal of Energy Research*, 16(9), 897–916.

Kim, W.-H., Park, I.S., Park, J. (2006). Acceptor-free alcohol dehydrogenation by recyclable ruthenium catalyst. *Organic Letters*, 8(12), 2543–2545.

Kim, K., Oh, J., Kim, T.W., Park, J.H., Han, J.W., Suh, Y.-W. (2017). Different catalytic behaviors of Pd and Pt metals in decalin dehydrogenation to naphthalene. *Catalysis Science & Technology*, 7(17), 3728–3735.

Kim, T.W., Park, S., Oh, J., Shin, C.-H., Suh, Y.-W. (2018). Hydrogenation of the LOHC compound monobenzyl toluene over ZrO_2-supported Ru nanoparticles: A consequence of zirconium hydroxide's surface hydroxyl group and surface area. *ChemCatChem*, 10(16), 3406–3410.

Kim, T.W., Kim, M., Kim, S.K., Choi, Y.N., Jung, M., Oh, H., Suh, Y.-W. (2021). Remarkably fast low-temperature hydrogen storage into aromatic benzyltoluenes over MgO-supported Ru nanoparticles with homolytic and heterolytic H_2 adsorption. *Applied Catalysis B: Environmental*, 286, 119889.

Kim, C.H., Lee, M.-W., Jang, J.S., Lee, S.H., Lee, K.-Y. (2022). Enhanced activity of a WOx-incorporated Pt/Al_2O_3 catalyst for the dehydrogenation of homocyclic LOHCs: Effects of impregnation sequence on Pt-WOx interactions. *Fuel*, 313, 122654.

Kim, T.W., Jo, Y., Jeong, K., Yook, H., Han, J.W., Jang, J.H., Han, G.B., Park, J.H., Suh, Y.-W. (2023). Tuning the isomer composition is a key to overcome the performance limits of commercial benzyltoluene as liquid organic hydrogen carrier. *Journal of Energy Storage*, 60, 106676.

Kirumakki, S.R., Shpeizer, B.G., Sagar, G.V., Chary, K.V.R., Clearfield, A. (2006). Hydrogenation of naphthalene over $NiO/SiO_2Al_2O_3$ catalysts: Structure activity correlation. *Journal of Catalysis*, 242(2), 319–331.

Kline, M.J., Karunarathne, S.A., Schwartz, T.J., Wheeler, M.C. (2022). Hydrogenation of 2-methylnaphthalene over bifunctional Ni catalysts. *Applied Catalysis A: General*, 630, 118462.

Klvana, D., Chaouki, J., Kusohorsky, D., Chavarie, C., Pajonk, G.M. (1988). Catalytic storage of hydrogen: Hydrogenation of toluene over a nickel/silica aerogel catalyst in integral flow conditions. *Appl. Catal.*, 42, 121–130.

Kothandaraman, J., Kar, S., Sen, R., Goeppert, A., Olah, G.A., Prakash, G.K.S. (2017). Efficient reversible hydrogen carrier system based on amine reforming of methanol. *Journal of the American Chemical Society*, 139(7), 2549–2552.

Kurhe, D.K., Fernandes, T.A., Deore, T.S., Jayaram, R.V. (2015). Oxidant free dehydrogenation of alcohols using chitosan/polyacrylamide entrapped Ag nanoparticles. *RSC Advances*, 5(58), 46443–46447.

Kwak, Y., Moon, S., Ahn, C.-I., Kim, A.-R., Park, Y., Kim, Y., Sohn, H., Jeong, H., Nam, S.W., Yoon, C.W. et al. (2021). Effect of the support properties in dehydrogenation of biphenyl-based eutectic mixture as liquid organic hydrogen carrier (LOHC) over Pt/Al_2O_3 catalysts. *Fuel*, 284, 119285.

Lee, S., Kim, T., Han, G., Kang, S., Yoo, Y.-S., Jeon, S.-Y., Bae, J. (2021a). Comparative energetic studies on liquid organic hydrogen carrier: A net energy analysis. *Renewable and Sustainable Energy Reviews*, 150, 111447.

Lee, S., Lee, J., Kim, T., Han, G., Lee, J., Lee, K., Bae, J. (2021b). Pt/CeO_2 catalyst synthesized by combustion method for dehydrogenation of perhydro-dibenzyl-toluene as liquid organic hydrogen carrier: Effect of pore size and metal dispersion. *International Journal of Hydrogen Energy*, 46(7), 5520–5529.

Leinweber, A. and Müller, K. (2018). Hydrogenation of the liquid organic hydrogen carrier compound monobenzyl toluene: Reaction pathway and kinetic effects. *Energy Technology*, 6(3), 513–520.

Li, Z., Chen, C., Zhan, E., Ta, N., Li, Y., Shen, W. (2014a). Crystal-phase control of molybdenum carbide nanobelts for dehydrogenation of benzyl alcohol. *Chemical Communications*, 50(34), 4469–4471.

Li, Z., Chen, C., Zhan, E., Ta, N., Shen, W. (2014b). Mo_2N nanobelts for dehydrogenation of aromatic alcohols. *Catalysis Communications*, 51, 58–62.

Li, Y., Nielsen, M., Li, B., Dixneuf, P.H., Junge, H., Beller, M. (2015). Ruthenium-catalyzed hydrogen generation from glycerol and selective synthesis of lactic acid. *Green Chemistry*, 17(1), 193–198.

Li, L., Yang, M., Dong, Y., Mei, P., Cheng, H. (2016). Hydrogen storage and release from a new promising liquid organic hydrogen storage carrier (LOHC): 2-methylindole. *International Journal of Hydrogen Energy*, 41(36), 16129–16134.

Li, X., Shen, P., Han, X., Wang, Y., Zhu, Y., Wu, Z. (2021). Dehydrogenation mechanisms of liquid organic hydrogen carriers over Pt, Pd, Rh, and Ni surfaces: Cyclo-hexane as a model compound. *Applied Surface Science*, 543, 148769.

Li, J., Tong, F., Li, Y., Liu, X., Guo, Y., Wang, Y. (2022). Dehydrogenation of dodecahydro-N-ethylcarbazole over spinel supporting catalyst in a continuous flow fixed bed reactor. *Fuel*, 321, 124034.

Lin, S.D. and Vannice, M.A. (1993). Hydrogenation of aromatic hydrocarbons over supported Pt catalysts. I. Benzene hydrogenation. *Journal of Catalysis*, 143(2), 539–553.

Lindfors, L.P., Salmi, T., Smeds, S. (1993). Kinetics of toluene hydrogenation on Ni/Al_2O_3 catalyst. *Chemical Engineering Science*, 48(22), 3813–3828.

Liu, C., Wang, Y., Chu, H., Qiu, M., Niu, L., Bai, G. (2014). Continuous dehydrogenation of n-pentanol over a Cr modified Cu/γ-Al_2O_3-La_2O_3 catalyst. *Chinese Journal of Chemistry*, 32(12), 1281–1286.

Liu, W., Sun, J., Zhang, X., Wei, M. (2018). Supported Ag catalysts on Mg-Al oxides toward oxidant-free dehydrogenation reaction of benzyl alcohol. *Industrial & Engineering Chemistry Research*, 57(46), 15606–15612.

Liu, C., Li, T., Dai, X., Zhao, J., He, D., Li, G., Wang, B., Cui, X. (2022). Catalytic activity enhancement on alcohol dehydrogenation via directing reaction pathways from single- to double-atom catalysis. *Journal of the American Chemical Society*, 144(11), 4913–4924.

Makaryan, I.A., Sedov, I.V., Maksimov, A.L. (2020). Hydrogen storage using liquid organic carriers. *Russian Journal of Applied Chemistry*, 93(12), 1815–1830.

Markiewicz, M., Zhang, Y.-Q., Empl, M., Lykaki, M., Thöming, J., Steinberg, P., Stolte, S. (2019). Hazard assessment of quinaldine-, alkylcarbazole-, benzene- and to-luene-based liquid organic hydrogen carrier (LOHCs) systems. *Energy & Environmental Science*, 12(1), 366–383.

van Meerten, R.Z.C., de Graaf, T.F.M., Coenen, J.W.E. (1977). Gas phase benzene hydrogenation on a nickel-silica catalyst: III. Low-field magnetization measurements on hydrogen, benzene, cyclohexene and cyclohexane adsorption, and benzene hydrogenation. *Journal of Catalysis*, 46(1), 1–12.

Metaxas, K.C. and Papayannakos, N.G. (2006). Kinetics and mass transfer of benzene hydrogenation in a trickle-bed reactor. *Industrial & Engineering Chemistry Research*, 45(21), 7110–7119.

Mevawala, C., Brooks, K., Bowden, M.E., Breunig, H.M., Tran, B.L., Gutiérrez, O.Y., Autrey, T., Müller, K. (2023a). The ethanol-ethyl acetate system as a biogenic hydrogen carrier. *Energy Technology*, 11(1), 2200892.

Mevawala, C., Autrey, T., Brooks, K., Bowden, M., Tran, B.L., Müller, K. (2023b). 1,4-Butanediol as a hydrogen carrier: Liquid- versus gas-phase dehydrogenation. *Energy & Fuels*, 37(1), 560–566.

Milstein, D., Hu, P., Fogler, E., Rao, J.A.G.N. (2017). Ruthenium complexes and their uses as catalysts in processes for formation and/or hydrogenation of esters, amides and related reactions. Report, US20170283447.

Mitsudome, T., Mikami, Y., Funai, H., Mizugaki, T., Jitsukawa, K., Kaneda, K. (2008). Oxidant-free alcohol dehydrogenation using a reusable hydrotalcite-supported silver nanoparticle catalyst. *Angewandte Chemie International Edition*, 47(1), 138–141.

Modisha, P. and Bessarabov, D. (2020). Stress tolerance assessment of dibenzyltoluene-based liquid organic hydrogen carriers. *Sustainable Energy & Fuels*, 4(9), 4662–4670.

Modisha, P.M., Jordaan, J.H.L., Bösmann, A., Wasserscheid, P., Bessarabov, D. (2018). Analysis of reaction mixtures of perhydro-dibenzyltoluene using two-dimensional gas chromatography and single quadrupole gas chromatography. *International Journal of Hydrogen Energy*, 43(11), 5620–5636.

Modisha, P.M., Ouma, C.N.M., Garidzirai, R., Wasserscheid, P., Bessarabov, D. (2019a). The prospect of hydrogen storage using liquid organic hydrogen carriers. *Energy & Fuels*, 33(4), 2778–2796.

Modisha, P., Gqogqa, P., Garidzirai, R., Ouma, C.N.M., Bessarabov, D. (2019b). Evaluation of catalyst activity for release of hydrogen from liquid organic hydrogen carriers. *International Journal of Hydrogen Energy*, 44(39), 21926–21935.

Mrusek, S., Preuster, P., Müller, K., Bösmann, A., Wasserscheid, P. (2021). Pressurized hydrogen from charged liquid organic hydrogen carrier systems by electrochemical hydrogen compression. *International Journal of Hydrogen Energy*, 46(29), 15624–15634.

Müller, K. (2022). Acceptorless dehydrogenation of amines to nitriles for hydrogen storage: Reality or wishful thinking? *Energy Technology*, 10(9), 2200468.

Müller, K., Stark, K., Müller, B., Arlt, W. (2012). Amine borane based hydrogen carriers: An evaluation. *Energy & Fuels*, 26(6), 3691–3696.

Müller, K., Völkl, J., Arlt, W. (2013). Thermodynamic evaluation of potential organic hydrogen carriers. *Energy Technology*, 1(1), 20–24.

Müller, K., Stark, K., Emel'yanenko, V.N., Varfolomeev, M.A., Zaitsau, D.H., Shoifet, E., Schick, C., Verevkin, S.P., Arlt, W. (2015). Liquid organic hydrogen carriers: Thermophysical and thermochemical studies of benzyl- and dibenzyl-toluene derivatives. *Industrial & Engineering Chemistry Research*, 54(32), 7967–7976.

Murthy, S.S.N., Paikaray, A., Arya, N. (1995). Molecular relaxation and excess entropy in liquids and their connection to the structure of glass. *The Journal of Chemical Physics*, 102(20), 8213–8220.

Naseem, M., Usman, M., Lee, S. (2021). A parametric study of dehydrogenation of various liquid organic hydrogen carrier (LOHC) materials and its application to methanation process. *International Journal of Hydrogen Energy*, 46(5), 4100–4115.

Nazir, H., Muthuswamy, N., Louis, C., Jose, S., Prakash, J., Buan, M.E., Flox, C., Chavan, S., Shi, X., Kauranen, P. et al. (2020). Is the H_2 economy realizable in the foreseeable future? Part II: H_2 storage, transportation, and distribution. *International Journal of Hydrogen Energy*. 45(41), 20693–20708.

Nguyen, D.H., Morin, Y., Zhang, L., Trivelli, X., Capet, F., Paul, S., Desset, S., Dumeignil, F., Gauvin, R.M. (2017). Oxidative transformations of biosourced alcohols catalyzed by earth-abundant transition metals. *ChemCatChem*, 9(14), 2652–2660.

Nicolau, G., Tarantino, G., Hammond, C. (2019). Acceptorless alcohol dehydrogenation catalysed by Pd/C. *ChemSusChem*, 12(22), 4953–4961.

Nie, J., Zhu, Z., Liao, Y., Xiao, X., Mauriello, F., Zhang, Z. (2022). Highly efficient and selective hydrogenation of quinolines at room temperature over Ru@NC-500 catalyst. *Molecular Catalysis*, 526, 112381.

Nielsen, M., Kammer, A., Cozzula, D., Junge, H., Gladiali, S., Beller, M. (2011). Efficient hydrogen production from alcohols under mild reaction conditions. *Angewandte Chemie International Edition*, 50(41), 9593–9597.

Nielsen, M., Junge, H., Kammer, A., Beller, M. (2012). Towards a green process for bulk-scale synthesis of ethyl acetate: Efficient acceptorless dehydrogenation of ethanol. *Angewandte Chemie International Edition*, 51(23), 5711–5713.

Niermann, M., Beckendorff, A., Kaltschmitt, M., Bonhoff, K. (2019). Liquid organic hydrogen carrier (LOHC) assessment based on chemical and economic properties. *Int. J. Hydrog. Energy*, 44(13), 6631–6654.

Niermann, M., Timmerberg, S., Drünert, S., Kaltschmitt, M. (2021). Liquid organic hydrogen carriers and alternatives for international transport of renewable hydrogen. *Renewable and Sustainable Energy Reviews*, 135, 110171.

Oh, J., Bathula, H.B., Park, J.H., Suh, Y.-W. (2019). A sustainable mesoporous palladium–alumina catalyst for efficient hydrogen release from N-heterocyclic liquid organic hydrogen carriers. *Communications Chemistry*, 2(1), 1–10.

Oh, J., Jo, Y., Kim, T.W., Bathula, H.B., Yang, S., Baik, J.H., Suh, Y.-W. (2022). Highly efficient and robust Pt ensembles on mesoporous alumina for reversible H_2 charge and release of commercial benzyltoluene molecules. *Applied Catalysis B: Environmental*, 305, 121061.

Okada, Y., Sasaki, E., Watanabe, E., Hyodo, S., Nishijima, H. (2006). Development of dehydrogenation catalyst for hydrogen generation in organic chemical hydride method. *International Journal of Hydrogen Energy*, 31(10), 1348–1356.

Okada, Y., Imagawa, K., Yasui, M. (2018). Development of novel dehydrogenation catalyst for hydrogen carrier system. In *Proceedings of TOCAT8*, Yokohama [Online]. Available at: https://www.shokubai.org/tocat8/pdf/Plenary/PL9.pdf.

Ombaka, L.M., Ndungu, P., Nyamori, V.O. (2013). Usage of carbon nanotubes as platinum and nickel catalyst support in dehydrogenation reactions. *Catalysis Today*, 217, 65–75.

Oner, O. and Khalilpour, K. (2022). Evaluation of green hydrogen carriers: A multicriteria decision analysis tool. *Renewable and Sustainable Energy Reviews*, 168, 112764.

Ouma, C.N.M., Obodo, K.O., Modisha, P.M., Rhyman, L., Ramasami, P., Bessarabov, D. (2021). Effect of chalcogen (S, Se and Te) surface additives on the dehydrogenation of a liquid organic hydrogen carrier system, octahydroindole-indole, on a Pt (111) surface. *Applied Surface Science*, 566, 150636.

Ouyang, M., Cao, S., Yang, S., Li, M., Flytzani-Stephanopoulos, M. (2020). Atomically dispersed Pd supported on zinc oxide for selective nonoxidative ethanol dehydrogenation. *Industrial & Engineering Chemistry Research*, 59(6), 2648–2656.

Pandey, P., Dutta, I., Bera, J.K. (2016). Acceptorless alcohol dehydrogenation: A mechanistic perspective. *Proceedings of the National Academy of Sciences, India Section A: Physical Sciences*, 86(4), 561–579.

Pawelczyk, E., Lukasik, N., Wysocka, I., Rogala, A., Gebicki, J. (2022). Recent progress on hydrogen storage and production using chemical hydrogen carriers. *Energies*, 15(14), 4964.

Peña-López, M., Neumann, H., Beller, M. (2015). Iron(II) pincer-catalyzed synthesis of lactones and lactams through a versatile dehydrogenative domino sequence. *ChemCatChem*, 7(5), 865–871.

Peters, W., Seidel, A., Herzog, S., Bösmann, A., Schwieger, W., Wasserscheid, P. (2015a). Macrokinetic effects in perhydro-N-ethylcarbazole dehydrogenation and H_2 productivity optimization by using egg-shell catalysts. *Energy & Environmental Science*, 8(10), 3013–3021.

Peters, W., Eypasch, M., Frank, T., Schwerdtfeger, J., Körner, C., Bösmann, A., Wasserscheid, P. (2015b). Efficient hydrogen release from perhydro-N-ethylcarba-zole using catalyst-coated metallic structures produced by selective electron beam melting. *Energy Environ. Sci.*, 8(2), 641–649.

Peters, R., Deja, R., Fang, Q., Nguyen, V.N., Preuster, P., Blum, L., Wasserscheid, P., Stolten, D. (2019). A solid oxide fuel cell operating on liquid organic hydrogen carrier-based hydrogen: A kinetic model of the hydrogen release unit and system performance. *International Journal of Hydrogen Energy*, 44(26), 13794–13806.

Pez, G.P., Scott, A.R., Cooper, A.C., Cheng, H. (2006). Hydrogen storage by reversible hydrogenation of pi-conjugated substrates. Report, US7101530B2.

Pez, G.P., Scott, A.R., Cooper, A.C., Cheng, H. (2008). Hydrogen storage by reversible hydrogenation of pi-conjugated substrates. Report, US7429372B2.

Polukeev, A.V., Wallenberg, R., Uhlig, J., Hulteberg, C.P., Wendt, O.F. (2022). Iridium-catalyzed dehydrogenation in a continuous flow reactor for practical on-board hydrogen generation from liquid organic hydrogen carriers. *ChemSusChem*, 15(8), e202200085.

Preuster, P., Papp, C., Wasserscheid, P. (2017). Liquid organic hydrogen carriers (LOHCs): Toward a hydrogen-free hydrogen economy. *Accounts of Chemical Research*, 50(1), 74–85.

Qi, S.-C., Wei, X.-Y., Zong, Z.-M., Wang, Y.-K. (2013). Application of supported metallic catalysts in catalytic hydrogenation of arenes. *RSC Advances*, 3(34), 14219–14232.

Quesada, J., Faba, L., Díaz, E., Ordóñez, S. (2018). Copper-basic sites synergic effect on the ethanol dehydrogenation and condensation reactions. *ChemCatChem*, 10(16), 3583–3592.

Ranjekar, A.M. and Yadav, G.D. (2021). Steam reforming of methanol for hydrogen production: A critical analysis of catalysis, processes, and scope. *Industrial & Engineering Chemistry Research*, 60(1), 89–113.

Rao, P.C. and Yoon, M. (2020). Potential liquid-organic hydrogen carrier (LOHC) systems: A review on recent progress. *Energies*, 13(22), 6040.

Rao, P.C., Kim, Y., Kim, H., Son, Y., Choi, Y., Na, K., Yoon, M. (2023). Methyl-benzyl naphthalene: Liquid organic hydrogen carrier for facile hydrogen storage and release. *ACS Sustainable Chemistry & Engineering*, 11(34), 12656–12666.

Rautanen, P.A., Lylykangas, M.S., Aittamaa, J.R., Krause, A.O.I. (2002). Liquid-phase hydrogenation of naphthalene and tetralin on Ni/Al_2O_3: Kinetic modeling. *Industrial & Engineering Chemistry Research*, 41(24), 5966–5975.

Rautenbach, R. and Albrech, R. (1989). *Membrane Processes.* John Wiley and Sons, New York.

Rüde, T., Dürr, S., Preuster, P., Wolf, M., Wasserscheid, P. (2022). Benzyltoluene/perhydro benzyltoluene pushing the performance limits of pure hydrocarbon liquid organic hydrogen carrier (LOHC) systems. *Sustainable Energy Fuels*, 6(6), 1541–1553.

Rüde, T., Lu, Y., Anschütz, L., Blasius, M., Wolf, M., Preuster, P., Wasserscheid, P., Geißelbrecht, M. (2023). Performance of continuous hydrogen production from perhydro benzyltoluene by catalytic distillation and heat integration concepts with a fuel cell. *Energy Technology*, 11(3), 2201366.

Ryabchuk, P., Agapova, A., Kreyenschulte, C., Lund, H., Junge, H., Junge, K., Beller, M. (2019). Heterogeneous nickel-catalysed reversible, acceptorless dehydrogenation of N-Heterocycles for hydrogen storage. *Chemical Communications*, 55(34), 4969–4972.

Sachtler, J.W.A. and Somorjai, G.A. (1984). Cyclohexane dehydrogenation catalyzed by bimetallic Au-Pt(111) single-crystal surfaces. *Journal of Catalysis*, 89(1), 35–43.

Saeys, M., Reyniers, M.-F., Neurock, M., Marin, G.B. (2005). Ab Initio reaction path analysis of benzene hydrogenation to cyclohexane on Pt(111). *The Journal of Physical Chemistry B*, 109(6), 2064–2073.

Safronov, S.P., Vostrikov, S.V., Samarov, A.A., Wasserscheid, P., Müller, K., Verevkin, S.P. (2023). Comprehensive thermodynamic study of substituted indoles/perhydro indoles as potential liquid organic hydrogen carrier system. *Fuel*, 331, 125764.

Santacesaria, E., Carotenuto, G., Tesser, R., Di Serio, M. (2012). Ethanol dehydrogenation to ethyl acetate by using copper and copper chromite catalysts. *Chemical Engineering Journal*, 179, 209–220.

Sapianik, A.A., Kovalenko, K.A., Samsonenko, D.G., Barsukova, M.O., Dybtsev, D.N., Fedin, V.P. (2020). Exceptionally effective benzene/cyclohexane separation using a nitro-decorated metal-organic framework. *Chemical Communications*, 56(59), 8241–8244.

Schoenmaker-Stolk, M.C., Verwijs, J.W., Don, J.A., Scholten, J.J.F. (1987). The catalytic hydrogenation of benzene over supported metal catalysts: I. Gas-phase hydrogenation of benzene over ruthenium-on-silica. *Applied Catalysis*, 29(1), 73–90.

Scola, D.A., Adams, J.S., Tewksbury, C.I., Wineman, R.J. (1965). Path of partial hydrogenation of o-, m-, and p-terphenyl. *J. Org. Chem.*, 30(2), 384–388.

Sebastián, D., Bordejé, E.G., Calvillo, L., Lázaro, M.J., Moliner, R. (2008). Hydrogen storage by decalin dehydrogenation/naphthalene hydrogenation pair over platinum catalysts supported on activated carbon. *International Journal of Hydrogen Energy*, 33(4), 1329–1334.

Sekine, Y. and Higo, T. (2021). Recent trends on the dehydrogenation catalysis of liquid organic hydrogen carrier (LOHC): A review. *Topics in Catalysis*, 64, 470–480.

Shan, J., Janvelyan, N., Li, H., Liu, J., Egle, T.M., Ye, J., Biener, M.M., Biener, J., Friend, C.M., Flytzani-Stephanopoulos, M. (2017). Selective non-oxidative dehy-drogenation of ethanol to acetaldehyde and hydrogen on highly dilute NiCu alloys. *Applied Catalysis B: Environmental*, 205, 541–550.

Shao, Z., Li, Y., Liu, C., Ai, W., Luo, S.-P., Liu, Q. (2020). Reversible interconversion between methanol-diamine and diamide for hydrogen storage based on manganese catalyzed (de)hydrogenation. *Nature Communications*, 11(1), 591.

Sharma, B.L., Gupta, S., Tandon, S., Kant, R. (2008). Physico-mechanical properties of naphthalene-acenaphthene eutectic system by different modes of solidification. *Materials Chemistry and Physics*, 111(2), 423–430.

Shiau, C.-Y., Chen, S., Tsai, J.C., Lin, S.I. (2000). Effect of zinc addition on copper catalyst in isoamyl alcohol dehydrogenation. *Applied Catalysis A: General*, 198(1), 95–102.

Shibata, M., Nagata, R., Saito, S., Naka, H. (2017). Dehydrogenation of primary aliphatic alcohols by Au/TiO_2 photocatalysts. *Chem. Lett.*, 46(4), 580–582.

Shin, B.S., Yoon, C.W., Kwak, S.K., Kang, J.W. (2018). Thermodynamic assessment of carbazole-based organic polycyclic compounds for hydrogen storage applications via a computational approach. *International Journal of Hydrogen Energy*, 43(27), 12158–12167.

Shirai, M., Rode, C.V., Mine, E., Sasaki, A., Sato, O., Hiyoshi, N. (2006). Ring hydrogenation of naphthalene and 1-naphthol over supported metal catalysts in supercritical carbon dioxide solvent. *Catalysis Today*, 115(1), 248–253.

Shylesh, S., Kim, D., Ho, C.R., Johnson, G.R., Wu, J., Bell, A.T. (2015). Non-oxidative dehydrogenation pathways for the conversion of C_2-C_4 alcohols to carbonyl compounds. *ChemSusChem*, 8(23), 3959–3962.

Sisáková, K., Podrojková, N., Oriňaková, R., Oriňak, A. (2021). Novel catalysts for dibenzyltoluene as a potential liquid organic hydrogen carrier use. A mini-review. *Energy & Fuels*, 35(9), 7608–7623.

Song, H., Kang, B., Hong, S.H. (2014). Fe-catalyzed acceptorless dehydrogenation of secondary benzylic alcohols. *ACS Catalysis*, 4(9), 2889–2895.

Speight, J.G. (2017). Industrial organic chemistry. In *Environmental Organic Chemistry for Engineers*. Butterworth-Heinemann, Oxford.

Stark, K., Keil, P., Schug, S., Müller, K., Wasserscheid, P., Arlt, W. (2016). Melting points of potential liquid organic hydrogen carrier systems consisting of N-Al-kylcarbazoles. *Journal of Chemical & Engineering Data*, 61(4), 1441–1448.

Steinhauer, J., Bachmann, P., Freiberger, E.M., Bauer, U., Steinrück, H.-P., Papp, C. (2020). Model catalytic studies of liquid organic hydrogen carriers: Indole/indoline/octahydroindole on Ni(111). *The Journal of Physical Chemistry C*, 124(41), 22559–22567.

Stepanenko, S.A., Shivtsov, D.M., Koskin, A.P., Koskin, I.P., Kukushkin, R.G., Yeletsky, P.M., Yakovlev, V.A. (2022). N-Heterocyclic molecules as potential liquid organic hydrogen carriers: Reaction routes and dehydrogenation efficacy. *Catalysts*, 12(10), 1260.

Sun, D., Misu, T., Yamada, Y., Sato, S. (2019). Advantages of using Cu/SiO_2 catalyst for vapor-phase dehydrogenation of 1-decanol into decanal. *Applied Catalysis A: General*, 582, 117109.

Sung, J.S., Choo, K.Y., Kim, T.H., Tarasov, A.L., Tkachenko, O.P., Kustov, L.M. (2008). A new hydrogen storage system based on efficient reversible catalytic hydrogenation/ dehydrogenation of terphenyl. *International Journal of Hydrogen Energy*, 33(11), 2721–2728.

Tang, C., Fei, S., Lin, G.D., Liu, Y. (2020). Natural liquid organic hydrogen carrier with low dehydrogenation energy: A first principles study. *International Journal of Hydrogen Energy*, 45(56), 32089–32097.

Taube, M., Rippin, D.W.T., Cresswell, D.L., Knecht, W. (1983). A system of hydrogen-powered vehicles with liquid organic hydrides. *Int. J. Hydrogen Energy*, 8, 213–225.

Tran, B.L., Johnson, S.I., Brooks, K.P., Autrey, S.T. (2021). Ethanol as a liquid organic hydrogen carrier for seasonal microgrid application: Catalysis, theory, and engineering feasibility. *ACS Sustainable Chemistry & Engineering*, 9(20), 7130–7138.

Trincado, M., Bösken, J., Grützmacher, H. (2021). Homogeneously catalyzed acceptor-less dehydrogenation of alcohols: A progress report. *Coordination Chemistry Reviews*, 443, 213967.

Tseng, K.-N.T., Rizzi, A.M., Szymczak, N.K. (2013). Oxidant-free conversion of primary amines to nitriles. *Journal of the American Chemical Society*, 135(44), 16352–16355.

Tuo, Y., Meng, Y., Chen, C., Lin, D., Feng, X., Pan, Y., Li, P., Chen, D., Liu, Z., Zhou, Y. et al. (2021). Partial positively charged Pt in $Pt/MgAl_2O_4$ for enhanced dehydrogenation activity. *Applied Catalysis B: Environmental*, 288, 119996.

Van Der Steen, P.J. and Scholten, J.J.F. (1990). Selectivity to cyclohexene in the gas phase hydrogenation of benzene over ruthenium, as influenced by reaction modifiers: II. Catalytic hydrogenation of benzene to cyclohexene and cyclohexane. *Applied Catalysis*, 58(1), 291–304.

Ventura-Espinosa, D., Marzá-Beltrán, A., Mata, J.A. (2016). Catalytic hydrogen production by ruthenium complexes from the conversion of primary amines to nitriles: Potential application as a liquid organic hydrogen carrier. *Chemistry – A European Journal*, 22(49), 17758–17766.

Verevkin, S.P. (1997). Thermochemistry of substituted benzenes. Experimental standard molar enthalpies of formation of o-,m-, and p-terphenyls and 1,3,5-triphenylbenzene. *The Journal of Chemical Thermodynamics*, 29(12), 1495–1501.

Völter, J., Hermann, M., Heise, K. (1968). Comparative hydrogenation and adsorption of benzene and methylbenzenes on cobalt and rhodium catalysts. *Journal of Catalysis*, 12(3), 307–313.

Wan, C., An, Y., Chen, F., Cheng, D., Wu, F., Xu, G. (2013). Kinetics of N-ethylcarba-zole hydrogenation over a supported Ru catalyst for hydrogen storage. *International Journal of Hydrogen Energy*, 38(17), 7065–7069.

Wang, Y. and Bai, X. (2023). Efficient catalytic hydrogen storage of N-ethylcarbazole over RuNi alloy nanoparticles loaded on SBA-15 prepared by electrostatic adsorption. *Fuel*, 331(1), 125709.

Wang, Z.-T., Xu, Y., El-Soda, M., Lucci, F.R., Madix, R.J., Friend, C.M., Sykes, E.C.H. (2017). Surface structure dependence of the dry dehydrogenation of alcohols on Cu(111) and Cu(110). *The Journal of Physical Chemistry C*, 121(23), 12800–12806.

Weitkamp, A.W. (1968). Stereochemistry and mechanism of hydrogenation of naphthalenes on transition metal catalysts and conformational analysis of the products. *Advances in Catalysis*, 18, 1–110.

Wunsch, A., Mohr, M., Pfeifer, P. (2018). Intensified LOHC-dehydrogenation using multi-stage microstructures and Pd-based membranes. *Membranes*, 8(4), 112.

Wunsch, A., Berg, T., Pfeifer, P. (2020). Hydrogen production from the LOHC per-hydro-dibenzyl-toluene and purification using a 5 μm PdAg-membrane in a coupled microstructured system. *Materials*, 13(2), 277.

Xia, Z., Lu, H., Liu, H., Zhang, Z., Chen, Y. (2017). Cyclohexane dehydrogenation over Ni-Cu/SiO_2 catalyst: Effect of copper addition. *Catalysis Communications*, 90, 39–42.

Yadav, V., Sivakumar, G., Gupta, V., Balaraman, E. (2021). Recent advances in liquid organic hydrogen carriers: An alcohol-based hydrogen economy. *ACS Catal.*, 11(24), 14712–14726.

Yamaguchi, R., Ikeda, C., Takahashi, Y., Fujita, K.-I. (2009). Homogeneous catalytic system for reversible dehydrogenation-hydrogenation reactions of nitrogen heterocycles with reversible interconversion of catalytic species. *Journal of the American Chemical Society*, 131(24), 8410–8412.

Yang, M., Dong, Y., Fei, S., Pan, Q., Ni, G., Han, C., Ke, H., Fang, Q., Cheng, H. (2013). Hydrogenation of N-propylcarbazole over supported ruthenium as a new prototype of liquid organic hydrogen carriers (LOHC). *RSC Advances*, 3(47), 24877–24881.

Yang, M., Xing, X., Zhu, T., Chen, X., Dong, Y., Cheng, H. (2020). Fast hydrogenation kinetics of acridine as a candidate of liquid organic hydrogen carrier family with high capacity. *Journal of Energy Chemistry*, 41, 115–119.

Yang, Y., Yao, J., Wang, H., Yang, F., Wu, Z., Zhang, Z. (2022). Study on high hydrogen yield for large-scale hydrogen fuel storage and transportation based on liquid organic hydrogen carrier reactor. *Fuel*, 321, 124095.

Ye, X., An, Y., Xu, G. (2011). Kinetics of 9-ethylcarbazole hydrogenation over Raney-Ni catalyst for hydrogen storage. *Journal of Alloys and Compounds*, 509(1), 152–156.

Yi, J., Miller, J.T., Zemlyanov, D.Y., Zhang, R., Dietrich, P.J., Ribeiro, F.H., Suslov, S., Abu-Omar, M.M. (2014). A reusable unsupported rhenium nanocrystalline catalyst for acceptorless dehydrogenation of alcohols through γ-C-H activation. *Angewandte Chemie International Edition*, 53(3), 833–836.

Yolcular, S. and Olgun, O. (2008). Ni/Al_2O_3 catalysts and their activity in dehydrogenation of methylcyclohexane for hydrogen production. *Catal. Today*, 138, 198–202.

Zhang, G., Vasudevan, K.V., Scott, B.L., Hanson, S.K. (2013). Understanding the mechanisms of cobalt-catalyzed hydrogenation and dehydrogenation reactions. *Journal of the American Chemical Society*, 135(23), 8668–8681.

Zhang, M., Zhao, Y., Liu, Q., Yang, L., Fan, G., Li, F. (2016). A La-doped Mg-Al mixed metal oxide supported copper catalyst with enhanced catalytic performance in transfer dehydrogenation of 1-decanol. *Dalton Transactions*, 45(3), 1093–1102.

Zhang, Y.-Q., Stolte, S., Alptekin, G., Rother, A., Diedenhofen, M., Filser, J., Markiewicz, M. (2020). Mobility and adsorption of liquid organic hydrogen carriers (LOHCs) in soils – Environmental hazard perspective. *Green Chemistry*, 22(19), 6519–6530.

Zhou, L., Sun, L., Xu, L., Wan, C., An, Y., Ye, M. (2020). Recent developments of effective catalysts for hydrogen storage technology using N-ethylcarbazole. *Catalysts*, 10(6), 648.

Zhukova, A.I., Chuklina, S.G., Maslenkova, S.A. (2021). Study of Cu modified Zr and Al mixed oxides in ethanol conversion: The structure-catalytic activity relationship. *Catalysis Today*, 379, 159–165.

Zou, Y.-Q., Zhou, Q.-Q., Diskin-Posner, Y., Ben-David, Y., Milstein, D. (2020). Synthesis of oxalamides by acceptorless dehydrogenative coupling of ethylene glycol and amines and the reverse hydrogenation catalyzed by ruthenium. *Chemical Science*, 11(27), 7188–7193.

Zubar, V., Borghs, J.C., Rueping, M. (2020). Hydrogenation or dehydrogenation of N-containing heterocycles catalyzed by a single manganese complex. *Organic Letters*, 22(10), 3974–3978.

4

Liquid Storage: Ammonia

Nicolas BION[1], Fabien CAN[1], Charlotte CROISÉ[1,2], Xavier COURTOIS[1] and Mohamed El Amine KRIBÉCHE[3,4]

[1]*IC2MP, CNRS, Université de Poitiers, France*

[2]*ICMCB, CNRS, Université de Bordeaux, Pessac, France*

[3]*IEM, CNRS, Université de Montpellier, France*

[4]*ETE France, Saint-Laurent-de-la-Salanque, France*

4.1. Introduction

Today, the vast majority of the world's energy needs are met by fossil fuels, such as coal, oil and natural gas. According to the International Energy Agency (IEA), their production accounts for 85% of the global energy demand. Despite the warning given by the Intergovernmental Panel on Climate Change (IPCC), global carbon dioxide emissions from fossil fuels continue to rise.

According to the IEA, the rise in global CO_2 emissions in 2021, estimated at over two billion tonnes, was the largest in history, largely offsetting the decline induced by the previous year's pandemic. Faced with these challenges, the decarbonizing industry is now a priority for geopolitical, economic and ecological reasons. The climatic consequences of the greenhouse effect and atmospheric pollution induced by anthropogenic carbon dioxide emissions call for a reduction in the use of fossil fuels of at least 70% to keep the rise in global temperatures below 2°C by 2100 (the objective of the Paris Agreement). The need to find substitutes is therefore a major challenge.

Green hydrogen is a promising option that is widely regarded as an alternative energy source that could play a vital role in the energy transition. It is a clean,

Hydrogen Storage,
coordinated by Patricia DE RANGO and Fermin CUEVAS.

environmentally friendly energy carrier (nonemitting greenhouse gases). With a very high net calorific value (NCV) (120.9 kJ.g^{-1} vs. approximately 44.5 kJ.g^{-1} for octane, for example), hydrogen is also characterized by a low energy density at ambient pressure and temperature (9.76 kJ.L^{-1} (gas) vs. 31,094 kJ.L^{-1} for liquid octane). This characteristic is a real handicap when it comes to storage and transport, as it requires very large volumes. It is therefore necessary to store hydrogen at a high density to make this molecule a viable fuel. Several methods of dense storage of hydrogen in the molecular state, in the form of compressed gas at high pressure (700 bar) or in liquefied form (T_{liq} = −253°C at 1 bar), are being considered, but they still face major problems of safety and energy costs linked to compression and low temperature.

Storing hydrogen in the form of ammonia is a less restrictive, more efficient and more environmentally friendly alternative. Ammonia (NH_3) contains 17.6% hydrogen by mass and is carbon free. It is easily liquefied. Its H_2 storage capacity makes it one of the most promising hydrogen carriers. Figure 4.1 shows the storage volume for 1 kg of hydrogen using the different techniques mentioned above.

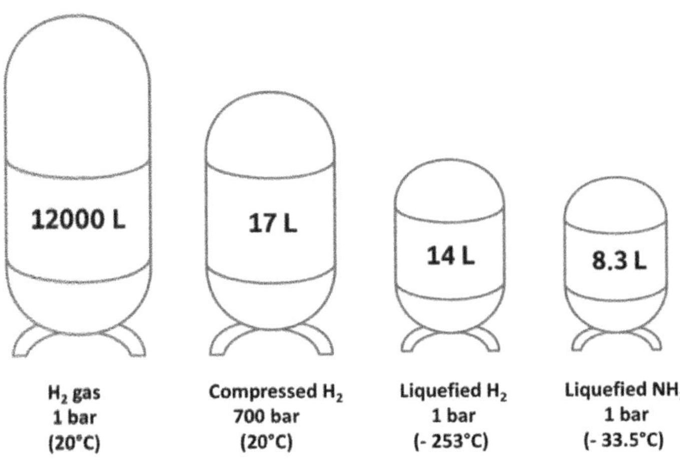

Figure 4.1. *Storage volume for 1 kg of hydrogen using different techniques*

4.2. Ammonia as an energy carrier

Ammonia is a particularly interesting molecule for mobile energy applications, either as a chemical carrier for hydrogen storage or directly as a fuel. It is one of the only potential fuels that contains no carbon and therefore generates no carbon dioxide during combustion. The molecular structure of NH_3 clearly indicates advantageous properties for hydrogen storage. With a hydrogen density of 136 kg.m^{-3}, liquid ammonia is one of the most hydrogen-dense fuels (17.6 wt%) (Itoh et al. 2014; Meng and Tsuru 2016). It is easily liquefied (by compression to 9.8 atm at 25°C or without compression to 1 atm at –33.35°C) and can therefore be transported in inexpensive, low-pressure containers. Although toxic and corrosive, ammonia is not explosive, except in a restricted mixing range of NH_3 and air (15%–28%). It is also nonflammable.

Ammonia has been abundantly produced by the industry for over a century using the Haber–Bosch process, which is based on the reaction between H_2 and N_2 at high temperatures and pressures (400–600°C; 20–40 MPa). This process currently supplies 180 Mt/year of ammonia, which is used mainly for agricultural applications, whether in the form of ammonia, ammonium nitrate or urea (Humphreys et al. 2021). The global consumption of ammonia is rising steadily as the population increases to ensure affordable food production. In the usual Haber–Bosch process, fossil fuels such as natural gas and coal are used as energy sources and hydrogen providers, releasing millions of tonnes of CO_2 into the atmosphere (Humphreys et al. 2021). Despite high energy consumption (28–166 GJ per tonne of NH_3, representing more than 1% of global energy production), process complexity and CO_2 emissions (1.87 tonnes per tonne of NH_3), there is still no industrial alternative to the Haber–Bosch process (Liu 2014; Foster et al. 2018).

Although other promising technologies based on the use of low-carbon renewable energy sources have been proposed to produce green ammonia (electrochemical and photocatalytic synthesis), the Haber–Bosch process remains the dominant technology for industrial ammonia synthesis in the medium term (Humphreys et al. 2021). From this standpoint, the development of better catalysts for a more virtuous synthesis is essential.

With an extensive distribution and storage network already in place (pipelines, trucks, ships, warehouses, etc.), ammonia is an attractive candidate for solving the problems of storing, transporting and distributing hydrogen for mobile or stationary use at multiple scales. It could be used as a primary energy source with zero CO_2 emissions or as a source of hydrogen for power generation; land, sea and air transport; and possibly for heating.

For a transition to a carbon-neutral energy system, ammonia production must be decarbonized. According to a report entitled “Green Ammonia: An Alternative Fuel”, published by FutureBridge, the cost of producing green ammonia from renewable energy sources ranges from \$700 to \$1,400 per tonne. By 2030, it is expected to fall to \$480 per tonne, then to \$310 per tonne by 2050. Moreover, the average cost of fossil-based ammonia has increased from €300 to €350/t over the past decade. Thus, green ammonia could become competitive on the markets and replace fossil-based ammonia on a global scale (Fasihi et al. 2021).

Furthermore, geopolitics can accelerate the phenomenon. A BloombergNEF analysis, echoed by a number of specialist media, concluded that the sharp rise in natural gas prices following the Russian invasion of Ukraine had made the cost of green ammonia more competitive than that of conventional ammonia (gray ammonia) by March 2022.

In a nutshell, ammonia is globally nonflammable and is easily obtained in liquid form without the need for costly and complex refrigeration technology, making it easy to handle. Ammonia therefore offers significant cost and convenience advantages because of its high hydrogen density and ease of storage and distribution. Table 4.1 shows the energy density values of conventional fuels that can be used to generate hydrogen.

Fuel	H_2 mass %	Energy density by mass (Wh/kg)	Volumetric energy density (Wh/L)
Ammonia	17.6	4,318	4,325
Methanol	12.5	6,400	4,600
Ethanol	13.0	7,850	6,100
Hydrogen (700 bar)	100	39,000	1,305
Diesel	15.8	12,200	9,700

Table 4.1. *Dihydrogen mass content and energy density of conventional fuels capable of generating hydrogen (Damle 2009; Chiuta et al. 2013)*

The hydrogen produced by decomposition of the ammonia carrier is intended for use in fuel cells to produce electrical energy or directly in internal combustion engines for mobile and stationary applications.

4.3. Compatibility with fuel cells

The decarbonization of our energy sources undoubtedly requires the replacement of gray hydrogen derived from fossil fuels with green hydrogen produced from renewable energies. It is therefore necessary to examine the operational obstacles that could hinder the prospects of hydrogen production via ammonia decomposition. The challenges ahead are considerable.

The low tolerance of proton exchange membrane (PEM) fuel cells to residual ammonia (0.1 ppm) may be the main operational challenge for this technology. Specifically, a residual ammonia concentration above the specified limit presents a high risk of poisoning the catalyst used at the anode, with a consequent loss of cell performance. Halseid et al. (2006) observed a significant drop in the performance of a PEM fuel cell when supplied with hydrogen containing 10 ppm ammonia. The poisoning process was relatively slow, taking over 24 h to reach the steady-state performance of the cell. On the other hand, Halseid et al. (2006) reported that this loss of performance was reversible in most cases but only after operation with pure hydrogen for several days.

Although no study has proven this hypothesis, high-temperature fuel cells (HT-PEMs) operating over a temperature range from 80°C to 200°C would be characterized by greater resistance to ammonia poisoning than low-temperature cells operating at temperatures below 80°C (Halseid et al. 2004). If this proves to be true, HT-PEM cells could be adapted to the use of hydrogen produced by ammonia decomposition. These cells have already demonstrated significant resistance to carbon monoxide (CO) poisoning. They operate with fuel containing up to 5% CO, which is 5,000 times greater than the 10 ppm CO limit applicable to low-temperature fuel cells (Das et al. 2009).

On the other hand, alkaline fuel cells (AFCs) can operate efficiently with hydrogen from ammonia decomposition without any need for purification (Hejze et al. 2008). These behaviors significantly reduce the threat posed by residual ammonia when NH_3 decomposition is implemented for fuel cell applications (Chiuta et al. 2013). In addition, some commercially available adsorbents, such as AmmoSorb™, have been shown to significantly reduce the residual ammonia concentration to 1 ppb (Choudhary et al. 2001; Chellappa et al. 2002; Alagharu et al. 2010).

However, the potential lifetime of these adsorbents for fuel cell operation has not yet been studied (Chiuta et al. 2013). Finally, solid oxide fuel cells (SOFCs) appear to be the most suitable and efficient for operating directly with ammonia (Giddey et al. 2017). The advantage lies in the high operating temperature (800°C) of these cells, which eliminates the need for prior ammonia decomposition (Rathore et al. 2021).

4.4. Ammonia synthesis

4.4.1. *The Haber–Bosch process*

From 1787, when C.L. Berthollet discovered that ammonia was composed of nitrogen and hydrogen, a great deal of research went into synthesizing the molecule. In the 19th century, ammonia was obtained from natural saltpetre deposits in Chile or recovered as a by-product of the coal industry. In the 20th century, several methods for fixing nitrogen from the air, such as the electric arc method or the cyanamide process, were tested to meet the growing demand for ammonia. Ultimately, the catalytic synthesis of ammonia from dinitrogen and dihydrogen was the most cost-effective. Currently, approximately 85% of the world's total ammonia production comes from this process (Aziz et al. 2020).

The synthesis of ammonia from dihydrogen and dinitrogen is exothermic and proceeds according to the following reaction (see equation [4.1]):

$$N_2 + 3\,H_2 \leftrightarrows 2\,NH_3 \qquad [4.1]$$

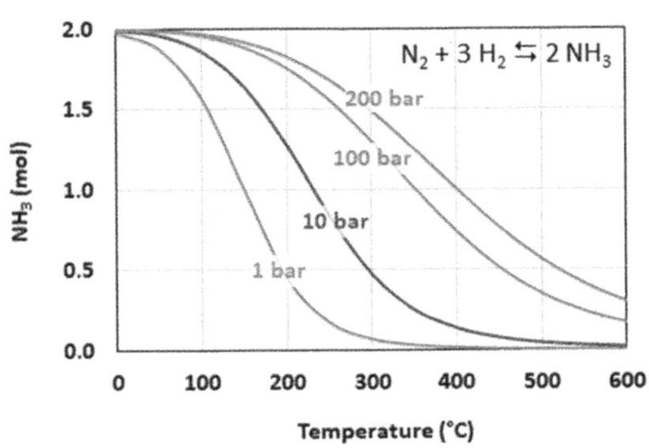

Figure 4.2. *Thermodynamic equilibrium of the $N_2 + 3H_2 \leftrightarrows 2\,NH_3$ reaction with ammonia as a function of temperature and pressure. For a color version of this figure, see www.iste.co.uk/derango-cuevas/hydrogen.zip*

As illustrated in Figure 4.2, this is a reversible reaction that is thermodynamically favored at low temperature and high pressure (Wang et al. 2019). However, the reaction rate at room temperature tends to be zero. With respect to industrial application, it is therefore necessary to increase the reaction kinetics by working in the presence of a catalyst at higher temperatures and consequently at higher pressures to achieve acceptable conversion rates (Aziz et al. 2020).

In 1908, the German chemist Fritz Haber developed a method for the synthesis of ammonia from dinitrogen and dihydrogen under high pressure. To overcome the problems of low conversion rates, Haber abandoned the then-majority-used static reaction method for a dynamic reaction (Tamaru 1991; Liu 2014).

In February 1908, Haber signed an agreement with Baden Aniline and Soda Company (BASF), which entrusted the task of industrial development to Carl Bosch. Industrialization was a major challenge, since a high-performance catalyst had to be developed that could operate at the lowest possible temperatures and pressures. Carl Bosch and his team succeeded in developing a commercial process in less than five years (Liu 2013; Humphreys et al. 2021). Fritz Haber and Carl Bosch were awarded a Nobel Prize in Chemistry for their work in 1918 and 1931, respectively.

The catalyst had to be economical, efficient and durable, resisting not only the temperature and pressure ranges used but also any impurities present in the feed gases. Fritz Haber's initial work had been carried out with osmium and uranium, but these two elements are in short supply and therefore difficult to obtain. Mittasch and Frankenburg (1950) carried out studies on catalysts, and in 1909, Mittasch hypothesized that "the winning catalyst must be a multicomponent system". Numerous tests were necessary to determine the best catalyst: between 1909 and 1922, 5,000 catalysts were synthesized, and 20,000 tests were carried out. Mittasch worked in particular on an iron-based catalyst. It was subsequently demonstrated that the addition of small quantities of dopants had a positive effect on catalyst performance. Currently, industrial catalysts for ammonia synthesis are still based on this principle.

The first ammonia production plant was built in 1911 in Oppau, Germany. In 1913, the plant was equipped with a 285 mm diameter reactor and 90 L of catalyst, with an ammonia synthesis reaction carried out at 200 bar. Initially, ammonia production was approximately 3–5 tonnes per day. The production scale rapidly expanded, reaching 30 tons of ammonia per day in 1915 and 230 tons in 1917 (Tamaru 1991; Liu 2013).

After a century of development, the catalytic synthesis of ammonia has progressed enormously. In the current Haber–Bosch industrial process, catalytic synthesis is achieved by passing a gaseous mixture of dihydrogen and dinitrogen over an iron-based catalyst promoted with K_2O and Al_2O_3, which is kinetically active at high temperatures (between 400 and 600°C). As thermodynamics are not favorable at these temperatures, high pressures (100–200 bar) are used to obtain sufficient ammonia yields (Nørskov et al. 2016; Marakatti and Gaigneaux 2020). An ammonia production furnace contains 100 tonnes of catalyst in the form of 1.5–20 mm grains, with a service life of up to 10 years. The typical composition of an industrial catalyst is described in section 4.4.2.

Many advances have been made in the catalytic synthesis of ammonia, but the process in modern ammonia production plants is much the same as that developed by Haber and Bosch. The main steps are as follows (Liu 2013; Amhamed et al. 2022):

– Hydrogen production: natural gas is the essential source of dihydrogen for ammonia synthesis. The hydrogen production stages are as follows:

- First, the sulfur compounds present in the natural gas are removed to avoid poisoning the catalysts for H_2 production and NH_3 synthesis. Catalytic hydrogenation is first carried out at 350–400°C in the presence of a molybdenum–cobalt or molybdenum–nickel catalyst to transform all of the sulfur compounds into gaseous hydrogen sulfide (H_2S). The hydrogen sulfide is then fixed at 400°C on zinc oxide, where it is converted into solid zinc sulfide.

- Natural gas is used to produce synthesis gas (syngas, a mixture of dihydrogen and carbon monoxide) by steam-reforming natural gas at 800–900°C and 3.3 MPa in the presence of a nickel oxide catalyst supported on alumina according to equation [4.2]:

$$CH_4 + H_2O \rightarrow CO + 3H_2 \quad [4.2]$$

- After reforming, the syngas mixture contains between 5% and 11% by volume of unconverted methane.

- The carbon monoxide in the syngas is then converted catalytically into carbon dioxide with additional production of dihydrogen according to the water–gas shift reaction, equation [4.3]:

$$CO + H_2O \rightarrow CO_2 + H_2 \quad [4.3]$$

- This water–gas shift reaction is carried out either at 320–370°C with an iron oxide/chromium oxide catalyst with copper metal addition or at 200–240°C with a copper oxide/chromium oxide/zinc oxide catalyst on alumina. A gas containing 70% hydrogen is then obtained.

- Carbon dioxide is usually removed by dissolving it in an amine solution at 35 bar.

- The last residual traces of carbon monoxide or carbon dioxide are eliminated by catalytic methanation according to equations [4.4] and [4.5]:

$$CO + 3H_2 \rightarrow CH_4 + H_2O \qquad [4.4]$$

$$CO_2 + 4H_2 \rightarrow CH_4 + 2H_2O \qquad [4.5]$$

- Secondary reforming or postcombustion is then carried out. At this stage, the syn-gas contains 5–10% unconverted methane. For this stage, air is added and the oxygen in the air oxidizes the remaining methane using a nickel oxide catalyst.

– Nitrogen production: nitrogen is either taken directly from the air used in the previous reaction and remaining in the gas mixture for ammonia synthesis or from an air separation unit.

– Compression of the gas mixture: the gas, composed of dihydrogen and nitrogen, is compressed by piston compressors to the pressure used in the synthesis unit.

– Ammonia synthesis and separation: the gas mixture is introduced into the reactor (T = 400–600°C, P = 100–200 bar) to form ammonia using the catalyst. This gas has previously been heated by the hot gases from the reactor. After the reaction, the ammonia-containing gas reaches the heat exchanger. The ammonia is condensed into a liquid (critical point: 132°C, 113 bar; the condensation temperature at 100 bar is approximately 120°C) and separated from the reaction mixture. Since conversion in a single pass does not exceed 20%, a large proportion of the gas must be recycled and returned to the synthesis loop.

The catalytic ammonia synthesis technology invented by Haber and Bosch is still used for ammonia production. Since 1913, enormous progress has been made in ammonia production plants, with the most recent units capable of producing between 1,000 and 3,200 tonnes per day (Liu 2014), even reaching over 10,000 tonnes per day for the largest ammonia production complex located in Donaldsonville, Louisiana (CF industries n.d.). The reaction conditions have also been somewhat improved. However, they are still energy intensive, and the high global demand for ammonia means that its production is responsible for 1–2% of total energy consumption worldwide and 1.6% of global CO_2 emissions. Saving energy and reducing gas emissions have always been, and still are, major challenges for the ammonia industry. The main difficulty in synthesizing ammonia is the activation of nitrogen molecules. Indeed, the dissociation energy of the N≡N triple bond is very high, at 942 $kJ.mol^{-1}$.

4.4.2. *Ammonia synthesis catalysts*

4.4.2.1. *Iron catalysts*

Initially, the iron catalyst selected in 1913 by Haber was chosen from approximately 5,000 tested formulations. It consists of molten Fe_3O_4 iron oxide with the addition of Al_2O_3 to increase the surface area and is promoted by basic oxides such as K_2O and CaO, which are electron donors. The evolution of the catalysts used in the Haber–Bosch process has been the subject of ongoing research throughout the 20th century. In the 1980s, another important development was the use of $Fe_{1-x}O$ iron oxide with a Wüstite-type structure. A typical industrial catalyst composition contains 40.5$\%_{at}$ Fe, 0.35$\%_{at}$ K, 2.0$\%_{at}$ Al, 1.7$\%_{at}$ Ca and 53.2$\%_{at}$ O (Ertl 1983).

During reducing, the catalyst takes the form of core–shell particles, with a core of magnetite (Fe_3O_4), a shell of Wüstite (FeO) and an outer layer of metallic iron, as shown in Figure 4.3 (Schlögl 1991). Thus, modern industrial catalysts are based on nanostructured iron with a metastable structure formed via complex synthesis from an oxide precursor (Holme et al. 1997).

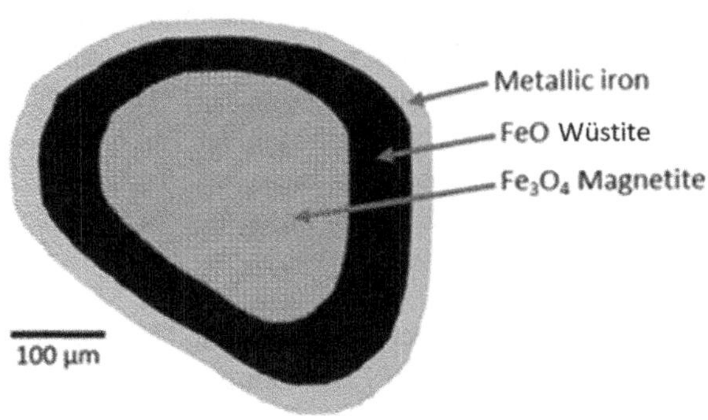

Figure 4.3. *Schematic representation of an iron catalyst for ammonia synthesis. Adapted from Schlögl (1991)*

The 1980s also saw the advent of cobalt-containing cocatalysts developed by the British company ICI (European patent application No. 78300276.9). This represents a significant development over traditional Fe_3O_4-based catalysts.

However, even after a century of research, the process using an iron-based catalyst is still commercially exploited. Despite some improvements, the lack of very significant progress for this family of materials highlighted the need to develop completely different alternative catalysts. At the end of the 20th century, a new family of ruthenium-based catalysts, classically supported on carbon, emerged.

These materials can be considered second-generation catalysts for NH_3 production (Raróg-Pilecka et al. 2006).

4.4.2.2. *Ruthenium catalysts*

The development of ruthenium-based catalysts for ammonia synthesis has a long history (Bielawa et al. 2001). The first works were published in 1917, although it was not until the 1970s that interest in ruthenium reappeared (Aika et al. 1972). Ruthenium-based catalysts are conventionally supported on carbon, and their activity can be considerably enhanced by doping with alkaline (Cs) or alkaline-earth (Ba) oxides (Aika et al. 1985; Aika 1986), which have electrodonic characteristics. This doping results in electron enrichment of the ruthenium, promoting electron transfer to the antibonding π^* orbitals of N_2, leading to a weakening of the N≡N bond and favoring its breakage.

A process was industrialized in 1992, using a ruthenium-based catalyst deposited by carbonyl sublimation on a carbon graphite support doped with barium and potassium (Forni et al. 1999). According to U.S. Patent No. 4250057, the catalyst must contain 5–10 wt% ruthenium.

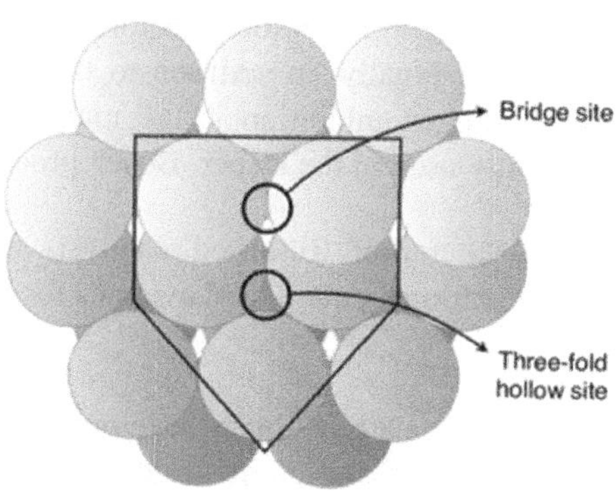

Figure 4.4. *Schematic representation of ruthenium B5 sites active for N≡N bond disruption at low temperatures. Reproduced with permission from Fernández et al. (2016). Copyright 2016 Elsevier Inc.*

Detailed studies have shown that the B_5 sites on Ru nanoparticles (Figure 4.4) are the active centers for N≡N bond breaking at low temperatures and control the reaction rate (Honkala et al. 2005; Raróg-Pilecka et al. 2005; Fernández et al. 2016). As a result, NH_3 synthesis can then be considered structure sensitive, meaning that a reaction for which particular sites on the catalyst surface will have a few orders of magnitude higher activity than other sites (Jacobsen et al. 2000).

The fraction of B_5 sites depends mainly on the ruthenium crystal size and support morphology, but the effect of the metal's crystal structure remains debated. In all supported catalysts, ruthenium crystallizes in a compact hexagonal structure rather than in a metastable face-centered cubic structure. Recently, with suitable supports such as $Ca(NH_2)_2$, a face-centered cubic crystal structure could be obtained, favoring the formation of more efficient B_5 sites for NH_3 synthesis (Inoue et al. 2016).

Controlling the face-centered cubic structure of supported ruthenium with different solutions in terms of synthesis methods is another research topic of interest (Kusada et al. 2013; Ghosh and Jagirdar 2018). High-surface-area (83 m^2/g) Ru nanosponges with a face-centered cubic crystal structure were developed using a particular reducing agent (sodium borohydride) and showed excellent activities for hydrogenation reactions (Ghosh and Jagirdar 2018).

As a result, a wide variety of doped ruthenium catalysts have been developed (Saadatjou et al. 2015). Among them, Cs-Ru/MgO is considered a reference catalyst with excellent yield in NH_3 synthesis (Szmigiel et al. 2002; Larichev et al. 2007).

However, these ruthenium catalysts suffer from severe hydrogen poisoning at the metal surface under the high reaction pressures required to achieve a satisfactory NH_3 yield (Kitano et al. 2012). Finally, in addition to the high cost of ruthenium compared with iron, this hydrogen inhibition is the main reason why iron-based catalysts are still predominantly used in the Haber–Bosch process today.

Although these second-generation Ru-based catalysts are more active than iron-based catalysts, their limitations remain, and avenues for improvement continue to be explored.

With this in mind, new materials have been developed in recent years, such as electrides, hydrides, nitrides, oxides and oxyhydride-nitrides, always with the aim of milder reaction conditions. Moreover, until recently, only Ru B_5 sites were considered active for N_2 activation.

However, ruthenium-based catalysts with particular electronic and geometrical structures but without B_5 sites have also shown significant activity for NH_3 synthesis. These include compounds such as Ru_2Y or LaRuSi. Structures in which ruthenium is in the form of isolated atoms, such as an exchanged zeolite or even a Ru-substituted ABO_3 perovskite-type structure, are also being developed, highlighting that the presence of the B_5 site is not always essential (Wang et al. 2018; Qiu et al. 2019).

4.4.2.3. *Nitride phases*

Owing to their N_2 adsorption properties, nitrides are natural candidates for ammonia synthesis. Over 20 years ago, ternary nitrides have been shown to be more active and stable than binary nitrides. These inexpensive materials have the advantage of being able to activate and exchange N_2 from the gas phase with the nitrogen in their crystal structure, as shown in Figure 4.5 (Hunter et al. 2013), opening new catalytic prospects (Hargreaves 2014). The most common formulation is Co_3Mo_3N, for which approximately 20% of the nitrogen atoms in the lattice can be involved in NH_3 formation (Hunter et al. 2010). Cesium doping limits NH_3 poisoning of the material (Boisen et al. 2002). Furthermore, activity in NH_3 synthesis can be promoted by supporting Co_3Mo_3N nanoparticles on CeO_2 (a 20-fold increase in activity compared with that of mass Co_3Mo_3N) (Tsuji et al. 2016).

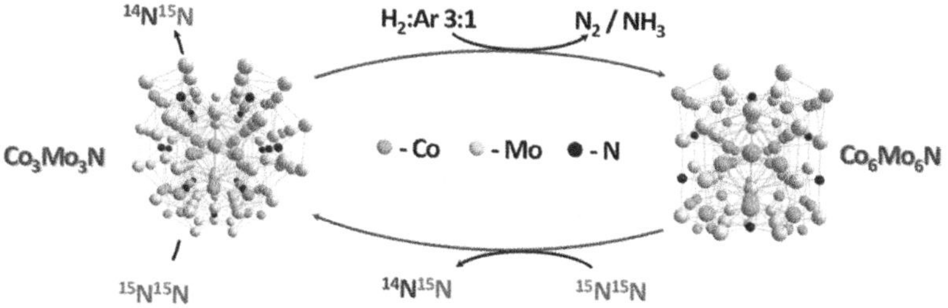

Figure 4.5. *Schematic representation of the lability of the –N atoms in the crystal lattice of the Co_3Mo_3N nitride phase. These atoms are exchangeable with molecular nitrogen at 400°C and can revert with H_2 to form NH_3. Adapted from Hunter et al. (2013). For a color version of this figure, see www.iste.co.uk/derango-cuevas/hydrogen.zip*

Several quaternary and ternary nitrides containing, for example, nickel or iron have also been newly developed, but they have not achieved the performance of Co_3Mo_3N. New methodologies for the synthesis of ternary nitrides, integrated into a support with a nanometric size, could improve the properties of these catalysts compared with those of bulk nitrides, with a new quantum confinement effect (Tsuji et al. 2016). The synthesis conditions for bulk Co_3Mo_3N are such that it is difficult to obtain this nitride phase with a specific surface area greater than 10 $m^2\ g^{-1}$. The confinement of these materials in mesoporous matrices is one of the interesting strategies for increasing the active surface area. The SBA-15 support, for example, was chosen for its good thermal stability, which resists the ammonolysis reaction, its large specific surface area and the uniformity of its mesopores. CoMo-N/SBA-15 catalysts containing 10, 20 or 30 wt% Co_3Mo_3N can thus achieve NH_3 yields of 1714, 1429 and 810 μmol $g_{phase\ active}^{-1}\ h^{-1}$, respectively, compared with

259 μmol g^{-1} h^{-1} for the unsupported phase under the same reaction conditions (Sfeir et al. 2023).

An important advancement in our understanding of how nitride catalysts work comes from theoretical calculations showing that nitrogen vacancy sites present on the surface would enable N_2 activation via a Mars–van Krevelen mechanism (Hunter et al. 2013; Zeinalipour–Yazdi et al. 2016). These results have inspired studies into the development of a bifunctional nitride catalyst whose vacancies enable N_2 activation, whereas another nearby site facilitates dissociative H_2 activation.

This is the case with the Ni/LaN compound, which has the advantage of containing no noble metal while exhibiting catalytic performance at 400°C and 1 bar that is far superior to all other Ni catalysts and most Ru catalysts (Ye et al. 2020). This synergistic effect is also present in Ni/CeN and Ni/YN (Ye et al. 2020). At 400°C and 9 bar, both N_2 and H_2 reagents could be activated on the vacancies of the CeN, LaN and YN supports, generating substantial amounts of NH_3. A concerted mechanism combining associative and dissociative mechanisms is then proposed for the Co/CeN catalyst, where N_2 can be activated on both Co and CeN vacancies (Ye et al. 2021).

4.4.2.4. *Electride phases*

Electride materials have specific properties linked to the existence of zones of high electron density within them, as shown in particular by theoretical approaches. We also speak of anionic electrons in the sense that an electride is a solid in which electrons, which have become detached from their original atom without being directly transferred to another atom as in an ionic bond, occupy interstitial sites between cations. These excess electrons are thus relatively free to diffuse. They are then easy to extract, resulting in solids with low work functions (Liu et al. 2020). The first of these materials to show interesting performance in ammonia synthesis is the C12A7:e^- compound. It is obtained from the inorganic oxide $12CaO \cdot 7Al_2O_3$ (C12A7), which consists of 12 cages per mesh, within which O^{2-} anions are present to compensate for the positive charge of the structure.

During chemical reduction, oxygen is extracted from the structure in the form of molecular oxygen, leaving excess electrons in the cages and giving the solid its electride character. Thus, in 2012, the association of Ru with this compound increased the activity of ruthenium catalysts (Cs@Ru/MgO or Ru/activated carbon) for the ammonia synthesis reaction by an order of magnitude (Kitano et al. 2012). The activation energy of the reaction is greatly decreased (< 60 kJ/mol) because of the retro-donation of anionic electrons to the ruthenium particles, weakening the N≡N bond and facilitating its breakage. Concomitantly, the material's cages are able to accommodate chemisorbed H adatoms, thus freeing the active metal surface from poisoning. The understanding of the mechanisms involved has shown that it is not the activation of the N≡N bond that would then be the limiting step but the formation of the N–H bond.

These results have opened the way for new research prospects for the development of new catalysts with electride properties. Thus, Ru/Ca_2N:e^- (Kitano et al. 2016), Ru/Y_5Si_3 (Lu et al. 2016), Ru/LaScSi (Wu et al. 2017) and Ru/LaNiSi (Mizoguchi et al. 2019) catalysts have recently been reported. The Y_5Si_3 support is no longer more active than Ca-based electrides but has the advantage of being more resistant to the presence of water vapor (3% $_{vol}$) in the reactant stream. The LaScSi compound was the first ternary intermetallic to be combined with ruthenium to synthesize NH_3.

In this phase, there are two types of interstitial sites formed by La_4 tetrahedra and La_2Sc_4 octahedra, where electrons are located (electride character), which can also accommodate hydrogen atoms, resulting in the formation of hydrides. A quantity of 1.5 H atoms per form unit can thus be stored. Combined with Ru at 8.3% by mass, this catalyst produces 5,300 μmol g^{-1} h^{-1} at 400°C and 1 bar, that is, 12 times that produced by the reference Ru-Cs/MgO catalyst under the same conditions. LaScSi belongs to the ternary intermetallic compounds of the RTX type, where R = yttrium or rare earth, T = transition metal and X = p-type element. These compounds were developed for their hydrogen storage capacity, similar to the NdScSi phase, which also has two preferred interstitial sites for the insertion of 1.5 H atoms per form unit (Tencé et al. 2016).

The effect of the rare earth composition of these phases on catalytic performance has recently been studied. A series of catalysts composed of Ru nanoparticles supported on RScSi intermetallic phases (with R = La, Ce, Pr, Nd, Sm or Gd) has been evaluated. The catalytic performance of these materials follows the lanthanide series (from La to Gd) and appears to be correlated with hydride phase formation. Cerium-based intermetallic compounds (Ru/CeScSi) show remarkable catalytic activity under moderate conditions (580 μmol g^{-1} h^{-1} at 300°C and 1 bar), combined with reversible hydrogen storage/loading properties (Croisé et al. 2022).

A study combining theoretical calculations using DFT with isotope exchange and hydrogen thermodesorption experiments also shows the beneficial role of a partial hydridation rate in Ru/LaScSi on the electride character of the materials and the dissociation activity of dinitrogen (Croisé et al. 2023). The disadvantage of these intermetallic compounds is their low specific surface area (<1 m^2 g^{-1}), which is related to their melt synthesis mode and the consequent difficulty in obtaining good dispersion of ruthenium particles.

Furthermore, the Ru deposition methods commonly used for these substrates, such as vapor deposition, do not lead to strong interactions between the metal and the substrate, which favors coalescence of the active sites during the reaction. Strategies such as incorporating Ru directly into the intermetallic phase, as in LaRuSi (Wang et al. 2019), and post treating this compound with ethylene diamine tetraacetic acid (EDTA) to increase the number of accessible active sites have been developed (Li et al. 2019). The specific surface areas of these catalysts remain low; however, the amount of Ru present in these phases raises questions for industrial application.

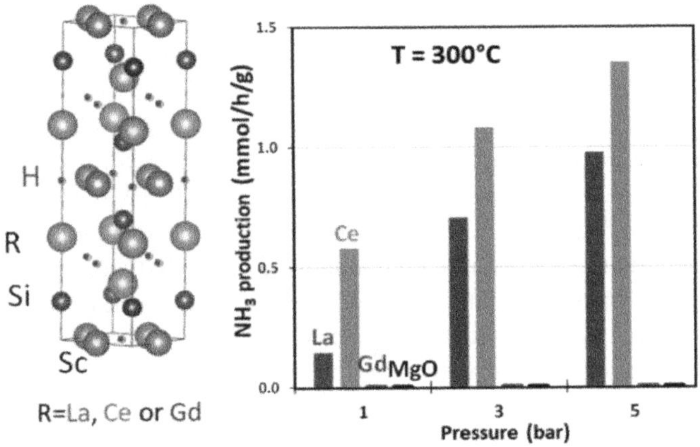

Figure 4.6. *NH_3 production rate at 300°C and moderate pressures of Ru/RScSi catalysts with R = La, Ce or Gd. Adapted from Croisé et al. (2022). For a color version of this figure, see www.iste.co.uk/derango-cuevas/hydrogen.zip*

4.4.2.5. *Hydride phases*

Research into hydride-type compounds for ammonia synthesis gained momentum following the development of a two-active-center strategy that combines a transition metal (TM) (V, Cr, Mn, Fe, Co, or Ni) with lithium hydride LiH (Wang et al. 2017; Gao et al. 2019). The advantage of this combination is that it overcomes the difficulty of being able at low temperatures both to strongly activate the main reactant (N_2) and to keep the reaction intermediates weakly bound so that they can react and desorb.

This is the case, for example, with TMs such as Fe or Ru, on which NH_x intermediates ($x = 0.1$ and 2) are strongly adsorbed and for which a high temperature is required to achieve NH_3 formation (Nørskov et al. 2009). In an Fe(N)-LiH tandem catalyst, the negatively charged H atom of the lithium hydride acts as a reducing agent to activate the –N atom or NH_x species of the iron or iron nitride. The $LiNH_2/Li_2NH$ species can then be hydrogenated to release NH_3. The role of lithium hydride has been particularly highlighted in its association with manganese.

"Conventional" catalysts based on this TM are particularly inactive in NH_3 synthesis, but the Mn_4N–LiH combination shows performance close to that of a ruthenium catalyst (Chang et al. 2018). The nature of the alkali or alkaline-earth element used to form the hydride is also essential. In combination with Mn_4N, the comparison of activities per gram of TM followed the order $BaH_2 > LiH > KH > CaH_2 > NaH$. This ranking is markedly different from the electronic promotion often proposed for alkaline or alkaline earths.

More broadly, approaches involving combining binary nitrides with hydrides to form nitride-hydrides have overcome the problems of deactivation of metal nitrides by N_2 and are emerging as an alternative route for exploring ternary nitrides in this direction (Wang et al. 2017; Kitano et al. 2019).

Even in the absence of a TM, hydride-type materials can be very active at low temperatures for NH_3 synthesis. This is the case for solid $KH_{0,19}C_{24}$, which consists of potassium hydrides confined within graphene sheets (Chang et al. 2022). The role of the hydride is in the hydrogenation step to form NH_3.

Despite their interesting performance, many of these compounds are highly sensitive to air and humidity. This complicates the characterization required to elucidate the reaction mechanisms and makes their industrial application difficult to understand.

4.4.2.6. *Other types of catalysts*

Early research into the use of perovskites (ABO_3 structural materials) as supports for ruthenium catalysts showed little interest and limited activity (Yang et al. 2010). Recently, the formation of oxy-hydrides and oxy-nitride-hydrides in perovskites ($ABO_{3-x}H_yN_z$) has generated significant interest for application in NH_3 synthesis. $ABO_{3-x}H_yN_z$ materials follow a specific mechanistic pathway involving a network of hydride anions (H^-) and nitride anions (N^{3-}) via the Mars–van Krevelen mechanism.

Compositions of particular interest in terms of catalytic performance are $BaCeO_{3-x}H_yN_z$ and $BaTiO_{3-x}H_x$ (Tang et al. 2018; Kitano et al. 2019). In the case of oxy-hydride, $BaTiO_{3-x}H_x$ acts as a support for low Co, Fe or Ru deposition. The high activity of these compositions is linked both to a Mars–van Krevelen mechanism with respect to hydrogen, which limits poisoning by this element, and to the electron-donating character of the hydride formed. The predominance of one of these two properties depends on the metal used. The wide possible variety of A and B cations embedded in the perovskite structure enables the design of diverse materials, which could lead to the development of a new family of oxy-nitride-hydride catalysts for NH_3 synthesis (Royer et al. 2014).

4.4.3. *Mechanisms and limiting steps in ammonia synthesis*

Despite intensive research over many years, the activation of nitrogen on iron-based catalysts is still the subject of controversy, whether by dissociative adsorption or molecular (associative) adsorption. These two routes (a and b) are presented below (* represents adsorption sites on the catalyst surface).

a) Dissociative adsorption of N_2:

(1): $N_2 + 2^* \leftrightarrows 2N^*$

(2): $H_2 + 2^* \leftrightarrows 2H^*$

(3): $N^* + H^* \leftrightarrows NH^* + {}^*$

(4): $NH^* + H^* \leftrightarrows NH_2^* + {}^*$

(5): $NH_2^* + H^* \leftrightarrows NH_3 + 2^*$

$N_2 + 3H_2 \leftrightarrows 2NH_3$

b) Molecular adsorption of N_2:

(1): $N_2 + 2^* \leftrightarrows N_2^*$

(2): $N_2^* + H_2^* \leftrightarrows N_2H_2^*$

(3): $N_2H_2^* + H_2 \leftrightarrows N_2H_4^*$

(4): $N_2H_4^* + H_2 \leftrightarrows 2NH_3 + {}^*$

$N_2 + 3H_2 \leftrightarrows 2NH_3$

Although most studies support the dissociative adsorption of N_2 (Por et al. 1991), many studies have asserted that molecular adsorption instead (Mortensen et al. 1999) is needed for NH_3 synthesis. These two reaction pathways could compete at the surface of the iron catalyst. In the dissociative pathway, the N≡N bond is broken prior to any hydrogen addition. In the associative pathway, the N≡N bond is broken upon release of the first NH_3, as illustrated in Figure 4.7.

In the associative mechanism, the hydrogenation process could occur via two pathways: the alternating pathway (A) and the distal pathway (D). The alternating pathway involves the alternating addition of H to the two N atoms on the catalyst surface. The distal route involves sequential additions of H to the N atom farthest from the catalyst surface, while the other N atom begins to be hydrogenated until the first ammonia molecule is released.

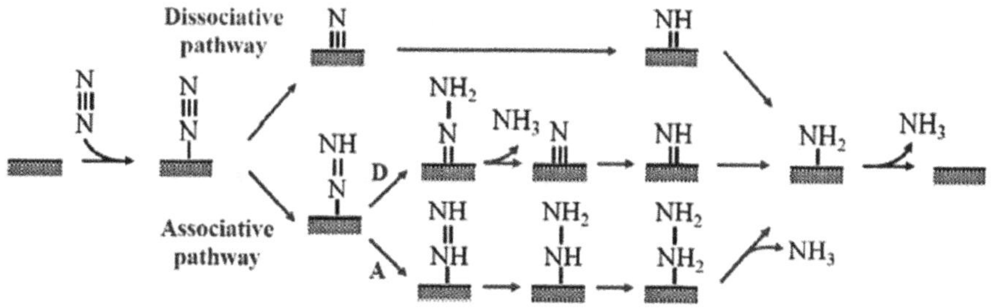

Figure 4.7. *Illustration of possible reactions for the reduction of N_2 to NH_3 (according to Wang et al. (2019)). For a color version of this figure, see www.iste.co.uk/derango-cuevas/hydrogen.zip*

In recent years, computational techniques have made it possible to predict the complex behavior of surface adsorbates (N, H and NH_X) and the reaction mechanisms of iron nanoparticle face orientations. These advances have considerably improved our understanding of the NH_3 synthesis reaction on these catalysts.

Thus, the determination of the configuration of the Fe active sites reveals that the breaking of the N_2 triple bond requires the presence of two neighboring surface-bridged sites, enabling a multistep mechanism where N_2 (i) first binds in a σ mode (retaining a triple N≡N bond) perpendicular to the surface on an Fe atom in the top layer, (ii) crosses an energy barrier to adsorb onto an iron site in the second layer, (iii) switches to a π-bonding mode (with a double N=N bond) horizontally on two Fe atoms in the second layer (and on one Fe atom in the third layer), (iv) forms two additional bonds with the surface, leaving a single N–N bond and (v) dissociates into two bridged N atoms (Fuller et al. 2022).

Overall, three different steps can be proposed as limiting for the entire NH_3 synthesis process: (i) dissociative adsorption of dinitrogen, (ii) surface reaction of adsorbed species and (iii) desorption of reaction products.

For iron-based materials, dissociative adsorption of N_2 (step (1), route (a)) remains the limiting step. The subsequent hydrogenation steps are energetically challenging, but the energy differences involved can be easily overcome at the reaction temperatures used in industrial ammonia synthesis (450–500°C). Because of these differences in activation energy, it appears that the kinetically limiting step would switch from dissociative adsorption of N_2 to hydrogenation of the adsorbed nitrogen species (steps 2–4) if the temperature were sufficiently lowered. Thus, in new-generation materials (see sections 4.4.2.4 and 4.4.2.5), where dissociative activation of N_2 is favored, the hydrogenation or even desorption steps for NH_X species become decisive.

With the advent of new materials, the limiting steps for ammonia synthesis are being revisited. Theoretical calculations have explored the possibility of the catalytic synthesis of NH_3 at low pressure and low temperature on TM surfaces. This work revealed scaling relationships between the transition state energy for N_2 dissociation (activation barrier) and the adsorption energies (binding energies) for all intermediates (Vojvodic et al. 2014; Medford et al. 2015).

Thus, an optimum performance, reflecting a compromise between two opposing properties, can be achieved to obtain the best activity: a low barrier to activation of the N≡N bond and a low rate of surface coverage of nitrogen atoms during the reaction. However, these two conditions require strong and weak N-surface interaction, respectively, as in the case of Co_3Mo_3N formulations. Using the same approach, systems based on Re in Cu have recently been proposed (Singh et al. 2018).

For MXene-based materials (2D materials with the general formula $M_{n+1}X_nT_x$ with M = transition metal, X = C and/or N, T = end groups of anionic character, see also section 4.5), it has also been reported that the 2D character of the surface facilitates N_2 dissociation by elongation of the N–N chemical bond (up to 20% on W_2N) (Gouveia et al. 2020). For these materials, the limiting step then becomes the hydrogenation step rather than the dissociative adsorption of N_2.

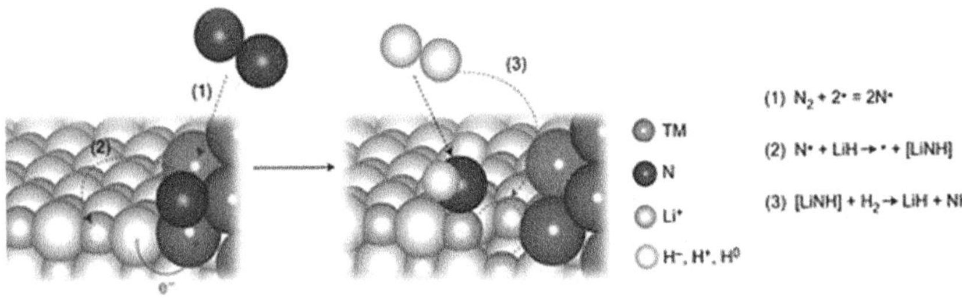

Figure 4.8. *Proposed mechanism for NH_3 synthesis on TMN@LiH materials (according to Wang et al. (2017)). For a color version of this figure, see www.iste.co.uk/derango-cuevas/hydrogen.zip*

Hybrid-composite materials that combine nitride and hydride functions have recently been proposed for NH_3 synthesis. These systems involve group 3*d*TM nitrides (TMNs) associated with lithium hydride. The synergy between TMN and LiH enables TMN-LiH composites to exhibit unprecedented catalytic activities at low temperatures. These results can be explained by a three-step mechanism, as depicted in Figure 4.8 (Wang et al. 2017). This model represents the activation of N_2 (reaction (1)) on the active sites of the TM nitrides, followed by the transfer of N with hydrogenation carried out by LiH (reaction (2)) and the heterolytic splitting of H_2 into (LiNH) to form NH_3 and regenerate the LiH hydride (reaction (3)).

In addition, numerous kinetic studies have been carried out on this reaction. The activation energies, as well as the reaction orders, also provide information on the decisive steps in NH_3 synthesis. Most standard Ru-based catalysts have activation energies in the range of 80–130 kJ/mol (Kobayashi et al. 2017) compared with 40–60 kJ/mol for electrides (Kitano et al. 2012). Furthermore, the reaction orders in N_2, H_2 and NH_3 (α, β and γ, respectively) provide insight into the catalytic behavior. For most conventional catalysts, α is close to 1, which is related to the fact that the rate-limiting step in the reaction is the unimolecular breakup reaction of an adsorbed N_2 species. For hydride or electride materials, the order reported for N_2 is closer to 0.5, reflecting easier dissociation of N_2. The order of the reactions on H_2, β, reflects the degree of hydrogen poisoning. In conventional materials, a negative order (from –0.5 to –1) is observed, whereas this order is rather positive (close to 1) for the latest generation of electride catalysts.

4.5. Alternative ammonia synthesis processes

Even if advances in materials make it possible to imagine the industrialization of new catalysts capable of producing ammonia under milder conditions than those currently used for the Haber–Bosch process in the not-too-distant future, it seems very difficult to design a process of this nature operating below 300°C at atmospheric pressure. Indeed, thermodynamic limitations remain severe at these temperatures, making it impossible to achieve high conversions without drastically increasing pressure (Figure 4.2). Research is therefore underway to circumvent this thermodynamic problem by proposing other concepts.

This is the case, for example, for the chemical looping concept applied to ammonia synthesis (Wang and Shen 2022; Gao et al. 2023). The general principle of a chemical loop is to decompose a given reaction into several subreactions to overcome the thermodynamic and kinetic limitations of the main reaction.

In this way, NH_3 generation can be achieved in two stages: a first stage in which N is attached to a "reservoir" material and a second stage in which NH_3 is formed by hydrogenation of the material. Good "reservoir" materials are, for example,

TM nitride compounds. In this case, the two reactions of fixation (see equation [4.6]) and hydrogenation (see equation [4.7]) can be written as follows:

$$2M_xN_{y-1} + N_2 \rightarrow 2M_xN_y \text{ (with M = transition metal)} \quad [4.6]$$

$$2M_xN_y + 3H_2 \rightarrow 2M_xN_{y-1} + 2NH_3 \quad [4.7]$$

These reactions are carried out in two separate reactors, enabling the operating conditions for the two stages to be optimized independently. Another advantage is that the adsorption competition between N_2 and H_2 is eliminated. The Co–Mo–N systems studied for the classical NH_3 synthesis reaction, and for which a Mars–van Krevelen-type mechanism is suspected, are good candidates for the thermochemical loop. However, these compounds need to be optimized to facilitate their ability to discharge and recharge nitrogen.

For example, Co_3Mo_3N nitride can destock half of its lattice N atoms by topotactic transformation to Co_6Mo_6N under H_2 flow at 700°C, with regeneration of the initial phase achieved at the same temperature (Gregory et al. 2011). These high temperatures increase the energy cost of the process and raise the issue of material stability for efficient cycling. One strategy for lowering the temperature of these steps is to associate an alkali or alkaline-earth hydride with the nitride to facilitate the N_2 activation step and, consequently, the temperature of the nitrogen fixation step.

As a result, N_2 is bound by the hydride BaH_2 to form BaNH in the first stage. In the second stage, BaNH is hydrogenated to produce NH_3 and regenerate BaH_2. By adding a Ni-based catalyst to the hydride, ammonia can be produced as early as 100°C and 1 bar. This Ni-BaH_2 composition supported on Al_2O_3 far exceeds the performance of Cs-Ru/Al_2O_3 at 300°C (Gao et al. 2018). The hydrogenation step can be replaced by a hydrolysis step multiplying the chemical loop possibilities for ammonia synthesis, as illustrated in Figure 4.9.

Another concept has attracted a great deal of interest, but published results are still vigorously debated. This involves the electroreduction of N_2 to NH_3, often referred to as the NRR for the Nitrogen Reduction Reaction (Montoya et al. 2015).

The idea is to produce, in small local units operating at ambient temperature and pressure, low concentrations of ammonia and thus avoid the considerable expense involved in transporting, distributing and spreading fertilizers from ammonia production. While the aim of this production method is less concerned with hydrogen storage, the resulting technological breakthroughs may be of interest in the development of catalysts for larger-scale processes.

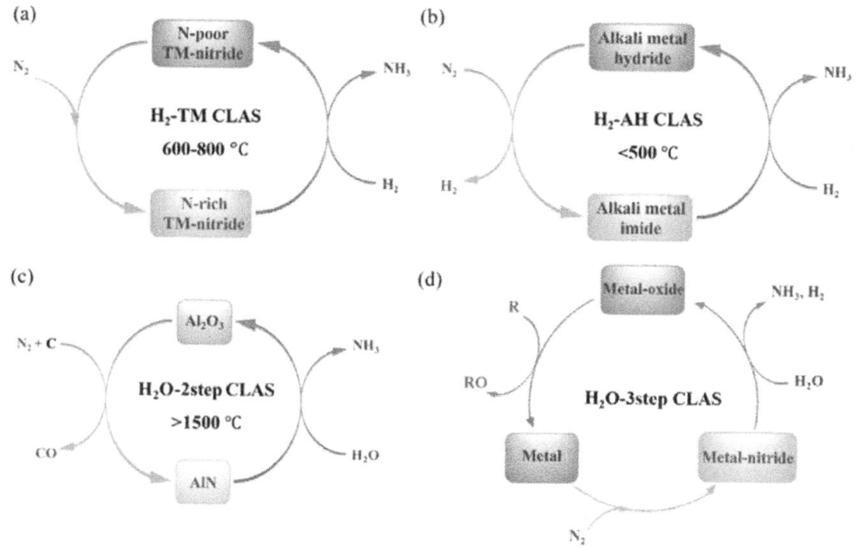

Figure 4.9. *Schematic of the thermochemical loop for NH_3 synthesis: (a) transition metal nitride-assisted; (b) alkali hydride-assisted; (c) H_2O as the hydrogen source in two steps; (d) H_2O as the hydrogen source in two steps. Reproduced with permission from Wang and Shen (2022). Copyright 2022 American Chemical Society. For a color version of this figure, see www.iste.co.uk/derango-cuevas/hydrogen.zip*

This is the case for MXene carbide, nitride or carbo-nitride phases (Kok et al. 2021). 2D MXenes are described by the general formula $M_{n+1}X_nT_x$ (M = transition metal, X = C and/or N, T = end groups) and are obtained by selective acid exfoliation of the A layer of $M_{n+1}AX_n$ (A = group 13 or 14 elements, such as Al or Si). T_x end groups, notably O, -OH, -F and -Cl, whose nature can be partially controlled by the exfoliation agent, play a major role in surface properties (Benchakar et al. 2020). The possible combinations of compositions are therefore innumerable. An initial theoretical study showing the interest in the Ti_3C_2Tx composition for the NRR has paved the way for many experimental studies on this promising material (Luo et al. 2019).

However, thermodynamics suggest that the electrochemical reduction of N_2 takes place at negative potentials similar to those required for the reaction to reduce water to dihydrogen (HER). In the aqueous phase, there is therefore competition between the HER and NRR reactions, which very often favor H_2 production. Strategies that completely inhibit the HER have the immediate consequence of inhibiting the NRR, which requires the same types of active sites (Chen et al. 2021).

Solutions such as controlling surface hydrophobicity to promote the NRR reaction are being developed (Yang et al. 2022).

Moreover, the quantities of NH_3 produced under NRR conditions are so small that it is sometimes very difficult to guarantee the reliability of the results. The nitrogen atom present in the NH_3 molecule produced does not always come from the N_2 molecule but from nitrogenous molecules present as impurities in the reagents used (e.g. nitrates) or from nitrogenous species present on the electrocatalysts (Marinho et al. 2023). A drastic and costly protocol, requiring the use of $^{15}N_2$-labeled nitrogen, should be generalized to limit the publication of "false positives", but recent literature shows that it is not always followed (Andersen et al. 2019).

4.6. Ammonia decomposition

From an economic standpoint, generating hydrogen through ammonia decomposition is a more advantageous process than is methanol reforming (Metkemeijer and Achard, 1994; Yin et al. 2004), and this technology could address issues related to hydrogen purification. For hydrogen production to be viable, the process must be competitive, clean and energy efficient. To date, two primary methods exist for producing hydrogen via ammonia decomposition: catalytic thermocracking and ammonia electrolysis (Chiuta et al. 2013).

4.6.1. *High-temperature ammonia cracking*

4.6.1.1. *Thermodynamic aspects*

Although ammonia is stable at room temperature, it can decompose at elevated temperatures into dihydrogen and dinitrogen in the presence of catalysts such as iron, nickel, zinc and uranium. This decomposition begins at temperatures as low as 300°C and is nearly complete between 500°C and 600°C (Boucet et al. 1999).

Producing hydrogen via the thermal decomposition of ammonia requires a thorough understanding of the reaction's thermodynamics. Ammonia decomposition proceeds according to the endothermic reaction shown in equation [4.8]:

$$2NH_3 \rightarrow N_2 + 3H_2,\ \Delta H^{\circ} = +\,46.19\ \text{kJ.mol}^{-1} \qquad [4.8]$$

The yield of this reaction depends mainly on temperature, as ammonia decomposition is thermodynamically limited at lower temperatures. This is illustrated in Table 4.2, which presents the theoretical decomposition rate of ammonia over a temperature range from 250°C to 700°C (Chiuta et al. 2013) (see also Figure 4.2).

T (°C)	250	300	350	400	450	500	600	700
NH_3 conversion %	89.2	95.7	98.1	99.1	99.5	99.7	99.9	99.95

Table 4.2. *Equilibrium conversion of ammonia as a function of temperature at a pressure of 1 bar (Chiuta et al. 2013)*

The decomposition reaction therefore requires relatively high temperatures to achieve satisfactory conversion rates (Yin et al. 2004). As shown in Table 4.2, temperatures above 400°C are necessary to achieve nearly complete ammonia decomposition (> 99%) (Yin et al. 2004). Above 450°C, the conversion of ammonia to hydrogen becomes virtually independent of temperature.

4.6.1.2. *Ammonia cracking material and efficiency*

The development of high-performance solid catalysts is essential for reducing energy consumption and enhancing the cost-effectiveness of hydrogen production via the thermal cracking of ammonia. A typical catalyst consists of two components: an active phase (a metal, for example) and a catalyst support material. The catalytic activity largely depends on the interaction between these two components.

Although research over the past decade has focused primarily on the development of ruthenium-based catalysts (Yin et al. 2004), numerous other studies have also evaluated the suitability of various metals and alloys for the thermal decomposition of ammonia.

These studies have investigated catalysts based on iron (Yin et al. 2004), nickel (Choudhary et al. 2001; Abashar et al. 2002; Yin et al. 2004), platinum (Yin et al. 2004), ruthenium (Bradford et al. 1997; Hashimoto et al. 2000; Choudhary et al. 2001; Yin et al. 2004), iridium (Choudhary et al. 2001), palladium (Yin et al. 2004) and rhodium (Yin et al. 2004), as well as alloy systems such as Ni/Pt (Chellappa et al. 2002), Ni/Ru (Kordesch et al. 2003) and Pd/Pt/Ru/La (Kordesch et al. 2003). In particular, Yin et al. (2004) compared the catalytic activity of various metals supported on carbon nanotubes (CNTs; catalyst particle size between 177 and 250 µm, molar charges of 4.75×10^{-4} mol/g_{cat}, corresponding to a mass content of 5% in the case of ruthenium).

These tests were carried out over several tens of hours and demonstrated stable performance for all of the metals studied. The results presented in Table 4.3 illustrate the influence of temperature on process efficiency. Regardless of the active metal used, increasing the reaction temperature leads to a sharp increase in the rate of H_2 formation, reaching 28.35 mmol/(min·g_{cat}) at 773 K compared with 2.11 mmol/(min g_{cat}) at 623 K in the case of ruthenium.

At 623 K, the hydrogen production rates for the various active metals follow the order: Ru > Rh ≈ Ni > Pt ≈ Pd > Fe, clearly indicating that ruthenium is the most active metal for hydrogen production via thermocatalytic decomposition of ammonia (Yin et al. 2004).

	H_2 formation rate (mmol/(min g_{cat}))					
Temperature (K)	Ru	Rh	Pt	Pd	Ni	Fe
623	2.11	0.2	0.05	0.03	0.10	–
643	3.42	0.34	0.10	0.07	0.17	0.02
673	5.73	0.6	0.21	0.15	0.33	–
703	9.51	1.38	0.37	0.3	0.74	0.07
723	14.64	1.94	0.61	0.57	1.11	0.17
743	–	–	–	–	–	0.34
773	28.35	4.19	1.41	1.43	2.90	0.65

Table 4.3. *Rate of H_2 production by ammonia decomposition catalyzed by various metals supported on carbon nanotube (CNT) supports (Yin et al. 2004)*

The difference in activity among these materials is commonly attributed to the reaction kinetics of the limiting step. In the case of the thermal decomposition of NH_3, it is generally accepted that nitrogen desorption is the limiting step. Ruthenium has high catalytic efficiency in the absence of ruthenium nitride formation.

To conclude, ruthenium remains the metal of choice for achieving conversions above 95% at temperatures below 500°C. This was clearly demonstrated in the report by Lamb et al. (2019). Inexpensive TMs such as nickel can also serve as catalysts. However, to reach 99% conversion, higher temperatures in the range of 600–800°C are needed (Le et al. 2021).

In addition to the nature of the metal, its content has also been reported to influence the ammonia-to-hydrogen conversion rate. For a ruthenium-based catalyst, varying the metal content between 0% and 35% revealed that the maximum ammonia conversion rate is achieved at a content of 15% (Yin et al. 2004). The decrease in activity at higher contents is attributed to the limited accessibility of the underlying ruthenium layers in materials with high metal contents.

Yin et al. (2004) also investigated the influence of a ruthenium support by comparing the catalytic activity of ruthenium supported on CNTs with that of catalysts supported on other materials, including Al_2O_3, activated carbon (AC), MgO, TiO_2 and ZrO_2. The hydrogen formation rates obtained with these different catalysts at various reaction temperatures are summarized in Table 4.3 (Yin et al. 2004).

Yin et al. (2004) reported that the rate of hydrogen formation follows the order: Ru-NTCs > Ru-MgO > Ru-TiO_2 = Ru-Al_2O_3 = Ru-ZrO_2 > Ru-CA. Among all of the supports studied, CNTs appear to be the most suitable support for ruthenium, providing the highest activity for hydrogen production via thermocatalytic decomposition of ammonia (Yin et al. 2004). From an economic perspective, less expensive TMs such as nickel can be used as catalysts to reduce the cost of producing hydrogen via the thermocatalytic decomposition of ammonia. However, to achieve 99% conversion, higher temperatures in the range of 600–800°C are needed. At these temperatures, ammonia becomes more corrosive, more susceptible to oxidation, more diffusive and therefore significantly more hazardous. As a result, implementing this process may necessitate a high-level and consequently more expensive safety infrastructure (CAPEX), which limits its applicability to stationary systems.

4.6.2. *Electro-oxidation of ammonia*

4.6.2.1. *Overview*

Electro-oxidation in aqueous alkaline media has also been explored in recent years as a method for producing hydrogen from ammonia. In 2016, Modisha and Bessarabov demonstrated that ammonia diluted in an alkaline medium (5 M KOH), in the presence of a platinum-iridium electrocatalyst, undergoes electrochemical oxidation at a potential of 0.9 V versus the standard hydrogen electrode (SHE) at 55°C, producing hydrogen with a purity of 86% (Modisha and Bessarabov 2016).

The electro-oxidation of ammonia in an alkaline medium is considered one of the most promising methods for hydrogen production. This technique is entirely CO_2-free, generating only nitrogen and water as reaction products. This technique eliminates the need for the thermal energy required in the thermocracking of ammonia (section 4.6.1). The principle of ammonia electro-oxidation for hydrogen production is illustrated by equation [4.11]. More specifically, the concept involves coupling the electro-oxidation reaction of ammonia (see equation [4.9]) with the hydrogen evolution reaction (HER), that is, the electrochemical reduction of water at the cathode (see equation [4.10]) to produce hydrogen in an alkaline electrolytic cell.

$$2NH_3\ (aq) + 6OH^- \rightarrow N_2(g) + 6H_2O + 6e^-\ (-0.77\ V/SHE) \qquad [4.9]$$

$$6H_2O + 6e^- \rightarrow 3H_2(g) + 6OH^-\ (-0.83\ V/SHE) \qquad [4.10]$$

$$2NH_3\ (aq) \rightarrow N_2(g) + 3H_2(g)\ (-0.06V\ /SHE) \qquad [4.11]$$

The mechanism of ammonia oxidation by alkaline electrolysis was first described by Gerischer and Mauere in 1970. The process begins with the absorption of ammonia onto the anode surface. This is followed by successive dehydrogenation steps leading to NH and/or NH_2 intermediates, which subsequently recombine to form adsorbed species such as N_2H_2, N_2H_4 and the radical N_2H_3. Finally, nitrogen is produced through a parallel reaction involving hydroxide ions (Gerischer and Mauerer 1970).

At 25°C, the oxidation potential of ammonia is –0.77 V (SHE), which is only 0.06 V less negative than the potential of the HER in alkaline solution, which is –0.83 V (SHE) (Simons et al. 1969). These thermodynamic values explain why hydrogen production coupled with ammonia oxidation is more attractive than hydrogen generation via water electrolysis, which requires a theoretical cell voltage of 1.223 V.

Thus, both thermodynamic considerations and experimental results clearly demonstrate the superiority of ammonia electrolysis over water electrolysis. In terms of energy performance, Chen et al. (2021) recently showed that a platinum-ruthenium alloy (Pt_6Ru) in the form of nanocubes is capable of electrolyzing ammonia in an alkaline medium (1 M KOH) at a cell voltage of 0.72 V, which is much lower than the 1.74 V required for water electrolysis (Xue et al. 2021). This finding was confirmed by another research group, who reported that with a conventional platinum-based electrocatalyst, ammonia electrolysis yields 734 L $(H_2).kWh^{-1}$ of hydrogen, compared with only 242 L $(H_2).kWh^{-1}$ for water electrolysis (Yang et al. 2021).

The theoretical energy consumption of an ammonia electrolyzer can be quantified from the standard cell potential. Ammonia electrolysis requires 1.55 Wh $g(H_2)^{-1}$, whereas water electrolysis requires at least 33 Wh $g(H_2)^{-1}$ under standard conditions. This finding indicates that, theoretically, ammonia electrolysis consumes approximately 95% less energy than water electrolysis (Vitse et al. 2005).

The second advantage of ammonia electrolysis lies in its ease of integration with renewable energy sources such as wind and solar power. Owing to its low energy consumption, an ammonia electrolyzer could be powered by renewable electricity or even by a portion of the energy generated by a hydrogen fuel cell if the electrolytic cell operates close to its theoretical potential. This approach would reduce the need for hydrogen storage, as hydrogen could be produced on demand (Vitse et al. 2005).

4.6.2.2. *Ammonia electro-oxidation materials and performance*

Since the kinetics of the ammonia electro-oxidation reaction remain the main limitation of the process, further research is needed to develop new catalysts capable of increasing the hydrogen production rate. Several studies have suggested that the presence of noble metals improves the efficiency of ammonia electrolysis (Vitse et al. 2005). In a slightly different context, Marincic and Leitz (1978) investigated the feasibility of this process at ambient temperature for the removal of low concentrations of ammonia from wastewater. Catalysts of different compositions were tested, including platinum, graphite, titanium and titanium coated with platinum black at different catalyst loadings (0.5–20 $mg.cm^{-2}$), and under various experimental conditions (pH 5.55–12.8, $[NH_3] = 2–170\ mg.L^{-1}$, different supporting electrolytes, and different current densities). The results showed that only platinum electrodes were able to ensure the electro-oxidation of ammonia at a reasonable rate.

As early as 1972, Giner and Moser revealed that platinum-group catalysts, such as Pt and Ir, enhance the kinetics of the ammonia electro-oxidation reaction (Giner and Moser 1972). This was later confirmed by McKee et al. (1973), who also proposed the use of Pt-Ir as an ammonia dissociation catalyst for ammonia fuel cells. Another bimetallic system based on a Pt–Rh pair was investigated by Copper and Botte (2006). These authors showed that, for comparable catalyst loading, the platinum–rhodium combination exhibited superior electrocatalytic activity compared with that of platinum alone. The observed synergistic effect was attributed to the ability of rhodium to promote OH^- adsorption, which appears to be a limiting factor in the reaction.

Owing to competitive adsorption between ammonia and hydroxide ions on the surface of platinum electrodes, blocking of active platinum sites has been observed (Cooper and Botte 2006). In line with these findings, several studies have reported that metal alloys, metal oxides and bimetallic catalysts exhibit higher electrocatalytic activity than monometallic catalysts for the ammonia oxidation reaction. This has been demonstrated, for example, with Ag–Pb alloys (Ge and Johnson 1995) and ruthenium oxides (Donten et al. 1997).

More recently, an in situ infrared spectroscopy study of ammonia oxidation on a platinum electrode in alkaline solution revealed that the metal is also easily poisoned by adsorbed nitrogen species, which significantly limits the rate of hydrogen production (Matsui et al. 2015).

Other studies have shown that not only does the composition of the catalysts affect the yield of the electrochemical reaction but that their microstructure also plays a critical role in the kinetics of ammonia oxidation (Vidal-Iglesias et al. 2003).

Xue et al. (2021) recently demonstrated that a Pt_6Ru platinum–ruthenium alloy in the form of nanocubes exhibits the highest activity for ammonia electrolysis in alkaline media. This enhanced performance was attributed to crystallographic effects and Pt-Ru bimetallic interactions.

With the aim of reducing the cost of hydrogen production via ammonia electro-oxidation, several studies have evaluated the use of various substrates for noble metal deposition, including nickel (Cooper and Botte 2006) and carbon fibers (Bonnin et al. 2008). The results revealed that carbon fiber is a more effective substrate than nickel because of its uniform noble metal coverage, the absence of surface blocking effects and its low mass density (Bonnin et al. 2008).

More recently, Yang et al. (2021) showed that a catalyst consisting of platinum electrodeposited in a flower-like morphology on a nickel mesh efficiently produces high-purity hydrogen with faradic yields exceeding 90%.

The electrocatalytic activity of the platinum electrodes synthesized by cyclic voltammetric deposition (CV-Pt) and chronopotentiometric deposition (CP-Pt) was evaluated and compared using cyclic voltammetry in alkaline media. The results showed that the CP-Pt and CV-Pt (100 cycles) electrodes exhibited similar current densities.

In contrast, increasing the number of cycles led to a rise in the current density of the CV-Pt electrodes, which reached a maximum after 500 cycles. The crystalline structure and surface morphology of the CV-Pt electrodes evolved with the number of cycles. The CV-Pt electrode prepared with 500 cycles demonstrated higher efficiency than did commercial Pt/C catalysts, with a mass activity of 4.37 A mg^{-1} (Yang et al. 2021).

Finally, the major challenge in hydrogen production via ammonia electro-oxidation is related to the reaction kinetics. Although the process is thermodynamically favorable, its slow kinetics remain a significant technological barrier.

Consequently, the development and industrial implementation of this technique require the development of catalysts specifically engineered to increase the reaction kinetics.

4.7. Overview and outlook

Green hydrogen is a key energy vector in the energy transition aimed at decarbonizing this sector. However, hydrogen remains a difficult molecule to store. In addition to being dangerous, liquefying or hyperbarically compressing hydrogen requires high energy consumption. However, there are other possible storage methods, such as chemical storage in the form of ammonia.

However, the usual process for synthesizing ammonia from nitrogen and dihydrogen (Haber–Bosch process) is particularly energy intensive (+/– 30 GJ/tonne) owing to the high temperatures and pressures required despite the use of a catalyst (typically 200 bar, 600°C). Moreover, the development of green hydrogen, obtained by the electrolysis of water fuelled by renewable energy, calls for a complete rethinking of the current industrialized process. Indeed, unlike the Haber–Bosch process designed for large-scale centralized production (> 1,000 tNH_3/day), the conversion of green hydrogen into ammonia seems better suited to small, flexible, decentralized units capable of withstanding the intermittent nature of renewable energies. These small units will be economically viable if the synthesis reaction can be carried out under much more moderate temperature and pressure conditions (300–350°C and 10–50 bar, for example). The race is therefore afoot to design new catalytic formulations capable of producing ammonia under these moderate conditions.

Research prior to 2010 focused mainly on promoted iron oxides and carbon-supported Ru catalysts, while ternary nitrides were in the early stages of development, with little knowledge gained (Panov and Kharitonov 1985; Saadatjou et al. 2015). The discovery of electrides, hydrides, nitrides and oxyhydrides, as well as their unique and promising reversible electron storage properties, has significantly improved the efficiency of ruthenium-based catalysts, inhibiting H_2 poisoning of the metal at high pressure (Kitano et al. 2012; Liu et al. 2020). These innovative supports have opened new avenues in the search for high-performance catalysts for NH_3 synthesis (Hara et al. 2017), and the Ru-Cs/MgO formulation is certainly no longer the catalyst of choice.

Each of these new catalysts, however, has its own advantages and drawbacks in terms of large-scale synthesis, stability and cost to market. For example, studies on Ru/Ca12A7:e- have revealed that the limiting step is no longer the activation of the

N≡N bond but the formation of the N–H bond (Hara et al. 2017). In this material, the turnover frequency (TOF) relative to the reaction rate per catalytic site increases with pressure up to 10 bar, after which the activity becomes pressure independent, limiting their application to relatively low working pressures (Kitano et al. 2012). The activity, stability and storage capacity of hydrides and electrides at high pressures (above 10 bar) remain to be clarified for commercial application in a reactor under moderate conditions (20–50 bar), and specific research is needed.

Nevertheless, knowledge of new materials and their mechanisms should make it possible to develop a suitable catalyst for large-scale NH_3 synthesis under mild conditions in the future. It has taken decades to develop alternatives to iron-based catalysts, but only 8–10 years for electrides and a few years for new oxy-nitride-hydride catalysts (Kitano et al. 2019).

Moreover, other routes are being explored for ammonia synthesis, such as the chemical looping process, which overcomes certain thermodynamic limitations, or electrosynthesis, even if the yields achieved today remain modest.

Moreover, although ammonia can be used as an energy carrier in a number of applications, using this molecule as a hydrogen carrier requires the ability to carry out NH_3 decomposition reactions to obtain H_2 molecules. For this reaction, two main routes are possible: thermocatalytic decomposition and electrooxidation of ammonia.

However, there are still challenges ahead, such as optimizing hydrogen production to make it more cost-effective and reducing the harmful emissions associated with the thermocatalytic decomposition of ammonia. The production of green hydrogen by electrolysis of ammonia is a very promising process.

The use of this technique remains limited, however, by reaction kinetics that are too slow. Platinum catalysts are the most active (Huang et al. 2020), but their high cost and sensitivity to poisoning by nitrogen species (Matsui et al. 2015) limit their industrial potential. The development of suitable catalytic materials that combine electrocatalytic activity, long-term stability and affordability is therefore crucial to enable the industrial implementation of this technique (Hanada et al. 2020). Finally, the increased use of hydrogen produced by ammonia electrolysis could have a greater impact than the thermocatalytic decomposition of ammonia on the transition to more sustainable and environmentally friendly energy.

Current research efforts are also focused on the economic and logistical aspects of the concept of producing green ammonia from renewable sources and its role as a hydrogen storage molecule. These technoeconomic and life-cycle analyses developed for each storage and transport route (compressed H_2, liquefied H_2, LOHC and solid hydrides) should enable us to compare and determine the most efficient solution depending on the storage duration and the distance separating the places where the hydrogen is produced and used (Crandall et al. 2023). Ammonia seems to be particularly suitable for longer storage periods. The environmental impact of low emissions and the toxicity of the molecule must also be carefully considered (Valera-Medina et al. 2018).

4.8. References

Abashar, M.E.E., Al-Sughair, Y.S., Al-Mutaz, I.S. (2002). Investigation of low-temperature decomposition of ammonia using spatially patterned catalytic membrane reactors. *Appl. Catal. Gen.*, 236(1–2), 35–53. doi: 10.1016/S0926-860X(02)00272-7.

Aika, K. (1986). Heterogeneous catalysis of ammonia synthesis at room temperature and atmospheric pressure. *Angew. Chem. Int. Ed. Engl.*, 25(6), 558–559. doi: 10.1002/anie.198605581.

Aika, K., Hori, H., Ozaki, A. (1972). Activation of nitrogen by alkali metal promoted transition metal I. Ammonia synthesis over ruthenium promoted by alkali metal. *J. Catal.*, 27(3), 424–431. doi: 10.1016/0021-9517(72)90179-0.

Aika, K., Ohya, A., Ozaki, A., Inoue, Y., Yasumori, I. (1985). Support and promoter effect of ruthenium catalyst: II. Ruthenium/alkaline earth catalyst for activation of dinitrogen. *J. Catal.*, 92(2), 305–311. doi: 10.1016/0021-9517(85)90265-9.

Alagharu, V., Palanki, S., West, K.N. (2010). Analysis of ammonia decomposition reactor to generate hydrogen for fuel cell applications. *J. Power Sources*, 195(3), 829–833. doi: 10.1016/j.jpowsour.2009.08.024.

Amhamed, A.I., Shuibul Qarnain, S., Hewlett, S., Sodiq, A., Abdellatif, Y., Isaifan, R.J., Alrebei, O.F. (2022). Ammonia production plants – A review. *Fuels*, 3(3), 408–435. doi: 10.3390/fuels3030026.

Andersen, S.Z., Čolić, V., Yang, S., Schwalbe, J.A., Nielander, A.C., McEnaney, J.M., Enemark-Rasmussen, K., Baker, J.G., Singh, A.R., Rohr, B.A. et al. (2019). A rigorous electrochemical ammonia synthesis protocol with quantitative isotope measurements. *Nature*, 570(7762), 504–508. doi: 10.1038/s41586-019-1260-x.

Aziz, M., Wijayanta, A.T., Nandiyanto, A.B.D. (2020). Ammonia as effective hydrogen storage: A review on production, storage and utilization. *Energies*, 13(12), 3062. doi: 10.3390/en13123062.

Benchakar, M., Loupias, L., Garnero, C., Bilyk, T., Morais, C., Canaff, C., Guignard, N., Morisset, S., Pazniak, H., Hurand, S. et al. (2020). One MAX phase, different MXenes: A guideline to understand the crucial role of etching conditions on Ti_3C_2Tx surface chemistry. *Appl. Surf. Sci.*, 530, 147209. doi: 10.1016/j.apsusc.2020.147209.

Bielawa, H., Hinrichsen, O., Birkner, A., Muhler, M. (2001). The ammonia-synthesis catalyst of the next generation: Barium-promoted oxide-supported ruthenium. *Angew. Chem. Int. Ed.*, 40(6), 1061–1063.

Boisen, A., Dahl, S., Jacobsen, C.J.H. (2002). Promotion of binary nitride catalysts: Isothermal N_2 adsorption, microkinetic model, and catalytic ammonia synthesis activity. *J. Catal.*, 208(1), 180–186. doi: 10.1006/jcat.2002.3571.

Bonnin, E.P., Biddinger, E.J., Botte, G.G. (2008). Effect of catalyst on electrolysis of ammonia effluents. *J. Power Sources*, 182(1), 284–290. doi: 10.1016/j.jpowsour.2008.03.046.

Bradford, M.C.J., Fanning, P.E., Vannice, M.A. (1997). Kinetics of NH_3 decomposition over well dispersed Ru. *J. Catal.*, 172(2), 479–484. doi: 10.1006/jcat.1997.1877.

CF Industries (n.d.). Ammonia production [Online]. Available at: https://www.cfindustries.com/what-we-do/ammonia-production [Accessed 14 September 2023].

Chang, F., Guan, Y., Chang, X., Guo, J., Wang, P., Gao, W., Wu, G., Zheng, J., Li, X., Chen, P. (2018). Alkali and alkaline earth hydrides-driven N_2 activation and transformation over Mn nitride catalyst. *J. Am. Chem. Soc.*, 140(44), 14799–14806. doi: 10.1021/jacs.8b08334.

Chang, F., Tezsevin, I., de Rijk, J.W., Meeldijk, J.D., Hofmann, J.P., Er, S., Ngene, P., de Jongh, P.E. (2022). Potassium hydride-intercalated graphite as an efficient heterogeneous catalyst for ammonia synthesis. *Nat. Catal.*, 5(3), 222–230. doi: 10.1038/s41929-022-00754-x.

Chellappa, A.S., Fischer, C.M., Thomson, W.J. (2002a). Ammonia decomposition kinetics over Ni-Pt/Al_2O_3 for PEM fuel cell applications. *Appl. Catal. Gen.*, 227(1–2), 231–240. doi: 10.1016/S0926-860X(01)00941-3.

Chellappa, A.S., Powell, M.R., Fountain, M., Call, C.J., Godshall, N.A. (2002b). Compact fuel processors for PEM fuel cells. *Fuel Chemistry Division Preprints*, 47(1), 713–715.

Chen, J., Cheng, H., Ding, L.-X., Wang, H. (2021). Competing hydrogen evolution reaction: A challenge in electrocatalytic nitrogen fixation. *Mater. Chem. Front.*, 5(16), 5954–5969. doi: 10.1039/D1QM00546D.

Chiuta, S., Everson, R.C., Neomagus, H.W.J.P., van der Gryp, P., Bessarabov, D.G. (2013). Reactor technology options for distributed hydrogen generation via ammonia decomposition: A review. *Int. J. Hydrog. Energy*, 38(35), 14968–14991. doi: 10.1016/j.ijhydene.2013.09.067.

Choudhary, T.V., Sivadinarayana, C., Goodman, D.W. (2001). Catalytic ammonia decomposition: COx-free hydrogen production for fuel cell applications. *Catal. Lett.*, 72(3–4), 197–201.

Cooper, M. and Botte, G.G. (2006). Hydrogen production from the electro-oxidation of ammonia catalyzed by platinum and rhodium on raney nickel substrate. *J. Electrochem. Soc.*, 153(10), A1894. doi: 10.1149/1.2240037.

Crandall, B.S., Brix, T., Weber, R.S., Jiao, F. (2023). Techno-economic assessment of green H_2 carrier supply chains. *Energy Fuels*, 37(2), 1441–1450. doi: 10.1021/acs.energyfuels.2c03616.

Croisé, C., Alabd, K., Tencé, S., Gaudin, E., Villesuzanne, A., Courtois, X., Bion, N., Can, F. (2022). Influence of the rare earth (R) element in Ru-supported RScSi electride-like intermetallic catalysts for ammonia synthesis at low pressure: Insight into NH_3 formation mechanism. *ChemCatChem*, 15(3), e202201172. doi: 10.1002/cctc.202201172.

Croisé, C., Alabd, K., Villesuzanne, A., Can, F., Courtois, X., Gaudin, E., Tencé, S., Bion, N. (2023). Role of hydride ion within Ru/LaScSi and Ru/CeTiGe catalysts for NH3 synthesis: A combination of DFT and experimental nitrogen isotopic exchange studies. *Catal. Commun.*, 179, 106689. doi: 10.1016/j.catcom.2023.106689.

Damle, A.S. (2009). Hydrogen production by reforming of liquid hydrocarbons in a membrane reactor for portable power generation. Experimental studies. *J. Power Sources*, 186(1), 167–177. doi: 10.1016/j.jpowsour.2008.09.059.

Das, S.K., Reis, A., Berry, K.J. (2009). Experimental evaluation of CO poisoning on the performance of a high temperature proton exchange membrane fuel cell. *J. Power Sources*, 193(2), 691–698. doi: 10.1016/j.jpowsour.2009.04.021.

Donten, M., Hyk, W., Ciszkowska, M., Stojek, Z. (1997). Electrooxidation of ammonia and simple amines at titanium electrodes modified with a mixture of ruthenium and titanium dioxides. *Electroanalysis*, 9(10), 751–754. doi: 10.1002/elan.1140091004.

Ertl, G. (1983). Kinetics of chemical processes on well-defined surfaces. In *Catalysis: Science and Technology*, Anderson, J.R. and Boudart, M. (eds). Springer, Berlin. doi: 10.1007/978-3-642-93229-8_3.

European Patent Office (n.d.). European publication server [Online]. Available at: https://data.epo.org/publication-server/document?iDocId=238&iFormat=2 [Accessed 3 February 2014].

Fasihi, M., Weiss, R., Savolainen, J., Breyer, C. (2021). Global potential of green ammonia based on hybrid PV-wind power plants. *Appl. Energy*, 294, 116170. doi: 10.1016/j.apenergy.2020.116170.

Fernández, C., Bion, N., Gaigneaux, E.M., Duprez, D., Ruiz, P. (2016). Kinetics of hydrogen adsorption and mobility on Ru nanoparticles supported on alumina: Effects on the catalytic mechanism of ammonia synthesis. *J. Catal.*, 344, 16–28. doi: 10.1016/j.jcat.2016.09.013.

Forni, L., Molinari, D., Rossetti, I., Pernicone, N. (1999). Carbon-supported promoted Ru catalyst for ammonia synthesis. *Appl. Catal. Gen.*, 185(2), 269–275. doi: 10.1016/S0926-860X(99)00144-1.

Foster, S.L., Bakovic, S.I.P., Duda, R.D., Maheshwari, S., Milton, R.D., Minteer, S.D., Janik, M.J., Renner, J.N., Greenlee, L.F. (2018). Catalysts for nitrogen reduction to ammonia. *Nat. Catal.*, 1, 490–500. doi: 10.1038/s41929-018-0092-7.

Fuller, J., An, Q., Fortunelli, A., Goddard, W.A.I. (2022). Reaction mechanisms, kinetics, and improved catalysts for ammonia synthesis from hierarchical high throughput catalyst design. *Acc. Chem. Res.*, 55(8), 1124–1134. doi: 10.1021/acs.accounts.1c00789.

Gao, W., Guo, J., Wang, P., Wang, Q., Chang, F., Pei, Q., Zhang, W., Liu, L., Chen, P. (2018). Production of ammonia via a chemical looping process based on metal imides as nitrogen carriers. *Nat. Energy*, 3(12), 1067–1075. doi: 10.1038/s41560-018-0268-z.

Gao, W., Guo, J., Chen, P. (2019). Hydrides, amides and imides mediated ammonia synthesis and decomposition. *Chin. J. Chem.*, 37(5), 442–451. doi: 10.1002/cjoc.201800586.

Gao, W., Wang, R., Feng, S., Wang, Y., Sun, Z., Guo, J., Chen, P. (2023). Thermodynamic and kinetic considerations of nitrogen carriers for chemical looping ammonia synthesis. *Discov. Chem. Eng.*, 3(1). doi: 10.1007/s43938-023-00019-4.

Ge, J. and Johnson, D.C. (1995). Electrocatalysis of anodic oxygen-transfer reactions: Oxidation of ammonia at anodized Ag-Pb eutectic alloy electrodes. *J. Electrochem. Soc.*, 142(10), 3420–3423. doi: 10.1149/1.2049997.

Gerischer, H. and Mauerer, A. (1970). Untersuchungen Zur anodischen oxidation von Ammoniak an Platin-Elektroden. *Journal of Electroanalytical Chemistry*, 25(3), 421–433.

Ghosh, S. and Jagirdar, B.R. (2018). Effect of the crystallographic phase of ruthenium nanosponges on arene and substituted-arene hydrogenation activity. *ChemCatChem*, 10(14), 3086–3095. doi: 10.1002/cctc.201800287.

Giddey, S., Badwal, S.P.S., Munnings, C., Dolan, M. (2017). Ammonia as a renewable energy transportation media. *ACS Sustainable Chem. Eng.*, 5(11), 10231–10239. doi: 10.1021/acssuschemeng.7b02219.

Giner, J.D. and Moser, J.R. (1972). Electrochemical cell with platinum-ruthenium electrode and method of using with ammonia. US Patent, 3,650,838.

Gouveia, J.D., Morales-García, Á., Viñes, F., Gomes, J.R.B., Illas, F. (2020). Facile heterogeneously catalyzed nitrogen fixation by MXenes. *ACS Catal.*, 10(9), 5049–5056. doi: 10.1021/acscatal.0c00935.

Gregory, D.H., Hargreaves, J.S.J., Hunter, S.M. (2011). On the regeneration of Co_3Mo_3N from Co_6Mo_6N with N_2. *Catal. Lett.*, 141(1), 22–26. doi: 10.1007/s10562-010-0464-3.

Halseid, R., Vie, P.J.S., Tunold, R. (2006). Effect of ammonia on the performance of polymer electrolyte membrane fuel cells. *J. Power Sources*, 154(2), 343–350. doi: 10.1016/j.jpowsour.2005.10.011.

Hanada, N., Kohase, Y., Hori, K., Sugime, H., Noda, S. (2020). Electrolysis of ammonia in aqueous solution by platinum nanoparticles supported on carbon nanotube film electrode. *Electrochimica Acta*, 341, 136027. doi: 10.1016/j.electacta.2020.136027.

Hara, M., Kitano, M., Hosono, H. (2017). Ru-loaded C12A7:e. Electride as a catalyst for ammonia synthesis. *ACS Catal.*, 7(4), 2313–2324. doi: 10.1021/acscatal.6b03357.

Hargreaves, J.S.J. (2014). Nitrides as ammonia synthesis catalysts and as potential nitrogen transfer reagents. *Appl. Petrochem. Res.*, 46(4), 3–10. doi: 10.1007/s13203-014-0049-y.

Hejze, T., Besenhard, J.O., Kordesch, K., Cifrain, M., Aronsson, R.R. (2008). Current status of combined systems using alkaline fuel cells and ammonia as a hydrogen carrier. *J. Power Sources*, 176(2), 490–493. doi: 10.1016/j.jpowsour.2007.08.117.

Holme, B., Skaugset, P., Taftø, J. (1997). Microstructure of reduced magnetite as a function of aluminum content and reduction temperature. *Appl. Catal. Gen.*, 162(1–2), 149–159. doi: 10.1016/S0926-860X(97)00092-6.

Honkala, K., Hellman, A., Remediakis, I.N., Logadottir, A., Carlsson, A., Dahl, S., Christensen, C.H., Nørskov, J.K. (2005). Ammonia synthesis from first-principles calculations. *Science*, 307(5709), 555–558. doi: 10.1126/science.1106435.

Huang, J., Cai, J., Wang, J. (2020). Nanostructured wire-in-plate electrocatalyst for high-durability production of hydrogen and nitrogen from alkaline ammonia solution. *ACS Appl. Energy Mater.*, 3(5), 4108–4113. doi: 10.1021/acsaem.9b02545.

Humphreys, J., Lan, R., Tao, S. (2021). Development and recent progress on ammonia synthesis catalysts for Haber–Bosch process. *Adv. Energy Sustain. Res.*, 2(1), 2000043. doi: 10.1002/aesr.202000043.

Hunter, S.M., Mckay, D., Smith, R.I., Hargreaves, J.S.J., Gregory, D.H. (2010). Topotactic nitrogen transfer: Structural transformation in cobalt molybdenum nitrides. *Chem. Mater.*, 22(9), 2898–2907. doi: 10.1021/cm100208a.

Hunter, S.M., Gregory, D.H., Hargreaves, J.S.J., Richard, M., Duprez, D., Bion, N. (2013). A study of 15N/14N isotopic exchange over cobalt molybdenum nitrides. *ACS Catal.*, 3(8), 1719–1725. doi: 10.1021/cs400336z.

Inoue, Y., Kitano, M., Kishida, K., Abe, H., Niwa, Y., Sasase, M., Fujita, Y., Ishikawa, H., Yokoyama, T., Hara, M. et al. (2016). Efficient and stable ammonia synthesis by self-organized flat Ru nanoparticles on calcium amide. *ACS Catal.*, 6(11), 7577–7584. doi: 10.1021/acscatal.6b01940.

Itoh, N., Oshima, A., Suga, E., Sato, T. (2014). Kinetic enhancement of ammonia decomposition as a chemical hydrogen carrier in palladium membrane reactor. *Catal. Today*, 236(Part A), 70–76. doi: 10.1016/j.cattod.2014.02.054.

Jacobsen, C.J.H., Dahl, S., Hansen, P.L., Törnqvist, E., Jensen, L., Topsøe, H., Prip, D.V., Møenshaug, P.B., Chorkendorff, I. (2000). Structure sensitivity of supported ruthenium catalysts for ammonia synthesis. *J. Mol. Catal. Chem.*, 163(1–2), 19–26. doi: 10.1016/S1381-1169(00)00396-4.

Kitano, M., Inoue, Y., Yamazaki, Y., Hayashi, F., Kanbara, S., Matsuishi, S., Yokoyama, T., Kim, S.-W., Hara, M., Hosono, H. (2012). Ammonia synthesis using a stable electride as an electron donor and reversible hydrogen store. *Nat. Chem.*, 4(11), 934–940. doi: 10.1038/nchem.1476.

Kitano, M., Inoue, Y., Ishikawa, H., Yamagata, K., Nakao, T., Tada, T., Matsuishi, S., Yokoyama, T., Hara, M., Hosono, H. (2016). Essential role of hydride ion in ruthenium-based ammonia synthesis catalysts. *Chem. Sci.*, 7(7), 4036–4043. doi: 10.1039/C6SC00767H.

Kitano, M., Kujirai, J., Ogasawara, K., Matsuishi, S., Tada, T., Abe, H., Niwa, Y., Hosono, H. (2019). Low-temperature synthesis of perovskite oxynitride-hydrides as ammonia synthesis catalysts. *J. Am. Chem. Soc.*, 141(51), 20344–20353. doi: 10.1021/jacs.9b10726.

Kobayashi, Y., Tang, Y., Kageyama, T., Yamashita, H., Masuda, N., Hosokawa, S., Kageyama, H. (2017). Titanium-based hydrides as heterogeneous catalysts for ammonia synthesis. *J. Am. Chem. Soc.*, 139(50), 18240–18246. doi: 10.1021/jacs.7b08891.

Kok, S.H.W., Lee, J., Tan, L.-L., Ong, W.-J., Chai, S.-P. (2021). MXene. A new paradigm toward artificial nitrogen fixation for sustainable ammonia generation: Synthesis, properties, and future outlook. *ACS Mater. Lett.*, 4(2), 212–245. doi: 10.1021/acsmaterialslett.1c00673.

Kordesch, K., Hacker, V., Faleschini, G., Koscher, G., Cifrain, M. (2003). Ammonia as hydrogen source for an alkaline fuel cell–battery hybrid system. Report.

Kusada, K., Kobayashi, H., Yamamoto, T., Matsumura, S., Sumi, N., Sato, K., Nagaoka, K., Kubota, Y., Kitagawa, H. (2013). Discovery of face-centered-cubic ruthenium nanoparticles: Facile size-controlled synthesis using the chemical reduction method. *J. Am. Chem. Soc.*, 135(15), 5493–5496. doi: 10.1021/ja311261s.

Lamb, K.E., Dolan, M.D., Kennedy, D.F. (2019). Ammonia for hydrogen storage: A review of catalytic ammonia decomposition and hydrogen separation and purification. *Int. J. Hydrog. Energy*, 44(7), 3580–3593. doi: 10.1016/j.ijhydene.2018.12.024.

Larichev, Y.V., Moroz, B.L., Zaikovskii, V.I., Yunusov, S.M., Kalyuzhnaya, E.S., Shur, V.B., Bukhtiyarov, V.I. (2007). XPS and TEM studies on the role of the support and alkali promoter in Ru/MgO and Ru–Cs+/MgO catalysts for ammonia synthesis. *J. Phys. Chem. C*, 111(26), 9427–9436. doi: 10.1021/jp066970b.

Le, T.A., Do, Q.C., Kim, Y., Kim, T.-W., Chae, H.-J. (2021). A review on the recent developments of ruthenium and nickel catalysts for COx-free H_2 generation by ammonia decomposition. *Korean J. Chem. Eng.*, 38(6), 1087–1103. doi: 10.1007/s11814-021-0767-7.

Li, J., Wu, J., Wang, H., Lu, Y., Ye, T., Sasase, M., Wu, X., Kitano, M., Inoshita, T., Hosono, H. (2019). Acid-durable electride with layered ruthenium for ammonia synthesis: Boosting the activity via selective etching. *Chem. Sci.*, 10(22), 5712–5718. doi: 10.1039/C9SC01539F.

Liu, H. (2013). *Ammonia Synthesis Catalysts: Innovation and Practice.* World Scientific/Chemical Industry Press, Singapore. doi: 10.1142/8199.

Liu, H. (2014). Ammonia synthesis catalyst 100 years: Practice, enlightenment and challenge. *Chin. J. Catal.*, 35(10), 1619–1640. doi: 10.1016/S1872-2067(14)60118-2.

Liu, C., Nikolaev, S.A., Ren, W., Burton, L.A. (2020). Electrides: A review. *J. Mater. Chem. C*, 8(31), 10551–10567. doi: 10.1039/D0TC01165G.

Lu, Y., Li, J., Tada, T., Toda, Y., Ueda, S., Yokoyama, T., Kitano, M., Hosono, H. (2016). Water durable electride Y5Si3: Electronic structure and catalytic activity for ammonia synthesis. *J. Am. Chem. Soc.*, 138(12), 3970–3973. doi: 10.1021/jacs.6b00124.

Luo, Y., Chen, G.-F., Ding, L., Chen, X., Ding, L.-X., Wang, H. (2019). Efficient electrocatalytic N-2 fixation with MXene under ambient conditions. *Joule*, 3(1), 279–289. doi: 10.1016/j.joule.2018.09.011.

Makhloufi, C. and Kezibri, N. (2021). Large-scale decomposition of green ammonia for pure hydrogen production. *Int. J. Hydrog. Energy*, 46(70), 34777–34787. doi: 10.1016/j.ijhydene.2021.07.188.

Marakatti, V.S. and Gaigneaux, E.M. (2020). Recent advances in heterogeneous catalysis for ammonia synthesis. *ChemCatChem*, 12(23), 5838–5857. doi: 10.1002/cctc.202001141.

Marinčić, L. and Leitz, F.B. (1978). Electro-oxidation of ammonia in waste water. *J. Appl. Electrochem.*, 8, 333–345. doi: 10.1007/BF00612687.

Marinho, A.L.A., Comminges, C., Habrioux, A., Célérier, S., Bion, N., Morais, C. (2023). Reactivity of nitrogen atoms from Zif-8 structure deposited over Ti_3C_2 MXene in the electrochemical nitrogen reduction reaction. *Chem. Commun.*, 59(67), 10133–10136. doi: 10.1039/d3cc02693k.

Matsui, T., Suzuki, S., Katayama, Y., Yamauchi, K., Okanishi, T., Muroyama, H., Eguchi, K. (2015). *In situ* attenuated total reflection infrared spectroscopy on electrochemical ammonia oxidation over Pt electrode in alkaline aqueous solutions. *Langmuir*, 31(42), 11717–11723. doi: 10.1021/acs.langmuir.5b02330.

McKee, D.W., Hills, B., Scarpellino, A.J. (1973). Ammonia fuel cell with iridium catalyst and method of using the cell. US Patent, 3,730,774.

Medford, A.J., Vojvodic, A., Hummelshøj, J.S., Voss, J., Abild-Pedersen, F., Studt, F., Bligaard, T., Nilsson, A., Nørskov, J.K. (2015). From the Sabatier principle to a predictive theory of transition-metal heterogeneous catalysis. *J. Catal., Special Issue: The Impact of Haldor Topsøe on Catalysis*, 328, 36–42. doi: 10.1016/j.jcat.2014.12.033.

Meng, L. and Tsuru, T. (2016). Hydrogen production from energy carriers by silica-based catalytic membrane reactors. *Catal. Today*, 268, 3–11. doi: 10.1016/j.cattod.2015.11.006.

Metkemeijer, R. and Achard, P. (1994). Comparison of ammonia and methanol applied indirectly in a hydrogen fuel cell. *Int. J. Hydrog. Energy*, 19(6), 535–542. doi: 10.1016/0360-3199(94)90009-4.

Mittasch, A., Frankenburg, W. (1950). Early studies of multicomponent catalysts, In *Advances in Catalysis*, Frankenburg, W.G., Komarewsky, V.I., Rideal, E.K. (eds). Academic Press, Cambridge, MA. doi: 10.1016/S03600564 (08)60375-2.

Mizoguchi, H., Park, S.-W., Kishida, K., Kitano, M., Kim, J., Sasase, M., Honda, T., Ikeda, K., Otomo, T., Hosono, H. (2019). Zeolitic intermetallics: LnNiSi (Ln = La–Nd). *J. Am. Chem. Soc.*, 141(8), 3376–3379. doi: 10.1021/jacs.8b12784.

Modisha, P. (2016). Electrocatalytic process for ammonia electrolysis: A remediation technique with hydrogen co-generation. *Int. J. Electrochem. Sci.*, 11(8), 6627–6635. doi: 10.20964/2016.08.54.

Montoya, J.H., Tsai, C., Vojvodic, A., Nørskov, J.K. (2015). The challenge of electrochemical ammonia synthesis: A new perspective on the role of nitrogen scaling relations. *ChemSusChem*, 8(13), 2180–2186. doi: 10.1002/cssc.201500322.

Mortensen, J.J., Hansen, L.B., Hammer, B., Nørskov, J.K. (1999). Nitrogen adsorption and dissociation on Fe(111). *J. Catal.*, 182(2), 479–488. doi: 10.1006/jcat.1998.2364.

Nørskov, J., Bligaard, T., Rossmeisl, J., Christensen, C.H. (2009). Towards the computational design of solid catalysts. *Nat. Chem.*, 1(1), 37–46. doi: 10.1038/nchem.121.

Nørskov, J., Chen, J., Miranda, R., Fitzsimmons, T., Stack, R. (2016). Sustainable ammonia synthesis. Exploring the scientific challenges associated with discovering alternative, sustainable processes for ammonia production. Report, US DOE Office of Science. doi: 10.2172/1283146.

Panov, G.I. and Kharitonov, A.S. (1985). Catalytic properties of nitrides in ammonia synthesis. *React. Kinet. Catal. Lett.*, 29, 267–274. doi: 10.1007/BF02067978.

Por, E., Haase, G., Citri, O., Kosloff, R., Asscher, M. (1991). Effect of molecular energy content on the dissociative chemisorption of N2 on Re(0001). *Chem. Phys. Lett.*, 186(6), 553–560. doi: 10.1016/0009-2614(91)90466-M.

Qiu, J.-Z., Hu, J., Lan, J., Wang, L.-F., Fu, G., Xiao, R., Ge, B., Jiang, J. (2019). Pure siliceous zeolite-supported Ru single-atom active sites for ammonia synthesis. *Chem. Mater.*, 31(22), 9413–9421. doi: 10.1021/acs.chemmater.9b03099.

Raróg-Pilecka, W., Miśkiewicz, E., Szmigiel, D., Kowalczyk, Z. (2005). Structure sensitivity of ammonia synthesis over promoted ruthenium catalysts supported on graphitised carbon. *J. Catal.*, 231(1), 11–19. doi: 10.1016/j.jcat.2004.12.005.

Raróg-Pilecka, W., Miśkiewicz, E., Jodzis, S., Petryk, J., Łomot, D., Kaszkur, Z., Karpiński, Z., Kowalczyk, Z. (2006). Carbon-supported ruthenium catalysts for NH3 synthesis doped with caesium nitrate: Activation process, working state of Cs–Ru/C. *J. Catal.*, 239(2), 313–325. doi: 10.1016/j.jcat.2006.01.035.

Rathore, S.S., Biswas, S., Fini, D., Kulkarni, A.P., Giddey, S. (2021). Direct ammonia solid-oxide fuel cells: A review of progress and prospects. *Int. J. Hydrogen Energy*, 46(71), 35365–35384. doi: 10.1016/j.ijhydene.2021.08.092.

Royer, S., Duprez, D., Can, F., Courtois, X., Batiot-Dupeyrat, C., Laassiri, S., Alamdari, H. (2014). Perovskites as substitutes of noble metals for heterogeneous catalysis: Dream or reality. *Chem. Rev.*, 114(20), 10292–10368. doi: 10.1021/cr500032a.

Saadatjou, N., Jafari, A., Sahebdelfar, S. (2015). Ruthenium nanocatalysts for ammonia synthesis: A review. *Chem. Eng. Commun.*, 202(4), 420–448. doi: 10.1080/00986445.2014.923995.

Schlögl, R. (1991). Preparation and activation of the technical ammonia synthesis catalyst. In *Catalytic Ammonia Synthesis: Fundamentals and Practice, Fundamental and Applied Catalysis*, Jennings, J.R. (ed.). Springer, Boston. doi: 10.1007/978-1-4757-9592-9_2.

Sfeir, A., Teles, C.A., Ciotonea, C., Reddy, G.N.M., Marinova, M., Dhainaut, J., Löfberg, A., Dacquin, J.-P., Royer, S., Laassiri, S. (2023). Enhancing ammonia catalytic production over spatially confined cobalt molybdenum nitride nanoparticles in SBA-15. *Appl. Catal. B Environ.*, 325, 122319. doi: 10.1016/j.apcatb.2022.122319.

Simons, E.L., Cairns, E.J., Surd, D.J. (1969). The performance of direct ammonia fuel cells. *J. Electrochem. Soc.*, 116, 556. doi: 10.1149/1.2411961.

Singh, A.R., Montoya, J.H., Rohr, B.A., Tsai, C., Vojvodic, A., Nørskov, J.K. (2018). Computational design of active site structures with improved transition-state scaling for ammonia synthesis. *ACS Catal.*, 8(5), 4017–4024. doi: 10.1021/acscatal.8b00106.

Szmigiel, D., Bielawa, H., Kurtz, M., Hinrichsen, O., Muhler, M., Raróg, W., Jodzis, S., Kowalczyk, Z., Znak, L., Zieliński, J. (2002). The kinetics of ammonia synthesis over ruthenium-based catalysts: The role of barium and cesium. *J. Catal.*, 205(1), 205–212. doi: 10.1006/jcat.2001.3431.

Tamaru, K. (1991). The history of the development of ammonia synthesis. In *Catalytic Ammonia Synthesis: Fundamentals and Practice, Fundamental and Applied Catalysis*, Jennings, J.R. (ed.). Springer, Boston. doi: 10.1007/978-1-4757-9592-9_1.

Tang, Y., Kobayashi, Y., Masuda, N., Uchida, Y., Okamoto, H., Kageyama, T., Hosokawa, S., Loyer, F., Mitsuhara, K., Yamanaka, K. et al. (2018). Metal-dependent support effects of oxyhydride-supported Ru, Fe, Co catalysts for ammonia synthesis. *Adv. Energy Mater.*, 8(36), 1801772. doi: 10.1002/aenm.201801772.

Tencé, S., Mahon, T., Gaudin, E., Chevalier, B., Bobet, J.-L., Flacau, R., Heying, B., Rodewald, U.C., Pöttgen, R. (2016). Hydrogenation studies on NdScSi and NdScGe. *J. Solid State Chem.*, 242(Part 2), 168–174. doi: 10.1016/j.jssc.2016.02.017.

Tsuji, Y., Kitano, M., Kishida, K., Sasase, M., Yokoyama, T., Hara, M., Hosono, H. (2016). Ammonia synthesis over Co–Mo alloy nanoparticle catalyst prepared via sodium naphthalenide-driven reduction. *Chem. Commun.*, 52(100), 14369–14372. doi: 10.1039/C6CC08566K.

Valera-Medina, A., Xiao, H., Owen-Jones, M., David, W.I.F., Bowen, P.J. (2018). Ammonia for power. *Prog. Energy Combust. Sci.*, 69, 63–102. doi: 10.1016/j.pecs.2018.07.001.

Vidal-Iglesias, F.J., García-Aráez, N., Montiel, V., Feliu, J.M., Aldaz, A. (2003). Selective electrocatalysis of ammonia oxidation on Pt(100) sites in alkaline medium. *Electrochem. Commun.*, 5(1), 22–26. doi: 10.1016/S1388-2481(02)00521-0.

Viscardi, R. and Matteo, P. (2020). Design of an ammonia electrolyzer for hydrogen production. *Int. J. Recent Eng. Sci.*, 7(5), 18–21. doi: 10.14445/23497157/IJRES-V7I5P103.

Vitse, F., Cooper, M., Botte, G.G. (2005). On the use of ammonia electrolysis for hydrogen production. *J. Power Sources*, 142(1–2), 18–26. doi: 10.1016/j.jpowsour.2004.09.043.

Vojvodic, A., Medford, A.J., Studt, F., Abild-Pedersen, F., Khan, T.S., Bligaard, T., Nørskov, J.K. (2014). Exploring the limits: A low-pressure, low-temperature Haber–Bosch process. *Chem. Phys. Lett.*, 598, 108–112. doi: 10.1016/j.cplett.2014.03.003.

Wang, B. and Shen, L. (2022). Recent advances in NH_3 synthesis with chemical looping technology. *Ind. Eng. Chem. Res.*, 61(50), 18215–18231. doi: 10.1021/acs.iecr.2c02926.

Wang, P., Chang, F., Gao, W., Guo, J., Wu, G., He, T., Chen, P. (2017). Breaking scaling relations to achieve low-temperature ammonia synthesis through LiH-mediated nitrogen transfer and hydrogenation. *Nat. Chem.*, 9, 64–70. doi: 10.1038/nchem.2595.

Wang, X., Peng, X., Ran, H., Lin, B., Ni, J., Lin, J., Jiang, L. (2018). Influence of Ru substitution on the properties of $LaCoO_3$ catalysts for ammonia synthesis: XAFS and XPS studies. *Ind. Eng. Chem. Res.*, 57(51), 17375–17383. doi: 10.1021/acs.iecr.8b04915.

Wang, Q., Guo, J., Chen, P. (2019). Recent progress towards mild-condition ammonia synthesis. *J. Energy Chem.*, 36, 25–36. doi: 10.1016/j.jechem.2019.01.027.

Wu, J., Gong, Y., Inoshita, T., Fredrickson, D.C., Wang, J., Lu, Y., Kitano, M., Hosono, H. (2017). Tiered electron anions in multiple voids of LaScSi and their applications to ammonia synthesis. *Adv. Mater.*, 29(36), 1700924. doi: 10.1002/adma.201700924.

Xue, Q., Zhao, Y., Zhu, J., Ding, Y., Wang, T., Sun, H., Li, F., Chen, P., Jin, P., Yin, S., Chen, Y. (2021). PtRu nanocubes as bifunctional electrocatalysts for ammonia electrolysis. *J. Mater. Chem. A*, 9(13), 8444–8451. doi: 10.1039/D1TA00426C.

Yang, X.-L., Zhang, W.-Q., Xia, C.-G., Xiong, X.-M., Mu, X.-Y., Hu, B. (2010). Low temperature ruthenium catalyst for ammonia synthesis supported on $BaCeO_3$ nanocrystals. *Catal. Commun.*, 11(10), 867–870. doi: 10.1016/j.catcom.2010.03.008.

Yang, Y., Kim, J., Jo, H., Seong, A., Lee, M., Min, H.-K., Seo, M., Choi, Y., Kim, G. (2021). A rigorous electrochemical ammonia electrolysis protocol with *in operando* quantitative analysis. *J. Mater. Chem. A*, 9(19), 11571–11579. doi: 10.1039/D1TA00363A.

Yang, Z., Li, Q., Zhang, Y., Chen, Z.-N., Zhang, L., Yang, Y. (2022). Microenvironment regulation of the $Ti_3C_2T_X$ MXene surface for enhanced electrochemical nitrogen reduction. *Acs Appl. Mater. Interfaces*, 14(50), 56344–56352. doi: 10.1021/acsami.2c17092.

Ye, T.-N., Park, S.-W., Lu, Y., Li, J., Sasase, M., Kitano, M., Hosono, H. (2020). Contribution of nitrogen vacancies to ammonia synthesis over metal nitride catalysts. *J. Am. Chem. Soc.*, 142(33), 14374–14383. doi: 10.1021/jacs.0c06624.

Ye, T.-N., Park, S.-W., Lu, Y., Li, J., Wu, J., Sasase, M., Kitano, M., Hosono, H. (2021). Dissociative and associative concerted mechanism for ammonia synthesis over Co-based catalyst. *J. Am. Chem. Soc.*, 143(32), 12857–12866. doi: 10.1021/jacs.1c06657.

Yin, S.F., Xu, B.Q., Zhou, X.P., Au, C.T. (2004a). A mini-review on ammonia decomposition catalysts for on-site generation of hydrogen for fuel cell applications. *Appl. Catal. Gen.*, 277(1–2), 1–9. doi: 10.1016/j.apcata.2004.09.020.

Yin, S.F., Zhang, Q.-H., Xu, B.-Q., Zhu, W.-X., Ng, C.-F., Au, C.-T. (2004b). Investigation on the catalysis of CO^x-free hydrogen generation from ammonia. *J. Catal.*, 224(2), 384–396. doi: 10.1016/j.jcat.2004.03.008.

Zeinalipour-Yazdi, C.D., Hargreaves, J.S.J., Catlow, C.R.A. (2016). DFT-D3 study of molecular N-2 and H-2 activation on Co_3Mo_3N surfaces. *J. Phys. Chem. C*, 120(38), 21390–21398. doi: 10.1021/acs.jpcc.6b04748.

5

Reversible Hydrogen Storage: Intermetallic Compounds and Mg-based Materials

Ghofrane FEDLOUK, Hugo BENET, Valérie PAUL-BONCOUR, Judith MONNIER and Junxian ZHANG
ICMPE, CNRS, UPEC, Thiais, France

5.1. Introduction

Hydrogen energy will undoubtedly be one of the main energy vectors in the future, but transportation and hydrogen storage are crucial issues that still need improvement. The criterion for storing hydrogen is to be safe and efficient and to be used anywhere and at any time. The solid-state storage method has the highest specific volumetric energy density. It can be divided into two categories: physisorption and chemisorption. Physisorption is based on van der Waals interactions between hydrogen molecules and materials, and normally, the storage capacity is directly correlated with the specific surface. Chemisorption consists of a chemical reaction between hydrogen atoms and elements, and the chemical bonds can be ionic, covalent, or metallic. Ionic (alkaline metal hydrides) or covalent (water or ammonia) hydrides form very stable compounds that cannot be used for reversible hydrogen storage under ambient conditions. The metal hydrides present metallic bonding between metal atoms and hydrogen atoms, with the latter occupying the interstitial sites. There are binary, ternary and multielement hydrides. The periodic table contains elements that form very stable hydrides, called A-elements, such as rare earths, alkali metals, alkali earth metals or early transition

Hydrogen Storage,
coordinated by Patricia DE RANGO and Fermin CUEVAS.

metals (YH_2 releases hydrogen at temperatures higher than 1000 K); the other elements, called B-elements, usually last for transition metals, and p-elements such as Al, Si and Sn, form very unstable hydrides ($NiH_{0.6}$ needs a pressure of 500 MPa to be formed). By combining these elements, it is possible to form alloys or intermetallic compounds that display intermediate thermodynamic stabilities.

Early works on intermetallic hydrides, which appeared in the 1950s, have already shown that intermetallic compounds containing *A*- (Zr) and *B*- (Ni) elements can reversibly absorb a large amount of hydrogen (Libowitz et al. 1958). Even though the hydrogen content in $ZrNiH_3$ is quite high (~2 wt%), a high temperature (523 K for $ZrNiH_3$ to ZrNiH and 773 K for ZrNiH at ambient pressure) is needed to release hydrogen (Burch and Mason 1979). The "accidental" discovery of the $SmCo_5H_{2.8}$ metal hydride (Zijlstra and Westendorp 1969), which is able to release hydrogen at ambient temperature, opened an avenue for the use of metal hydrides as hydrogen storage materials. The later discovery of reversible hydrogen sorption in $LaNi_5$ and TiFe opened new possibilities for hydrogen storage developments (Reilly and Wiswall 1974; Willems and Buschow 1987). Metal hydrides are the most compact way to store hydrogen; for example, it is possible to store up to 150 g H_2/L in $LaNi_5$, which has a two-time larger volumetric density than liquid hydrogen (71 g H_2/L).

Metal hydrides exhibit a reversible interaction between hydrogen and metals (elements such as Mg, body centered cubic (BCC) V-based alloys or intermetallic compounds such as $LaNi_5$). Hydride formation or hydrogen absorption is exothermic, whereas hydride decomposition or hydrogen desorption is endothermic. The heat released during hydrogen absorption, that is, the enthalpy of hydride formation ΔH_f, is closely related to the crystal structure and chemistry of the metals. The hydrogen desorption temperature and pressure depend directly on the dehydrogenation enthalpy, $\Delta H \sim -\Delta H_f$. To fulfill the temperature (293 K $\leq T \leq$ 373 K) and pressure (0.1 MPa $\leq P \leq$ 1 MPa) criteria to supply a fuel cell, the enthalpies must be between 25 and 50 kJ/mol H_2.

The first part of this chapter is devoted to materials that store hydrogen reversibly under the abovementioned conditions, that is, BCC V-based alloys, *AB*, AB_2, AB_5 and intergrowth AB_x compounds. Mg based materials do not fulfill all these prerequisites because of the high stability ($\Delta H_f \sim -75$ kJ/mol H_2) and sluggish kinetics of MgH_2. However, Mg is one of the most abundant elements in the Earth's crust, and its low cost and high hydrogen content (7.6 wt%) make it very attractive for various applications. Therefore, many studies have been performed to improve the kinetic by nanostructuring and tuning the thermodynamic properties of Mg by alloying effects. These studies will be developed in section 5.3.

5.2. Intermetallic compounds

5.2.1. *BCC-type alloys*

Body-centered cubic (Space Group, *s.g.* $Im\bar{3}m$) metals and alloys have been studied owing to their high hydrogen storage capacity (Akiba and Iba 1998). Vanadium is a typical BCC metal used for hydrogen storage (Akiba and Iba 1998). It has a storage capacity of close to 4 wt%, which is among the highest for metal hydrides after Mg (7.6 wt%) (Kumar et al. 2017). However, it has many disadvantages, such as the presence of two plateaus at very different equilibrium pressures (0.1 Pa and 0.3 MPa at room temperature) separated by a slope, which can be problematic for various applications (Figure 5.1). It also exhibits slow kinetic and difficulties in the activation process (Akiba and Iba 1998).

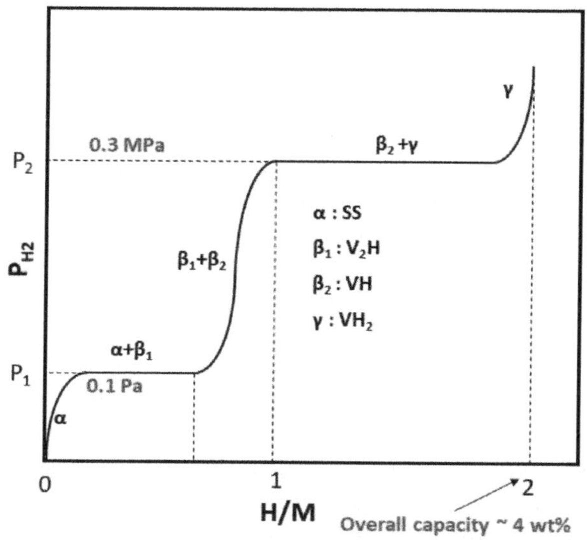

Figure 5.1. *Schematic PCI curve for Vanadium metal at room temperature, M stands for a metallic atom (Kumar et al. 2017). For a color version of this figure, see www.iste.co.uk/derango-cuevas/hydrogen.zip*

To modify and improve these properties, the alloying of other metals with *BCC*-type structure was considered. For binary alloys, Ti-V is the first system to have been studied. However, since its equilibrium pressure is directly related to the

V/Ti ratio, a decrease in the V content results in a very low undetectable plateau. Therefore, to accurately measure the PCI of V_xTi_{1-x} ($x = 0.4; 0.5$), high temperatures (383–623 K) are needed, which can be problematic in the case of applications (Matsuda and Akiba 2013). However, other systems, such as V-Al, V-Mn and V-Ru, have also been investigated (Liu et al. 2018). The alloying of V with a small amount (5 at%) of these elements improved the hydrogenation–dehydrogenation kinetics under moderate conditions and improved the reversible capacity. Alloying with other *3d*, *4d* and *5d* transition metals, such as Cr, Mo, Ta and Nb, was also studied. The addition of all these elements leads to less stable hydrides (Kumar et al. 2015). However, alloying with Nb is reported to lower the capacity and increase the absorption/desorption hysteresis (Lynch et al. 1984). Moreover, these alloys have poor cycling abilities; therefore, the addition of a third element was considered.

Among the studied ternary systems (Ti-V-Fe, Ti-V-Mn, Ti-V-Co, Ti-V-Ni, etc.), Ti-V-Cr is the most investigated and promising system. The addition of Cr improves the system's cycling ability (up to 100 cycles), while its hydrogen storage capacity remains stable (Kumar et al. 2017; Itoh et al. 2005). By varying the alloy composition, the storage capacity can be modified. The highest and most effective hydrogen storage capacity was obtained for $V_{60}Ti_{15}Cr_{25}$ at 2.6 wt% at room temperature and an applied pressure of 3.3 MPa (Tsukahara 2011). The addition of other elements, such as Fe and Al, to the Ti-V-Cr system was also studied. For example, when the Fe content increases (from 1 to 3 at%), the hydrogen storage capacity of these alloys decreases (from 2.3 to 0.75 wt%), and their plateau equilibrium pressure increases (from 0.3 to 1 MPa) at 298 K (Wu et al 2021b).

The Ti-V-Fe system has also shown promising results. The largest absorption capacity of 3.9 wt%, for an applied pressure of 3 MPa, was obtained for $Ti_{43.5}V_{49}Fe_{7.5}$ at 253 K, and a reversible capacity of 2.4 wt% was obtained at 573 K (Nomura and Akiba 1995; Kumar et al. 2017). However, increasing the percentage of iron in the Ti-V-Fe system decreases the formation enthalpy, which results in a destabilized hydride (Massicot 2009). Other elements (such as Mn, Ni and Zr) were then introduced into the same system, and as such, the addition of Mn and Ni can flatten the plateau region and enhance the hydrogenation-dehydrogenation kinetics (Kumar et al. 2017).

BCC-type metals and alloys are therefore very promising materials with interesting hydrogen storage capacities, and much work has been done with multielement compounds. Currently, there are multiple new studies on BCC high-entropy alloys composed of at least five base elements with atomic percentages ranging from 5 to 35%. This topic is discussed in detail in the next chapter.

5.2.2. *AB* compounds

ZrNi is the first reported intermetallic compound to absorb hydrogen in a reversible way (Libowitz et al. 1958). *AB* compounds have become popular since the discovery of TiFe and TiNi compounds with CsCl-type structure (*s.g.* $Pm\bar{3}m$) and can be used as negative electrode in nickel-metal hydride (Ni-*M*H) batteries (Cuevas et al. 2001). For TiFe, *pressure-composition isotherms* (PCIs) at ambient temperature are characterized by two subsequent plateau pressures related to the consecutive formation of the monohydride, β-TiFeH (orthorhombic, *s.g.* $P222_1$), and the dihydride, γ-$TiFeH_2$ (orthorhombic, *s.g.* *Cmmm*), which represents a high storage capacity of 1.9 wt%, as shown in Figure 5.2 (Reilly and Wiswall 1974; Ćirić et al. 2012). It can withstand multiple cycles of hydrogenation–dehydrogenation with no visible amorphization (Zadorozhnyy et al. 2012). However, as surface oxidation of TiFe alloys blocks hydrogen penetration, activation at high temperature (673 K) and pressure (5 MPa) is often needed (Bououdina et al. 2006; Dematteis et al. 2021). Furthermore, their poor hydrogenation kinetics and high equilibrium pressure can be problematic for reversible hydrogen storage.

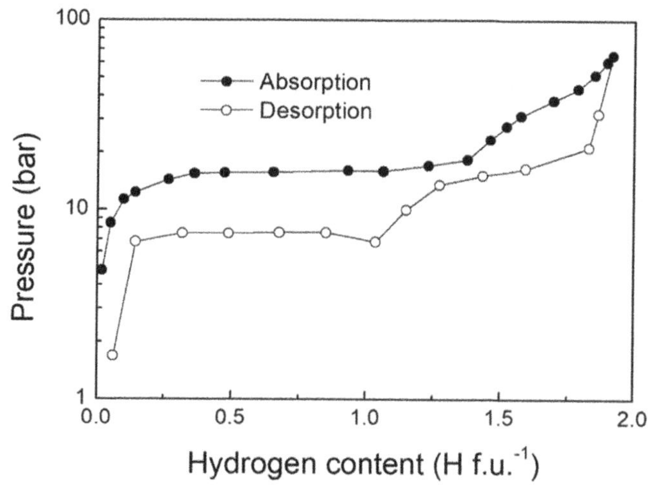

Figure 5.2. *PCI for TiFe at 40°C (Reilly and Wiswall 1974)*

The partial substitution of Fe in TiFe by Mn, Ni, Cr, V and even Ti can modify the activation conditions and decrease the equilibrium pressure (Rusman and Dahari 2016). Among the listed substitutions, that of Fe by Mn (Dematteis et al. 2021; Zheng et al. 2022) leads to an improvement in the performance of the alloys toward hydrogen storage by increasing their resistance to gas impurities and reducing the activation temperature from 673 to 423 K (Yuan et al. 2021). For $TiFe_{0.8}Mn_{0.2}$, the hydrogen storage capacity is 1.79 wt% at temperature of 303 K up to 3 MPa (Yuan et al. 2021).

The substitution of Fe by Ni has also been widely investigated, and it was found that it lowers the equilibrium pressure and slightly improves the maximum storage capacity and the kinetics of the whole process (Ćirić et al. 2012). The increase in the quantity of Ni in $TiFe_{1-x}Ni_x$ at a low Ni content lowered the equilibrium pressure of the first plateau (from 0.2 MPa for $x = 0.1$–0.03 MPa for $x = 0.2$) at 323 K (Lee and Perng 1999), which increased its hydrogenation process kinetics and eliminated the need for an activation process (Lee and Perng 1999; Bououdina et al. 2006; Rusman and Dahari 2016). However, a high amount of Ni results in an increase in the hydrogen desorption temperature and a decrease in the storage capacity; therefore, the most promising composition for this system is $TiFe_{0.8}Ni_{0.1}$, with a maximal storage capacity of 1.1 wt% and an equilibrium pressure of approximately 0.2 MPa at 323 K (Ćirić et al. 2012).

5.2.3. *AB*₂ compounds

AB_2 intermetallic compounds, often called Laves phases, can crystallize in three different types of crystal structures: *C*14 ($P6_3/mmc$), *C*15 ($Fd\bar{3}m$) and *C*36 ($P6_3/mmc$) (Figure 5.3). In these compounds, the hydrogen atoms can occupy three tetrahedral interstitial sites. Each of these sites presents a unique environment (Table 5.1).

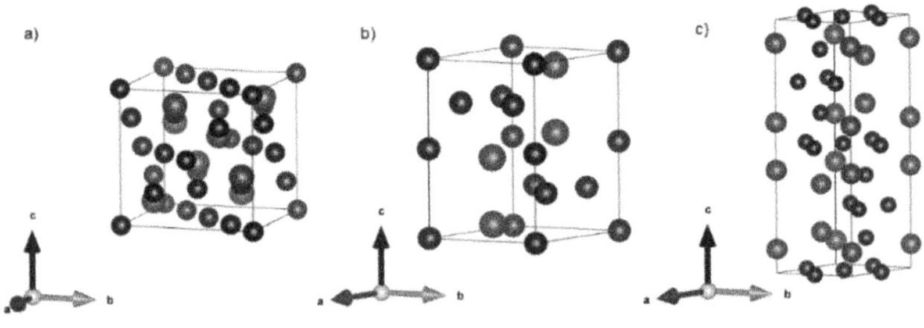

Figure 5.3. *(a) $MgCu_2$-type structure (C15), (b) $MgZn_2$-type structure (C14), and (c) $MgNi_2$-type structure (C36) (Mg in red, Cu, Zn and Ni in blue). For a color version of this figure, see www.iste.co.uk/derango-cuevas/hydrogen.zip*

AB_2 compounds crystallizing in *C*14 and *C*15 structures are known to have high hydrogen storage capacities and fast kinetics and can withstand multiple cycles

(Bououdina et al. 2006). One of the most effective *A*-elements for promoting the formation of Laves phases is zirconium (Zr) (Rusman and Dahari 2016). $ZrCo_2$ and $ZrFe_2$ (*C*15) have very low storage capacities, approximately 0.25 wt% for $ZrCo_2$ and 0.54 wt% for $ZrFe_2$, under ambient conditions because of their elevated equilibrium pressures (80 MPa) (Filipek et al. 2010). The partial substitution of Fe and Co by Al or Mn, which are elements with a relatively large atomic radius, decreases the alloy equilibrium pressure because of the increase in the cell volume (Yadav et al. 2012). However, Al substitution decreases the hydrogen storage capacity. The partial substitution of Fe by V in $ZrFe_2$ decreases the plateau pressure and increases the storage capacity to 1.68 wt%, which is obtained for $Zr(Fe_{0.5}V_{0.5})_2$ with an applied pressure of 6 MPa at room temperature (Yadav et al. 2012).

The capacity is greater for other *B*-elements, such as V, Cr and Mn: 2.76 wt% for ZrV_2 (*C*15-type structure), 2 wt% for $ZrCr_2$ and 1.8 wt% for $ZrMn_2$ (*C*14-type structure), at room temperature and under 0.1 MPa (Yartys and Lototskyy 2022). However, the formed hydrides are very stable (Bououdina et al. 2006). Therefore, pseudobinary compounds, such as $ZrCr_{2-x}M_x$ (M = Mn, Fe, Cu, Co, Ni and $x < 0.8$), were considered. They form less stable hydrides, are easier to desorb and are more suitable for applications in a large range of temperatures (296–473 K) and pressures (1–10 MPa) (Drasner 1990; Drasner and Blažina 1991; Bououdina et al. 2006; Yartys and Lototskyy 2022).

Atom	Environment	*C*14	*C*15
		Site	
H	*A2B2*	12*k*	96*g*
H	*AB3*	4*f*	32*e*
H	*B4*	4*e*	8*b*

Table 5.1. *Possible tetrahedral interstitial sites and their environments for each hydrogen atom in AB_2 (C14 and C15) compounds (Gesari et al. 2010)*

AFe$_2$ and AMn$_2$ (A = rare earth) compounds were also studied. YFe_2 can absorb up to 2.5 wt% hydrogen at room temperature and 0.8 MPa (approximately 5 H/formula unit, f.u.) (Ropka et al. 2009). However, the absorption of a large amount of hydrogen can cause a complete change in its initial structure, and it is difficult to desorb completely at room temperature (RT) (Leblond et al. 2009; Paul-Boncour et al. 2017). YMn_2 has a maximum reversible storage capacity of approximately 2.2 wt% at ambient temperature and pressure (4.4 H/f.u.) (Przewoznik et al. 1995). In various papers, deuterium was used instead of hydrogen to study the crystallographic structure of the hydride by neutron diffraction (Paul-Boncour et al. 2005).

Other AB_2 compounds have been compiled in excellent reviews and books (Paul-Boncour et al. 2018; Lys et al. 2020; Téliz et al. 2020; Wu et al 2021a; Yartys and Lototskyy 2022).

5.2.4. AB_5 compounds

AB_5 intermetallic compounds are also known as Haucke phases. They are one of the most studied families for hydrogen storage. They have a unique structure, which is a $CaCu_5$ hexagonal crystal structure described in the *P6/mmm* space group (Figure 5.4). As observed for AB_2, the hydrogen atoms occupy specific tetrahedral interstices that have different environments (Table 5.2).

Atom	Environment	Site
H	*B4*	4*h*
H	*A2B2*	6*m*
H	*AB3*	12*n*
H	*AB3*	12*o*

Table 5.2. *Possible tetrahedral interstitial sites and their environments for each hydrogen atom in AB₅ compounds (Percheron-Guégan et al. 1980)*

These compounds have a simple activation process and fast absorption/desorption kinetics under ambient conditions, but have a limited capacity of 1.5 wt% (approximately 6.5 H/f.u.) at room temperature and an applied pressure of 2 MPa (Percheron-Guégan et al. 1985; Lys et al. 2020). The most interesting and studied compounds are $LaNi_5$ and their substituted derivatives. $LaNi_5$ has a typical capacity of 1.4 wt% under ambient conditions. In addition to its good storage capacity, the pressure of its absorption plateau is near ambient pressure (approximately 0.2 MPa) at room temperature, which makes it a promising material for applications (Joubert et al. 2021). Substitutions of Ni with other 3*d*, 4*d* and 5*d* metals, such as Fe, Al, Co, Sn, Cu and Mn, were performed (Table 5.3). They do not improve the storage capacity, but these substitutions tailor the thermodynamic properties of the hydride and make it suitable for various applications. The use of AB_5-type alloys as negative electrodes in Ni-*M*H batteries is one of the largest applications of metal hydrides (Joubert et al. 2021).

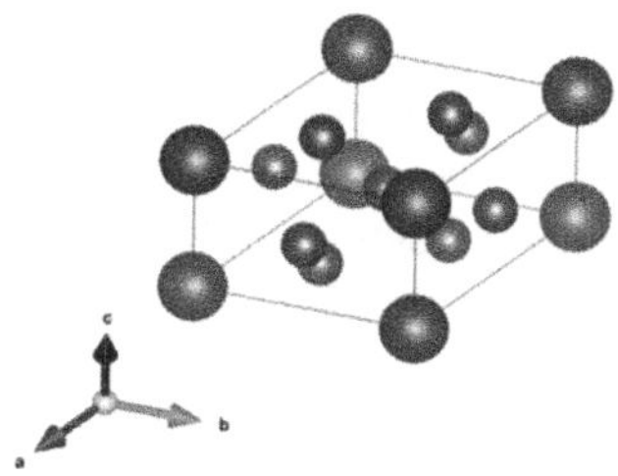

Figure 5.4. *$CaCu_5$-type structure (Ca in red, Cu in blue). For a color version of this figure, see www.iste.co.uk/derango-cuevas/hydrogen.zip*

Other compounds, such as ACo_5, ANi_5 and AFe_5 (A = Y, La, Ce, Pr, Nd, Sm, Gd, Ca), have also been studied and compiled in several excellent reviews (Paul-Boncour et al. 2018; Lys et al. 2020; Joubert et al. 2021).

Compound	a (Å)	c (Å)	V (Å^3)	References
$LaNi_5$	5.019	3.981	86.85	(Joubert et al. 2003)
$LaNi_{4.5}Mn_{0.5}$	5.045	4.022	88.65	(Lartigue et al. 1980)
$LaNi_4Mn$	5.089	4.082	91.55	(Lartigue et al. 1980)
$LaNi_{4.5}Fe_{0.5}$	5.0372	4.0008	87.91	(Nakamura et al. 2000)
$LaNi_4Fe$	5.0372	4.0008	88.64	(Pasturel et al. 1982)
$LaNi_4Co$	5.037	3.983	87.52	(Colinet et al. 1987)
$LaNiCo_4$	5.085	3.988	89.30	(Colinet et al. 1987)
$LaNi_4Cu$	5.033	4.007	87.90	(Pasturel et al. 1984)
$LaNiCu_4$	5.136	4.069	92.95	(Pasturel et al. 1984)
$LaNi_{4.5}Al_{0.5}$	5.043	4.025	88.65	(Diaz et al. 1979)
$LaNi_4Al$	5.061	4.07	90.28	(Diaz et al. 1979)
$LaNi_{4.5}Sn_{0.5}$	5.104	4.074	91.91	(Joubert et al. 1999)

Table 5.3. *Lattice parameters of $LaNi_{5-x}M_x$ (M = Mn, Fe, Co, Cu, Al, Sn) compounds with $CaCu_5$-type structure and the P6/mmm space group*

5.2.5. *AB_x* compounds (2 < x < 5)

AB_x (A = rare earth, Mg, Ca; B = Ni, Co, Fe, Mn and Al) intermetallic compounds are also called intergrowth compounds since their structure is the stacking of two units previously described, [A_2B_4] and [AB_5], along the crystallographic axis *c*. These compounds have two different types of structure depending on the stacking sequence: a hexagonal structure (2*H*) in the $P6_3/mmc$ space group and a rhombohedral structure (3*R*) in the $R\bar{3}m$ space group (Crivello et al. 2011). The hexagonal structure (*2H*) is formed by stacking of elemental units alternating two polymorphs of the unit [A_2B_4] (*C*15 and *C*14) and [AB_5], forming two elemental bricks. The polymorphic rhombohedral structure (3*R*) is formed by three identical elemental bricks using only the *C*15 polymorph for the unit [A_2B_4] (Figure 5.5). The structure of AB_x can be described by the stacking of the units [A_2B_4] and [AB_5] according to the following formula:

$$AB_x = \boldsymbol{p}\,[A_2B_4] + \boldsymbol{n}\,[AB_5] \tag{5.1}$$

with $x = \frac{5n+4p}{n+2p}$, where $\boldsymbol{p}$ and $\boldsymbol{n}$ are integers.

The most studied compounds are AB_3 (p = 1, n = 1), A_2B_7 (p = 1, n = 2) and A_5B_{19} (p = 1, n = 3) because of their interesting hydrogenation properties. In addition to these compounds, a few other less reported compounds exist, such as A_5B_{13}, A_7B_{23}, A_6B_{24} and A_7B_{29}. They will be discussed later in section 5.2.5.4.

Many high-quality reviews on rare earth- and Mg-Ni-based AB_x alloys have been published (Liu et al. 2011; Ouyang et al. 2017; Jiang et al. 2021). For this section, we emphasize compounds other than Mg-containing compounds.

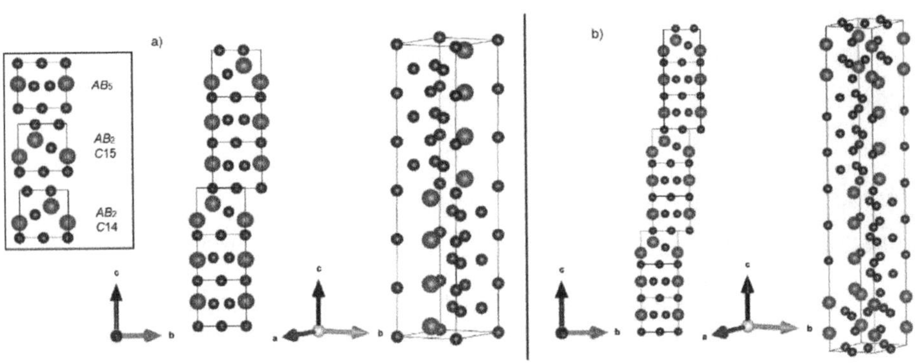

Figure 5.5. *Schematic diagram showing the stacking and structure of Ce_2Ni_7-type A_2B_7 compounds (2H) (a) and Gd_2Co_7-type A_2B_7 compounds (3R) (b) (A: elements in red; B: elements in blue). For a color version of this figure, see www.iste.co.uk/derango-cuevas/hydrogen.zip*

5.2.5.1. AB_3 compounds

As mentioned previously, AB_3 compounds have two different structures. They can crystallize either in the $PuNi_3$-type rhombohedral structure or in the $CeNi_3$-type hexagonal structure. These structures are formed by the stacking of one unit of [A_2B_4] and one unit of [AB_5] along the *c* axis.

Most rare earth elements and calcium as *A*-elements can form binary AB_3 compounds with B = Ni, Fe and Co. These alloys can absorb large quantities of hydrogen. Multiple studies on different systems, such as ACo_3, AFe_3 and ANi_3 (A = Ce, Ca, Dy, Er, Gd, Ho, Mm, Nd, Pr, Sm, Tb), have shown variable capacities ranging from 0.6 wt.% for YCo_3 at a temperature of 313 K and an equilibrium pressure of 60.8 kPa to 2 wt% for $CaNi_3$ at a temperature of 293 K and an equilibrium pressure of 40.5 kPa (Chen et al. 2000b; Paul-Boncour et al. 2018).

Rare earth-based binary AB_3 compounds are not completely reversible, and they suffer severe degradation of their initial structure after a hydrogenation/ dehydrogenation cycle. Subunits [AB_5] and [A_2B_4] both contain six atoms and possess similar volumes. However, [A_2B_4] is slightly larger than [AB_5], as it contains an extra larger "*A*" atom. This difference induces a mismatch leading to microstrains between the two subunits, resulting in structural instability.

Therefore, substitutions of both "*A*" and "*B*" elements were conducted to resolve this issue. The most effective alternative is rare earth substitution by magnesium, as Mg occupies only the rare earth site in [A_2B_4], which drastically reduces the volume of [A_2B_4] subunits (Crivello et al. 2011). The particularity of AB_3 compounds is that the rare earth atoms in [A_2B_4] can be completely substituted by Mg, which is not observed in other compounds with stacking structures, resulting in AMg_2Ni_9 compounds (which will be discussed in section 5.3).

Another strategy is to substitute large rare earth elements (La) with smaller rare earth elements such as Y in $LaNi_3$ (Boussami et al. 2013; Belgacem et al. 2015) or to substitute Ni with Mn in YNi_3 and $LaNi_3$. Recent studies have shown that the structures of the pseudobinary alloys $(La,Y)(Ni,Mn)_3$ are preserved after 30 hydrogenation–dehydrogenation cycles (Yan et al. 2017; Shi et al. 2019).

In a recent work, a series of AB_3 intermetallics, including $PrNi_3$, $NdNi_3$, $SmNi_3$, $SmNi_{2.67}Mn_{0.33}$, and $Sm_{0.9}Mg_{0.1}Ni_3$, was studied (Fang et al. 2019). The investigation of their hydrogen storage and crystal structural properties revealed a correlation between the subunit volume and reversibility: the [A_2B_4] and [AB_5] subunits should have critical volumes below 89.2 Å^3 and 88.3 Å^3, respectively. This hypothesis opens a door to investigate the Mg-free AB_3 compounds as hydrogen storage materials.

5.2.5.2. A_2B_7 compounds

As previously mentioned, A_2B_7 compounds also exist in two different structures. A hexagonal Ce_2Ni_7-type structure and a rhombohedral Gd_2Co_7-type structure are formed by the stacking of two [AB_5] units and one [A_2B_4] unit along the *c* axis (Figure 5.5). These two polymorph structures are often observed in A_2B_7-type compounds.

The most studied alloys are based on A = La, Y, Sm, Gd, or Mg and B = Ni, Mn, or Al (Crivello et al. 2011; Wei et al. 2011; Charbonnier et al. 2015, 2016; Xiong et al. 2017; Yan et al. 2017; Guo et al. 2021; Zhao et al. 2021; Zhang et al. 2022). For binary alloys, such as La_2Ni_7 and Y_2Ni_7, good storage capacities (1.6 wt% and 1.5 wt%) were measured at 298 K at an applied pressure of 10 MPa (Zhang et al. 2021). However, they exhibit multiplateau behavior, and H absorption in La_2Ni_7 is not fully reversible (Figure 5.6).

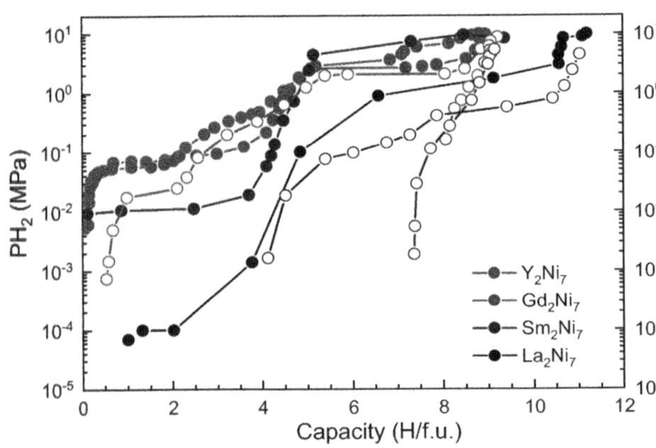

Figure 5.6. *PCI curves for A_2Ni_7 compounds (A = La, Sm, Gd, Y) measured at 298 K (Zhang et al. 2021). For a color version of this figure, see www.iste.co.uk/derango-cuevas/hydrogen.zip*

The most studied A-side substitution is the partial substitution of different A-elements by Mg, which noticeably improves their reversibility, reduces the number of plateaus and makes them suitable for practical applications (Zhang et al. 2021). The highest amount of Mg that can be introduced into A_2Ni_7 is 5.5 at%, corresponding to the composition $A_{1.5}Mg_{0.5}Ni_7$. This structure can be described as the stacking of 2 [ANi_5] + [$AMgNi_4$] subunits, with Mg located only in the [A_2B_4] subunits and occupying only half of the A-sites (Crivello et al. 2011; Liu et al. 2016). For example, for $La_{1.5}Mg_{0.5}Ni_7$, Mg substitution lowers the equilibrium pressure of La_2Ni_7 from 2 MPa to 0.02 MPa at 298 K, so its 2 plateaus turn into a unique flat plateau. Mg remarkably improved the alloy's reversibility but lowered its maximum storage capacity from 1.6 wt% to 1.47 wt% (Denys et al. 2008) (Figure 5.7).

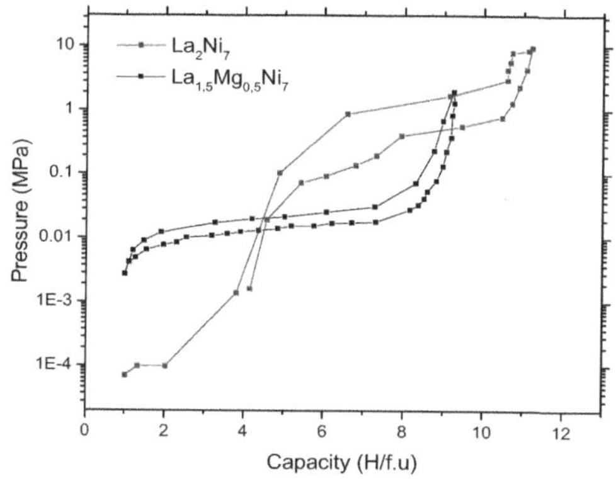

Figure 5.7. *PCI curves of $La_{2-x}Mg_xNi_7$ (x = 0; 0.5) at 298 K adapted from (Denys et al. 2008; Zhang et al. 2021). For a color version of this figure, see www.iste.co.uk/derango-cuevas/hydrogen.zip*

However, owing to the very different melting points and high evaporation pressure of Mg, the synthesis of Mg-containing intermetallic compounds faces security issues, especially at the industrial scale. For AB_3 compounds, the subunit volume rule is also valid for the A_2B_7 phase. Recently, Mg-free substitutions of both elements A (A = Y, La, Sm, Pr) and B (B = Ni, Mn, Al, Co), such as Y-La-Ni, Sm-Ni-Mn, La-Y-Ni-Mn, La-Y-Ni-Mn-Al and La-Y-Sm-Ni-Mn-Al, have demonstrated improved hydrogen storage capacities and better reversibility (Table 5.4) (Xiong et al. 2017; Fersi et al. 2018; Madern et al. 2020; Guo et al. 2021; Zhang et al. 2021, 2022).

For example, for $LaY_2Ni_{10.5-x}Mn_x$, increasing the Mn content from x = 0 to 1 increased its hydrogen storage capacity from 0.91 to 1.49 wt%, with a flatter plateau and smaller hysteresis (Xiong et al. 2017). Furthermore, increasing the Y content in $La_{3-x}Y_xNi_{9.7}Mn_{0.5}Al_{0.3}$ from x = 1 to 2.5 improved the cycling ability of the alloy (Zhao et al. 2022). The effects of Mn, Al and Zr substitutions on the structural and hydrogen storage properties of $LaY_2Ni_{10.5}$-based A_2B_7 superlattice alloys have been studied in this work. Y occupies the A-site in the [A_2B_4] subunit, Mn only replaces the Ni in the [AB_5] subunit, Al mainly replaces the Ni at the 6*h* position between the two [AB_5] subunits, while Zr substitutes only the Y at the 4*f* position in the [A_2B_4] subunit.. The preferential occupation of Y, Mn, Al and Zr results in a gradual decrease in the volume difference between the [A_2B_4] and [AB_5] subunits. As a result, the $LaY_2Ni_{9.7}Mn_{0.5}Al_{0.3}$ alloy shows high hydrogen storage capacity under ambient conditions. Indeed, it was used as a negative electrode in a Ni-*M*H battery, with an electrochemical capacity of 384 mAh g^{-1} and a capacity retention of 76.1% after 200 cycles (Zhao et al. 2022).

A_2B_7	Type structure	Space group	*a* (Å)	*c* (Å)	*V* (Å³)	References
La_2Ni_7	Ce_2Ni_7	$P6_3/mmc$	5.065	24.71	549.2	(Iwase et al. 2010)
Y_2Ni_7	Gd_2Co_7	$R\bar{3}m$	4.9465	36.260	768.19	(Zhang et al. 2021)
Sm_2Ni_7	Ce_2Ni_7	$P6_3/mmc$	4.969	24.35	520.68	(Buschow 1970)
	Gd_2Co_7	$R\bar{3}m$	4.969	36.53	781.12	
Gd_2Ni_7	Ce_2Ni_7	$P6_3/mmc$	4.9621	24.230	516.67	(Zhang et al. 2021)
	Gd_2Co_7	$R\bar{3}m$	4.9608	36.357	774.86	
Pr_2Co_7	Ce_2Ni_7	$P6_3/mmc$	5.068	24.463	544.14	(Fersi et al. 2018)
Nd_2Co_7	Ce_2Ni_7	$P6_3/mmc$	5.066	24.484	544.18	(Iwase et al. 2022)
	Gd_2Co_7	$R\bar{3}m$	5.066	36.621	813.94	
$LaY_2Ni_{10}Mn_{0.5}$	Ce_2Ni_7	$P6_3/mmc$	5.010	24.31	529.01	(Xiong et al. 2017)
	Gd_2Co_7	$R\bar{3}m$	5.010	36.46	793.86	
$La_{1.5}Mg_{0.5}Ni_7$	Ce_2Ni_7	$P6_3/mmc$	5.0285	24.22	530.42	(Denys et al. 2008)
$Sm_{1.97}Ni_{6.51}Mn_{0.49}$	Ce_2Ni_7	$P6_3/mmc$	5.0052	24.389	529.14	(Madern et al. 2020)
	Gd_2Co_7	$R\bar{3}m$	5.0051	36.570	793.41	

Table 5.4. *Lattice parameters of some A_2B_7 compounds*

5.2.5.3. *A_5B_{19} compounds*

As observed for AB_3 and A_2B_7, A_5B_{19} compounds also crystallize in two different structures: a hexagonal Sm_5Co_{19}-type structure and a rhombohedral Ce_5Co_{19}-type structure, which are formed by the stacking of three units of [AB_5] and one unit of [A_2B_4] along axis *c*.

The most popular system that has been studied is La_5Ni_{19} with a 2*H* structure, which presents a maximum capacity of 1.4 wt% at 298 K for an applied pressure of 1 MPa (Férey et al. 2009). The most studied substitution of La for Mg in La_5Ni_{19} is La_4MgNi_{19}. This substitution not only favors the 3*R*-type A_5B_{19} structure but also influences the relative sizes between the [A_2B_4] and [AB_5] subunits (Zhang et al. 2010). Similar to A_2B_7 compounds, the maximum Mg content corresponds to half of the *A*-site occupation in the [A_2B_4] subunits. La_4MgNi_{19} shows a large flat plateau at a pressure of 0.08 MPa for absorption, at 0.06 MPa for desorption and a reversible hydrogen capacity of 1.75 wt% at 303 K (Zhao et al. 2018), 0.045 MPa for absorption at 0.02 MPa for desorption and a reversible capacity of 1.56 wt% at 298 K for an applied pressure of 1 MPa (Férey et al. 2009). The substitution of Ni for Mn in $La_5Ni_{19-x}Mn_x$ resulted in a capacity increase of up to 1.68 wt% for $x = 2$ at an applied pressure of 4 MPa with an equilibrium pressure of 0.03 MPa at 298 K. However, increasing the Mn content above $x = 2$ decreases the storage capacity (Zhong et al. 2011).

Other systems, such as Pr_5Co_{19}, which has a maximum storage capacity of 0.55 wt% at 293 K, an equilibrium pressure of 1 MPa and complete reversibility, have also been explored (Bouzidi et al. 2020). Pr_5Ni_{19} and Nd_5Ni_{19} showed maximum storage capacities of 1.33 wt% and 1.17 wt%, respectively, at 298 K and under 10 MPa (Figure 5.8). However, these capacities decreased following each hydrogenation/dehydrogenation cycle, which indicates compound degradation during cycling (Lemort et al. 2011; Bouzidi et al. 2020). Finally, Yan et al. (2017) studied the La-Y-Ni system, including the substitution of both Mn and Al for Ni, for $LaY_2Ni_{10.6}Mn_{0.5}Al_{0.3}$ (A_5B_{19}). The compound shows reversible plateau-like behavior with a capacity in the order of 1.4 wt% at 313 K and 2 MPa. This improvement is attributed to both the substitution of La by Y and Ni by Mn and Al.

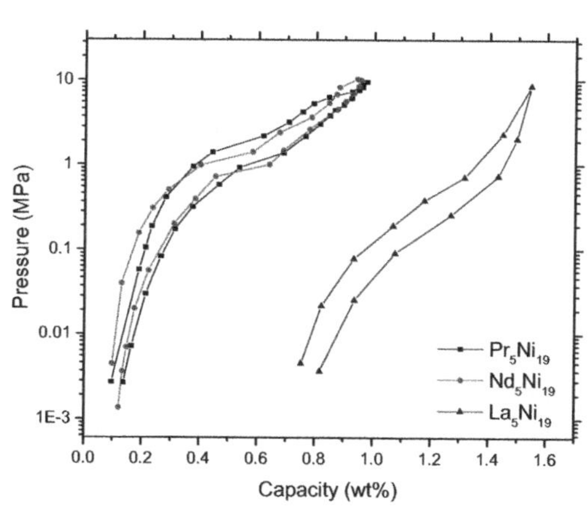

Figure 5.8. *PCI of La_5Ni_{19}, Nd_5Ni_{19} and Pr_5Ni_{19} at 298 K (adapted from Férey et al. (2009) and Lemort et al. (2011)). For a color version of this figure, see www.iste.co.uk/derango-cuevas/hydrogen.zip*

AB_3, A_2B_7 and A_5B_{19} are the most studied intermetallic compounds for the purpose of hydrogen storage in addition to AB_2 and AB_5. However, other types of intergrowth compounds, such as A_5B_{13}, A_7B_{23}, A_6B_{23} and A_6B_{24}, are less investigated.

5.2.5.4. *Other intermetallic compounds*

5.2.5.4.1. A_7B_{23} compounds

Li et al. (2019) characterized a new type of stacking described by the following formula: $[AB_5]$-$[A_2B_4]$-$[AB_5]$-$[AB_5]$-$[A_2B_4]$, which corresponds to $p = 2$, $n = 3$ in equation [5.1], a less common stacking sequence (Figure 5.9).

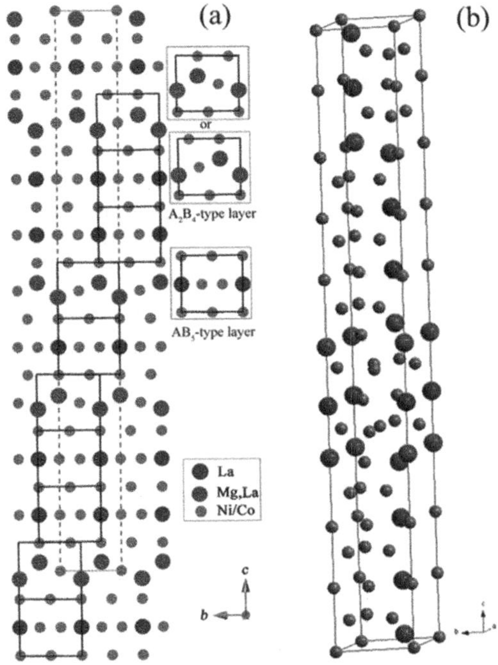

Figure 5.9. *(a) Layered structure stacking along the c-axis and (b) cell unit of the novel A_7B_{23} structure (Li et al. 2019). For a color version of this figure, see www.iste.co.uk/derango-cuevas/hydrogen.zip*

This new structure was first observed in the La-Mg-Ni-Co system for the $La_2MgNi_{7.2}Co_{1.8}$ multiphase alloy. The maximum hydrogen storage capacity of this multiphase alloy is 1.61 wt% at an applied pressure of 5 MPa at 298 K. Furthermore, it has a good reversible hydrogenation/dehydrogenation capacity of 1.57 wt% at room temperature up to 2 MPa. This new structure shows promising properties for hydrogen storage and as an anode material for Ni-*M*H batteries (Li et al. 2019; Liu et al. 2021).

5.2.5.4.2. A_6B_{24} compounds

A_6B_{24} compounds, also often referred to as AB_4 compounds, are the result of the stacking of one unit of [A_2B_4] and four units of [AB_5] (p = 1, n = 4 in equation [5.1]) (Figure 5.10) (Crivello et al. 2011; Zhang et al. 2012, 2018a).

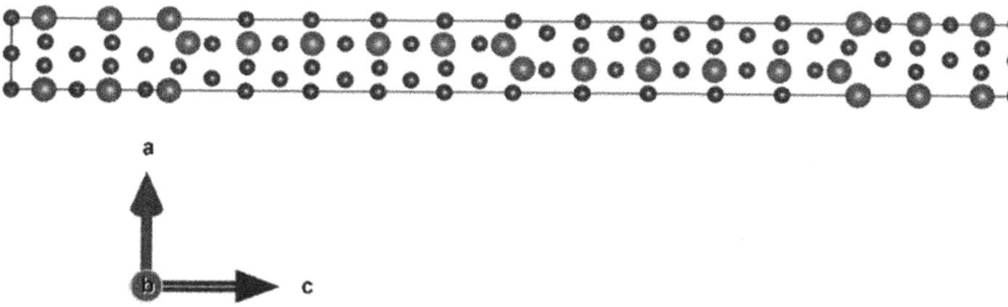

Figure 5.10. *La_5MgNi_{24}-type in the space group $R\bar{3}m$ (La and Mg in red and Ni in blue). For a color version of this figure, see www.iste.co.uk/derango-cuevas/hydrogen.zip*

The most popular compound used to obtain this structure is La_5MgNi_{24}. The alloys synthesized can be single phase AB_4 or contain other structures such as AB_5 and A_5B_{19}.

The $La_{0.85}Mg_{0.15}Ni_{3.8}$ multiphase alloy has a storage capacity of approximately 1.4 wt% at a temperature of 313 K, an equilibrium pressure of 0.1 MPa and a reversible capacity of 1.3 wt% in the pressure range of 0.01–1 MPa (Zhang et al. 2012). $La_{0.78}Mg_{0.22}Ni_{3.9}$ is stored at 1.5 wt% at 298 K, with a reversible capacity of 1.3 wt% and a desorption pressure at the middle plateau of 0.05 MPa (Zhao et al. 2016). $La_{0.8}Mg_{0.2}Ni_{3.6}$ is stored at 1.62 wt% at an applied pressure of 6 MPa and a temperature of 298 K, and it has fast kinetics (full absorption in less than 5 min at room temperature and 3 MPa of pressure). However, after a few hydrogenation/dehydrogenation cycles, its capacity decreases to 1.45 wt% at a pressure of 3 MPa (Wang et al. 2020).

Substituting Ni for Al in $La_{0.78}Mg_{0.22}Ni_{3.77-x}Al_x$ (x = 0.1) leads to a single-phase compound with an AB_4-type structure and provides a hydrogen storage capacity of 1.5 wt% at 298 K under 2 MPa. However, its reversible capacity decreased to 90% after 20 hydrogenation/dehydrogenation cycles (Zhang et al. 2018a). The Mg-free compound $La_{1.5}Y_{1.5}Ni_{12}$ with an AB_4-stacking structure has been reported, and substituting Ni for Mn in $La_{1.5}Y_{1.5}Ni_{12-x}Mn_x$ slightly improved its storage capacity from 1.3 to 1.4 wt% for $x = 1$ at 313 K and its reversibility (Wan et al. 2022).

5.2.5.4.3. A_5B_{13} and A_6B_{23} compounds

A_5B_{13} compounds are obtained by stacking one unit of [AB_5] and two units of [A_2B_4] ($p = 2$, $n = 1$ in equation [5.1]) in space group $R\bar{3}m$. This structure has been observed only in the Ca-Mg-Ni-based compound $Ca_3Mg_2Ni_{13}$. It has a maximum capacity of 1.8 wt% at 298 K under 3 MPa. However, complete reversibility was not observed (Zhang et al. 2009).

A_6B_{23} compounds usually crystallize in a face-centered cube within the $Fm\bar{3}m$ space group instead of stacking [AB_5] and [A_2B_4] subunits (Westlake 1983). They can absorb up to 1 H/M (Paul-Boncour et al. 2018). The most popular studied alloys are Y_6Mn_{23} and Th_6Mn_{23}. They exhibited reasonable storage capacities of approximately 1.3 wt% at 295 K and 1.1 wt% at 296 K. Other compounds, such as Ho_6Mn_{23}, Er_6Mn_{23} and A_6Fe_{23} (A = Y, Er, Ho, Lu), were also studied (Paul-Boncour et al. 2018). However, owing to their high rare earth weight, they are less interesting for H_2 storage.

5.3. Mg-based materials

Magnesium is very interesting for hydrogen storage applications for several reasons. It is a light element with high hydrogen absorption capacity up to 7.6 wt%, it is abundant, non-toxic and it has a low manufacturing cost (Zhu et al. 2013). Nevertheless, it has sluggish absorption kinetics, is very reactive to air, and has a high temperature for hydrogen desorption. These drawbacks are inconvenient and should be overcome for practical applications (Cui et al. 2008; Paul-Boncour et al. 2018). Many studies have been conducted to improve the kinetics, such as size reduction and catalyst addition (Crivello et al. 2016). Identifying rare earth and transition metals that can be associated with magnesium to change its thermodynamic properties is also an interesting approach to using magnesium for reversible hydrogen storage (Zhu et al. 2013).

5.3.1. *Magnesium hydride MgH_2*

5.3.1.1. *Structure and polymorphisms*

Metallic magnesium displays a hexagonal structure ($P6_3/mmc$ space group), but MgH_2 hydride crystallizes in different crystallographic structures depending on the temperature and pressure. X-ray diffraction (XRD) and neutron diffraction techniques were used to determine the different crystallographic structures, including the sites occupied by hydrogen (Bastide et al. 1980; Bortz et al. 1999; Cui et al. 2008). The structural properties of the MgH_2 hydride phases under various pressures are reported in Table 5.5.

Phase	Formation pressure (GPa)	Type structure	Space group	Lattice parameters (Å)	Cell volume (Å³)	References
α - MgH_2	Ambient	TiO_2	$P4_2/mnm$	$a = 4.5025$ $c = 3.0123$	61.07	(Zachariasen et al. 1963)
δ - MgH_2	0.39	SrO_2	$Pbc2_1$	$a = 4.8604$ $b = 4.6354$ $c = 4.7511$	107.04	(Vajeeston et al. 2006)
γ - MgH_2	2.00	α-PbO_2	$Pbcn$	$a = 4.5051$ $b = 5.4197$ $c = 4.9168$	120.06	(Bortz et al. 1999)
β - MgH_2	3.84	Modified CaF_2	$Pa\bar{3}$	$a = 4.6655$	101.55	(Vajeeston et al. 2006)
ε - MgH_2	21.90	α-$PbCl_2$	$Pnma$	$a = 4.9536$ $b = 2.9453$ $c = 5.6677$	82.69	(Moriwaki et al. 2006)

Table 5.5. *Structural parameters of the MgH_2 hydride phases*

5.3.1.2. *Thermodynamic and kinetic properties of MgH_2*

The enthalpy of α-MgH_2 formation is very negative ($\Delta H_f = -75$ kJ/mol H_2) when the entropy is assumed to be −135 J/K.mol H_2, so MgH_2 is a very stable hydride that requires high desorption temperatures (555 K at ambient pressure) (Bogdanović et al. 1999). The Mg/MgH_2 *PCI* exhibits a low-pressure plateau at room temperature; in addition, its sluggish kinetic requires temperatures above 676 K for *PCI* measurement in reasonable time (Figure 5.11(a)). As previously mentioned, both hydrogenation and dehydrogenation of micrometric Mg have slow kinetics (Zaluska 1999; Crivello et al. 2016). Indeed, Mg is very reactive to air, and under ambient conditions, MgO forms at the surface of the particles, inhibiting hydrogen dissociation on the surface. Moreover, the diffusion of hydrogen in MgH_2 is very slow (Zhu et al. 2013) compared with that in Mg: the diffusion coefficient of hydrogen in MgH_2 is 1.5×10^{-16} m^2/s versus 4×10^{-13} m^2/s for Mg at 673 K (Teichmann et al. 2018). The dehydrogenation process is complicated by the difficulty of breaking Mg–H bonds and sluggish hydrogen atom recombination on the Mg surface (Martin et al. 1996; Vajeeston et al. 2006; Tanniru et al. 2010).

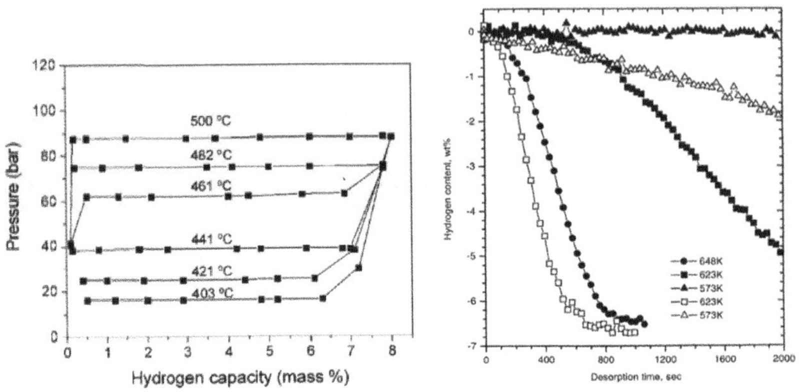

Figure 5.11. *(a) PCI of the Mg powder, particle size 25–40 μm, in the temperature range 670–770 K (Bogdanović et al. 1999); (b) hydrogen desorption curves of unmilled MgH_2 (filled marks) and ball-milled (hollow marks) MgH_2 under a hydrogen pressure of 0.015 MPa (Huot et al. 1999)*

5.3.1.3. *Nanostructuration*

Experimentally, no hydrogen desorption has been observed for micrometric magnesium hydride, whereas the desorption kinetics are improved at all temperatures owing to the nanostructuration produced by ball milling (Zaluska 1999) (Figure 5.11(b)). Ball milling is one of the most common methods used to prepare nanocrystalline Mg-based hydrides but is not the only method used. Severe plastic deformation (SPD), which consists of applying large mechanical strains to increase defaults and obtain "bulk" nanograined material, is also used in combination with other methods, such as cold rolling and cold forging (Skryabina et al. 2015; Wen et al. 2020; Botta et al. 2022). Nanosizing the material reduces the length of diffusion paths for H atoms and provides a highly specific surface available for H chemisorption. Ball milling improves both hydrogenation and dehydrogenation properties for several reasons. It creates defects in the structure that aid hydride nucleation and the presence of grain boundaries provides a pathway for hydrogen diffusion. The particle size and average crystallite size, also referred to as the grain size inside the powder particles, are also important parameters. Zaluska et al. (1999) measured the average particle size by scanning electron microscopy (SEM) and the grain size inside the particles by transmission microscopy. They have shown the best hydrogenation properties (kinetic and capacity) is obtained with the smallest grain size (30 nm) by diminishing the diffusion lengths (Figure 5.12(a)).

The drawback of nanostructuration is the increase in specific surface area which causes a reaction with air and results in an increase of the MgO concentration after ball milling (0.6 wt% before ball milling and 1.4 wt% after 20 h of ball milling; Huot et al. 1999). Ball milling can also induce the formation of Mg hydroxide, which can decompose upon activation by heating (Zaluska 1999). However, nanosized materials do not show good kinetics and reversibility upon cycling because of the grain size growing after cycling. One way to overcome this issue is the synthesis of nanocomposites, which limit grain size growth. To improve the kinetic and cycle stability, nanocomposites are usually synthesized with catalysts, as discussed in sections 5.3.1.4 and 5.3.1.5.

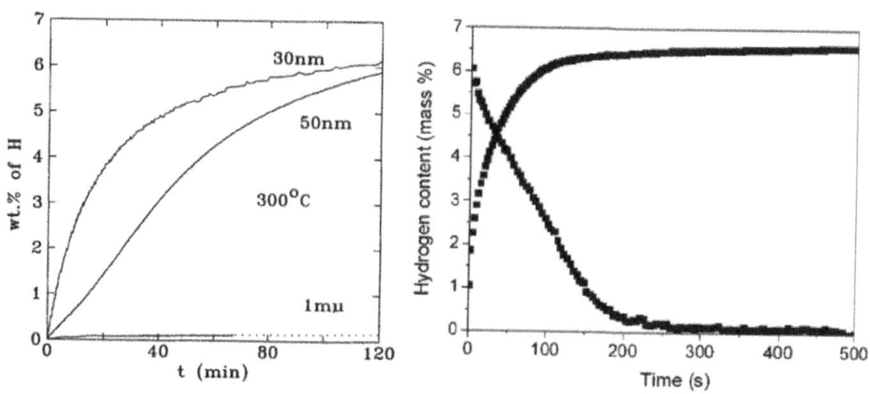

Figure 5.12. *(a) Effect of grain size on the hydrogen absorption of ball-milled magnesium powders (Zaluska 1999); (b) hydrogen absorption/desorption kinetics of MgH_2 milled with 10 wt% diamond powder on a Fritsch planetary mill for 10 h. The measurements were performed at 570 K at 1 MPa of hydrogen for absorption and 10^{-4} MPa for desorption (Aguey-Zinsou and Ares-Fernández 2010)*

5.3.1.4. *Catalyst addition*

Another method used to improve the kinetics of hydrogenation and dehydrogenation consists in the addition of catalysts. These catalysts are often transition metals or oxides of transition metals added by ball milling or other chemical methods. The nanometric catalytic particles are usually distributed uniformly on the metal surface. The advantage of adding catalysts is that the Mg surface does not need to be activated, as the catalyst particles dissociate and recombine with hydrogen (Zaluska 1999).

For example, the addition of several metals, such as Ti, Nb, V, Co, Mo, Li, Al, Mn or Fe, in small proportions with a nanometric grain size, to Mg by powder metallurgy can improve the kinetics of hydrogenation and dehydrogenation (Zaluska 1999; Cui et al. 2014). Cui et al. compared the impact of adding transition metals such as Ti, Nb, V, Co, Mo and Ni. They improved the kinetics and reduced the hydrogenation temperature (to below 500 K at 2 MPa) while maintaining good capacity. The beneficial catalytic effects of those metals on kinetics can be ranked in the following order, from more effective to less effective: Ti, Nb, Ni, V, Co, and Mo. We can observe that the less electronegative element (Ti) provides the greatest improvement in the hydrogenation properties. Zaluska (1999) has observed that mixtures of transition metal additives such as Zr + Ti or Mn + Zr improved the kinetics better while maintaining a capacity of up to 6 wt% hydrogen for both absorption and desorption.

The Pd effect as a catalyst has been studied and has greatly improved the kinetics. Callini et al. (2009) synthesized Mg nanoparticles with partial decoration of their oxide shells with Pd. The advantage of Pd is that it helps H_2 dissociation and thus increases hydrogen sorption and prevents oxidation (Callini et al. 2009; Liu et al. 2020). With 8 wt% Pd, they managed to reversibly absorb 6.1 wt% hydrogen in 16 min at 610 K under 0.8 MPa for absorption and 0.06 MPa for desorption instead of 0.7 wt% hydrogen in 125 min for Mg nanoparticles without Pd.

Metal oxide nanoparticles such as Sc_2O_3, V_2O_5, Cr_2O_3, Mn_2O_3, Fe_3O_4, CuO, Al_2O_3 and SiO_2 were added to Mg by high-energy ball milling to form nanocomposites. They have a significant catalytic effect on hydrogen sorption and desorption kinetics and can sometimes lower the desorption temperature (Oelerich et al. 2001; Dehouche et al. 2002; Barkhordarian et al. 2006; Sadhasivam et al. 2013; Pukazhselvan et al. 2017). Oelerich et al. (2001) have shown that among those metal oxides, Cr_2O_3 has the fastest absorption kinetics, whereas Fe_3O_4 and V_2O give the fastest desorption kinetics. This was attributed to the high defect density at the MgH_2/Me_xO_y (Me_xO_y = Sc_2O_3, TiO_2, V_2O_5, Cr_2O_3, Mn_2O_3, Fe_3O_4, CuO, Al_2O_3, SiO_2) interface introduced by high-energy ball milling, which acts as a gateway for hydrogen. Later, Barkhordarian et al. (2006) discovered that Nb_2O_5 leads to a higher capacity and faster kinetic than the other metal oxide nanoparticles mentioned before. With 1 mol% Nb_2O_5, they managed to desorb approximately 7 wt% in 2 min at 573 K. According to their study, the reasons for the improvement in desorption kinetic are the high affinity of Nb with hydrogen and the high valence state of Nb_2O_5, which permits the exchange of more electrons than other metal oxides (NbO, for example).

5.3.1.5. *Nanocomposite synthesis*

MH_x/MgH_2 (MH_x: metal hydride) nanocomposites can be synthesized by reactive ball milling under a hydrogen atmosphere and then used as catalysts.

Rizo-Acosta et al. (2018) studied the impact of TiH_2 on the kinetic and thermodynamic properties of MgH_2 (de)hydrogenation. They reported that the diffusion of hydrogen atoms is faster in TiH_2 than in MgH_2. Ponthieu et al. (2014) suggested that TiH_2 acts as a gateway for hydrogen and favors hydrogen dissociation and diffusion. Shahi et al. (2015) compared the dehydrogenation/hydrogenation properties of the $MgH_2/ZrFe_2H_x$, $MgH_2/ZrFe_2$ and pure MgH_2 nanocomposites. They reported that a mixture of $MgH_2/ZrFe_2H_x$ with 5 wt% $ZrFe_2H_x$ gives the best capacity and kinetic (6.5 wt% hydrogen in 60 min), decreases the desorption temperature (from 680 to 640 K) and diminishes the activation energy for rehydrogenation from 61 to 30 kJ/mol H_2. The improvement in the absorption and desorption kinetics is attributed to the smaller grain size of $ZrFe_2H_x$ than that of $ZrFe_2$.

Allotropic forms of carbon (graphite, fullerene, graphene and nanotubes) can also be used as catalysts (Zhang et al. 2018b). Indeed, nanocomposites of carbon and Mg with organic additives such as benzene, cyclohexene or cyclohexane have been studied in the context of hydrogen storage materials. For example, Aguey-Zinsou and Ares-Fernández (2010) have shown that the addition of 10 wt% diamond powder with Mg can improve MgH_2 hydrogenation kinetic and enable good cyclability (Figure 5.12(b)). High-energy ball milling of Mg with graphite leads to the creation of dangling carbon bonds that act as sites where hydrogen can be taken up (Imamura et al. 2003). The effect of adding organic additives to Mg/graphite nanocomposites on the microstructure and dehydrogenation/hydrogenation properties has been studied by Imamura et al. (2003). XRD measurements have shown that milling Mg/graphite nanocomposites without organic additives leads to rapid amorphization, which inhibits Mg–graphite interactions (Imamura et al. 2000). These findings suggest that organic additives enhance the Mg–graphite solid phase interaction during milling, thereby influencing the microstructure of the nanocomposites and their hydrogenation/dehydrogenation properties. Indeed, Mg/graphite nanocomposites with additives can be milled for a long time without losing their structure and show better improvement of hydrogenation/dehydrogenation properties than without additives.

It seems that ball milling of Mg with graphene and carbon nanotubes does not cause amorphization, as in the case of graphite. For example, Liu et al. (2013) have synthesized nanocomposites composed of MgH_2 and graphene nanosheets by ball milling to study their dehydrogenation/hydrogenation properties. They compared the impact of the ball milling time and reported that the longest ball milling time (20 h) led to the best absorption kinetics (6.6 wt% in 25 min at 570 K) and lowered the desorption temperature to 600 K under 5.10^{-3} MPa of hydrogen pressure with 5 wt% graphene nanosheets. MgH_2/single-walled carbon nanotube (SWNT) nanocomposites were studied by Wu et al. (2006). They reported that MgH_2 with

5 wt% SWNTs milled for 10 h can absorb 6.7 wt% of hydrogen at 570 K in 2 min under 2 MPa of hydrogen pressure, and desorb 6 wt% of hydrogen at 620 K in 5 min at less than 10^{-6} MPa. Wu et al. (2010) then investigated the hydrogenation/dehydrogenation properties of MgH_2 with 5 wt% Ni-coated carbon nanofibers. These nanocomposites can absorb 6 wt% hydrogen in 5 min at 470 K and 5 wt% hydrogen in 10 min at 420 K under the same conditions (2 MPa for absorption and less than 10^{-6} MPa for desorption).

5.3.2. *Binary Mg-based compounds*

5.3.2.1. *Mg-rich intermetallic compounds*

$MgNi_2$

$MgNi_2$ is the second intermetallic compound of the Mg–Ni binary phase after Mg_2Ni. The $MgNi_2$ compound crystallizes in the C36 Laves-type structure ($P6_3/mmc$ space group, see Figure 5.13(a) (Komura and Tokunaga 1980; ASM Handbook 1992) with $a = 4.8256$ and $c = 15.8323$ Å. $MgNi_2H_{3.2}$ trihydride can be synthesized only under a hydrogen pressure in the order of GPa (Kataoka et al. 2007; Yartys et al. 2015); therefore, it is not a good candidate for reversible hydrogen storage near ambient conditions of pressure and temperature.

Mg_6Pd

Mg_6Pd is an intermetallic compound that crystallizes in the FCC structure ($F\bar{4}3m$ space group; see Figure 5.13(b)) with a lattice parameter of $a = 20.10$ Å (Ferro 1959; Sampson 1972; Kume and Weiss 1987). This phase is called the ρ-phase and has the highest Mg content among the intermetallic phases of the binary Mg–Pd system. In addition, this phase has a low peritectic temperature (973 K), which facilitates its synthesis (Dufour and Huot 2007). In Huot et al. (2009), Mg_6Pd was synthesized via ball milling and then exposed to a maximum hydrogen pressure of 4 MPa at 573 K to measure a full isotherm (absorption and desorption). They obtained a maximum capacity of 4.3 wt% with 3 equilibrium plateaus. The first plateau was measured at 0.2 MPa and corresponds to the reaction described in equation [5.2]. The second plateau at 0.24 MPa and the third plateau at 1.5 MPa correspond to the reactions described in equations [5.3] and [5.4], respectively.

$$Mg_6Pd + 2.35H_2 \rightarrow Mg_{3.65}Pd + 2.35MgH_2 \qquad [5.2]$$

$$Mg_{3.65}Pd + 2.35MgH_2 + 1.15H_2 \rightarrow 0.5Mg_5Pd_2 + 3.5MgH_2 \qquad [5.3]$$

$$0.5Mg_5Pd_2 + 3.5MgH_2 + 1.5H_2 \rightarrow MgPd + 5MgH_2 \qquad [5.4]$$

Compound	Type structure	Space group	Lattice parameters (Å)	Cell volume (Å^3)	References
Mg_2FeH_6	K_2PtCl_6	$Fm\bar{3}m$	$a = 6.4458$	267.82	(Retuerto et al. 2015)
Mg_2Ni	Mg_2Ni	$P6_222$	$a = 5.2066$ $c = 13.2543$	311.17	(Poletaev et al. 2012)
Mg_2NiH_4	K_2PtCl_6	$C2/c$	$a = 14.3667$ $b = 6.4067$ $c = 6.4941$	547.44	(Martínez-Coronado et al. 2013)
Mg_2CoH_5	Mg_2CoH_5	$P4/nmm$	$a = 4.463$ $c = 6.593$	131.32	(Zolliker et al. 1985)
Mg_6Pd	$Mg_{83}(Mg_{0.51}Pd_{0.49})_4Pd_{12}$	$F\bar{4}3m$	$a = 20.10$	8120.6	(Kume and Weiss 1987)
$MgNi_2$	$MgNi_2$	$P6_3/mmc$	$a = 4.82565$ $c = 15.8323$	319.29	(Yartys et al. 2015)

Table 5.6. *Structural parameters of Mg_2T (T = Fe, Ni, Co)H_x, Mg_6Pd and $MgNi_2$*

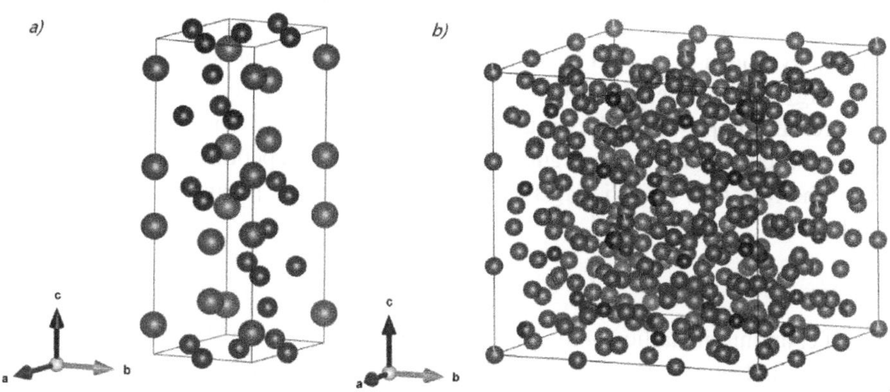

Figure 5.13. *(a) $MgNi_2$; (b) Mg_6Pd structure with Mg in red and Ni and Pd in blue. For a color version of this figure, see www.iste.co.uk/derango-cuevas/hydrogen.zip*

Besides its high desorption temperature, the disadvantage of Mg_6Pd is the cost and heavy weight of Pd; therefore, the substitution of Pd with other transition metals, such as Ni (Ponthieu et al. 2013), has been studied. A study of Ni/Pd substitution in Mg_6Pd revealed that $Mg_6Pd_{0.25}Ni_{0.75}$ can reversibly absorb 5.6 wt% hydrogen at 620 K under 3 MPa of hydrogen pressure (Ponthieu et al. 2013). The

PCI at 620 K exhibits a flat plateau pressure at approximately 0.75 MPa and a second sloppy plateau between 1.25 and 1.75 MPa. These two pressure plateaus are also present in desorption with a hysteresis effect. The $Mg_6Pd_{0.25}Ni_{0.75}$ de/hydrogenation temperature remains high; it is a promising material for reversible hydrogen storage, as nickel reduces the cost and weight of Mg_6Pd and increases its weight capacity.

5.3.2.2. Mg_2TH_x (*T* = Fe, Co and Ni)

To decrease the desorption temperature of MgH_2, it is necessary to tune its thermodynamic properties, which can be achieved by alloying Mg with other elements. Many studies have been conducted with Mg_2TH_x (T = Fe, Co and Ni) complex hydrides to lower the enthalpy for dehydrogenation ΔH_d compared with that of Mg (ΔH = 75 kJ/mol H_2). The reduction in the enthalpy for dehydrogenation depends on the enthalpy of formation for the alloys or compounds formed (Crivello et al. 2016). Therefore, finding new Mg-TH_x hydrides with lower ΔH_d values is crucial for the use of Mg-based compounds for reversible hydrogen storage applications.

Notably, the Mg_2TH_x (T = Fe, Co and Ni, x = 6, 5 and 4, respectively) hydrides presented in this part are complex hydrides where hydrogen creates covalent bonds with the transition metal of the structure, which is different from metallic hydrides where the hydrogen is located inside the interstitial sites of the structure, such as in $MgNi_2$. Mg_2TH_x (T = Fe, Co and Ni, x = 6, 5 and 4, respectively) complex hydrides, which are all composed of the metallic cation Mg^{2+} and the complex anion TH_x^{4-}, and their structures (shown in Figure 5.14) are different depending on the transition metals. The Mg_2TH_x, Mg_2Ni, Mg_6Pd and $MgNi_2$ structural parameters are reported in Table 5.6.

Mg_2FeH_6

Mg_2Fe, an intermetallic compound, does not exist because of the immiscibility of magnesium and iron in the binary Mg–Fe phase diagram (Williams 1984; Brutti et al. 2018). However, Mg_2FeH_6 is a very stable ternary complex hydride that has been widely studied in the context of reversible hydrogen storage (Williams 1984; Huot et al. 1997; Polanski et al. 2011; Asselli et al. 2012; Li et al. 2017; Brutti et al. 2018; Thiangviriya et al. 2021). Brutti et al. (2018) synthesized Mg_2FeH_6 by reactive ball milling under 5 MPa H_2 pressure and by mixing $2MgH_2$ and Fe. Mg_2FeH_6 synthesized from a simple mixture of 2 MgH_2 and Fe was able to store 3.8 wt% at 610 K and 4.9 wt% at 750 K with a pressure ranging from 0.01 to 10 MPa. For the sample synthesized by reactive ball milling (RBM), 5.4 wt% hydrogen was stored as expected for Mg_2FeH_6. Both samples contain two phases: MgH_2 and Mg_2FeH_6. The

PCI of those samples exhibited two pressure plateaus associated with the hydrogenation and dehydrogenation reactions of Mg_2FeH_6 and MgH_2 (equations [5.5] and [5.6]). The enthalpy of dissociation of Mg_2FeH_6 was calculated with the van't Hoff law, and the authors reported that $\Delta H_{dis} = 77 \pm 2$ kJ/mol H_2, which is in good agreement with early work (Bogdanović et al. 2002).

$$2Mg(s) + Fe(s) + 3H_2 \leftrightarrow Mg_2FeH_6 \quad [5.5]$$

$$Mg(s) + H_2 \leftrightarrow MgH_2(s) \quad [5.6]$$

With metal boron hydrides, Mg_2FeH_6 forms reactive hydride composites (RHCs). $M(BH_4)_n$ (M = Mg, Li, Na, K and Ca) borohydrides have high theoretical hydrogen capacities (up to 18.5 wt% for Li) because of the light weight of the elements used in borohydrides. Moreover, they have a less negative enthalpy than Mg_2FeH_6; thus, they can enhance their desorption properties by decreasing the desorption temperature to 540 K for KBH_4/Mg_2FeH_6 RHC (Chaudhary et al. 2015).

Li et al. (2017) studied $x\,Mg_2FeH_6 + (1\text{-}x)\,LiBH_4$ nanocomposites with $x = 0.5$ and reported full hydrogenation and dehydrogenation of 6 wt% hydrogen at 670 K; notably, this capacity accounts for the capacity of $LiBH_4$. After four cycles, this compound maintained its maximum absorption and desorption capacity. The dehydrogenation of $x\,Mg_2FeH_6 + (1 - x)\,LiBH_4$ can be described by equation [5.7], and the enthalpy of this reaction is 64 kJ/mol H_2, which is lower than that of pure Mg_2FeH_6.

$$LiBH_4 + Mg_2FeH_6 \rightarrow LiH + 2MgH_2 + FeB + 2.5H_2 \quad [5.7]$$

The hydrogen sorption properties of $Mg_2FeH_6/LiNH_2$ composites have been investigated (Langmi et al. 2012). This RHC was synthesized by ball milling with a 3:10 ratio of Mg_2FeH_6 and $LiNH_2$ respectively. Differential Scanning Calorimetry (DSC) measurements were performed on this sample in a flow of N_2. They observed a broad endothermic peak between 520 K and 620 K associated with Mg_2FeH_6-$LiNH_2$ desorption. However, this compound cannot absorb hydrogen after the first dehydrogenation.

Notably, borohydride presents the risk of releasing B_2H_6, which is a toxic gas; therefore, it can be a problem for practical applications. For amide-hydride compounds such as $LiNH_2$, small amounts of NH_3 can be released with hydrogen (Langmi et al. 2012).

Mg_2CoH_5

Mg_2Co does not exist in the Mg-Co phase diagram; however, the Mg_2CoH_5 hydride can be synthesized by RBM from Mg, Co metal or MgH_2 and Co powder

(Zolliker et al. 1985; Fernández et al. 2007, 2008; Zhang et al. 2011; Verón et al. 2013; Zepon et al. 2015, 2017). Mg_2CoH_5 was also studied for reversible hydrogen storage. The hydrogenation/dehydrogenation of ball-milled $2MgH_2$-Co and $3MgH_2$-Co samples has been studied (Verón et al. 2013). The reactions occurring during the RBM of $2MgH_2$-Co and $3MgH_2$-Co can be expressed as equations [5.8] and [5.9], respectively:

$$MgH_2(s) + Co(s) + 0.5H_2(g) \rightarrow Mg_2CoH_5(s) \qquad [5.8]$$

$$3MgH_2(s) + Co(s) + 0.5H_2(g) \rightarrow Mg_2CoH_5(s) + MgH_{2\,\mathrm{unreacted}}(s) \qquad [5.9]$$

Three phases were detected by XRD in the $2MgH_2$-Co sample (Mg_2CoH_5, MgH_2 and Co), and four phases were detected in the $3MgH_2$-Co sample (Mg_2CoH_5, $Mg_6Co_2H_{11}$, MgH_2 and Co). The $2MgH_2$-Co sample was able to absorb 2.5 and 2.8 wt% hydrogen under 2 MPa at 523 K and 573 K respectively but only desorbed 2.6 wt% hydrogen at 573 K under 2×10^{-2} MPa of hydrogen pressure. The $3MgH_2$-Co sample absorbed 3.4 and 3.7 wt% at 2 MPa at 250 and 570 K, respectively, and desorbed 3.7 wt% at 570 K under 2×10^{-2} MPa of hydrogen pressure. The two samples exhibit fast kinetics, which can be explained by the catalytic effect of unreacted metallic Co contained in the sample. XRD measurements were performed on the two samples after 30 cycles of hydrogenation/dehydrogenation. For the $2MgH_2$-Co sample, the main phases detected at 570 K are MgH_2 and Co, with a minor amount of Mg_2CoH_5. For the $3MgH_2$-Co sample, the main phases are Mg_2CoH_5 and $Mg_6Co_2H_{11}$ in the presence of MgH_2 and Co. These results show that reactions [5.8] and [5.9] are not fully reversible at 570 K. In another study, Fernández et al. (2007), Mg_2CoH_5 was synthesized by a milling and sintering method under 6 MPa at 683 K for seven days. A full *PCI* at 673 K shows that it can reversibly store 3 wt% with two equilibrium plateaus with a large hysteresis. The first plateau at 0.67 MPa and the second one with a sloppy shape between 1.6 and 3 MPa are associated with the formation of Mg_2CoH_5 and $Mg_6Co_2H_{11}$ respectively. DSC measurements of Mg_2CoH_5 under argon flow revealed two endothermic peaks: one at 476 K, which corresponds to the change in the structure of Mg_2CoH_5 (from tetragonal to cubic), and the other at 663 K, which corresponds to Mg_2CoH_5 decomposition. The sample shows fast kinetics, as it can absorb 2 wt% in only 2 minutes. Moreover, good reversibility after 50 cycles at 570 K was observed in this study. Another synthesis method can be used to synthesize Mg_2CoH_5, which consists of the mechanical milling of a 2Mg-Co mixture under argon followed by RBM (Fernández et al. 2008). In this work, a 2:1 atomic ratio mixture of Mg and Co was milled for 12,000 min (premilling) under 0.1 MPa of argon and then milled under 0.5 MPa of hydrogen for 5,400 min. The RBM method decreases the powder size and increases the density of defects, but absorption/desorption cycling increases the crystallite size. Therefore, after several cycles, there is no difference in the kinetics between the sintered and RBM samples. However, the RBM method reduces the initial decomposition temperature of Mg_2CoH_5 to 478 K. This is attributed to the catalytic effect of the small cobalt particles observed by the MEB in this sample. Zepon et al. (2015) synthesized Mg_2CoH_5 by RBM under H_2 pressure for different durations. They studied the hydrogen sorption properties of a sample milled for 720 min. The theoretical capacity of Mg_2CoH_5 (4.48 wt%) was obtained in 30 min at 620 K under 2 MPa. This sample released 4.35 wt% under a pressure of 0.03 MPa, which represents a reaction yield of 97%. These results are attributed to the appropriate milling parameters and duration. SEM and transmission electron microscopy (TEM) characterization revealed a particle size of 2 μm and a crystallite size below 15 nm, which can improve the hydrogen sorption properties.

Mg_2NiH_4

Unlike Mg_2Fe and Mg_2Co, Mg_2Ni exists as an intermetallic compound. Mg_2Ni is also an interesting compound because the dissociation energy of its hydride Mg_2NiH_4 is $\Delta H_d = 64.5$ kJ/mol H_2, which is lower than those of MgH_2, Mg_2FeH_6 and Mg_2CoH_5 (Reilly and Wiswall 1968; Palade et al. 2006). Mg_2NiH_4 can also be synthesized by mechanical milling, hydride combustion synthesis, or induction melting of Mg_2Ni (Zaluski et al. 1995; Li et al. 2000, 2001; Hsu et al. 2010; Martínez-Coronado et al. 2013; Hou et al. 2014; Emami et al. 2016; Ma et al. 2022) followed by hydrogenation. One of the first studies on Mg_2Ni showed that it can absorb approximately 3.5 wt% between 570 K and 620 K under 2.5 MPa of hydrogen pressure, which is close to the theoretical value of Mg_2NiH_4 (3.6 wt%) (Reilly and Wiswall 1968).

Si et al. (2017) have studied the hydrogenation properties of the $Mg_2Ni_{1-x}Cu_x$ system (x = 0, 0.05, 0.1 and 0.15). After four cycles of activation at 620 K and 3 MPa, the samples were hydrogenated at 550 K, 570 K and 590 K. The authors reported that for all these temperatures the samples can absorb and desorb about 3.5 wt% of hydrogen in the first cycle with a single plateau from 0.2 MPa at 550 K to 0.5 MPa at 590 K corresponding to the Mg_2Ni de/hydrogenation described in equation [5.10]. The same behavior was observed for x = 0.05 and 0.1, with a lower desorption temperature when $x \leq 0.1$ (from 520 K for Mg_2NiH_4 to 500 K for $Mg_2Ni_{0.9}Cu_{0.1}$). The reaction pathway is described in equation [5.11] for $x \leq 0.1$ and was found to have a lower enthalpy of dissociation than Mg_2NiH_4 ($\Delta H_d(Mg_2Ni_{0.9}Cu_{0.1}) = 59.8$ kJ/mol H_2), which means that the addition of Cu destabilizes Mg_2NiH_4. However, for $Mg_2Ni_{0.85}Cu_{0.15}$, a second plateau appears at 0.3 MPa at 550 K, and a sloppy peak appears at 0.5 MPa at 590 K. Moreover, DSC measurements under argon flow have shown an additional endothermic peak at approximately 600 K that was not observed for compounds without Cu. Similar results were obtained by Liang et al. (1998) and Hsu et al. (2010), which were interpreted by the decomposition of Mg_2Cu described in equation [5.12]. Although those reaction pathways are different and occur at different temperatures, they are all reversible.

$$Mg_2Ni + 2H_2 \leftrightarrow Mg_2NiH_4 \quad [5.10]$$

$$Mg_2(Ni_{1-x}Cu_x) + 2H_2 \leftrightarrow Mg_2(Ni_{1-x}Cu_x)H_4 \quad [5.11]$$

$$2Mg_2Cu + 3H_2 \leftrightarrow 3MgH_2 + MgCu_2 \quad [5.12]$$

The substitution of Ni by Co in Mg_2Ni was also studied. $Mg_2Ni_{0.5}Co_{0.5}H_{4.4}$ was synthesized via reactive ball milling, and the dehydrogenation properties were studied via thermal desorption spectroscopy (TDS) measurements under vacuum (Verbovytskyy et al. 2015). They synthesized $Mg_2Ni_{0.5}Co_{0.5}H_{4.4}$ by RBM at 7.3 MPa and reported that it desorbed from 490 to 530 K, which is significantly lower than the desorption rates of MgH_2, Mg_2NiH_4 and Mg_2CoH_5.

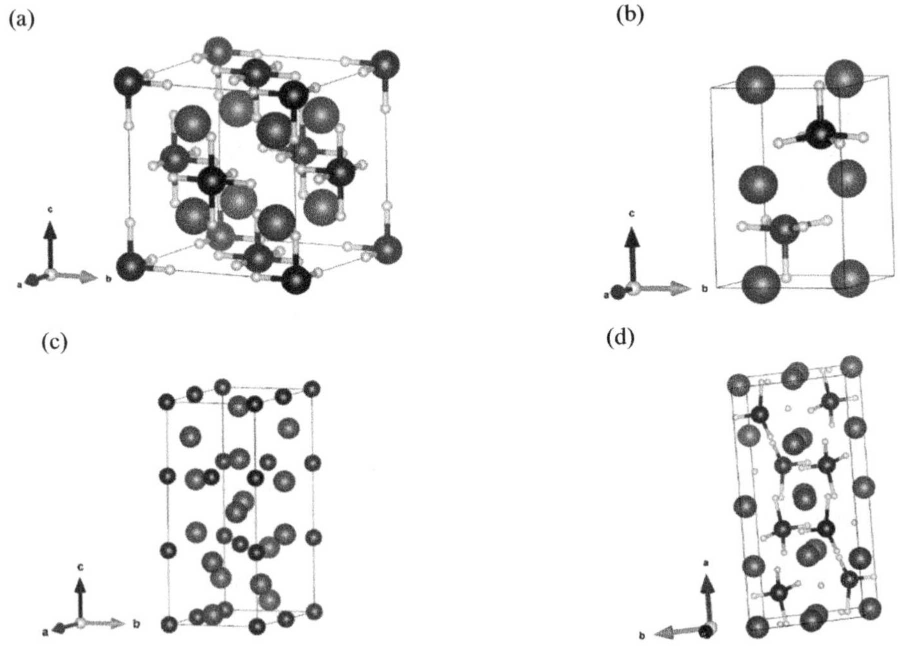

Figure 5.14. *(a) Mg_2FeH_6, (b) Mg_2CoH_5, (c) Mg_2Ni, and (d) Mg_2NiH_4 structures with Mg in red; Fe, Ni and Co in blue; and H in cyan. For a color version of this figure, see www.iste.co.uk/derango-cuevas/hydrogen.zip*

5.3.3. *Ternary Mg-based compounds*

Many efforts have been made to find new R_xMgT_y (R = rare earth, T = Ni, Co) ternary compounds with improved hydrogenation properties compared to MgH_2 and the binary Mg-based compounds. Indeed, the association of rare earth elements, which have a high affinity for hydrogen, with transition metals, which can be desorbed at low temperatures, could enhance the sorption properties of Mg-based compounds. In this section, ternary Mg-based compounds that could be good candidates for reversible hydrogen storage applications are presented.

5.3.3.1. *RMgT₄ (R = Y, Lanthanide, T = Ni, Co and Cu)*

RMgT_4 (R = Y, La, Pr, Nd, Sm and Gd, T = Ni, Co) compounds crystallize in a $MgCu_4Sn$-type structure (space group $F\bar{4}3m$), which is a superstructure of the C15 Laves phase, where the R and Mg atoms are occupied at two different sites, 4a and 4c, respectively (Table 5.7). The *PCI* of all RMgT_4 samples show a first plateau that corresponds to the formation of RMg$T_4H_{\approx 4}$, but in some cases, a second plateau is observed at higher pressure for the $LaMgNi_4$, $PrMgNi_4$ and RMg$Ni_{4-x}Co_x$ compounds ($x > 2$), which corresponds to the formation of RMg$T_4H_{\approx 6}$ (Sakaki et al. 2013; Shtender et al. 2014). Among RMgT_4 compounds, $YMgNi_4$ is an interesting material because it can reversibly absorb 1.13 wt% hydrogen at ambient temperature with good kinetics (35 min for absorption) (Shtender et al. 2014). Shtender et al. (2014) have studied YMgT_4 (T = Ni, Co and Cu) compounds and reported that $YMgCo_2Ni_2$ and $YMgCo_4$ can absorb reversibly 1.4 wt% and 1.92 wt% hydrogen, respectively, at ambient temperature under a hydrogen pressure of 2 MPa. $YMgCu_4$ and $NdMgCu_4$ do not interact with hydrogen even at temperatures and pressures up to 690 K and 15 MPa (Shtender et al. 2015a).

RMgNi_4 (R = La, Gd, Sm, Nd and Pr) hydrogen sorption properties have been investigated by Sakaki et al. (2012). They explained that $LaMgNi_4$ can absorb 1.5 wt% at 298 K under 9 MPa of hydrogen pressure but only desorbs 0.5 wt%. $GdMgNi_4$ is able to absorb reversibly 1 wt% at 293 K at a hydrogen pressure of 4 MPa. They also reported that $SmMgNi_4$ and $NdMgNi_4$ can absorb approximately 1 wt% hydrogen reversibly at 293 K.

The $NdMgNi_{4-x}Co_x$ compounds have been studied by Shtender et al. (2015a), who reported that $NdMgNi_4$ can reversibly absorb approximately 1 wt% at 320 K and 2 MPa, as observed by Guénée et al. (2003). $NdMgNi_2Co_2$ and $NdMgNiCo_3$ reversibly absorb 1.39 and 1.54 wt% hydrogen, respectively, since their molar capacities increase from 4 to 6 H/f.u. due to Co substitution. In addition to the fact that the Co content in $NdMgNi_{4-x}Co_x$ increases the capacity, it also decreases the plateau pressure and the hysteresis ratio P_{abs}/P_{des} (1.25 for $NdMgNiCo_3$ instead of 1.79 for $NdMgNi_4$).

$RMgT_4$	a (Å)	V (Å^3)	References
$LaMgNi_4$	7.1722	368.94	(Sakaki et al. 2012)
$PrMgNi_4$	7.0901	356.41	(Sakaki et al. 2013)
$NdMgNi_4$	7.0877	356.05	(Sakaki et al. 2012)
$NdMgNi_2Co_2$	7.1184	360.07	(Shtender et al. 2015a)
$NdMgNiCo_3$	7.1286	362.25	(Shtender et al. 2015a)
$Nd_{1.2}Mg_{0.8}Ni_3Co$	7.1108	359.54	(Verbovytskyy et al. 2021)
$Nd_{0.8}Mg_{1.2}Ni_3Co$	7.0472	349.98	(Verbovytskyy et al. 2021)
$SmMgNi_4$	7.0532	350.88	(Sakaki et al. 2012)
$GdMgNi_4$	7.0243	346.58	(Sakaki et al. 2012)
$YMgNi_4$	7.0129	344.90	(Shtender et al. 2014)
$YMgNi_2Co_2$	7.0247	346.64	(Shtender et al. 2014)
$YMgCo_4$	7.0596	351.84	(Shtender et al. 2014)

Table 5.7. *Structural parameters of $RMgT_4$ compounds of the MgCu4Sn type structure and space group $F\bar{4}3m$*

$R_{2-x}Mg_xNi_{4-y}Co_y$ (R = Pr, Nd; x = 0.8–1.2; y = 0–2) compounds were studied by Verbovytskyy et al. (2021), who reported that $Nd_{1.2}Mg_{0.8}Ni_3Co$ and $Nd_{0.8}Mg_{1.2}Ni_3Co$ can reversibly absorb 1.3 and 1 wt% of hydrogen, respectively, at 320 K. These two compounds have interesting thermodynamic properties, as their desorption enthalpies are low (ΔH_d ($Nd_{1.2}Mg_{0.8}Ni_3Co$) = 41.3 kJ/mol H_2 and ΔH_d ($Nd_{0.8}Mg_{1.2}Ni_3Co$) =36.8 kJ/mol H_2; Verbovytskyy et al. 2021).

5.3.3.2. $R_{3-x}Mg_xT_9$ (R = Ca, rare earth; T = Ni and Co)

$R_{3-x}Mg_xT_9$ are *AB_3-type* compounds with a *$PuNi_3$*-type structure (S.G $R\bar{3}m$) (Table 5.8). They seem to be promising compounds for reversible hydrogen storage. As previously mentioned in section 5.1.5.1, this structure can be considered a stacking of [$R_{2-x}Mg_xT_4$] and [RT_5] subunits. The feature of these compounds is that Mg can fully substitute R in the [R_2T_4] units (instead of half for other stacking compounds A_2B_7, A_5B_{19}, etc.), leading to stacking of [RT_5] + [Mg_2T_4] = RMg_2T_9. Among the RMg_2T_9 compounds (T = Ni, Co), RMg_2Ni_9 is the most promising for reversible hydrogen storage.

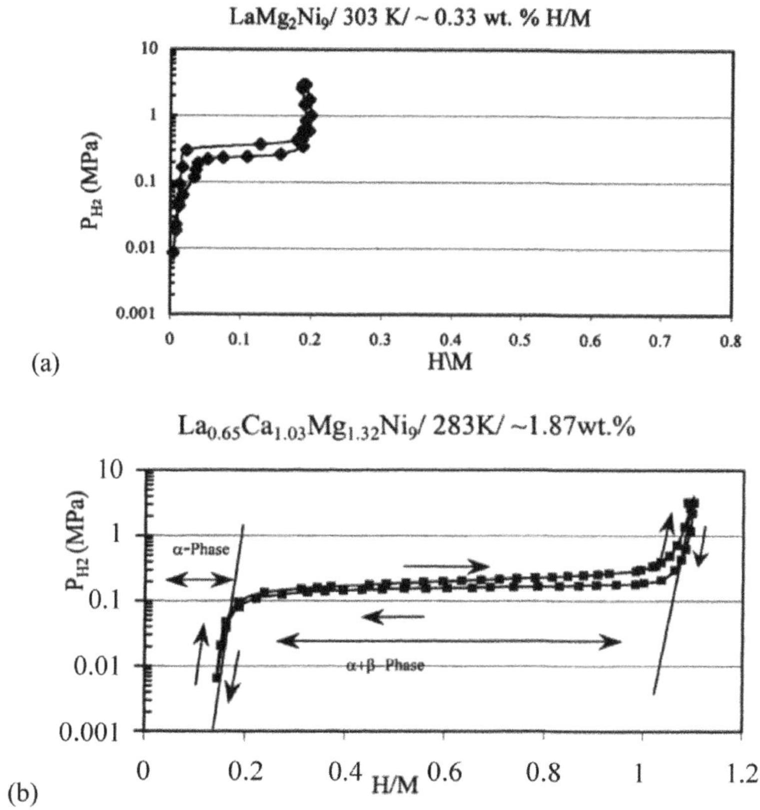

Figure 5.15. *(a) Hydrogen isotherms at 303 K for $LaMg_2Ni_9$ in the hydrogen pressure range of 0.01–3.3 MPa; (b) hydrogen isotherms at 283 K for $La_{0.65}Ca_{1.03}Mg_{1.32}Ni_9$ in the hydrogen pressure range of 0.01–3.3 MPa (Kadir et al. 2000). For a color version of this figure, see www.iste.co.uk/derango-cuevas/hydrogen.zip*

As $LaMg_2Ni_9$ has a low hydrogen absorption capacity (0.33 wt% at 300 K and 3.3 MPa), Kadir et al. have replaced La with Ca and reported that $CaMg_2Ni_9$ can absorb 1.48 wt% hydrogen at 273 K and 3 MPa, which is almost four times the capacity of $LaMg_2Ni_9$ (Kadir et al. 1999b). A full *PCI* was measured, and the equilibrium pressures for absorption and desorption were 0.03 and 0.02 MPa, respectively, at 283 K. Kadir et al. (2000a) investigated the substitution of Mg by La and Ca and reported that $La_{0.65}Ca_{1.03}Mg_{1.32}Ni_9$ can reversibly absorb 1.87 wt% hydrogen at 283 K and 3.3 MPa. The *PCI* curves of $LaMg_2Ni_9$ and $La_{0.65}Ca_{1.03}Mg_{1.32}Ni_9$ at 303 K and 283 K are shown in Figure 5.15a and b. Considering the *PCI* profile of $LaMg_2Ni_9$, a second plateau might exist at higher pressures, as is the case for other AB_3 compounds, for example, YNi_3 (Madern 2018) or $La_{0.5}Ca_{1.5}MgNi_9$ (Chen et al. 2000b), but it was not measured in this study. For $La_{0.65}Ca_{1.03}Mg_{1.32}Ni_9$, a flat plateau was observed, as was a drastic increase in capacity. Kadir et al. (1999c) have also studied the hydrogenation properties of $Y_{0.5}Ca_{1.5}MgNi_9$, which has a maximum capacity of 1.98 wt% at 263 K with a sloped plateau.

Denys and Yartys (2011) and Denys et al. (2012b) have also studied the substitution of La by Mg in the $La_{3-x}Mg_xNi_9$ system and reported that $La_{1.5}Mg_{1.5}Ni_9$, La_2MgNi_9 and $La_{2.3}Mg_{0.7}Mg_9$ can reversibly absorb 1.47, 1.58 and 1.6 wt%, respectively, at 293 K. The low desorption enthalpy of those compounds obtained in this study ($\Delta H_d = 29.5 \pm 0.5$; 35.9 ± 0.3 and 40.3 kJ/mol H_2 for $La_{1.5}Mg_{1.5}Ni_9$, La_2MgNi_9 and $La_{2.3}Mg_{0.7}Ni_9$, respectively) can explain the fully reversible hydrogenation at only 293 K. Notably, the La_2MgNi_9 *PCI* plateaus at 0.009 MPa at 293 K, which can be shifted near atmospheric pressure (0.077 MPa) when hydrogenation is performed at 340 K but diminishes the capacity to 1.21 wt%. Neutron diffraction experiments were performed on the $La_2MgNi_9D_{13.1}$ deuteride to determine the hydrogen position in the crystal structure. Figure 5.16 shows the stacking of $[LaNi_5D_{5.2}]$ and $[LaMgNi_4D_{7.9}]$ units, where the amount of hydrogen in each subunit roughly corresponds to the capacity of individual compounds $LaNi_5H_6$ and $LaMgNi_4H_6$ (Figure 5.16(b)).

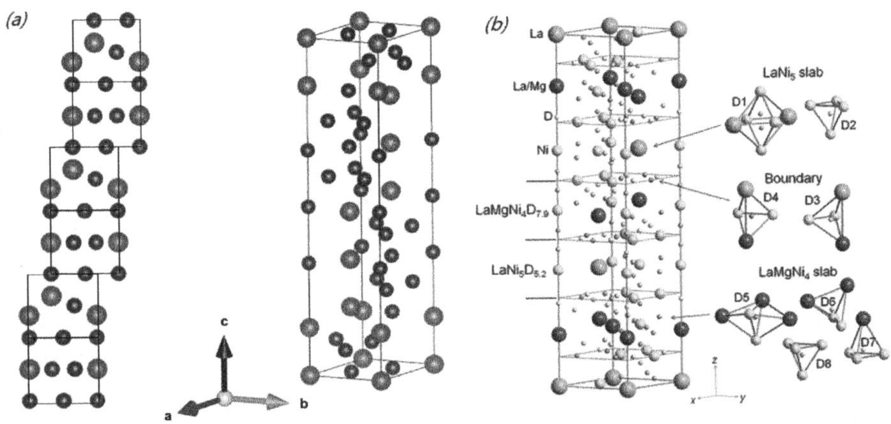

Figure 5.16. *(a) Crystal structure (PuNi$_3$-type) of AB_3 compounds: A in red and B in blue; (b) crystal structure of $La_2MgNi_9D_{13.1(3)}$ showing the stacking of $LaNi_5D_{5.2(1)}$ (CaCu$_5$ type) and $LaMgNi_4D_{7.9(2)}$ (C15 Laves type) (Denys et al. 2012b). For a color version of this figure, see www.iste.co.uk/derango-cuevas/hydrogen.zip*

Si et al. (2009) studied $Ca_{3-x}Mg_xNi_9$ (x = 0.5, 1.0, 1.5 and 2) compounds. $Ca_{2.5}Mg_{0.5}Ni_9$ absorbs 0.71 wt% hydrogen at 298 K, but the hydride is too stable to desorb all the hydrogen. Ca_2MgNi_9 can absorb 1.57 wt% at 293 K with a very sloppy plateau and can only desorb 1.23 wt%. For $Ca_{1.5}Mg_{1.5}Ni_9$ and $CaMg_2Ni_9$, an increase in the Mg proportion leads to the formation of less stable hydrides, thus improving the reversibility with an equilibrium pressure close to atmospheric pressure. The shift of the equilibrium plateau to higher pressure is due to the decrease in the cell volume induced by Mg for Ca substitution. The decrease in the cell volume increases the enthalpy of hydride formation from –39 to –32 kJ/mol H_2 for $Ca_{1.5}Mg_{1.5}Ni_9$ and $CaMg_2Ni_9$, respectively (Si et al. 2009). As a drawback, the shift in the equilibrium plateau leads to a decrease in the capacity with increasing Mg content, from 1.71 to 1 wt% for $Ca_{1.5}Mg_{1.5}Ni_9$ and $CaMg_2Ni_9$, respectively.

Among $R_{3-x}Mg_xCo_9$, Ce_2MgCo_9 can reversibly absorb 1.35 wt.% at 293 K with an equilibrium pressure equal to 2.8 MPa and a large hysteresis between absorption and desorption (Denys et al. 2012a). Nd_2MgCo_9 can also absorb reversibly 1.35 wt% at 293 K with an equilibrium pressure of 0.03 MPa, which can be shifted to atmospheric pressure at 320 K with a slight decrease in capacity (from 1.35 wt.% at 293 K to 1.27 wt% at 320 K) (Shtender et al. 2017). Pr_2MgCo_9 and Y_2MgCo_9 are also interesting compounds because they can reversibly absorb 1.54 wt% and 1.36 wt%, respectively, at 293 K (Shtender et al. 2022).

$R_{3-x}Mg_xT_9$	a (Å)	c (Å)	V (Å^3)	References
$CaMg_2Ni_9$	4.9048	23.7868	495.57	(Kadir et al. 1999b)
$Ca_{1.5}Mg_{1.5}Ni_9$	4.9223	23.798	499.35	(Si et al. 2009)
Ca_2MgNi_9	4.9545	23.911	504.77	(Si et al. 2009)
$Ca_{2.5}Mg_{0.5}Ni_9$	5.0008	24.177	519.97	(Si et al. 2009)
$LaMg_2Ni_9$	4.9235	23.866	501.02	(Kadir et al. 1997)
$La_{1.5}Mg_{1.5}Ni_9$	4.9840	24.006	512.83	(Denys and Yartys 2011)
La_2MgNi_9	5.0314	24.302	529.07	(Denys and Yartys 2011)
$La_{2.3}Mg_{0.7}Ni_9$	5.0488	24.498	537.03	(Denys and Yartys 2011)
$La_{0.65}Ca_{1.03}Mg_{1.32}Ni_9$	4.9523	23.907	507.77	(Kadir et al. 2000a)
$Y_{0.5}Ca_{1.5}MgNi_9$	4.9277	23.8887	502.36	(Kadir et al. 1999c)
Ce_2MgCo_9	4.9524	24.258	511.66	(Denys et al. 2012a)
Pr_2MgCo_9	5.0415	24.2405	533.57	(Shtender et al. 2022)
Y_2MgCo_9	4.9850	24.087	518.39	(Shtender et al. 2022)

Table 5.8. *Structural parameters of $R_{3-x}Mg_xT_9$ compounds with the Mg_2LaNi_9 space group $R\bar{3}m$*

Other families of ternary Mg-based compounds, such as RMg_2T_2 and R_4MgT, have been studied. Among the RMg_2T_2 compounds, only $LaMg_2Cu_2$ absorbs reversibly hydrogen up to 1.2 wt% at 440 K (Kadir et al. 1999a). Due to the large amount of rare earths with high affinity towards hydrogen in R_4MgT, these compounds form very stable hydrides and therefore high temperature (up to 1220 K) is required to achieve complete desorption (Yartys et al. 1997; Shtender et al. 2015b).

5.3.3.3. *Mg-T-T'*

Mg can also be associated with two transition metals (*T* and *T'*) to improve the hydrogenation properties of pure Mg. Mg_3MnNi_2 and $Mg_{88}Mn_4Ni_8$ have been synthesized by high-energy ball milling, and their structural and hydrogenation properties have been investigated (Denys et al. 2010). XRD measurements and neutron diffraction revealed that these two compounds crystallize in a Ti_2Ni structure (SG $Fd\bar{3}m$, a = 11.56 Å). These authors reported that Mg_3MnNi_2 has a reversible capacity of 1 wt% at room temperature under 2 MPa with an equilibrium desorption pressure of 0.03 MPa. Its dissociation enthalpy is equal to 41 kJ/mol H_2. $Mg_{88}Mn_4Ni_8$ shows better hydrogenation properties, as it can absorb approximately 5.4 wt% at temperatures ranging from 520 to 620 K.

5.4. Conclusion

In this chapter, we have reviewed the structural and hydrogenation properties of the most promising alloys and intermetallic compounds for reversible hydrogen storage. In the first part, the hydrogenation properties of BCC-type alloys and *AB*, AB_2, AB_5 and AB_x ($2 < x < 5$) compounds were reported, while the second part was dedicated to Mg-rich-based compounds. In all these systems, the chemical composition must be adjusted to optimize their hydrogenation properties, such as their weight capacities and enthalpies of absorption/desorption, as well as to improve their activation process, kinetics, and stability during cycling.

Among the V-based BCC alloys, $Ti_{43.5}V_{49}Fe_{7.5}$ has the largest absorption capacity of 3.9 wt% at 253 K, and a reversible capacity of 2.4 wt% was obtained at 573 K for an applied pressure of 3 MPa. The interesting *AB* compounds are $TiFe_{0.8}Mn_{0.2}$, with a reversible capacity of approximately 1.8 wt% at 303 K up to 3 MPa, and $TiFe_{0.8}Ni_{0.2}$, with a capacity of 1.1 wt% at 323 K and 0.02 MPa, which both improve the kinetic and activation processes compared with those of TiFe. For the AB_2 system, $ZrCr_{2-x}M_x$ compounds (M = Mn, Fe, Cu, Co, Ni and $x < 0.8$) have been selected because their capacity is close to 2 wt%, and their equilibrium pressure (1–10 MPa in a broad *T* range between 296 and 473 K) can be easily adapted by appropriate chemical substitutions. $LaNi_5$-based compounds, which have been widely commercialized, display many suitable properties for hydrogen storage, including easy activation, fast kinetics, tunable equilibrium pressures near RT, and good cyclability; however, their weight capacity is limited to approximately 1.5 wt%. To improve the capacity, while retaining the other advantages of AB_5 compounds, AB_x intergrowth compounds composed of the stacking of [A_2B_4] and [AB_5] subunits along the crystallographic axis *c* have attracted broad interest. AB_3, A_2B_7 and A_5B_{19} are the most studied families. For binary compounds, several plateau pressures can be observed in their *PCI*, but they can be tuned into one unique flat plateau with small hysteresis and good reversibility with appropriate substitution, such as in $La_{1.5}Mg_{0.5}Ni_7$. Multiple substituents on both the *A* and *B* sides, such as La-Y-Sm-Ni-Mn-Al, improved the hydrogen storage capacity and reversibility. The A_5Ni_{19} stoichiometry also provides interesting systems such as La_4MgNi_{19} (1.8 wt% at 303 K and an applied pressure of 3 MPa with a plateau pressure of 0.07 MPa) or $La_5Ni_{17}Mn_2$ (1.7 wt% at 298 K and an applied pressure of 4 MPa with a plateau pressure of 0.03 MPa). Other *B*-rich compounds with A_7B_{23} stoichiometries, such as $La_2MgNi_{7.2}Co_{1.8}$, present a reversible capacity of 1.6 wt.% at 298 K. $Ca_3Mg_2Ni_{13}$ (A_5Ni_{13}) also has a large capacity of 1.8 wt% at 298 K under 3 MPa, but it is not fully reversible.

Mg hydride has a large weight capacity and is inexpensive and abundant but presents kinetic limitations (surface oxidation, low H diffusion) and thermodynamic drawbacks because it is highly stable and needs temperatures above 573 K to be desorbed. The activation procedure and kinetics can be improved by nanostructuring stabilized with appropriate additives such as TiH_2 to maintain good kinetics and capacity during cycling. The addition of nanoparticles of metals or oxides, hydride-forming alloys or allotropic forms of carbon (graphite, fullerene, graphene, and carbon nanotubes) as catalysts can significantly improve the kinetics of hydrogen dissociation at the surface and limit oxidation.

However, to tune the thermodynamics and obtain reversible metal hydrides near room temperature, an alloying effect is necessary. Several binary and ternary Mg alloys have been investigated. Complex binary hydrides MgT_4H_x, Mg_2Ni or Mg_6Pd have been extensively studied. These compounds present interesting reversible capacities: approximately 2.2 wt% for the $MgNi_2H_{3.2}$ hydride, 3.5 wt% for Mg_2NiH_4 and 5.6 wt% for $Mg_6Pd_{0.25}Ni_{0.75}$ (multistep reactions). However, the desorption temperatures, although lower than those for MgH_2 even with additives, remain elevated at approximately 473 K. Investigations of ternary hydrides were therefore undertaken to obtain reversible Mg hydrides with absorption and desorption temperatures closer to ambient temperature.

The most studied ternary compounds contain, in addition to Mg and transition metals, a third *A*-type element, such as rare earth or Ca. $RMgT_4$ (T = Ni, Co) has been widely investigated, and the largest capacity obtained for $YMgCo_4H_6$ is 1.9 wt% at ambient temperature and 2 MPa. The second interesting family is $R_{3-x}Mg_xT_9$ AB_3-*type* compounds (R = Ca, rare earth, T = Ni and Co). For example, $La_{0.65}Ca_{1.03}Mg_{1.32}Ni_9$ can reversibly absorb 1.9 wt% at 283 K and 3.3 MPa, and $Y_{0.5}Ca_{1.5}MgNi_9$ has a reversible capacity of 2 wt% at 263 K. Other ternary systems without rare earth elements, such as Mg-Ni-Mn, are interesting but have a lower capacity (1 wt%). They can be used to form composites with Mg, but in this case, the thermodynamic is dominated by those of Mg.

A plot of the desorption temperature under ambient pressure versus the reversible weight percentage of selected representative compounds is displayed in Figure 5.17. These findings indicate that compounds with weight capacities above 2 wt%, such as V-, Mg- or Mg-complex hydrides, require elevated desorption temperatures above 500 K, whereas metal hydrides can reversibly store hydrogen near room temperature but have reversible capacities below 2 wt%.

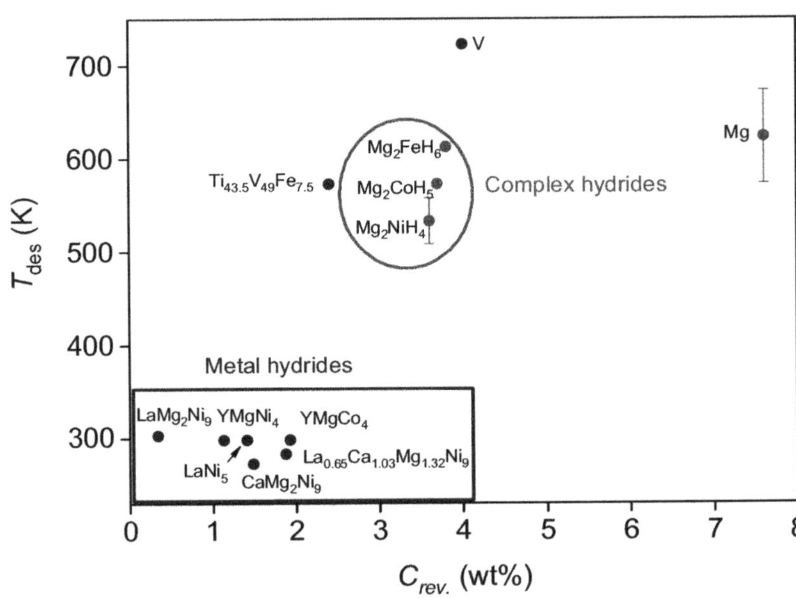

Figure 5.17. *Desorption temperature versus reversible capacity of the selected compounds studied in this book. For a color version of this figure, see www.iste.co.uk/derango-cuevas/hydrogen.zip*

To conclude, a safe compact and reversible hydrogen storage material can be obtained with many alloys and intermetallic compounds by adjusting their chemical composition. The difficulty remains that a compromise should be made between the best weight capacity and the adjustment of the equilibrium pressure under ambient conditions, in addition to all the other important hydrogenation parameters, such as kinetics, stability upon cycling and cost of the alloy. This means that it is not possible to determine one unique and universal material for hydrogen storage but that it should be chosen among all the possible optimized systems and adapted for a given practical application.

This work was funded by the French National Research Agency (ANR), reference project "ANR-22-PEHY-0007", as part of the France 2030 programme.

5.5. References

Aguey-Zinsou, K.-F. and Ares-Fernández, J.-R. (2010). Hydrogen in magnesium: New perspectives toward functional stores. *Energy Environ. Sci.*, 3, 526.

Akiba, E. and Iba, H. (1998). Hydrogen absorption by Laves phase related BCC solid solution. *Intermetallics*, 461–470.

ASM Handbook (1992). ASM International alloy phase diagram and the handbook committees. Report, ASM.

Asselli, A.A.C., Leiva, D.R., Jorge, A.M., Ishikawa, T.T., Botta, W.J. (2012). Synthesis and hydrogen sorption properties of Mg_2FeH_6–MgH_2 nanocomposite prepared by reactive milling. *J. Alloys Compd.*, 536, S250–S254.

Barkhordarian, G., Klassen, T., Bormann, R. (2006). Catalytic mechanism of transition-metal compounds on Mg hydrogen sorption reaction. *J. Phys. Chem. B*, 110(22), 11020–11024.

Bastide, J., Bonnetot, B., Létoffé, J.-M., Claudy, P. (1980). Polymorphisme de l'hydrure de magnesium sous haute pression. *Materials Research Bulletin*, 15, 1779.

Belgacem, Y.B., Khaldi, C., Lamloumi, J., Takenouti, H. (2015). Effect of the discharge rate on the electrochemical properties of LaY_2Ni_9 hydrogen storage alloy. *J. Alloys Compd.*, 631, 7–14.

Bogdanović, B., Bohmhammel, K., Christ, B., Reiser, A., Schlichte, K., Vehlen, R., Wolf, U. (1999). Thermodynamic investigation of the magnesium–hydrogen system. *J. Alloys Compd.*, 282(1/2), 84–92.

Bortz, M., Bertheville, B., Bottger, G., Yvon, K. (1999). Structure of the high pressure phase g-MgH_2 by neutron powder diffraction. *J. Alloys Compd.*, 287, L4–L6.

Botta, W.J., Zepon, G., Ishikawa, T.T., Leiva, D.R. (2022). Metallurgical processing of Mg alloys and MgH_2 for hydrogen storage. *J. Alloys Compd.*, 897, 162798.

Bououdina, M., Grant, D., Walker, G. (2006). Review on hydrogen absorbing materials–structure, microstructure, and thermodynamic properties. *Int. J. Hydrog. Energy*, 31(2), 177–182.

Boussami, S., Khaldi, C., Lamloumi, J., Mathlouthi, H., Takenouti, H., Vivier, V. (2013). The impedance response of LaY_2Ni_9 negative electrode materials after activation. *J. Phys. Chem. Solids*, 74(10), 1369–1374.

Bouzidi, W., Patout, L., Charai, A., Mliki, N., Bessais, L. (2020). Crystal structure of nanocrystalline Pr_5Co_{19} compound and its hydrogen storage properties. *Int. J. Hydrog. Energy*, 45(19), 11190–11198.

Brutti, S., Farina, L., Trequattrini, F., Palumbo, O., Reale, P., Silvestri, L., Panero, S., Paolone, A. (2018). Extremely pure Mg_2FeH_6 as a negative electrode for lithium batteries. *Energies*, 11(8), 1952.

Burch, R. and Mason, N.B. (1979). Absorption of hydrogen by titanium–cobalt and titanium–nickel intermetallic alloys. Part 1.–Experimental results. *J. Chem. Soc., Faraday Trans. 1*, 75, 561.

Buschow, K.H.J. (1970). The crystal structure of rare earth nickel compounds of the type R_2Ni_7. *Journal of the Less Common Metals*, 22(4), 419-428. doi: 10.1016/0022-5088(70)90129-3

Callini, E., Pasquini, L., Piscopiello, E., Montone, A., Antisari, M.V., Bonetti, E. (2009). Hydrogen sorption in Pd-decorated Mg–MgO core-shell nanoparticles. *Appl. Phys. Lett.*, 94(22), 221905.

Charbonnier, V., Zhang, J., Monnier, J., Goubault, L., Bernard, P., Magén, C., Serin, V., Latroche, M. (2015). Structural and hydrogen storage properties of Y_2Ni_7 deuterides studied by neutron powder diffraction. *J. Phys. Chem. C*, 119(22), 12218–12225.

Charbonnier, V., Monnier, J., Zhang, J., Paul-Boncour, V., Joiret, S., Puga, B., Goubault, L., Bernard, P., Latroche, M. (2016). Relationship between H_2 sorption properties and aqueous corrosion mechanisms in A_2Ni_7 hydride forming alloys (A = Y, Gd or Sm). *J. Power Sources*, 326, 146–155.

Chaudhary, A.-L., Li, G., Matsuo, M., Orimo, S., Deledda, S., Sørby, M.H., Hauback, B.C., Pistidda, C., Klassen, T., Dornheim, M. (2015). Simultaneous desorption behavior of M borohydrides and Mg_2FeH_6 reactive hydride composites (M = Mg, then Li, Na, K, Ca). *Appl. Phys. Lett.*, 107(7), 073905.

Chen, J., Kuriyama, N., Takeshita, H.T., Tanaka, H., Sakai, T., Haruta, M. (2000a). Hydrogen storage alloys with $PuNi_3$-type structure as metal hydride electrodes. *Electrochem. Solid-State Lett.*, 3(6), 249.

Chen, J., Takeshita, H.T., Tanaka, H., Kuriyama, N., Sakai, T., Uehara, I., Haruta, M. (2000b). Hydriding properties of $LaNi_3$ and $CaNi_3$ and their substitutes with $PuNi_3$-type structure. *J. Alloys Compd.*, 302(1), 304–313.

Ćirić, K.D., Kocjan, A., Gradišek, A., Koteski, V.J., Kalijadis, A.M., Ivanovski, V.N., Laušević, Z.V., Stojić, D.L.J. (2012). A study on crystal structure, bonding and hydriding properties of Ti–Fe–Ni intermetallics – Behind substitution of iron by nickel. *Int. J. Hydrog. Energy*, 37(10), 8408–8417.

Colinet, C., Pasturel, A., Percheron-Guegan, A., Achard, J.C. (1987). Enthalpies of formation and hydrogenation of $La(Ni_{1-x}Co_x)_5$ compounds. *J. Less-Common Met.*, 134, 109–122.

Crivello, J.-C., Zhang, J., Latroche, M. (2011). Structural stability of AB_y phases in the (La,Mg)–Ni system obtained by density functional theory calculations. *J. Phys. Chem.*, 115(51), 25470–25478.

Crivello, J.-C., Dam, B., Denys, R.V., Dornheim, M., Grant, D.M., Huot, J., Jensen, T.R., de Jongh, P., Latroche, M., Milanese, C. et al. (2016). Review of magnesium hydride-based materials: Development and optimisation. *Appl. Phys. A*, 122(2), 97.

Cuevas, F., Joubert, J.-M., Latroche, M., Percheron-Guégan, A. (2001). Intermetallic compounds as negative electrodes of Ni/*M*H batteries. *Appl. Phys. A*, 72(2), 225–238.

Cui, S., Feng, W., Hu, H., Feng, Z., Wang, Y. (2008). Structural phase transitions in MgH_2 under high pressure. *Solid State Commun.*, 148, 403–405.

Cui, J., Liu, J., Wang, H., Ouyang, L., Sun, D., Zhu, M., Yao, X. (2014). Mg–*TM* (*TM*: Ti, Nb, V, Co, Mo or Ni) core–shell like nanostructures: synthesis, hydrogen storage performance and catalytic mechanism. *J. Mater. Chem. A*, 2(25), 9645–9655.

Dehouche, Z., Klassen, T., Oelerich, W., Goyette, J., Bose, T.K., Schulz, R. (2002). Cycling and thermal stability of nanostructured MgH_2–Cr_2O_3 composite for hydrogen storage. *J. Alloys Compd.*, 347(1–2), 319-323. doi: 10.1016/S0925-8388(02)00784-3

Dematteis, E.M., Dreistadt, D.M., Capurso, G., Jepsen, J., Cuevas, F., Latroche, M. (2021). Fundamental hydrogen storage properties of TiFe-alloy with partial substitution of Fe by Ti and Mn. *J. Alloys Compd.*, 874, 159925.

Denys, R.V. and Yartys, V.A. (2011). Effect of magnesium on the crystal structure and thermodynamics of the $La_{3-x}Mg_xNi_9$ hydrides. *J. Alloys Compd.*, 509, S540–S548.

Denys, R.V., Riabov, A.B., Yartys, V.A., Sato, M., Delaplane, R.G. (2008). Mg substitution effect on the hydrogenation behaviour, thermodynamic and structural properties of the La_2Ni_7–H(D)$_2$ system. *J. Solid State Chem.*, 181(4), 812–821.

Denys, R.V., Zavaliy, I.Y., Paul-Boncour, V., Berezovets, V.V., Koval'chuk, I.V., Riabov, A.B. (2010). New Mg–Mn–Ni alloys as efficient hydrogen storage materials. *Intermetallics*, 18(8), 1579–1585.

Denys, R.V., Riabov, A.B., Černý, R., Koval'chuk, I.V., Zavaliy, I.Y.U. (2012a). New $CeMgCo_4$ and Ce_2MgCo_9 compounds: Hydrogenation properties and crystal structure of hydrides. *J. Solid State Chem.*, 187, 1–6.

Denys, R.V, Yartys, V.A., Webb, C.J. (2012b). Hydrogen in $La_2MgNi_9D_{13}$: The role of magnesium. *Inorg. Chem.*, 51, 4231–4238.

Diaz, H., Percheron-Guégan, A., Achard, J.C. (1979). Thermodynamic and structural properties of $LaNi_{5-y}Al_y$ compounds and their related hydrides. *Int. J. Hydrog. Energy*, 4, 445–454.

Drasner, A. (1990). On the structural and hydrogen desorption properties of the $Zr(Cr_{1-x}Ni_5)_5$ alloys. *J. Less-Common Met.*, 163, 151–157.

Drasner, A. and Blažina, Z. (1991). On the structural and hydrogen desorption properties of the $Zr(Cr_{1-x}Cu_5)_5$ alloys. *J. Less-Common Met.*, I75, 103–108.

Dufour, J. and Huot, J. (2007). Study of Mg_6Pd alloy synthesized by cold rolling. *J. Alloys Compd.*, 446(447), 147–151.

Emami, H., Edalati, K., Staykov, A., Hongo, T., Iwaoka, H., Horita, Z., Akiba, E. (2016). Solid-state reactions and hydrogen storage in magnesium mixed with various elements by high-pressure torsion: Experiments and first-principles calculations. *RSC Adv.*, 6(14), 11665–11674.

Fang, F., Chen, Z., Wu, D., Liu, H., Dong, C., Song, Y., Sun, D. (2019). Subunit volume control mechanism for dehydrogenation performance of AB_3-type superlattice intermetallics. *J. Power Sources*, 427, 145–153.

Férey, A., Cuevas, F., Latroche, M., Knosp, B., Bernard, P. (2009). Elaboration and characterization of magnesium-substituted La_5Ni_{19} hydride forming alloys as active materials for negative electrode in Ni-*M*H battery. *Electrochim. Acta*, 54(6), 1710–1714.

Fernández, I.G., Meyer, G.O., Gennari, F.C. (2007). Reversible hydrogen storage in Mg_2CoH_5 prepared by a combined milling-sintering procedure. *J. Alloys Compd.*, 446(447), 106–109.

Fernández, I.G., Meyer, G.O., Gennari, F.C. (2008). Hydriding/dehydriding behavior of Mg_2CoH_5 produced by reactive mechanical milling. *J. Alloys Compd.*, 464(1/2), 111–117.

Ferro, R. (1959). Research on the alloys of noble metals with the more electropositive elements: II. Micrographic and roentgenographic examination of the magnesium-palladium alloys. *J. Less-Common Met.*, 1, 424–438.

Fersi, R., Mliki, N., Bessais, L. (2018). Hydrogenation and the effect of H absorption on structural and magnetic properties of nanocrystalline Pr_2Co_7 alloys. *J. Magn. Magn. Mater.*, 465, 220–227.

Filipek, S.M., Paul-Boncour, V., Kuriyama, N., Takeichi, N., Tanaka, H., Liu, R.-S., Wierzbicki, R., Sato, R., Kuo, H.-T. (2010). Hydrides of Laves phases intermetallic compounds synthesized under high hydrogen pressure. *Solid State Ion.*, 181(5/7), 306–310.

Gesari, S.B., Pronsato, M.E., Visintin, A., Juan, A. (2010). Hydrogen storage in AB_2 laves phase (A = Zr, Ti; B = Ni, Mn, Cr, V): Binding energy and electronic structure. *J. Phys. Chem. C*, 114(39), 16832–16836.

Guenee, L., Favre-Nicolin, V., Yvon, K. (2003). Synthesis, crystal structure and hydrogenation properties of the ternary compounds $LaNi_4Mg$ and $NdNi_4Mg$. *J. Alloys Compd.*, 348, 129–137.

Guo, M., Yuan, H., Liu, Y., Jiang, L. (2021). Effect of Sm on the cyclic stability of La–Y–Ni-based alloys and their comparison with *RE*–Mg–Ni-based hydrogen storage alloy. *Int. J. Hydrog. Energy*, 46(10), 7432–7441.

Hou, X., Hu, R., Zhang, T., Kou, H., Song, W., Li, J. (2014). Hydrogen desorption performance of high-energy ball milled Mg_2NiH_4 catalyzed by multi-walled carbon nanotubes coupling with TiF_3. *Int. J. Hydrog. Energy*, 39(34), 19672–19681.

Hsu, F.-K., Hsu, C.-W., Chang, J.-K., Lin, C.-K., Lee, S.-L., Jiang, C.-E. (2010). Structure and hydrogen storage properties of $Mg_2Cu_{1-x}Ni_x$ (x = 0–1) alloys. *Int. J. Hydrog. Energy*, 35(24), 13247–13254.

Huot, J., Hayakawa, H., Akiba, E. (1997). Preparation of the hydrides Mg_2FeH_6, and Mg_2CoH_5, by mechanical alloying followed by sintering. *J. Alloys Compd.*, 248, 164–167.

Huot, J., Liang, G., Boily, S., Van Neste, A., Schulz, R. (1999). Structural study and hydrogen sorption kinetics of ball-milled magnesium hydride. *J. Alloys Compd.*, 293(295), 495–500.

Huot, J., Yonkeu, A., Dufour, J. (2009). Rietveld analysis of neutron powder diffraction of Mg_6Pd alloy at various hydriding stages. *J. Alloys Compd.*, 475, 168–172.

Imamura, H., Tabata, S., Takesue, Y., Sakata, Y., Kamazaki, S. (2000). Hydriding–dehydriding behavior of magnesium composites obtained by mechanical grinding with graphite carbon. *Int. J. Hydrog. Energy*, 25(9), 837–843.

Imamura, H., Kusuhara, M., Minami, S., Matsumoto, M., Masanari, K., Sakata, Y., Itoh, K., Fukunaga, T. (2003). Carbon nanocomposites synthesized by high-energy mechanical milling of graphite and magnesium for hydrogen storage. *Acta Mater.*, 51(20), 6407–6414.

Itoh, H., Arashima, H., Kubo, K., Kabutomori, T., Ohnishi, K. (2005). Improvement of cyclic durability of BCC structured Ti–Cr–V alloys. *J. Alloys Compd.*, 404(406), 417–420.

Iwase, K., Sakaki, K., Nakamura, Y., Akiba, E. (2010). Phase transformation and crystal structure of $La_2Ni_7H_x$ studied by in situ X-ray diffraction. *Inorg. Chem.*, 49(19), 8763–8768.

Iwase, K., Honjyo, Y., Mori, K. (2022). Phase transformation of hydride-phase Nd_2Co_7 with superlattice structure. *J. Solid State Chem.*, 316, 123575.

Jiang, W., Chen, Y., Hu, M., Zeng, C., Liang, C. (2021). Rare earth-Mg-Ni-based alloys with superlattice structure for electrochemical hydrogen storage. *J. Alloys Compd.*, 887, 161381.

Joubert, J.-M., Latroche, M., Černý, R., Bowman, R.C., Percheron-Guégan, A., Yvon, K. (1999). Crystallographic study of $LaNi_{5-x}Sn_x$ ($0.2 \leq x \leq 0.5$) compounds and their hydrides. *J. Alloys Compd.*, 293(295), 124–129.

Joubert, J.-M., Charton, J., Percheron-Guégan, A. (2003). Investigation of structural and hydrogen absorption properties in the $LaNi_{5-x}Pt_x$–H_2 system. *J. Solid State Chem.*, 173(2), 379–386.

Joubert, J.-M., Paul-Boncour, V., Cuevas, F., Zhang, J., Latroche, M. (2021). $LaNi_5$ related AB_5 compounds: Structure, properties and applications. *J. Alloys Compd.*, 862, 158163.

Kadir, K., Sakai, T., Uehara, I. (1997). Synthesis and structure determination of a new series of hydrogen storage alloys; RMg_2Ni_9 (R = La, Ce, Pr, Nd, Sm and Gd) built from $MgNi_2$ Laves-type layers alternating with AB_5 layers. *J. Alloys Compd.*, 257(1/2), 115–121.

Kadir, K., Tanaka, H., Sakai, T., Uehara, I. (1999a). Hydrogen behavior in the La–Mg–Cu system. *J. Alloys Compd.*, 289(1/2), 66–70.

Kadir, K., Kuriyama, N., Sakai, T., Uehara, I., Eriksson, L. (1999b). Structural investigation and hydrogen capacity of $CaMg_2Ni_9$: A new phase in the AB_2C_9 system iso-structural with $LaMg_2Ni_9$. *J. Alloys Compd.*, 284(1/2), 145–154.

Kadir, K., Sakai, T., Uehara, I. (1999c). Structural investigation and hydrogen capacity of YMg_2Ni_9 and $(Y_{0.5}Ca_{0.5})(MgCa)Ni_9$: New phases in the AB_2C_9 system isostructural with $LaMg_2Ni_9$. *J. Alloys Compd.*, 287(1/2), 264–270.

Kadir, K., Sakai, T., Uehara, I. (2000). Structural investigation and hydrogen storage capacity of $LaMg_2Ni_9$ and $(La_{0.65}Ca_{0.35})(Mg_{1.32}Ca_{0.68})Ni_9$ of the AB_2C_9 type structure. *J. Alloys Compd.*, 302(1/2), 112–117.

Kataoka, R., Goto, Y., Kamegawa, A., Takamura, H., Okada, M. (2007). High-pressure synthesis of novel hydride in Mg–Ni–H and Mg–Ni–Cu–H systems. *J. Alloys Compd.*, 446(447), 142–146.

Komura, Y. and Tokunaga, K. (1980). Structural studies of stacking variants in Mg-base Friauf–Laves phases. *Acta Cryst.*, 36(7), 1548–1554.

Kumar, S., Tiwari, G.P., Krishnamurthy, N. (2015). Tailoring the hydrogen desorption thermodynamics of V_2H by alloying additives. *J. Alloys Compd.*, 645, S252–S256.

Kumar, S., Jain, A., Ichikawa, T., Kojima, Y., Dey, G.K. (2017). Development of vanadium-based hydrogen storage material: A review. *Renew. Sust. Energ. Rev.*, 72, 791–800.

Kume, Y. and Weiss, A. (1987). On the interaction of hydrogen with the intermetallic phase Mg_6Pd. *J. Less-Common Met.*, 136(1), 51–54.

Langmi, H.W., McGrady, G.S., Newhouse, R., Rönnebro, E. (2012). Mg_2FeH_6–$LiBH_4$ and Mg_2FeH_6–$LiNH_2$ composite materials for hydrogen storage. *Int. J. Hydrog. Energy*, 37(8), 6694–6699.

Lartigue, C., Percheron-Guégan, A., Achard, J.C., Tasset, F. (1980). Thermodynamic and structural properties of $LaNi_{5-x}Mn_x$ compounds and their hydrides. *J. Less-Common Met.*, 75, 23–29.

Leblond, T., Paul-Boncour, V., Cuevas, F., Isnard, O., Fernández, J.F. (2009). Study of the multipeak deuterium thermodesorption in YFe_2D_x ($1.3 \leq x \leq 4.2$) by DSC, TD and in situ neutron diffraction. *Int. J. Hydrog. Energy*, 34(5), 2278–2287.

Lee, S.-M. and Perng, T.-P. (1999). Correlation of substitutional solid solution with hydrogenation properties of $TiFe_{1-x}M_x$ (M=Ni, Co, Al) alloys. *J. Alloys Compd.*, 291(1/2), 254–261.

Lemort, L., Latroche, M., Knosp, B., Bernard, P. (2011). Elaboration and characterization of unreported $(Pr,Nd)_5Ni_{19}$ hydrides. *J. Alloys Compd.*, 509, S823–S826.

Li, L., Akiyama, T., Yagi, J. (2000). Hydrogen storage alloy of Mg_2NiH_4 hydride produced by hydriding combustion synthesis from powder of mixture metal. *J. Alloys Compd.*, 308(1/2), 98–103.

Li, L., Akiyama, T., Yagi, J. (2001). Activation behaviors of Mg_2NiH_4 at different hydrogen pressures in hydriding combustion synthesis. *Int. J. Hydrog. Energy*, 26, 1035–1040.

Li, G., Matsuo, M., Takagi, S., Chaudhary, A.-L., Sato, T., Dornheim, M., Orimo, S. (2017). Thermodynamic properties and reversible hydrogenation of $LiBH_4$–Mg_2FeH_6 composite materials. *Inorganics*, 5(4), 81.

Li, Y., Liu, Z., Guofang, Z., Zhang, Y., Ren, H. (2019). Novel A_7B_{23}-type La-Mg-Ni-Co compound for application on Ni-*M*H battery. *J. Power Sources*, 441, 126667.

Liang, G.X., Boily, S., Huot, J., Van Neste, A., Schulz, R. (1998). Hydrogen storage properties of nanocrystalline $Mg_2Ni_xCu_{1-x}$ synthesized by mechanical alloying. *Mater. Sci. Forum.*, 269–272, 1049–1054.

Libowitz, G.G., Hayes, H.F., Gibb, T.R.P. (1958). The system zirconium–nickel and hydrogen. *J. Phys. Chem.*, 62(1), 76–79.

Liu, Y., Pan, H., Gao, M., Wang, Q. (2011). Advanced hydrogen storage alloys for Ni/MH rechargeable batteries. *J. Mater. Chem.*, 21(13), 4743–4755.

Liu, G., Wang, Y., Xu, C., Qiu, F., An, C., Li, L., Jiao, L., Yuan, H. (2013). Excellent catalytic effects of highly crumpled graphenenanosheets on hydrogenation/ dehydrogenation of magnesium hydride. *Nanoscale*, 5(3), 1074–1081.

Liu, J., Han, S., Li, Y., Zhang, L., Zhao, Y., Yang, S., Liu, B. (2016). Phase structures and electrochemical properties of La-Mg-Ni-based hydrogen storage alloys with superlattice structure. *International Journal of Hydrogen Energy,* 41(44), 20261 20275. doi: 10.1016/ j.ijhydene.2016.08.149

Liu, Z., Xiong, L., Li, J., Liu, S., Er, S. (2018). Effects of alloying elements (Al, Mn, Ru) on desorption plateau pressures of vanadium hydrides: An experimental and first-principles study. *Int. J. Hydrog. Energy*, 43(46), 21441–21450.

Liu, Y., Zhu, J., Liu, Z., Zhu, Y., Zhang, J., Li, L. (2020). Magnesium nanoparticles with Pd decoration for hydrogen storage. *Front. Chem.*, 7, 949.

Liu, J., Zhu, S., Chen, X., Xu, J., Zhang, L., Yan, K., Chen, W., Cheng, H., Han, S. (2021). Superior electrochemical performance of La-Mg-Ni-based alloys with novel A_2B_7-A_7B_{23} biphase superlattice structure. *J. Mater. Sci. Technol.*, 80, 128–138.

Lynch, J.F., Libowitz, G.G., Maeland, A.J. (1984). Hysteresis in the Nb-V-H system. *J. Less-Common Met.*, 103(1), 117–122.

Lys, A., Fadonougbo, J.O., Faisal, M., Suh, J.-Y., Lee, Y.-S., Shim, J.-H., Park, J., Cho, Y.W. (2020). Enhancing the hydrogen storage properties of A_xB_y intermetallic compounds by partial substitution: A short review. *Hydrogen*, 1(1), 38–63.

Ma, Z., Tang, Q., Ni, J., Zhu, Y., Zhang, Y., Li, H.-W., Zhang, J., Liu, Y., Ba, Z., Li, L. (2022). Synergistic effect of TiH_2 and air exposure on enhancing hydrogen storage performance of Mg_2NiH_4. *Chem. Eng. J.*, 433, 134489.

Madern, N. (2018). Propriétés de sorption de l'hydrogène et mécanismes de corrosion en milieu aqueux d'alliages AB_x (A = La, Sm, Y, Mg; B = Ni, Mn; $3 < x < 5$) pour le stockage de l'énergie. PhD Thesis, Université Paris-Est, Paris.

Madern, N., Charbonnier, V., Monnier, J., Zhang, J., Paul-Boncour, V., Latroche, M. (2020). Investigation of H sorption and corrosion properties of $Sm_2Mn_xNi_{7-x}$ ($0 \leq x < 0.5$) intermetallic compounds forming reversible hydrides. *Energy J.*, 13(13), 3470.

Martin, M., Gommel, C., Borkhart, C., Fromm, E. (1996). Absorption and desorption kinetics of hydrogen storage alloys. *J. Alloys Compd.*, 238(1/2), 193–201.

Martínez-Coronado, R., Retuerto, M., Torres, B., Martínez-Lope, M.J., Fernández-Díaz, M.T., Alonso, J.A. (2013). High-pressure synthesis, crystal structure and cyclability of the Mg_2NiH_4 hydride. *Int. J. Hydrog. Energy*, 38(14), 5738–5745.

Massicot, B. (2009). Étude du système Fe–Ti–V et de ses applications au stockage de l'hydrogène. PhD Thesis, UPE, Paris.

Matsuda, J. and Akiba, E. (2013). Lattice defects in V–Ti BCC alloys before and after hydrogenation. *J. Alloys Compd.*, 581, 369–372.

Moriwaki, T., Akahama, Y., Kawamura, H., Nakano, S., Takemura, K. (2006). Structural phase transition of rutile-type MgH_2 at high pressures. *J. Phys. Soc. Jpn*, 75(7), 074603.

Nakamura, Y., Oguro, K., Uehara, I., Akiba, E. (2000). X-ray diffraction peak broadening and lattice strain in $LaNi_5$-based alloys. *J. Alloys Compd.*, 298(1/2), 138–145.

Nomura, K. and Akiba, E. (1995). H_2 Absorbing-desorbing characterization of the TiVFe alloy system. *J. Alloys Compd.*, 231(1/2), 513–517.

Oelerich, W., Klassen, T., Bormann, R. (2001). Metal oxides as catalysts for improved hydrogen sorption in nanocrystalline Mg-based materials. *J. Alloys Compd.*, 315(1/2), 237–242.

Ouyang, L., Huang, J., Wang, H., Liu, J., Zhu, M. (2017). Progress of hydrogen storage alloys for Ni-*M*H rechargeable power batteries in electric vehicles: A review. *Mater. Chem. Phys.*, 200, 164–178.

Palade, P., Sartori, S., Maddalena, A., Principi, G., Lo Russo, S., Lazarescu, M., Schinteie, G., Kuncser, V., Filoti, G. (2006). Hydrogen storage in Mg–Ni–Fe compounds prepared by melt spinning and ball milling. *J. Alloys Compd.*, 415(1/2), 170–176.

Pasturel, A., Chatillon-Colinet, C., Guégan, A.P., Achard, J.C. (1982). Thermodynamic properties of $LaNi_4M$ compounds and their related hydrides. *J. Less-Common Met.*, 84, 73–78.

Pasturel, A., Liautaud, F., Colinet, C., Allibert, C., Percheron-Guégan, A., Achard, J.C., Mktallurgique, C. (1984). Thermodynamic study of the $LaNi_{5-x}Cu_x$ system. *J. Less-Common Met.*, 96, 93–97.

Paul-Boncour, V., Filipek, S.M., Dorogova, M., Bourée, F., André, G., Marchuk, I., Percheron-Guégan, A., Liu, R.S. (2005). Neutron diffraction study, magnetic properties and thermal stability of YMn_2D_6 synthesized under high deuterium pressure. *J. Solid State Chem.*, 178(1), 356–362.

Paul-Boncour, V., Guillot, M., Isnard, O., Ouladdiaf, B., Hoser, A., Hansen, T., Stuesser, N. (2017). Interplay between crystal and magnetic structures in $YFe_2(H_\alpha D_{1-\alpha})_{4.2}$ compounds studied by neutron diffraction. *J. Solid State Chem.*, 245, 98–109.

Paul-Boncour, V., Joubert, J.-M., Cuevas, F., Latroche, M. (2018). Hydrogen storage materials. In *Landolt-Börnstein: Numerical Data and Functional Relationships in Science and Technology*, Burzo, E. (ed.). Wiley, New York.

Percheron-Guégan, A., Lartigue, C., Achard, J.C., Germi, P., Tasset, F. (1980). Neutron and X-ray diffraction profile analyses and structure of $LaNi_5$, $LaNi_{5-x}Al_x$ and $LaNi_{5-x}Mn_x$ intermetallics and their hydrides (deuterides). *J. Less-Common Met.*, 74, 1–12.

Percheron-Guégan, A., Lartigue, C., Achard, J.C. (1985). Correlations between the structural properties, the stability and the hydrogen content of substituted $LaNi_5$ compounds. *J. Less-Common Met.*, 109(2), 287–309.

Polanski, M., Bystrzycki, J., Varin, R.A., Plocinski, T. (2011). Rapid hydrogenation at 30°C of magnesium (Mg) and iron (Fe) nanocomposite obtained through a decomposition of Mg_2FeH_6 precursor. *Int. J. Hydrog. Energy*, 36(1), 1059–1065.

Poletaev, A.A., Denys, R.V., Maehlen, J.P., Solberg, J.K., Tarasov, B.P., Yartys, V.A. (2012). Nanostructured rapidly solidified $LaMg_{11}Ni$ alloy: Microstructure, crystal structure and hydrogenation properties. *Int. J. Hydrog. Energy*, 37(4), 3548–3557.

Ponthieu, M., Fernández, J.F., Cuevas, F., Ares, J.R., Leardini, F., Bodega, J., Sánchez, C. (2013). Reversible hydrogen storage in the Ni-rich pseudo-binary $Mg_6Pd_{0.25}Ni_{0.75}$ intermetallic compound: Reaction pathway, thermodynamic and kinetic properties. *J. Alloys Compd.*, 548, 96–104.

Ponthieu, M., Calizzi, M., Pasquini, L., Fernández, J.F., Cuevas, F. (2014). Synthesis by reactive ball milling and cycling properties of MgH_2–TiH_2 nanocomposites: Kinetics and isotopic effects. *Int. J. Hydrog. Energy*, 39(18), 9918–9923.

Przewoznik, J., Paul-Boncour, V., Latroche, M., Percheron-Guégan, A. (1995). Structural study of YMn_2 hydrides. *J. Alloys Compd.*, 225(1/2), 436–439.

Pukazhselvan, D., Nasani, N., Sandhya, K.S., Singh, B., Bdikin, I., Koga, N., Fagg, D.P. (2017). Role of chemical interaction between MgH_2 and TiO_2 additive on the hydrogen storage behavior of MgH_2. *Appl. Surf. Sci.*, 420, 740–745.

Reilly, J.J. and Wiswall, R.H. (1968). Reaction of hydrogen with alloys of magnesium and nickel and the formation of Mg_2NiH_4. *Inorg. Chem.*, 7(11), 2254–2256.

Reilly, J.J. and Wiswall, R.H. (1974). Formation and properties of iron titanium hydride. *Inorg. Chem.*, 13(1), 218–222.

Retuerto, M., Alonso, J.A., Martínez, R., Jiménez-Villacorta, F., Sánchez-Benítez, J., Fernández-Díaz, M.T., Garcia-Ramos, C.A., Ruskov, T. (2015). Neutron powder diffraction, X-ray absorption and Mössbauer spectroscopy on Mg_2FeH_6. *Int. J. Hydrog. Energy*, 40(30), 9306–9313.

Rizo-Acosta, P., Cuevas, F., Latroche, M. (2018). Optimization of TiH_2 content for fast and efficient hydrogen cycling of MgH_2-TiH_2 nanocomposites. *Int. J. Hydrog. Energy*, 43(34), 16774–16781.

Ropka, J., Černý, R., Paul-Boncour, V., Proffen, T. (2009). Deuterium ordering in Laves-phase deuteride $YFe_2D_{4.2}$. *J. Solid State Chem.*, 182(7), 1907–1912.

Rusman, N.A.A. and Dahari, M. (2016). A review on the current progress of metal hydrides material for solid-state hydrogen storage applications. *Int. J. Hydrog. Energy*, 41(28), 12108–12126.

Sadhasivam, T., Sterlin Leo Hudson, M., Pandey, S.K., Bhatnagar, A., Singh, M.K., Gurunathan, K., Srivastava, O.N. (2013). Effects of nano size mischmetal and its oxide on improving the hydrogen sorption behaviour of MgH_2. *Int. J. Hydrog. Energy*, 38(18), 7353–7362.

Sakaki, K., Terashita, N., Tsunokake, S., Nakamura, Y., Akiba, E. (2012). Effect of rare earth elements and alloy composition on hydrogenation properties and crystal structures of hydrides in $Mg_{2-x}RE_xNi_4$. *J. Phys. Chem. C*, 116(36), 19156–19163.

Sakaki, K., Terashita, N., Kim, H., Proffen, T., Majzoub, E.H., Tsunokake, S., Nakamura, Y., Akiba, E. (2013). Crystal structure and local structure of $Mg_{2-x}Pr_xNi_4$ (x = 0.6 and 1.0) deuteride using in situ neutron total scattering. *Inorg. Chem.*, 52(12), 7010–7019.

Sampson, S. (1972). Complex cubic A_6B compounds. II. The crystal structure of Mg_6Pd. *Acta Cryst.*, 28(3), 936–945.

Shahi, R.R., Bhatanagar, A., Pandey, S.K., Shukla, V., Yadav, T.P., Shaz, M.A., Srivastava, O.N. (2015). MgH_2-$ZrFe_2H_x$ nanocomposites for improved hydrogen storage characteristics of MgH_2. *Int. J. Hydrog. Energy*, 40(35), 11506–11513.

Shi, Y., Leng, H., Wei, L., Chen, S., Li, Q. (2019). The microstructure and electrochemical properties of Mn-doped La-Y-Ni-based metal-hydride electrode materials. *Electrochim. Acta*, 296, 18–26.

Shtender, V.V., Denys, R.V., Paul-Boncour, V., Riabov, A.B., Zavaliy, I.Y.U. (2014). Hydrogenation properties and crystal structure of $YMgT_4$ (T=Co, Ni, Cu) compounds. *J. Alloys Compd.*, 603, 7–13.

Shtender, V.V., Denys, R.V., Paul-Boncour, V., Verbovytskyy, Yu.V., Zavaliy, I.Y.U. (2015a). Effect of Co substitution on hydrogenation and magnetic properties of Nd $MgNi_4$ alloy. *J. Alloys Compd.*, 639, 526–532.

Shtender, V.V., Paul-Boncour, V., Riabov, A.B., Denys, R.V., Zavaliy, I.Y.U. (2015b). Hydrogenation behavior of the R_4MgCo (R = Y, La, Nd, Tb) compounds. *J. Solid State Chem.*, 229, 135–140.

Shtender, V.V., Denys, R.V., Paul-Boncour, V., Zavaliy, I.Y.U., Verbovytskyy, Yu.V., Taylor, D.D. (2017). Crystal structure, hydrogen absorption-desorption behavior and magnetic properties of the $Nd_{3-x}MgCo_9$ alloys. *J. Alloys Compd.*, 695, 1426–1435.

Shtender, V.V., Paul-Boncour, V., Denys, R., Hedlund, D., Svedlindh, P., Zavaliy, I. (2022). Impact of the *R* and Mg on the structural, hydrogenation and magnetic properties of $R_{3-x}MgCo_9$ (R = Pr, Nd, Tb and Y) compounds. *Mater. Res. Bull.*, 156, 111981.

Si, T., Zhang, Q.A., Pang, G., Liu, D.M., Liu, N. (2009). Structural characteristics and hydrogen storage properties of $Ca_{3-x}Mg_xNi_9$ (x = 0.5, 1.0, 1.5 and 2.0) alloys. *Int. J. Hydrog. Energy*, 34, 1483–1488.

Si, T., Ma, Y., Li, Y., Liu, D. (2017). Solid solution of Cu in Mg_2NiH_4 and its destabilized effect on hydrogen desorption. *Mater. Chem. Phys.*, 193, 1–6.

Skryabina, N., Medvedeva, N., Gabov, A., Fruchart, D., Nachev, S., de Rango, P. (2015). Impact of severe plastic deformation on the stability of MgH_2. *J. Alloys Compd.*, 645, S14–S17.

Tanniru, M., Tien, H.-Y., Ebrahimi, F. (2010). Study of the dehydrogenation behavior of magnesium hydride. *Scr. Mater.*, 63(1), 58–60.

Teichmann, N., Hamm, M., Pundt, A. (2018). Fast lateral hydrogen diffusion in magnesium-hydride films on sapphire substrates studied by electrochemical hydrogenography. *Int. J. Hydrog. Energy*, 43(3), 1634–1642.

Téliz, E., Abboud, M., Faccio, R., Esteve, M., Zinola, F., Díaz, V. (2020). Hydrogen storage in AB_2 hydride alloys: Diffusion processes analysis. *J. Electroanal. Chem.*, 879, 114780.

Thiangviriya, S., Plerdsranoy, P., Hagenah, A., Le, T.T., Kidkhunthod, P., Utke, O., Dornheim, M., Klassen, T., Pistidda, C., Utke, R. (2021). Effects of Ni-loading contents on dehydrogenation kinetics and reversibility of Mg_2FeH_6. *Int. J. Hydrog. Energy*, 46(63), 32099–32109.

Tsukahara, M. (2011). Hydrogenation properties of vanadium-based alloys with large hydrogen storage capacity. *Mater. Trans.*, 52(1), 68–72.

Vajeeston, P., Ravindran, P., Hauback, B.C., Fjellvåg, H., Kjekshus, A., Furuseth, S., Hanfland, M. (2006). Structural stability and pressure-induced phase transitions in MgH_2. *Phys. Rev. B*, 73(22), 224102.

Verbovytskyy, Y., Zhang, J., Cuevas, F., Paul-Boncour, V., Zavaliy, I. (2015). Synthesis and properties of the $Mg_2Ni_{0.5}Co_{0.5}H_{4.4}$ hydride. *J. Alloys Compd.*, 645, S408–S411.

Verbovytskyy, Y., Oprysk, V., Paul-Boncour, V., Zavaliy, I., Berezovets, V., Lyutyy, P., Kosarchyn, Y. (2021). Solid gas and electrochemical hydrogenation of the selected alloys $(R',R'')_{2-x}Mg_xNi_{4-y}Co_y$ (R', R'' = Pr, Nd; x = 0.8–1.2; y = 0–2). *J. Alloys Compd.*, 876, 160155.

Verón, M.G., Condó, A.M., Gennari, F.C. (2013). Effective synthesis of Mg_2CoH_5 by reactive mechanical milling and its hydrogen sorption behavior after cycling. *Int. J. Hydrog. Energy*, 38(2), 973–981.

Wan, C., Zhao, S., Wang, H., Yang, L., Liu, J., Ouyang, L., Zhu, M. (2022). Annealing temperature-dependent phase structure and electrochemical hydrogen storage properties of AB_4-type $La_{1.5}Y_{1.5}Ni_{12-x}Mn_x$ (x = 0, 1.0) superlattice alloys. *International Journal of Hydrogen Energy,* 48(4), 1472-1481. doi: 10.1016/j.ijhydene.2022.09.280

Wang, W., Guo, W., Liu, X., Zhang, S., Zhao, Y., Li, Y., Zhang, L., Han, S. (2020). The interaction of subunits inside superlattice structure and its impact on the cycling stability of AB_4-type La–Mg–Ni-based hydrogen storage alloys for nickel-metal hydride batteries. *J. Power Sources*, 445, 227273.

Wei, W., Yungui, C., Chaoling, W. (2011). Hydrogen-induced amorphization in $LaNi_{2.5}M_{0.2}$ (*M*=Ni, Fe, Mn, Si) alloys. *Rare Metal Mat. and Eng.*, 40(12), 2080–2082.

Wen, J., de Rango, P., Allain, N., Laversenne, L., Grosdidier, T. (2020). Improving hydrogen storage performance of Mg-based alloys through microstructure optimization. *J. of Power Sources*, 480, 228823.

Westlake, D.G. (1983). Preferred hydrogen sites and hydride stoichiometry for A_6B_{23} compounds. *J. Mater. Sci.*, 18(2), 605–611.

Willems, J.J.G. and Buschow, K.H.J. (1987). From permanent magnets to rechargeable hydride electrodes. *J. Less-Common Met.*, 129, 13–30.

Williams, A.F. (1984). Dimagnesium iron(I1) hydride, Mg_2FeH_6, containing octahedral FeH_2-anions. *Inorg. Chem.*, 23, 1957–1967.

Wu, C.Z., Wang, P., Yao, X., Liu, C., Chen, D.M., Lu, G.Q., Cheng, H.M. (2006). Hydrogen storage properties of MgH_2/SWNT composite prepared by ball milling. *J. Alloys Compd.*, 420(1/2), 278–282.

Wu, C.Z., Yao, X.D., Zhang, H. (2010). Hydriding/dehydriding properties of MgH_2/5 wt.% Ni coated CNFs composite. *Int. J. Hydrog. Energy*, 35(1), 247–252.

Wu, Y., Peng, Y., Jiang, X., Zeng, H., Wang, Z., Zheng, J., Li, X. (2021a). Reversible hydrogenation of AB_2-type Zr–Mg–Ni–V based hydrogen storage alloys. *Prog. Nat. Sci.*, 31(2), 319–323.

Wu, Y., Zhao, W., Jiang, L., Li, Z., Guo, X., Ye, J., Yuan, B., Wang, S., Hao, L. (2021b). Effect of Fe and Al on hydrogen storage properties of 75 V-Ti-Cr alloys. *J. Alloys Compd.*, 887, 161181.

Xiong, W., Yan, H., Wang, L., Verbetsky, V., Zhao, X., Mitrokhin, S., Li, B., Li, J., Wang, Y. (2017). Characteristics of A_2B_7-type (La-Y-Ni)-based hydrogen storage alloys modified by partially substituting Ni with Mn. *Int. J. Hydrog. Energy*, 42(15), 10131–10141.

Yadav, T.P., Shahi, R.R., Srivastava, O.N. (2012). Synthesis, characterization and hydrogen storage behaviour of AB_2 ($ZrFe_2$, $Zr(Fe_{0.75}V_{0.25})_2$, $Zr(Fe_{05}V_2)_2$) type materials. *Int. J. Hydrog. Energy*, 37, 3689–3696.

Yan, H., Xiong, W., Wang, L., Li, B., Li, J., Zhao, X. (2017). Investigations on AB_3-, A_2B_7- and A_5B_{19}-type La-Y-Ni system hydrogen storage alloys. *Int. J. Hydrog. Energy*, 42(4), 2257–2264.

Yartys, V.A. and Lototskyy, M.V. (2022). Laves type intermetallic compounds as hydrogen storage materials: A review. *J. Alloys Compd.*, 916, 165219.

Yartys, V.A., Gutfleisch, O., Panasyuk, V.V., Harris, I.R. (1997). Desorption characteristics of rare earth (*R*) hydrides (*R* = Y, Ce, Pr, Nd, Sm, Gd and Tb) in relation to the HDDR behaviour of *R*–Fe-based-compounds. *J. Alloys Compd.*, 253–254, 128–133.

Yartys, V.A., Antonov, V.E., Beskrovnyy, A.I., Crivello, J.-C., Denys, R.V., Fedotov, V.K., Gupta, M., Kulakov, V.I., Kuzovnikov, M.A., Latroche, M. et al. (2015). Hydrogen-assisted phase transition in a trihydride $MgNi_2H_3$ synthesized at high H_2 pressures: Thermodynamics, crystallographic and electronic structures. *Acta Mater.*, 82, 316–327.

Yuan, Z., Qi, Z., Zhai, T., Wang, H., Wang, H.-Y., Zhang, Y. (2021). Effects of La substitution on microstructure and hydrogen storage properties of Ti–Fe–Mn-based alloy prepared through melt spinning. *Trans. Nonferrous Met.*, 31(10), 3087–3095.

Zachariasen, W.H., Holley, C.E., Stamper, J.F. (1963). Neutron diffraction study of magnesium deuteride. *Acta Cryst.*, 16(5), 352–353.

Zadorozhnyy, V., Klyamkin, S., Zadorozhnyy, M., Bermesheva, O., Kaloshkin, S. (2012). Hydrogen storage nanocrystalline TiFe intermetallic compound: Synthesis by mechanical alloying and compacting. *Int. J. Hydrog. Energy*, 37(22), 17131–17136.

Zaluska, A. (1999). Nanocrystalline magnesium for hydrogen storage. *J. Alloys Compd.*, 288, 217–225.

Zaluski, L., Zaluska, A., Ström-Olsen, J.O. (1995). Hydrogen absorption in nanocrystalline Mg_2Ni formed by mechanical alloying. *J. Alloys Compd.*, 217(2), 245–249.

Zepon, G., Leiva, D.R., Kaufman, M.J., Figueroa, S.J.A., Floriano, R., Lamas, D.G., Asselli, A.A.C., Botta, W.J. (2015). Controlled mechanochemical synthesis and hydrogen desorption mechanisms of nanostructured Mg_2CoH_5. *Int. J. Hydrog. Energy*, 40(3), 1504–1515.

Zepon, G., Leiva, D.R., Strozi, R.B., Terra, B.C.M., Figueroa, S.J.A., Floriano, R., Jorge, A.M., Botta, W.J. (2017). Structural characterization and hydrogen storage properties of MgH_2–Mg_2CoH_5 nanocomposites. *Int. J. Hydrog. Energy*, 42(21), 14593–14601.

Zhang, Q., Pang, G., Si, T.Z., Liu, D.M. (2009). Crystal structure and hydrogen absorption–desorption properties of $Ca_3Mg_2Ni_{13}$. *Acta Mater.*, 57(6), 2002–2009.

Zhang, Q., Fang, M., Si, T., Fang, F., Sun, D., Ouyang, L., Zhu, M. (2010). Phase stability, structural transition, and hydrogen absorption–desorption features of the polymorphic La_4MgNi_{19} compound. *J. Phys. Chem. C*, 114(26), 11686–11692.

Zhang, J., Cuevas, F., Zaïdi, W., Bonnet, J.-P., Aymard, L., Bobet, J.-L., Latroche, M. (2011). Highlighting of a single reaction path during reactive ball milling of Mg and *TM* by quantitative H_2 gas sorption analysis to form ternary complex hydrides (*TM* = Fe, Co, Ni). *J. Phys. Chem. C*, 115(11), 4971–4979.

Zhang, J., Villeroy, B., Knosp, B., Bernard, P., Latroche, M. (2012). Structural and chemical analyses of the new ternary La_5MgNi_{24} phase synthesized by spark plasma sintering and used as negative electrode material for Ni-*M*H batteries. *Int. J. Hydrog. Energy*, 37(6), 5225–5233.

Zhang, L., Wang, W., Rodríguez-Pérez, I.A., Zhao, Y., Jia, Z., Li, Y., Han, S., Wang, L.-M. (2018a). A new AB_4-type single-phase superlattice compound for electrochemical hydrogen storage. *J. Power Sources*, 401, 102–110.

Zhang, X., Leng, Z., Gao, M., Hu, J., Du, F., Yao, J., Pan, H., Liu, Y. (2018b). Enhanced hydrogen storage properties of MgH_2 catalyzed with carbon-supported nanocrystalline TiO_2. *J. Power Sources*, 398, 183–192.

Zhang, J., Charbonnier, V., Madern, N., Monnier, J., Latroche, M. (2021). Improvement of reversible H storage capacity by fine tuning of the composition in the pseudo-binary systems $A_{2-x}La_xNi_7$ (A = Gd, Sm, Y, Mg). *J. Alloys Compd.*, 852, 157008.

Zhang, W., Zhao, D., Zhang, Y., Li, J., Guo, S., Qi, Y., Gao, J. (2022). Effect of Y partially substituting La on the phase structure and hydrogen storage property of La–Mg–Ni alloys. *J. Phys. Chem. Solids*, 167, 110744.

Zhao, Y., Han, S., Li, Y., Liu, J., Zhang, L., Yang, S., Cao, J., Jia, Z. (2016). Structural phase transformation and electrochemical features of La–Mg–Ni-based AB_4-type alloys. *Electrochim. Acta*, 215, 142–149.

Zhao, Y., Zhang, S., Liu, X., Wang, W., Zhang, L., Li, Y., Han, S. (2018). Phase formation of Ce_5Co_{19}-type super-stacking structure and its effect on electrochemical and hydrogen storage properties of $La_{0.6}M_{0.2}Mg_{0.2}Ni_{3.8}$($M$ = La, Pr, Nd, Gd) compounds. *Int. J. Hydrog. Energy*, 43(37), 17809–17820.

Zhao, S., Wang, H., Hu, R., Jiang, W., Liu, J., Ouyang, L., Zhu, M. (2021). Phase transformation and hydrogen storage properties of $LaY_2Ni_{10.5}$ superlattice alloy with single Gd_2Co_7-type or Ce_2Ni_7-type structure. *J. Alloys Compd.*, 868, 159254.

Zhao, S., Wang, H., Yang, L., Liu, J., Ouyang, L., Zhu, M. (2022). Modulating superlattice structure and cyclic stability of Ce_2Ni_7-type $LaY_2Ni_{10.5}$-based alloys by Mn, Al, and Zr substitutions. *J. Power Sources*, 524, 231067.

Zheng, W., Song, W., Wu, T., Wang, J., He, Y., Lu, X.-G. (2022). Experimental investigation and thermodynamic modeling of the ternary Ti–Fe–Mn system for hydrogen storage applications. *J. Alloys Compd.*, 891, 161957.

Zhong, C., Chao, D., Chen, Y., Wang, W., Zhu, D., Wu, C. (2011). Microstructures and electrochemical properties of $LaNi_{3.8-x}Mn_x$ hydrogen storage alloys. *Electrochim. Acta*, 58, 668–673.

Zhu, M., Lu, Y., Ouyang, L., Wang, H. (2013). Thermodynamic tuning of Mg-based hydrogen storage alloys: A review. *Materials*, 6(10), 4654–4674.

Zijlstra, H. and Westendorp, F.F. (1969). Influence of hydrogen on the magnetic properties of $SmCo_5$. *Solid State Commun.*, 7(12), 857–859.

Zolliker, P., Yvon, K., Fischer, P., Schefer, J. (1985). Dimagnesium cobalt(I) pentahy-dride, Mg_2CoH_5, containing square-pyramidal pentahydrocobaltate (4$^-$) (CoH_5^{4-}) anions. *Inorg. Chem.*, 24(24), 4177–4180.

6

High-Entropy Alloys for Hydrogen Storage

Nayely PINEDA ROMERO[1], Claudia ZLOTEA[1], Kylia MARCUS[2]
[1]*ICMPE, CNRS, UPEC, Thiais, France*
[2]*Institute Néel, CNRS, Université Grenoble Alpes, France*

6.1. Context

Among the different methods for hydrogen storage, the formation of *metal hydrides* starting from a variety of alloys/intermetallics, as described in Chapter 5, is one of the safest and most promising routes because of the possibility of adjusting the kinetics and thermodynamics of the reaction with hydrogen by tuning the chemical composition of the host material. Hydrogen interacts with metals or alloys to form metallic bonds. The process initiates when hydrogen gas molecules are adsorbed onto the surface of the metal and dissociate into two atoms, which can diffuse into the volume of the metal and then occupy the interstitial sites of the lattice (Figure 6.1). An expansion of the lattice can occur up to 30% in volume upon hydrogen insertion into the metal/alloy.

The hydride formation that occurs directly from the hydrogenation of the metal/alloy follows the reaction given in equation [6.1] (Amirthan and Perera 2022; Usman 2022).

$$M + {}^{x}/_{2}H_2 \leftrightarrow MH_x + Q \qquad [6.1]$$

The reaction of hydrogen absorption is exothermic, whereas the opposite reaction, hydrogen desorption, is endothermic and involves the same amount of energy in ideal systems (Züttel 2004; Usman 2022). The formation of the hydride can be related to the affinity of the metal/alloy for hydrogen, and two parameters are usually used to describe it: the enthalpy of the solution at infinite dilution (ΔH_∞) when the dissolution of the atomic hydrogen through the lattice takes place, and the enthalpy of forming the concentrated hydride (ΔH_f). Hydride formation can occur when both values are negative and with large magnitude (Keith et al. 2023). Solid solution formation occurs when a metal/alloy dissolves a small amount of hydrogen in the range of $H/M < 0.1$, known as the α-phase. When the pressure of hydrogen increases to an equilibrium value at a given temperature, the nucleation and growth of a new hydride phase (β-phase) are initiated (Schlapbach and Züttel 2001; Amirthan and Perera 2022).

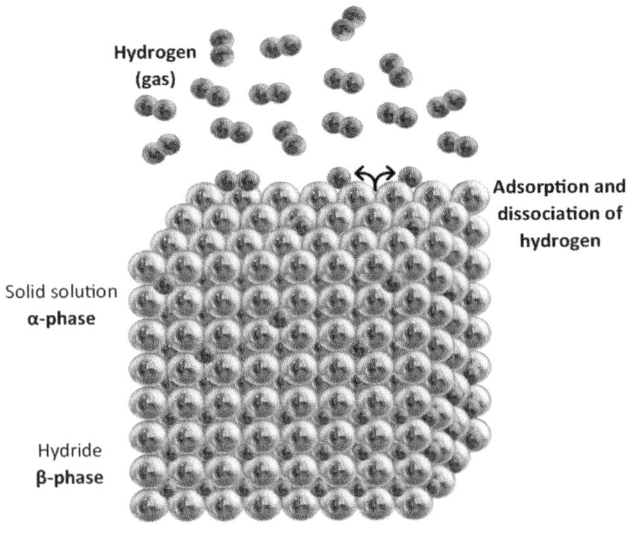

Figure 6.1. *Metal hydride formation. For a color version of this figure, see www.iste.co.uk/derango-cuevas/hydrogen.zip*

Furthermore, metal hydrides create metallic bonds that have high volumetric density (> 100 kg/m^3). Metal hydrides generally operate under mild conditions of temperature and pressure, which makes them safer than compressed and liquified hydrogen. Moreover, the reaction with H_2 is reversible. However, these materials may face slow absorption/desorption kinetics, high temperatures are required for hydrogen desorption, the materials used are heavy and the gravimetric capacities remain below the DOE targets (6 wt.% over the system, 80°C). One well-studied hydride is MgH_2, which has a very good gravimetric capacity of approximately 7 wt.%; it is a lightweight, abundant, and low-cost material, but its low kinetics and high operation temperature (above 300°C) required for desorption complicate its practical use (Tarhan and Çil 2021; Amirthan and Perera 2022; Usman 2022). Consequently, the development of new materials is important for overcoming the challenges involved in the use of metal hydrides.

6.2. High-entropy alloys

A new alloy design strategy was recently proposed for high-entropy alloys (HEAs) or multiprincipal element alloys (MPEAs). These materials focus on the center of the multicomponent phase diagrams, unlike traditional alloys, which emphasize the vertices, edges or faces (boundaries) of the phase diagrams. Therefore, HEAs allow exploration of an extensive composition space, which results in the possibility of having a variety of properties that could be useful in different applications. One application that has received particular attention is their use as structural materials, where the mechanical properties are the focus of study. Additionally, research on the functional properties of HEAs, such as their magnetic properties (e.g. saturation magnetization and coercivity), physical properties (e.g. electrical resistivity and thermal conductivity), thermoelectric properties, superconductivity and, more recently, hydrogen solid-state storage properties, has increased over the years (Gao et al. 2018).

6.2.1. *Definition of HEAs*

For many years, alloy development has been based on the following strategy: selection of a main component on the basis of a requirement for a primary property, followed by the addition of small amounts of other elements to impact the final properties of the material. However, starting in 2004, a new strategy for alloy production was proposed. This unconventional strategy has led to the development of a number of functional materials (Cantor 2021).

In 1979, Brian Cantor of the University of Sussex (UK) proposed a new metallurgical approach: alloys with an equiatomic mixture of different elements. Equimolar alloys of 16 and 20 elements were synthesized by melting, then cooling and solidifying in air. The two materials solidified with a complex microstructure and contained different phases. Energy dispersive X-ray analysis using a scanning electron microscope revealed that one of these phases consisted mainly of chromium, manganese, iron, cobalt and nickel in approximately equal proportions. The equimolar alloy of these five elements was then synthesized, and the X-ray diffractogram revealed that the alloy was a single phase with a face-centered cubic (*fcc*) structure. Surprisingly, when chromium and iron have a body-centered cubic (*bcc*) lattice, manganese has a complex face-centered cubic structure, cobalt has a hexagonal close-packed (*hcp*) structure and only nickel has a *fcc* structure (Barrett 1943).

Since this discovery, the $Cr_{20}Mn_{20}Fe_{20}Co_{20}Ni_{20}$ composition has been known as the Cantor alloy. Since 1995, on the basis of the original composition of $Cr_{20}Mn_{20}Fe_{20}Co_{20}Ni_{20}$, various alloys of five or more elements have been synthesized by adding titanium, vanadium, copper, germanium and niobium in equiatomic and non-equiatomic proportions. Some alloys are single-phase with *bcc* or *fcc* lattices or multiphase with these two types of structures (Cantor 2021). These results were first presented at a conference in Oxford in 2002 (Rapidly Quenched and Metastable Materials conference, RQ11) before being published in 2004: microstructural development in equiatomic multicomponent alloys (Cantor et al. 2004).

In 1995, independent of Brian Cantor, Jien-Wei Yeh developed the idea that high mixing entropy could promote the mixing of elements and reduce the number of phases. Together with his students, they synthesized 40 equimolar alloys by arc melting (AM). The microstructure, hardness and corrosion resistance of these alloys were analyzed before and after annealing. From this set of data, 20 alloys based on titanium, vanadium, chromium, iron, cobalt, nickel, copper, molybdenum, zirconium, palladium and aluminum were selected. These alloys are grouped into three equiatomic series of six to nine elements each: the Cu series, the Al series and the Mo series (Gao and Qiao 2018). In 2004, Jien-Wei Yeh published his first article on MPEAs (multiprincipal-element alloys), and a few months later, he was the first to explain the concept of high-entropy alloys (HEAs) with *fcc* $CuCoNiCrAl_{0.5}Fe$ and CuCoNiCrFe alloys (Yeh et al. 2004; Gao and Qiao 2018).

The term "high-entropy alloys" was proposed by Yeh in 2004. In the case of an equiatomic alloy in the liquid state or in the regular solid solution state, the configurational entropy per mole can be calculated from the following expression (Yeh et al. 2004):

$$\Delta S_{conf} = R \times \ln(n) \qquad [6.2]$$

where n is the number of components and R is the perfect gas constant (8.314 J/mol·K).

In this definition of configurational entropy, the atoms are considered in an ideal solution, that is, randomly distributed and all the same size. In the case of a real solution, an excess configurational term is added to the ideal configurational entropy. In addition to this configurational excess term, other terms participate in the total mixing entropy ΔS_{mix}, which therefore has four contributions: total configurational ($\Delta S_{conf} + \Delta S_{exces}$), vibrational, magnetic and electronic. The excess configurational entropy term can arise from the atomic size difference, which can lead to some uncertainty in atom positions (Miracle and Senkov 2017). The ideal configuration entropy ΔS_{conf} dominates the other three contributions, and the excess configuration term ΔS_{exces} is generally negative and disappears at very low atomic packing densities (Ye et al. 2016). For this reason, the configuration entropy ΔS_{conf} is often represented as the mixing entropy ΔS_{mix} (Miracle and Senkov 2017; Gao and Qiao 2018). The configuration entropy reduces the segregation and ordering tendency of atoms (De Boer et al. 1988; Yeh et al. 2004). Consequently, alloys containing many principal elements are more likely than intermetallic compounds to form solid solutions during solidification. As shown in Figure 6.2, the mixing entropy ΔS_{mix} of the alloy increases with the number of components, *N* (Zhang and Zhou 2007). To take full advantage of the maximum mixing entropy, high-entropy alloys are defined as those composed of five or more principal elements in equimolar proportions. To broaden the scope of this concept, the definition has been extended to include alloys with atomic concentrations between 5% and 35% (Yeh et al. 2004).

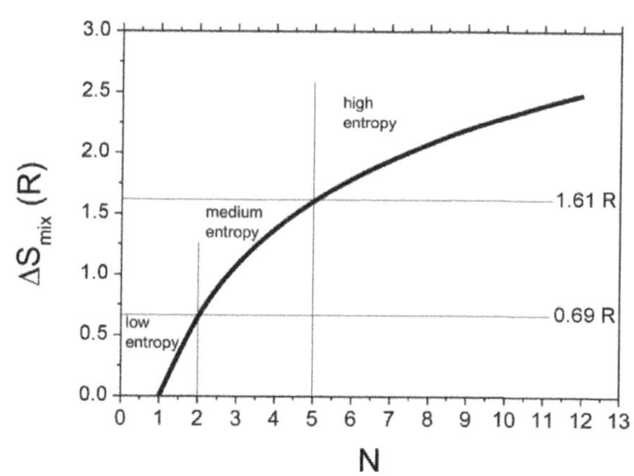

Figure 6.2. *The entropy of mixing of equiatomic alloys as a function of the number of components, N*

This terminology was subsequently criticized in the literature as explained by Cantor (2014, 2021), Murty et al. (2014, 2019), Gao and Qiao (2018), Tsai and Yeh (2014) and (Zhang et al. (2014) before being fully accepted, particularly with regard to the distinction between low, medium and high entropy. The term "medium entropy" refers to alloys with fewer than five elements, whereas "high entropy" refers to those with more than five elements. Other alternative names have emerged, such as multiple principal element alloys and compositionally complex and complex concentrated alloys.

Multi-principal element alloys have four main characteristics (Yeh 2013; Gao and Qiao 2018; Cantor 2021): stabilization of solid solutions due to the high entropy of mixing, significant lattice distortion due to size mismatch among all component atoms, slow and complex diffusion, and a "cocktail" effect responsible for improved properties compared with the sum of the individual properties of the constituents.

High-entropy mixing can inhibit the formation of intermetallic compounds in favor of solid solution formation since a multielement alloy has a high probability of having an average ΔH_{mix} due to the proportion of different atom pairs within the phase (Gao et al. 2016).

Significant lattice distortion results from the difference in size between all component atoms, which share the same crystallographic site. Each atom is surrounded by other atoms, resulting in asymmetric chemical bonds and electronic structure. As a result, the atom experiences greater lattice deformation and stress than do conventional alloys (He and Yang 2018).

Slow diffusion was proposed by Yeh et al. (2004) to explain the formation of nanoparticles with a *bcc* structure within the CuCoNiCrAlFe alloy. Similar crystalline nanostructures are rarely observed in conventional or amorphous alloys. This effect would depend on the specific components of the alloy, would not be directly proportional to the number of components and would only be "relative" to conventional materials at an equivalent temperature (Cantor 2021).

The cocktail effect proposed by Ranganathan (2003) refers to the synergy between elements in a multicomponent alloying system to enhance the properties of materials. The enhanced property comes not only from the basic properties of the component elements in the alloy following the mixture rule but also from the mutual interaction/synergy between all of the elements. Thus, the cocktail effect ranges from atomic-scale multi-principal element interaction to microscale multi-phase composite effect.

6.2.2. Thermodynamics of HEAs

The definition of high-entropy alloys remains controversial. This is mainly due to their microstructural complexity. The composition of the alloy and the cooling rate influence the final microstructure of the alloy. The most commonly accepted definition is that HEAs are single-phase solid solutions (Gao and Qiao 2018). However, a high configurational entropy greater than 1.5 R (where R is the perfect gas constant; Miracle et al. 2014) does not prevent the formation of multi-phase solid solutions (Otto et al. 2013), intermetallic compounds or amorphous solid solutions (Wang 2014). Physicochemical parameters such as the atomic size difference, mixing enthalpy, mixing entropy and melting temperature are used to predict the formation of different equilibrium phases for given compositions (Zhang et al. 2008; Guo and Liu 2011; Yang and Zhang 2012; Guo et al. 2013).

Phase formation is thermodynamically controlled by the Gibbs free energy, G:

$$\Delta G_{mix} = \Delta H_{mix} - T\Delta S_{mix} \qquad [6.3]$$

where ΔG_{mix} is the Gibbs free energy of mixing, ΔH_{mix} is the enthalpy of mixing, ΔS_{mix} is the entropy of mixing and T is the melting temperature of the alloy. The term ΔS_{mix} includes all sources of entropy (configurational, vibrational, electronic and magnetic). An estimate of the mixing enthalpy ΔH_{mix} from the mixing enthalpy of equiatomic binary alloys AB ΔH_{AB}^{mix} has been proposed by Zhang et al. (2008), where c_i and c_j are the atomic contents of the ith and jth components:

$$H_{mix} = \sum_{i=1, i\neq j}^{N} 4\,\Delta H_{AB}^{mix} c_i c_j \qquad [6.4]$$

The mixing entropy ΔS_{mix} is often assumed to be equal to the configuration entropy ΔS_{conf}:

$$\Delta S_{mix} = -R \sum_{i=1}^{N} c_i \ln c_i \qquad [6.5]$$

This simplification means that HEAs are considered to be in a liquid state or to form an ideal solid solution. The solid solution phase is the simplest phase to form and the most stable when its ΔG_{mix} is the lowest of all ΔG_{mix} of the possible phases. Calculating the ΔG_{mix} of all possible phases for a multi-component alloy is laborious, so another strategy is needed to predict the formation of solid solutions.

The mixing enthalpy ΔH_{mix} and the product of the mixing entropy with the melting temperature do not have the same influence on solid solution formation. Miedema et al. (1980) experimentally reported that the enthalpy of mixing ΔH_{mix} of a liquid alloy influences the formation of phases in a binary system. If $\Delta H_{mix} < 0$, one or more compounds are stable; if $\Delta H_{mix} > 0$, there is immiscibility in the liquid phase and consequently no stable compounds in the solid phase. For a multi-element alloy, the influence of ΔH_{mix} is similar (Yang and Zhang 2012): if $\Delta H_{mix} < 0$, the possibility of forming an intermetallic compound is greater. The more negative the value is, the greater the strength of the bonds between the elements; if $\Delta H_{mix} > 0$, the higher the value is, the lower the miscibility of the elements, resulting in segregation of the elements. In summary, a high absolute value of $|\Delta H_{mix}|$ prevents the formation of a solid solution, whereas a value close to zero leads to a stable solid solution.

In HEAs, ΔS_{mix} is high and therefore makes an important contribution to the formation of a solid solution. A high value of ΔS_{mix} decreases the Gibbs free energy ΔG_{mix} and leads to a random distribution of elements in the crystal lattice and a decrease in element segregation. The solid solution is therefore easier to form and more stable than intermetallic compounds are. The effect of entropy is even greater with increasing temperature. The $T\Delta S_{mix}$ value can be used to assess the driving force behind the formation of solid solutions. The Ω parameter characterizes the competition between $T_m\Delta S_{mix}$ and ΔH_{mix} and allows estimation of the formation of solid solutions and their stability (Yang and Zhang 2012; Zhang et al. 2012).

$$\Omega = \frac{T_m \Delta S_{mix}}{|\Delta H_{mix}|} \qquad [6.6]$$

$$T_m = \sum_{i=1}^{N} x_i (T_m)_i \qquad [6.7]$$

where T_m is the average melting temperature of the N-element alloy and $(T_m)_i$ is the melting temperature of the ith component of the alloy. Rules have been developed to account for an N-body system. They are added to the thermodynamic parameters to predict phase formation.

Other empirical parameters used to predict solid solution stability are the atomic size difference, the electronegativity difference and the valence electron concentration.

The atomic size difference δ is defined by (Zhang et al. 2008):

$$\delta\,\% = \sqrt{\sum_{i=1}^{n} x_i \left(1 - \frac{d_i}{\sum_{j=1}^{n} c_j d_j}\right)^2} \qquad [6.8]$$

where N is the number of elements, c_i and c_j are the compositions of the ith and jth elements, respectively, and d_i and d_j are the atomic diameters of the ith and jth elements, respectively.

The electronegativity difference $\Delta\chi$ is defined by (Guo and Liu 2011):

$$\Delta\chi = \sqrt{\sum_{i=1}^{n} c_i\left(\chi_i - \sum_{i=1}^{n} c_i\chi_j\right)^2} \quad [6.9]$$

The valence electron concentration (VEC) plays a critical role in the phase stability and physical properties of alloys. The total number of electrons, including the d-electrons on the valence layer, is used to calculate the VEC (Guo and Liu 2011):

$$VEC = \sum_{i=1}^{n} c_i(VEC)_i \quad [6.10]$$

where $(VEC)_i$ is the VEC of the ith element.

Yang related the difference in atomic size δ to the enthalpy of mixing ΔH_{mix}, the parameter Ω, the difference in electronegativity $\Delta\chi$ and the VEC on the basis of data from the literature and the previous equations.

The difference in the atomic size δ plays a crucial role in phase formation. A low value of δ ($\delta \leq 6.6\%$) favors the formation of a solid solution (Zhang et al. 2008; Yang et al. 2014). The three relationships proposed by Yang, $\delta - \Delta H_{mix}$, $\delta - \Omega$ and $\delta - \Delta\chi$, seem to be the most interesting to explain "empirically" the formation of the solid solution, which is favored for the following parameters:

– $-22 \leq \Delta Hmix \leq -5$ kJ/mol

– $\Omega \geq 1.1$

– $\Delta\chi \leq 0.175$

However, the parameters $\delta - \Delta H_{mix}$ and $\delta - \Omega$, which are based on the classical Hume-Rothery rules, have been the most widely used (Li et al. 2019).

Several statistical evaluations have been performed by relating the empirical parameters to investigate the formation of phases in HEAs. The variation in ΔH_{mix} as a function of the δ parameter shows that solid solution (SS) formation occurs when $\delta \leq 6.6\%$ and approximately $-11.6 \leq \Delta H_{mix} \leq 3.2$ kJ/mol. Moreover, amorphous phase (AM) formation occurs when $\delta \geq 6.4\%$ and with a more negative mixing enthalpy around $\Delta H_{mix} \leq -12.2$ kJ/mol. Furthermore, intermetallic compounds (IM) along with SS+IM can be formed under intermediate conditions (Guo et al. 2013). Another interesting relationship is Ω as a function of δ, and SS is obtained mainly when $\delta \leq 6.6\%$ and $\Omega \geq 1.1$, where the constituent elements of the HEAs have a small atomic size difference, the strain energy in the alloy caused by lattice distortion is lower, and the effect of ΔS_{mix} exceeds that of ΔH_{mix} (Yang and Zhang 2012). In addition, distinguishing between single and multiphase alloys via ΔH_{mix} as a function of the VEC relationship is not possible, but single-phase *bcc* and *fcc* formations appear in distinct zones that can be discerned by their VEC (Couzinié and Dirras 2019). Therefore, the VEC appears to be crucial in distinguishing the stability of solid solutions with a *bcc* lattice from those with an *fcc* structure (Yang et al. 2014).

A recent compilation of data showed that materials reported from 2010 to January 2018 crystallized in at least one *bcc* phase as a single phase or multiphase material, revealing their mechanical properties and information that can be used as a guide for further design of HEAs (Couzinié et al. 2018).

6.2.3. *Elaboration methods for HEAs*

HEAs synthesis methods can be classified according to their state of mixing as liquid, solid or gas. Within the classification as a liquid state, there are methods such as arc-melting (AM), laser-engineered net shaping (LENS), laser cladding and Bridgeman solidification. Mechanical alloying (MA) via high-energy ball milling (HEBM) is within the solid state, whereas gas mixing methods such as sputtering or pulsed laser deposition are considered. In general, the various preparation methods for HEAs are the same as those for classical alloys, with AM and high-energy ball milling being the most widely used methods (Shahi et al. 2022; Yadav et al. 2022), which will be described in more detail below.

The AM principal parts are the melting chamber with the copper crucible, the vacuum pumping system, the movable electrode of tungsten, the cooling water system and the unit power (Williams and Levingston 1961). High-purity metals of the desired composition are placed into a water-cooled copper crucible as individual elements and in the form of bulk pieces (Yang et al. 2022). Melting is performed under a protective atmosphere, such as vacuum or inert gas (argon and helium) to avoid oxidation of the sample and the introduction of impurities (Kong et al. 2023). The vacuum pressure is normally $\leq 1.33 \times 10^{-5}$ mbar, and the pressure of the inert gas during the melting process is ≤ 1.3 bar. The use of an oxygen getter such as Ti or

Zr could be adopted prior to the melting of the sample to trap contaminants (Williams and Levingston 1961). The method is based on the formation of an arc and plasma by applying high voltage and the discharge between the electrodes when the tungsten tip approaches the copper crucible. The movable electrode of tungsten acts as the cathode (–) and the bottom plate with the copper crucible as the anode (+). The pure metals are heated until a molten state is reached at 3000°C. Therefore, this is an ideal method to synthesize HEAs with refractory elements (Nb, Mo, Ti, etc.), but it is not suitable for metals with very low melting point and high vapor pressure (e.g. Mg, Zn and Mn). However, the bottom of the lump is not completely molten due to the action of the water-cooled crucible, and to improve the homogenization, remelting several times and flipping the sample between each melt are necessary (Williams and Levingston 1961; El-Hadad 2022; Shahi et al. 2022; Yang et al. 2022; Kong et al. 2023). After the melting process, the lump rapidly cooled and solidified in the copper crucible. A fast cooling rate is important to avoid the formation of any phase segregation. Therefore, this method allows the formation of alloys with minimum contamination and very fast cooling rates (Lai et al. 2019; Shahi et al. 2022; Kong et al. 2023).

HEBM involves different processes when a mixture of metallic powders is used to form an alloy/hydride under MA with inert gas (argon and He) or reactive milling (RM) by using a chemically reactive gas, for example, hydrogen, during the milling of powders (Lai et al. 2019). HEBM achieves the alloying of metallic powders by several deformations, cold welding and crushing. Furthermore, materials with high vapor pressures, such as Mg, are suitable for use in this alloying process (Kong et al. 2023). HEBM under an inert or reactive atmosphere has been successfully reported for the synthesis of alloys for hydrogen storage, especially for Mg-based alloys that cannot be prepared by direct melting alloying (Yang et al. 2022). As a result, the formation of a homogeneous solid solution is possible, but continuous deformation/fracture/welding could lead to amorphization and nanocrystallization (Yadav et al. 2022). Additionally, repeated compression, wear, shear and impact cause the formation of new surfaces and cracks, which simultaneously form defects and dislocations. This creates a high surface area, smaller crystallite size and greater number of boundaries. The latter is assumed to favor the kinetics of hydrogen diffusion during absorption/desorption (Lai et al. 2019). The ball milling alloying process has been shown to be influenced by the ball-to-powder weight ratio, speed of rotation, milling time, ball size, process control agent, powder hardness and type of grinding media (Shahi et al. 2022). Consequently, the powders are placed inside the vial together with spherical balls that impact them freely and constantly until a steady state between cold welding and fracture occurs. The spherical shape of the balls allows a greater grinding surface area and lower degree of wear. The size of the metallic powders decreases to the micrometer range during the process.

The brittleness and ductility of elements play important roles in mechanical alloying. The brittle powders are fragmented into small pieces that ultimately act as ductile powders. The ductile materials tend to be flattened and cold welded without further reduction in the particle size. In a mixture of brittle and ductile powders, small brittle particles are embedded in ductile materials. One type of high-energy mill is the planetary mill, where the centrifugal force controls the procedure and typically has the following milling conditions: the velocity of the balls is 2.5–4.0 m/s, the kinetic energy per impact is 0.01–0.4 J/hit, the shock frequency is approximately 100 Hz, and the power is 0.01–0.8 W/g/ball (Lai et al. 2019). Introducing high-energy milling increases the temperature of the vial because of the impact between ball-to-ball or ball-to-wall collisions. Other parameters to consider are the density of the balls, which should be high enough to have sufficient impact on the powders, and the milling media needs to be harder than the powders to avoid excessive contamination. The use of stainless steel in the vials could lead to more than 5 wt.% of Fe contamination in the milled material. Other factors that influence contamination are the duration of the milling process, the size of the balls and the free space inside the vial. Effective interaction between the ball/powder will not occur if the balls are too large compared with the capacity of the vial and if the vial is overfilled or underfilled with powder. The vial should be filled at 50%, with a ball-to-powder mass ratio of 10:1 in vials of approximately 50 mL (Lai et al. 2019).

6.2.4. *Prediction of HEAs phases and element selection*

As mentioned above, different empirical parameters are used to predict the stability of phases in HEAs and are based on the Hume–Rothery rules. According to these rules, the similarities in atomic size, electronegativity, valence state and crystalline structure favor the formation of solid solutions (Hume-Rothery and Powell 1935). After several experiments, the formation of solid solution phases is promoted when $\delta \leq 6.5\%$, -15 kJ/mol $\leq \Delta H_{mix} \leq 5$ kJ/mol, 12 J/K·mol $\leq \Delta S_{mix} \leq$ 17.5 J/K·mol and $\Omega \geq 1.1$ (Luo et al. 2024). Another parameter that has been useful in determining phase stability is VEC; VEC $\geq$ 8 favors *fcc* solid solution formation, whereas VEC < 6.87 favors *bcc* solid solutions (Guo and Liu 2011; Guo et al. 2011). The formation of mixed *fcc* plus *bcc* phases occurs when 6.87 $\leq$ VEC < 8 (Guo et al. 2011; El-Hadad 2022; Kaushik et al. 2022). Recent work revisited the VEC rule in Al-Co-Cr-Fe-Ni compositions, finding > 90% of compositions with *bcc* lattice when 5.7 $\leq$ VEC $\leq$ 7.2 and 100% of the *fcc* phase is formed when VEC $\geq$ 8.4 (Yang et al. 2020).

The selection of elements for the preparation of HEAs has been guided by conventional alloys for hydrogen storage. The conventional classification established that the A-type elements are hydride forming with a high affinity for hydrogen. Moreover, B-type elements are non-hydride forming elements with a low affinity for hydrogen and are considered to regulate the reversibility of hydrogen absorption and desorption (Luo et al. 2024).

The elements most commonly used for HEAs as hydrogen storage materials are Ti, V, Cr, Mn, Fe, Zr, Ni, Nb, Hf, Co and Mo, which have been used more than 20 times in different studies. However, four elements, V, Ti, Cr and Mn, have been used more than 100 times. In addition, Ti, V and Zr have been used in 80% of the HEAs studied. Therefore, 3*d* and 4*d* transition metals are more likely to be used, but to obtain high gravimetric capacity in HEAs, it is necessary to pay attention to the use of light elements such as Mg and Al (Marques et al. 2021; Luo et al. 2024).

Most reports on HEAs for hydrogen storage focus on single-phase materials; thus, the structure of HEAs is usually evaluated via X-ray powder diffraction (XRD).

The crystal structure is determined through the diffraction of photons, neutrons or electrons. XRD is the most commonly used technique for identifying the crystallinity and phase distribution in powder samples and is generated by the scattering of X-rays on the electron clouds of the analyzed sample (Kittel 2005). The incident X-ray beam is diffracted in specific directions following the well-known Bragg's law. This equation allows the calculation of the angle where constructive interference between the incident and diffracted X-rays via parallel planes of atoms will produce a diffraction peak.

6.2.5. *Microstructure of HEAs*

The microstructure and properties of HEAs are influenced by several factors, such as four core effects (high entropy, sluggish diffusion, lattice distortion and the cocktail effect), the method of preparation and the solidification process. A low cooling rate during the melting steps promotes the formation of dendrites and interdendritic regions. The former are enriched in elements with the highest melting points, whereas a higher solidification rate tends to cause the dendrites to disappear. The constituent elements with a large difference in enthalpy of mixing are prone to segregate during the solidification process. Furthermore, a sufficiently rapid cooling rate favors the formation of single solid solution HEAs because the formation of multiple phases (or intermetallic phases) tends to be avoided. This is due to the different atomic radii in HEAs, which increase the energy at the solid–liquid interface and make it more difficult for atoms to diffuse over the long range within the crystal lattice (Singh et al. 2011; Couzinié et al. 2014; Gao et al. 2016; El-Hadad 2022).

Consequently, the microstructure might play an important role in the hydrogenation properties of HEAs that can be refined, modified and homogenized via different processing methods, resulting in a microstructure without dendrite/interdrite zones or compositional differences. Scanning electron microscopy (SEM) has been extensively used to analyze the microstructure of HEAs, and energy-dispersive X-ray spectroscopy (EDS) has been used to investigate chemical homogeneity (Marques et al. 2021).

The principle of SEM is a narrow, collimated beam of electrons applied to a specimen under vacuum to avoid the scattering of the electrons in the air, and by obtaining the diffracted current from the specimen it is possible to construct an image (Korchinski et al. 2019). An elemental analysis of a sample can be obtained via energy dispersive X-ray spectroscopy (EDS) by measuring the energies of the characteristic X-rays, where the intensity of the X-rays is proportional to the amount of the element present in the specimen. Furthermore, X-ray mapping analysis can be performed to obtain a qualitative analysis of the elements from a certain area, observing their distribution throughout the area of analysis (JEOL 2020).

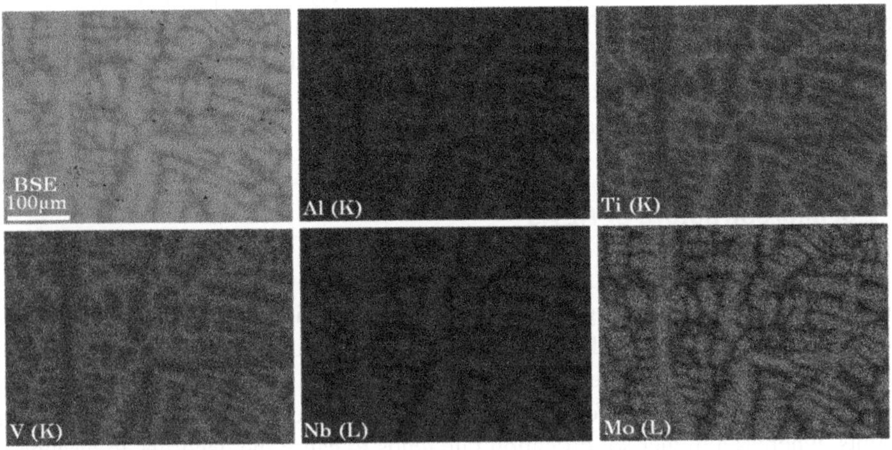

Figure 6.3. *SEM–EDX chemical mapping of the as-cast $Al_{0.05}(TiVNb)_{0.85}Mo_{0.10}$ composite. Figure taken from Pineda-Romero et al. (2023). For a color version of this figure, see www.iste.co.uk/derango-cuevas/hydrogen.zip*

An example of X-ray mapping of $Al_{0.05}(TiVNb)_{0.85}Mo_{0.10}$ is shown in Figure 6.3. The formation of Nb-enriched and Mo-enriched dendrites is shown. Moreover, the interdendritic zones are enriched with Al–Ti–V. This distribution is very common in HEAs produced by AM, where these chemical modulations can be attributed to the solidification process (as mentioned before), but overall, a single-phase *bcc* lattice is maintained.

6.2.6. *Large-scale instruments for the characterization of HEAs*

The most commonly used experimental large-scale platforms are synchrotron radiation and neutron sources, which are applied in a variety of topics related to materials science. These platforms provide the opportunity to work with highly specialized instrumentation and with specific experimental conditions that would not be possible to achieve in most normal laboratories (such as analyses in situ or ex situ, samples under certain environments or temperature conditions) (D'Ippolito and Rüling 2019). The experiments can be carried out under extreme conditions such as low or high temperatures or pressures and high magnetic fields. In HEAs, the use of synchrotron radiation and neutrons for powder diffraction under in situ and ex situ conditions has been proposed for several compositions to characterize the structural properties of the hydride/deuteride phases and their evolution.

In a synchrotron, electrons are accelerated, and electromagnetic waves are emitted when the electrons are deviated by magnetic fields in the circular accelerator. The light emitted is synchrotron radiation, and the most important advantage over laboratory X-rays is its brilliance (measured by the quality of the X-ray sources considering the beam divergence/dimensions, with units p.p.s./mm^2/mrad^2 within 0.1% bandwidth, p.p.s. = photons per second), which is approximately a billion times greater. This allows us to obtain more details of the material under study with greater precision and deeper penetration into the material (Fitch 2019; Kern 2019).

Alternatively, neutrons have the power to provide information that is impossible to obtain from another source of radiation. Furthermore, the scattering of neutrons can be calculated very precisely and can provide isotope contrast. Unlike X-rays, light atoms (hydrogen, deuterium, nitrogen, etc.) are visible or have cross sections comparable to or greater than those of heavier atoms, discerning neighboring atoms (Howard and Kisi 2019). The use of deuterium that scatters coherently has improved the resolution of diffraction patterns and even provides the possibility to determine the exact position of D atoms (Maréchal 2007; Huot and Černý 2016).

Moreover, a complementary analysis used to characterize the local structure is the pair distribution function (PDF) analysis of scattering data, which is based on powder diffraction data Fourier transformed to direct space from neutrons or synchrotron sources (Dove and Li 2022). From a PDF pattern, it is possible to obtain important information about the real space structure. The distance of atomic pairs is related to the peak position, and the abundance of relevant pairs is linked to the peak area. Moreover, the peak width indicates the disorder within the material, which can be structural or thermal vibrations. The peak at the maximum distance (r_{max}) gives the size of the coherent domain (Wang et al. 2020). A detailed study of the local structure in HEAs is important but difficult to obtain. Several good properties, such as high mechanical strength and superior hydrogen storage, have been attributed to lattice distortion, but there is still no experimental evidence that these properties are directly related. Combined PDF analysis from synchrotron and neutron diffraction diagrams could be a good option for assessing the local structure of HEAs (both metals and hydrogen networks), providing information on the interatomic distance and relating lattice distortion to the broadening and positional shifts of the peaks (Nygård et al. 2020).

6.2.7. *H_2 storage properties in HEAs*

Despite extensive research efforts, most hydrides still have hydrogen storage properties that are not suitable for application. The heat of reaction with hydrogen and the rate of hydrogen uptake/release need to be properly adjusted for final use under the targeted pressure and temperature conditions. Therefore, it is important to understand the interaction between hydrogen and materials. For some HEAs, weaker H–M bonds improve the thermodynamics and kinetics, whereas correct activation and proper operating conditions enable their viable application (Lai et al. 2019). Recently, it has been shown that HEAs may achieve better volumetric and gravimetric hydrogen uptake than other interstitial hydrides do (Keith et al. 2023). This can be achieved according to the criteria proposed to consider four key factors to increase hydrogen performance: the transformation of the preferred *bcc* (as-cast) phase to *fcc* dihydride, the VEC parameter, the degree of randomness of the element distribution, and the affinity for hydrogen (Keith et al. 2023). Therefore, different quantitative parameters have been proposed to study hydrogen storage properties, such as the maximum hydrogen capacity, kinetics, thermodynamics, stability of hydrogen absorption/desorption during cycling, and diffusivity of hydrogen into the metal matrix.

6.2.7.1. *Capacity of hydrogen*

The hydrogen storage ability of HEAs can be affected by their composition, crystal structure, atomic size of the constituent elements, defect concentration and surface. In addition, the hydrogen pressure, temperature and number of cycles can influence the capacity of hydrogen (Luo et al. 2024).

The capacity of hydrogen can be expressed as the ratio (H/M) of hydrogen (H) to the metal atom host (M) and as the gravimetric capacity (wt.%), which represents the amount of hydrogen absorbed per mass unit. In equations [6.11] and [6.12], n_H and n_M are the number of moles of hydrogen and metal host, respectively, and M_H and M_M are the molar masses of hydrogen and the metal host, respectively (Wei et al. 2017).

$$H/M = \frac{n_H}{n_M} = \frac{mass_H/M_H}{mass_M/M_M} \quad [6.11]$$

$$weight\ percent\ (wt.\%) = \frac{mass_H}{mass_M + mass_H} \times 100 \quad [6.12]$$

The technique most commonly used to measure the hydrogen storage properties of a material is the volumetric method, which is also known as the Sieverts or manometric method. The pressure–temperature–volume relationship is used for the calculation of the hydrogen storage capacity (Gross et al. 2012). The measurement is based on the change in pressure in a closed system with constant and known calibrated volumes. It is cost effective, easy to set up and reliable (Wei et al. 2017). Figure 6.4 shows a Sieverts system, consisting of a reservoir connected to a sample holder, both of which are maintained at constant temperature (not necessarily the same) with calibrated volumes. The change in pressure can be obtained via pressure transducers. For the method to be reliable, it is very important that the temperature is controlled over the entire system (tubbing and vessels) and the sample holder, as well as considering the gradients of temperature that may arise (Gross et al. 2012).

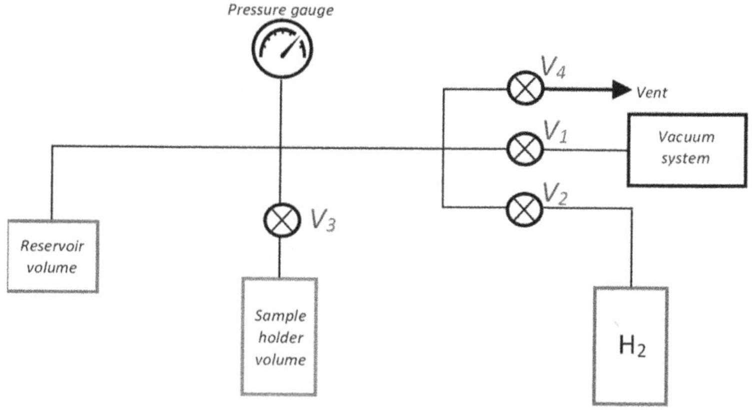

Figure 6.4. *Sieverts apparatus to measure hydrogen uptake*

The capacities of several HEAs have been investigated. Among all alloys, TiVZrNbHf has been reported to have a high capacity, and during the hydrogenation process, the material transforms from a *bcc* metal phase to a body-centered tetragonal (*bct*) hydride phase. The high capacity was attributed to the occupancy of deuterium atoms in both tetrahedral and octahedral interstitial sites, as revealed by in situ neutron diffraction at high temperatures (Sahlberg et al. 2016; Karlsson et al. 2018). In the case of $Mg_{0.10}Ti_{0.30}V_{0.25}Zr_{0.10}Nb_{0.25}$, the transformation during hydrogenation is from the *bcc* metal phase to the *fcc* dihydride phase (Montero et al. 2021b), as encountered in classical *bcc* alloys. Other studies have focused on the effect of the addition of Al to TiVNb, a very stable hydride-forming alloy. The addition of Al to the $Al_x(TiVNb)_{1-x}$ (x = 0, 0.05, 0.10, 0.175 and 0.25) system destabilizes the *fcc* dihydride but at the expense of capacity. Compared with TiVNb, $Al_{0.05}(TiVNb)_{0.95}$ was found to be the most promising material with a high initial capacity (2.96 wt.%), a less stable hydride, better desorption properties and enhanced reversible capacity during cycling (2.83 wt.%) (Pineda-Romero et al. 2022; Pineda-Romero and Zlotea 2022). Therefore, the addition of light elements can improve the hydrogenation properties of refractory HEAs.

6.2.7.2. *Kinetic properties*

The measurement of kinetics is very important for understanding the mechanism involved in the absorption and desorption of hydrogen. Kinetics is the rate at which the hydrogenation reaction occurs, that is, the speed of change in the concentration of reactants (H_2 + metal) and products (metal-hydride). The factors that may influence kinetics can be divided into two groups: intrinsic properties of the material and external factors. The *intrinsic properties* include surface interactions, mass transport (hydrogen diffusion in metals/hydrides), energy transport (heat transfer), bond breaking and bond forming mechanisms. The *external factors* are the temperature, driving force, properties of the sample (particle size, packing density, shape), shape of the sample holder and thermal contact between the sample and sample holder. However, a good comparison between different materials and different experimental setups is not possible. In some cases, the average kinetic rate has been described as 95% of the total material capacity reached in time. When comparing different materials, it is possible to obtain the same average kinetic rate but with a different shape in the kinetic curve (Gross et al. 2012).

Recently, the compositions of $Ti_{35}V_{35}Nb_{20}Cr_5Mn_5$, $Ti_{32}V_{32}Nb_{18}Cr_9Mn_9$ and $Ti_{27.5}V_{27.5}Nb_{20}Cr_{12.5}Mn_{12.5}$ were designed via thermodynamic calculations. $Ti_{27.5}V_{27.5}Nb_{20}Cr_{12.5}Mn_{12.5}$ showed the highest capacity (3.38 wt.%) at room temperature after the first cycle (considered activation), but the maximum capacity and kinetic properties decreased during cycling. However, $Ti_{32}V_{32}Nb_{18}Cr_9Mn_9$ showed the best kinetics after activation and the highest maximum capacity during cycling, with an increase in capacity after the second cycle (2.72 wt.%). This has been attributed to the decrease in the material during the absorption/desorption cycles, which creates a new surface that promotes the diffusion rates of hydrogen into the material (Serrano et al. 2023). The hydrogen sorption kinetics of three different TiVNbCr HEAs have been reported, with the same time to reach 90% of the maximum hydrogen uptake of approximately 110 s. The possible limitations for the kinetics are the dissociation of hydrogen on the metal surface, the diffusion of H in the bulk material and hydride formation at the metal/hydride interface. In addition, it was suggested that more lattice defects due to the mismatch between the initial and hydride phases can increase the nucleation rate of hydrides. The addition of transition metals (Fe, Mn and Ni) has been studied, leading to the formation of Laves phases, which have been reported to increase the kinetics of conventional alloys. However, for these HEAs, it was not possible to increase the kinetics but rather led to a decrease in the desorption temperature (Cheng et al. 2022).

6.2.7.3. *Thermodynamic properties*

For a reaction to occur, the Gibbs free energy difference (equation [6.13]) must be negative at a fixed temperature (T). Controlling it is crucial to the thermodynamics of hydrogen absorption, where ΔH is the change in enthalpy and ΔS is the change in entropy. For metal hydrides, ΔS is considered the entropy loss of gaseous hydrogen being absorbed into the metallic bulk at approximately 130 J/mol $H_2 \cdot K$ (Lai et al. 2019).

$$\Delta G = \Delta H - T\Delta S \tag{6.13}$$

Pressure–composition–isotherm (PCI) measurements (Figure 6.5, left) are a collection of data points that represent the evolution of the hydrogen concentration and phase transformation toward hydride formation when an equilibrium pressure is reached at a given temperature. When the occupancy of hydrogen within a metal is disordered, it is known as a solid solution or the α-phase. The hydride phase, also called the β-phase, is formed when the concentration of hydrogen is high enough that the H–H interactions become important so that an ordered arrangement of hydrogen atoms within the lattice is formed (Gross et al. 2012; Wei et al. 2017).

This process can be reversed under certain pressure and temperature conditions. The first region of a PCI at low pressure and a low hydrogen concentration corresponds to hydrogen diffusing into the volume of the metal and randomly occupying the interstitial sites of the lattice, forming the α-phase. The increase in pressure and hydrogen concentration, usually above 0.1 H/M, indicates that the α-phase begins to transform into the hydride β-phase. When equilibrium is reached between the hydrogen gas and the α and β phases, a plateau region is observed whose length and flatness determine how much hydrogen capacity can be reversibly stored. The thermodynamic behavior of this process follows the one-degree-of-freedom Gibbs phase rule, which means that under equilibrium conditions, the hydrogen pressure remains constant (plateau pressure, P_p) during the phase transition as the hydrogen concentration increases. The pressure increases again where the α-phase is completely converted to a β-phase single hydride, and the maximum storage capacity is reached (Gross et al. 2012; Wei et al. 2017). The ease of hydrogen reversibility depends on the thermodynamic stability and absorption/desorption kinetics of the material (Gross et al. 2012).

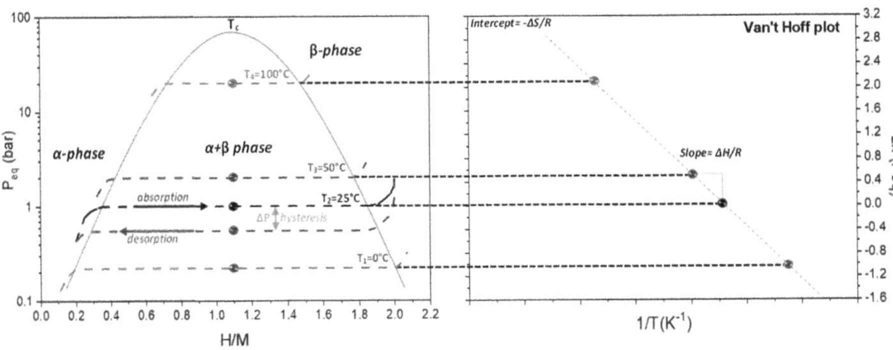

Figure 6.5. *Pressure composition isotherms (left) at different temperatures and the corresponding van't Hoff plot (right) for hydrogen absorption. For a color version of this figure, see www.iste.co.uk/derango-cuevas/hydrogen.zip*

Furthermore, thermodynamic information can be obtained from the P_p at different temperatures, where at a given temperature, there is a corresponding equilibrium plateau pressure related to a point on the van't Hoff plot for the absorption and desorption of hydrogen (Figure 6.5, right). The van't Hoff equation (equation [6.14]) relates the slope of the plot to the enthalpy of the reaction (ΔH = kJ/mol H_2) and the intercept at the origin (ln P_p) to the entropy change (ΔS = J/K·mol H_2). *R* is the gas constant (J/mol·K) and *T* is the temperature (K) (Gross et al. 2012; Wei et al. 2017).

$$\ln(P_p) = \frac{\Delta H}{RT} - \frac{\Delta S}{R} \qquad [6.14]$$

As mentioned before, all points of the van't Hoff plot are from the plateau region of the corresponding PCI, which must be taken at the same hydrogen concentration, usually the middle of the plateau pressure length. Therefore, sloped plateaus could increase thermodynamic calculation errors (Gross et al. 2012).

The hydrogen storage efficiency and capacity, thermodynamic stability, heat of reaction during the absorption/desorption of hydrogen and free energy are influenced by the thermodynamic properties of HEAs (Luo et al. 2024). The thermodynamic properties of $(TiVNb)_{100-x}Cr_x$ (x = 15, 20, 30, 35) were studied by performing PCI experiments and applying the van't Hoff method. The first tests at 24°C for $(TiVNb)_{100-x}Cr_x$ (x = 15, 25, 35) revealed that an increase in Cr increased the equilibrium plateau pressure without significant loss in capacity but rather pronounced hysteresis (Strozi et al. 2021b). Later, PCIs were collected at different temperatures for the $(TiVNb)_{100-x}Cr_x$ (x = 30, 35) alloys, which exhibited two plateaus. The first plateau occurred at low equilibrium pressures below the detection limit of the pressure transducer, which was related to the equilibrium of the *bcc* solid solution and *bcc* intermediate monohydride. A second plateau is observed at higher pressures, which is related to the equilibrium between the *bcc* intermediate monohydride and the *fcc* dihydride phase. Both compositions also presented hysteresis with one order of magnitude of difference between the PCIs of absorption and desorption. The experimental ΔH values of dihydride formation for $(TiVNb)_{100-x}Cr_x$ (x = 30, 35) were – 49 and – 39 kJ/mol H_2, respectively. Moreover, the experimental ΔS values of absorption were – 154 and – 134 J/K·mol H_2, respectively (Strozi et al. 2023a). In addition, the thermodynamic properties of $(TiVNb)_{100-x}Cr_x$ (x = 15, 20) were determined, and the enthalpy of hydride formation was – 57 and – 54 kJ/mol H_2, together with the entropy of hydride formation of – 151 and – 149 J/K·mol H_2, respectively. The high value of ΔS was attributed to the sloped plateaus of the PCIs. Furthermore, the destabilization of the *fcc* hydride was found to be almost linear with increasing Cr content, which is consistent with machine learning (ML) predictions. This thermodynamic destabilization was related not only to steric effects (smaller unit cell volume and volume of interstitial sites) but also to the low affinity of Cr for hydrogen (Strozi et al. 2023b). The equimolar composition of TiVNbCr also aligns with the linear destabilization trend with increasing Cr content, with ΔH = – 51.6 kJ/mol H_2 and ΔS = – 129.3 J/K·mol H_2 (Nygård et al. 2021).

Further studies of the thermodynamic properties of the $Al_x(TiVNb)_{1-x}$ (x = 0, 0.05, 0.10) system have been reported. The hydrogen absorption PCI measurements recorded at several temperatures (25–235°C) reveal the existence of two plateaus. The thermodynamic properties of dihydride formation have been determined from the variation in the P_p of the second plateau with temperature, that is, via the van't Hoff approach. The increase in the Al content in TiVNb led to a progressive decrease in the experimental enthalpy of dihydride formation from – 67 to – 52 and – 49 kJ/mol H_2 at x = 0, 0.05 and 0.10, respectively. Additionally, it was reported that the presence of Al flattens the plateaus of the PCI curves, a desirable attribute for practical applications, despite the reduction in storage capacity. The decrease in the experimental errors in the thermodynamic calculations with increasing Al content was reasonably explained by the presence of more flat plateaus in the case of the Al-containing alloys. In addition, the measured values of ΔS are close to the expected value for the entropy of H_2 free gas, that is, – 130 J/K·mol H_2. In this work, the ML-predicted values are also in good agreement with those obtained for the compositions with different Al contents, despite a large difference for the ternary alloy (Pineda-Romero et al. 2022; Pineda-Romero and Zlotea 2022). To further destabilize the dihydride phase and increase the system-level capacity, the addition of Mo to the previous compositions was reported, that is, $Al_{0.05}(TiVNb)_{0.95-x}Mo_x$ (x = 0.05, 0.10). PCIs suggest a two-step reaction for hydrogen absorption with the formation of a final *fcc* dihydride phase. The $Al_{0.05}(TiVNb)_{0.95-x}Mo_x$ (x = 0.05, 0.10) thermodynamic properties obtained were ΔH = – 45 and – 37 kJ/mol H_2 and ΔS = – 139 and – 125 J/K·mol H_2, respectively. The experiments demonstrate that increasing the Mo content results in significant hydride destabilization, which is consistent with predictions from a gradient boosting tree data-driven model for MH thermodynamics. The gravimetric capacities are 2.79 and 2.75 wt.%, respectively. Therefore, the simultaneous addition of non-forming hydrides, such as Al and Mo, to refractory HEAs in limited amounts can greatly improve hydride thermodynamics without significant capacity loss (Pineda Romero et al. 2023).

6.2.7.4. *Cycle–life properties*

A cycle is the process of complete absorption followed by full desorption of hydrogen from the material. Cycling tests evaluate the ability of a material to maintain a stable hydrogen capacity in each cycle, thus indicating the degradation behavior of the material. The efficiency of reversible hydrogen storage materials has been studied primarily for on-board applications. Capacity and hydrogen uptake/release kinetics over each cycle are often affected by the cycling process. The factors that can influence the cycle–life performance of metal hydrides can be divided into extrinsic and intrinsic factors. The *extrinsic factors* are the maximum operating temperature to ensure complete hydrogen absorption/desorption, the driving force, and impurities in the hydrogen gas or vacuum. In contrast, the *intrinsic factors* include the degree of decrepitation, number of plateaus in the sample, composition of the alloy, volume expansion, defect formation, kinetics, and phase disproportionation. After several cycles, the materials may undergo changes in their structure, crystallinity, microstructure, thermodynamics and phase stability. Therefore, it is important to analyze the materials after cycling via XRD and EDS-SEM to determine the state of the materials (Gross et al. 2012).

As mentioned above, the metal hydride is decomposed during each absorption/desorption cycle, which can affect the life cycle properties of the material. In addition, some metal alloys can undergo surface decomposition, where an element with a high affinity for hydrogen or oxygen can form stable hydrides or oxides that reduce the hydrogen storage capacity during cycling. This decomposition is known as disproportionation. In addition, impurities in gas can strongly influence the cycling performance of metal hydrides. They may delay the uptake/release of hydrogen by acting as a barrier if the inert gas is trapped in cracks or voids. In addition, the interaction may cause poisoning, which may result in retardation of the reaction rate or oxidation reactions (Gross et al. 2012).

The absorption/desorption reversibility during 20 cycles was investigated for $Ti_{0.30}V_{0.25}Zr_{0.10}Nb_{0.25}Ta_{0.10}$ (Montero et al. 2020). The maximum capacity was 2.0 H/M (2.52 wt.%), with a slight decrease in the capacity for the subsequent cycles, stabilizing around the eighth cycle and obtaining a reversible capacity of approximately 1.71 H/M (2.19 wt.%). The kinetics of absorption occurred between two and three minutes, which improved with increasing number of cycles. To exclude the possibility that the decrease in capacity was due to phase segregation, SEM investigations were carried out, which revealed the formation of numerous cracks in the material due to the multiple absorption/desorption cycles, but no phase segregation was found. This finding is in agreement with the XRD analysis, which confirmed the structural stability, recovery of the *bcc* phase after the desorption process and maintenance of the hydride phase in the cycled material. A broadening of the diffraction peaks after cycling was observed and explained as the progressive introduction of defects (Montero et al. 2020). For $Mg_{0.10}Ti_{0.30}V_{0.25}Zr_{0.10}Nb_{0.25}$ prepared by high-energy ball-milling, a slight decrease in the initial capacity was observed during the second cycle, but the capacity stabilized at approximately 2.4 wt.% in the following cycles, which demonstrated an improvement in the reversible hydrogen capacity in comparison with $Ti_{0.325}V_{0.275}Zr_{0.125}Nb_{0.275}$ (Montero et al. 2021b). Furthermore, the material is stable during cycling (no phase segregation and the same crystalline structures are present during the cycling process). The high homogeneity of the Mg-containing alloy obtained by ball milling may be responsible for the enhanced cycling properties (Montero et al. 2021b). The related $Al_{0.10}Ti_{0.30}V_{0.25}Zr_{0.10}Nb_{0.25}$ alloy exhibited good cycling properties and a higher reversible capacity than the material did without Al, and no phase decomposition occurred. The material has good structural stability as both a desorbed alloy and hydride after the cycling process. The initial capacity of the material was 1.6 H/M (2.6 wt.%), which slightly decreased until the capacity stabilized at approximately 1.5 H/M (2.45 wt.%). In terms of gravimetric capacity, the addition of Al increases the reversible gravimetric capacity in comparison to that of $Mg_{0.10}Ti_{0.30}V_{0.25}Zr_{0.10}Nb_{0.25}$ (2.40 wt.%) and $Ti_{0.325}V_{0.275}Zr_{0.125}Nb_{0.275}$ (2.25 wt.%) (Montero et al. 2021a).

A single-phase *bcc* $Ti_{0.30}V_{0.25}Zr_{0.10}Nb_{0.25}Mo_{0.10}$ alloy was synthesized via arc melting. During cycling, this alloy initially loses capacity from 1.8 H/M (2.7 wt.%) to 1.5 H/M (2.3 wt.%) and stabilizes at this value from the fifth cycle until the 20th cycle. The decrease in capacity was associated with both extrinsic and intrinsic factors. SEM–EDS analysis revealed a microstructure similar to that of the as-cast material with dendrites/interdendrites, and no phase segregations developed during cycling. In addition, the crystalline structures of the cycled material, as both hydride and desorbed phases, are stable during cycling (Bouzidi et al. 2021). Similar results were obtained for $Ti_{0.30}V_{0.25}Cr_{0.10}Zr_{0.10}Nb_{0.25}$ after 20 cycles, with a decrease in the gravimetric capacity from 3.0 wt.% to 2.44 wt.% (Bouzidi et al. 2022). In addition, first-principles calculations combined with experiments were used to design a $Ti_{0.4}Zr_{1.6}CrMnFeNi$ alloy. This material was found to have excellent cycling properties with an almost unchanged capacity of approximately 1.6 wt.% even after 1000 cycles, and the PCIs recorded for the different cycled materials (after the 4th, 30th, 100th and 1000th cycles) were almost identical (Mohammadi et al. 2022).

6.3. Review of high-entropy alloys

6.3.1. *bcc phase in HEAs*

Since 2014, reports of high-entropy alloys for hydrogen storage have increased because of their interesting properties, such as kinetics, cycling, thermodynamics, capacity and activation. In fact, these results suggest that the hydrogenation properties of *bcc* HEAs are even better than those of traditional *bcc* alloys. Within conventional alloys, a solid solution with a *bcc* structure has shown a high gravimetric capacity owing to the large number of interstitial sites. Like conventional *bcc* alloys, HEA/MPEAs can exhibit full occupancy of tetrahedral interstitial sites (eight interstitial sites per lattice) in the *fcc* structure (four metals per lattice), allowing a maximum capacity of up to 2 H/M. Given the molar mass of the transition metals, an interesting gravimetric capacity is expected (at least 1.5% by mass). 3*d* and 4*d* transition elements are the most commonly used, particularly Ti, V and Zr (Gao and Qiao 2018). Hydrogenation generally induces several phase transformations. HEA/MPEAs that crystallize in a *bcc* system may first transform to a *bcc* or distorted *bcc* (*bct*) monohydride followed by a second phase transformation into an *fcc* or *bct* dihydride after complete hydrogenation, similar to traditional *bcc* alloys. However, the useable capacity is only half of the total uptake since it is based solely on the second transition between the monohydride and dihydride phases. It has also been reported that some HEAs can undergo a single-step phase transformation from a *bcc* initial phase to a distorted *fcc* (*bct*) dihydride (Sahlberg et al. 2016). A single-step transformation is more suitable for application since it doubles the usable capacity compared with common two-step reactions.

In Table 6.1, a summary of several *bcc* HEA compositions reported as hydrogen storage materials is provided. These materials are synthesized via AM, laser engineered net-shaping (LENS), radiofrequency (RF) with induction melting, or ball-milling (BM).

Composition	Synthesis	As-cast phase	Max. H_2 capacity H/M or wt.% (25°C)	Hydride phase	Ref.
TiZrNbMoV	LENS 300 W	Mo + *bcc* + $NbTi_4$ type	1.78–2.3 (30–50°C)	*fcc* δTiHx + *bcc* $NBH_{0.4}$ + αZr-rich	(Kunce et al. 2014)
TiVZrNbHf	AM	*bcc*	2.7 (299°C)	*bct*	(Sahlberg et al. 2016)
$MgZrTiFe_{0.5}Mn_{0.5}Ni_{0.5}$	BM	bcc	1.2	*fcc*	(Zepon et al. 2018)
HfNbTiVZr	AM	*bcc*	1.9 (289°C)	*bct*	(Karlsson et al. 2018)
TiZrNbHfTa	AM	*bcc*	~2 H/M (300°C)	*fcc*	(Zlotea et al. 2019)
TiZrNbTa	AM	*bcc*	~1.4	–	(Zhang et al. 2019)
TiZrHfMoNb	AM	*bcc*	1.18	*fcc*	(Shen et al. 2019)
TiZrHfScMo	AM	*bcc*	2.14 (DFT)	–	(Hu et al. 2019)
TiVNbTa, TiVZrNb, $TiVZr_2Nb$	AM	*bcc*	1.86, 2.83, 2.66	*fcc* + *bct* *fcc* *fcc*	(Nygård et al. 2019a)
TiVCrNb	AM	*bcc*	1.96	*fcc*	(Nygård et al. 2019b)
$Ti_{0.325}V_{0.275}Zr_{0.125}Nb_{0.275}$	AM, BM, RBM	*bcc*	2.5, 2.5, 2.7	*bct*	(Montero et al. 2019)
$MgTiNbCr_{0.5}Mn_{0.5}Ni_{0.5}$	BM	*bcc*	1.6	*fcc* + Cr + Mg_2NiH_4	(Marques et al.2020)

Table 6.1. *Reported bcc HEAs and their hydrogen storage performances (LENS = laser engineered net shaping, AM = arc melting, bct = body-centered tetragonal, DFT = capacity obtained with density functional theory (DFT), BM = ball milling under argon, RBM = reactive ball milling under hydrogen, TG = thermogravimetric, D = deuterium capacity, FPC = first-principles calculations)*

Composition	Synthesis	As-cast phase	Max. H_2 capacity H/M or wt.% (25°C)	Hydride phase	Ref.
$Ti_{0.20}Zr_{0.20}Hf_{0.20}Mo_{0.10}Nb_{0.30}$	AM	*bcc*	1.54 (TG curves)	*fcc*	(Shen et al. 2020)
$Ti_{0.30}V_{0.25}Zr_{0.10}Nb_{0.25}Ta_{0.10}$	AM	*bcc*	2.5 (100°C)	*fcc*	(Montero et al. 2020)
TiVNb, TiVZrNb, TiVZrNbHf	AM	*bcc*	2 D/M	*fcc*	(Nygård et al. 2020)
MgAlTiFeNi	BM	bcc	1	*bcc* + TiH_2 + Mg_2FeH_6	(Cardoso et al. 2021)
TiVCrNb	AM	*bcc*	1.96	*fcc*	(Nygård et al. 2021)
TiVNb, TiVZrNb, TiVZrNbHf	AM	*bcc*	1.96, 1.98, 1.99 H/M	*fcc*	(Ek et al. 2021)
$Mg_{0.10}Ti_{0.30}V_{0.25}Zr_{0.10}Nb_{0.25}$	BM	*bcc*	2.70	*fcc*	(Montero et al. 2021b)
MgVAlCrNi	RBM	*bcc*	0.30	*bcc*	(Strozi et al. 2021a)
$(TiVNb)_{85}Cr_{15}$, $(TiVNb)_{95.3}Co_{4.7}$, $(TiVNb)_{96.2}Ni_{3.8}$	AM	*bcc*	3.18, 3.11 3.17	*fcc*	(Silva et al. 2021)
$Al_{0.10}Ti_{0.30}V_{0.25}Zr_{0.10}Nb_{0.25}$	AM	*bcc*	2.6	*bct*	(Montero et al. 2021a)
$(TiVNb)_{100-x}Cr_x$ (x = 15, 25, 35 at.%)	AM	*bcc* + minor *fcc* (Ti-rich)	3.0, 2.7, 2.8	*fcc* + Ti (15–25 at.%) *bcc* + Ti (35 at.%)	(Strozi et al. 2021b)
$Ti_{0.30}V_{0.25}Zr_{0.10}Nb_{0.25}Mo_{0.10}$	AM	*bcc*	2.8	*fcc*	(Bouzidi et al. 2021)
$TiHfZrNb_{1-x}V_{1+x}$	AM	*bcc* (x = 0) *bcc* + *fcc* + *hcp* (x = 0.1–0.4) *bcc* + *fcc* + *hcp* + C15 (x = 0.6) C15 + *bcc* (x = 1)	(X = 0.1–1) ≈2	*bct* + *fcc* + *hcp* (x = 0.1–0.4) *bct* + *fcc* + *hcp* + C15 (x = 0.6) C15 + *bct* (x = 1)	(Sleiman and Huot 2022)

Table 6.1. *Continued*

Composition	Synthesis	As-cast phase	Max. H_2 capacity wt.% (25°C)	Hydride phase	Ref.
$TiZrHfMoNbPt_{0.0025}$	AM	*bcc*	1.75	*fcc*	(Li et al. 2022)
$Al_x(TiVNb)_{1-x}$ (x = 0, 0.05, 0.10, 0.175, 0.25)	AM	*bcc*	3.12, 2.96, 2.63, 1.41, 1.25	*fcc*	(Pineda-Romero et al. 2022; Pineda-Romero and Zlotea 2022)
$Mg_{35}Al_{15}Ti_{25}V_{10}Zn_{15}$	BM RBM	*bcc* + Mg	2.5 (375°C) 2.75	MgH_2 + *fcc* + *bct* + $MgZn_2$ *fcc* + MgH_2	(Ferraz et al. 2022)
$Ti_{0.30}V_{0.25}Cr_{0.10}Zr_{0.10}Nb_{0.25}$	AM	*bcc*	3.0	*fcc*	(Bouzidi et al. 2022)
$V_{0.3}Ti_{0.3}Cr_{0.25}Mn_{0.1}Nb_{0.05}$	AM	*bcc* + hexagonal phase P63/mmc	3.45	*bct* + *fcc*	(Liu et al. 2022)
$Ti_{0.30}V_{0.25}Zr_{0.10}Nb_{0.25}M_{0.10}$ (M = Mn, Fe, Co, Ni, Cu and Zn)	AM (M=Mn) BM RBM	*bcc*	2.0 H/M 1.16 H/M 1.23 H/M 1.1 H/M 1.35 H/M 1.18 H/M	*fcc*	(Zlotea et al. 2022)
$(TiVNb)_{100-x}(CrAl)_x$ (x = 10, 20, 30, 40)	AM	*bcc*	3.0, 2.4, 1.2, 0.8	*fcc*, *bct* + *bcc*, *bcc*, *bcc*	(Ferreira et al. 2023)
$Ti_4V_3NbCr_2$ $Ti_3V_3Nb_2Cr_2$ $Ti_2V_3Nb_3Cr_2$ $Ti_4V_3NbCr_2Ni$ $Ti_4V_3NbCr_2Fe$ $Ti_4V_3NbCr_2Mn$	AM	*bcc* *bcc* *bcc* *bcc* + Ti-Ni *bcc* *bcc*	3.7 3.4 3.2 3.19 3.51 3.65	*fcc* *fcc* *fcc* + $NbCr_2$ *fcc* + $Ti_2NiH_{0.5}$ *fcc* *fcc*	(Cheng et al. 2022)
$(TiVNb)_{100-x}Cr_x$ (x = 30, 35, 40 at.%)	AM	mainly *bcc* ($Cr_{x=30,35}$) mainly C15 Laves (Cr_{40})	~3.06 ~2.8 ~0.95	*fcc* + *bcc* *bcc* + C15 C15 + *fcc*	(Strozi et al. 2023a)

Table 6.1. *Continued*

Composition	Synthesis	As-cast phase	Max. H_2 capacity wt.% (25°C)	Hydride phase	Ref.
$Ti_{31}V_{26}Nb_{26}Zr_{12}Fe_5$ $Ti_{31}V_{26}Nb_{26}Zr_{12}Co_5$ $Ti_{31}V_{26}Nb_{26}Zr_{12}Ni_5$	AM	*bcc* + C14	2.92 2.89 2.93	*fcc* + C14	(Chanchetti et al. 2023)
$Ti_{35}V_{35}Nb_{20}Cr_5Mn_5$ $Ti_{32}V_{32}Nb_{18}Cr_9Mn_9$ $Ti_{27.5}V_{27.5}Nb_{20}Cr_{12.5}Mn_{12.5}$	AM	*bcc* *bcc* *bcc* + C14 Laves	2.47 2.09 3.38	*fcc*	(Serrano et al. 2023)
$Mg_{0.10}Ti_{0.30}V_{0.25}Zr_{0.10}Nb_{0.25}$	FPC	*bcc*	3.16	*fcc*	(Deng et al. 2023)
$Al_{0.05}(TiVNb)_{0.95-x}Mo_x$ (x = 0.05, 0.10)	AM	*bcc*	2.79, 2.75	*fcc*	(Pineda Romero et al. 2023)

Table 6.1. *Continued*

Both empirical and computational methods have been proposed to improve the compositional design of *bcc*-HEAs. The chemical composition together with the preparation route have been identified as important parameters that influence the *bcc* HEAs properties for the targeted application (Kong et al. 2023).

6.3.1.1. *Empirical methods for the design of bcc HEAs*

As mentioned before, solid solution formation is favored when $\Omega \geq 1.1$ and $\delta \leq 6.6$ (Yang and Zhang 2012; Zhang et al. 2014). Moreover, VEC < 6.87 and, more specifically, when $5.7 \leq \text{VEC} \leq 7.2$ favor *bcc* solid solution formation (Guo et al. 2011; Yang et al. 2020). The VEC has also been identified as a parameter for predicting hydrogen storage properties. Nygård et al. (2019b) demonstrated that *bcc* HEAs with VEC ≤ 5.0 form *fcc* dihydrides with high capacity. Moreover, the onset temperature for hydrogen desorption (T_{onset}) decreases linearly with increasing VEC. When VEC > 4.75, the hydride spontaneously desorbs, and when VEC ≥ 5.0, the maximum hydrogen capacity decreases. TiVCrNb with VEC = 5.0 was identified as a promising material. Moreover, this publication highlights that the expansion of the volume per metal atom from the *bcc* HEA to the corresponding *fcc* hydride increases linearly with the VEC.

On the other hand, T_{onset} increases almost linearly with increasing VEC in the series of HEAs $Ti_{0.30}V_{0.25}Zr_{0.10}Nb_{0.25}M_{0.10}$ (M = Mg, Al, Cr, Mn, Fe, Co, Ni, Cu, Zn, Mo, Ta) and $Ti_{0.325}V_{0.275}Zr_{0.125}Nb_{0.275}$. The maximum hydrogen storage capacity also depends on VEC: a high capacity (1.5–2.0 H/M) was found for a low VEC value (< 4.9), whereas a decrease in capacity (1.1–3.5 H/M) was observed with a high VEC value (> 4.9) (Zlotea et al. 2022). In the Ti-V-Nb-Cr system, the highest capacity and the lowest thermal stability of hydrides are also found for VEC $\leq$ 5.0 (Cheng et al. 2022).

Furthermore, the probability of an interstitial site being unoccupied increases with the average VEC of the nearest-neighbor metals in the (TiVNb)-D system (Nygård et al. 2020). Three different compositions (TiVNb)-Cr/Co/Ni have been reported with the same VEC = 4.87 (Silva et al. 2021). The maximum capacity and thermodynamic properties might be related to this parameter, but the desorption properties seem to be affected by other factors, such as the microstructure, surface state and history of the samples.

Another empirical parameter studied is the lattice distortion (δ). The $TiZrHfNbMo_x$ alloys show an increase in δ and a decrease in the unit cell volume of the hydrides as the amount of Mo increases because of the smaller atomic radius of Mo. As a result, the reduced thermal stability of the hydrides is attributed to the smaller cell volume of the hydrides (Shen et al. 2020). In addition, different *bcc* HEAs with varying δ values have been reported, but no correlation with the H/M capacity was found. Furthermore, the authors noticed the expansion of the *bcc* and *fcc* unit cells with the zirconium-to-metal ratio [Zr]/[M]. An increase in the Zr concentration stabilizes hydrides (Nygård et al. 2019a). Similarly, the unit cell volume expansion per metal atom upon hydrogenation increases linearly with increasing VEC, and the T_{onset} of hydrogen desorption for the second event also decreases linearly with increasing VEC; however, no correlation with δ was found for any of the hydrogenation properties investigated (Ek et al. 2021). Therefore, the conclusion of these studies indicates that the use of empirical parameters still has some limitations and restrictions in designing HEAs for hydrogen storage because the combined influence of other factors is ignored.

6.3.1.2. *Computational methods for the design of bcc HEAs*

The CALPHAD model is a computational phase diagram tool that assists in the design of HEAs by calculating the phase equilibrium relationship on the basis of thermodynamic properties (Kong et al. 2023). Strozi et al. (2023a) reported TiVNbCr alloys with various Cr/TiVNb ratios and that the CALPHAD approach accurately predicted the formation of a major *bcc* structure in $(TiVNb)_{70}Cr_{30}$ and $(TiVNb)_{65}Cr_{35}$; moreover, a major C15 Laves phase was expected in $(TiVNb)_{60}Cr_{40}$. The experimental results are in good agreement with CALPHAD modeling. Similarly, phase prediction by CALPHAD modeling has been reported for TiVNbCr compositions, and a single *bcc* phase has been confirmed experimentally by XRD (Cheng et al. 2022).

The PCIs are important in determining the applicability of an alloy and are composition dependent, which is why Zepon et al. (2021) presented a thermodynamic model for *bcc* HEAs. Compared with the values predicted with the model, the experimental entropy (ΔS) and enthalpy (ΔH) of absorption/desorption agree well with each other and are powerful tools for alloy design. The hydrogen multicomponent system has been considered under paraequilibrium conditions, where the metal atoms are frozen without any mobility and only the hydrogen atoms are mobile. Therefore, in this equilibrium condition, the chemical potential of the co-existing phases is the same. The variation in the Gibbs free energy of each possible phase during dihydride formation (*bcc* solid solution → *bcc* or *bct* intermediate hydride → *fcc* dihydride) with increasing hydrogen content is important for determining the chemical potential of metal atoms and hydrogen. To describe the configurational entropy, the equation developed by Garcés (2010) for interstitial solid solution with site blocking effect was used, and the enthalpy of hydrogen mixing in the possible phases was found to vary linearly with the hydrogen content. Additionally, the partial molar enthalpy (which describes how the addition or removal of a specific substance affects the total enthalpy of a mixture) of a multicomponent phase was approximated by an ideal mixing rule of the components of the alloy.

An ML model developed by Witman et al. (2020) is useful for the rational selection of new materials with desired thermodynamic properties (i.e. low-stability hydrides) by several descriptors. The training of the ML model was performed by using the HydPARK experimental MH database. A gradient boosting tree regressor identified the strong dependence of the equilibrium pressure on a volume-based descriptor (average volume per atom in the solid, v_{pa}), which can be computed from solely the elemental composition of the alloy and the ground-state volume per atom of each component. The underlying structure–property relationship between ΔH and v_{pa} provides the thermodynamic basis for this correlation. To screen HEAs, the ML is constantly improved by adding newly reported training examples in the HydPARK database, for which thermodynamic properties have been reported. A second model improvement is achieved by the addition of new composition-derived features. This model screens the hydride stability of several *bcc* HEAs, which allows the selection of specific compositions for laboratory validation on the basis of thermodynamic properties. The results showed that the experimental values validated the ML model. However, ML requires more training beyond the dataset studied, but it could lead to the selection of new compositions that exhibit thermodynamic destabilization without affecting the maximum capacity of hydrogen (Witman et al. 2021).

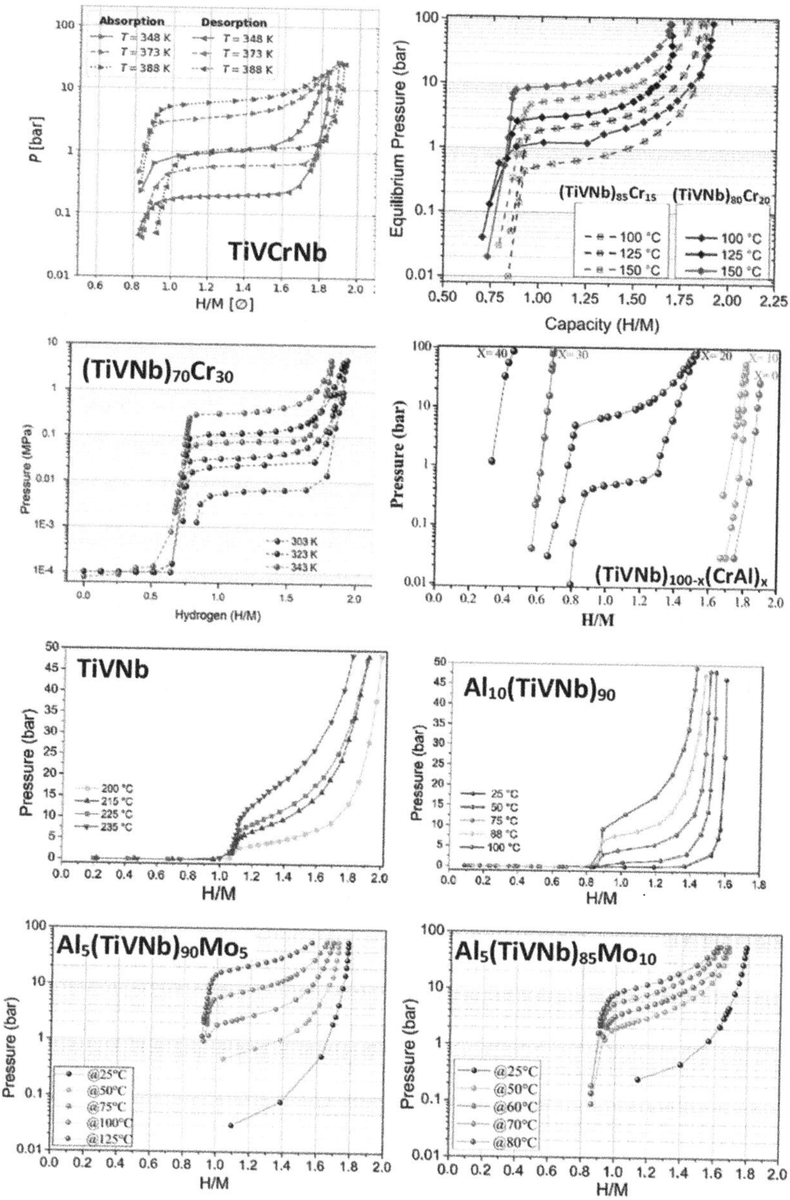

Figure 6.6. *PCIs of TiVCrNb (Nygård et al. 2021), $(TiVNb)_{100\text{-}x}Cr_x$ (x = 15, 20) (Strozi et al. 2023b), $(TiVNb)_{70}Cr_{30}$ (Strozi et al. 2023a), $(TiVNb)_{100\text{-}x}(CrAl)_x$ (x = 10, 20, 30, 40) (Ferreira et al. 2023), $Al_x(TiVNb)_{100\text{-}x}$ (x = 0, 10) (Pineda-Romero et al. 2022), and $Al_5(TiVNb)_{95\text{-}x}Mo_x$ (x = 5, 10) (Pineda Romero et al. 2023). For a color version of this figure, see www.iste.co.uk/derango-cuevas/hydrogen.zip*

The most common synthesis method used to obtain *bcc* single-phase alloys is the AM technique, and the maximum capacity achieved for the *bcc* HEAs studied to date is approximately 3 wt.% or 2 H/M, with the most common transformation to an *fcc* or *bct* dihydride phase (Table 6.1). However, hydrogenation properties such as capacity, activation, kinetics, thermodynamics and cycling stability still require modifications to obtain suitable values for practical applications.

For the TiZrNbTa HEA, a two-step activation mechanism was proposed to reduce surface oxides to suboxides and transform suboxides into subhydroxides. This process increases the size and concentration of vacancy clusters and can be responsible for the improvement in hydrogen absorption kinetics (Zhang et al. 2019).

One of the main challenges is the stability of hydrides; for hydrides that are too stable, high temperatures are needed to recover the stored hydrogen. Consequently, hydride destabilization has become an important research strategy. Several reported PCIs are plotted in Figure 6.6. Typically, two plateaus are observed, few compositions are unstable hydrides, and the reversible capacities remain modest. However, we can observe that a small variation in the chemical composition has a large effect on the thermodynamic properties. Therefore, much remains to be done to optimize the compositions and/or to identify new compositions with the desired thermodynamic properties.

6.3.2. *Intermetallic phases in HEAs*

The term Laves phases was implemented in 1939 for AB_2-type intermetallics, which are now recognized as important hydride formers with capacities of up to 2 wt.% and excellent stability during cycling processes. However, the greatest challenge that requires attention is adapting the stability of the hydride at room temperature. Refining the composition that can directly impact its properties is an important key to overcoming this problem. Laves phases, being a tetrahedrally close-packed structure, possess either A2B2, AB3 or B4 coordinated interstitial sites with four nearest neighbors. Their preferential occupancy by hydrogen atoms and their stability related to bonding energies are crucial.

Several studies have focused on lightweight AB_2 Laves phases with hydrogenation properties that are still not adequate in terms of applications. The most encouraging Laves phases include Zr and/or Ti as A-type elements and Co/Cr/Mn/V/Ni/Fe as B-type elements with capacities of up to 1.7 wt.% (Yadav et al. 2022). In general, HEAs containing Laves phases such as hexagonal C14 Laves are the most common in the literature. These materials have attracted considerable interest as hydrogen storage materials because they have shown high phase stability after hydrogenation and desorption and have greater flexibility in accommodating different atom sizes than cubic C15 Laves do (Stein and Leineweber 2021; Yang et al. 2022).

Composition	Synthesis	As-cast phase	H_2 capacity (wt. %)	Hydride phase	Ref.
$CoFeMnTi_{2.0}VZr$ $CoFeMnTiV_{2.6}Zr$ $CoFeMnTiVZr_{2.3}$	AM	C14 Laves	1.82 1.60 1.80	C14 Laves	(Kao et al. 2010)
$Zr_{0.163}Ti_{0.162}V_{0.222}Cr_{0.135}Fe_{0.158}Ni_{0.160}$	LENS	C14 Laves + α-Ti	1.81 (50°C)	$C14_{Laves}$ + $C14_{hydride}$	(Kunce et al. 2013)
$La_{0.1}Ni_{0.5}Fe_{0.1}V_{0.1}Mn_{0.2}$	LENS	La $(Ni,Mn)_5$ + *fcc* + ?	0.83 (35°C)	*fcc* + La $(Ni,Mn)_5$	(Kunce et al. 2017)
FeMnTiVZr, $CrFeMnTi_{2.0}VZr$, CrMnTiVZr, $CrFeMnTiVZr_{2.0}$	AM	C14 Laves	1.88, 1.85, 2.23 (5°C), 2.17 (5°C)	C14 Laves	(Chen et al. 2018)
TiZrCrMnFeNi	AM	C14 Laves+ cubic phase Pm-3m	1.7 (30°C)	C14 Laves	(Edalati et al. 2020)
TiZrNbFeNi	AM	C14 Laves 98.5% + *bcc* 0.8% + B2	1.64 (32°C)	C14 Laves 99%+*bcc*	(Floriano et al. 2020)
$Ti_{20}Zr_{20}Nb_5Fe_{40}Ni_{15}$		C14 Laves 99.1% + *bcc*	1.38 (32°C)	C14 Laves	
ZrTiVNiCrFe	AM + MS	C14 Laves	1.60 (27°C)	C14 Laves	(Zadorozhnyy et al. 2020)
CoFeMnTiVZr	AM + MS	C14 Laves	1.7	C14 Laves	(Sarac et al. 2020)
TiZrNbCrFe	AM	C14 Laves + *bcc*	1.95 (30°C) 1.90 (200°C)	C14 Laves + *fcc*	(Floriano et al. 2021)
ZrTiNbMoCr	AM	C14 Laves + bcc	~0.4 (300°C)	C14 Laves + bcc	(Huang et al. 2021)
$(ZrTiVFe)_{0.80}Al_{0.20}$	AM	C14 + hcp + tetragonal	1.40 (200°C)	C14 Laves + hcp	(Ma et al. 2022a)
ZrTiVFe ZrTiVAl	AM	C14 Laves + hcp	1.67 (200°C) 1.40 (200°C)	C14 Laves	(Ma et al. 2022b)
TiZrFeMnCrV	AM	C14 Laves	1.80 (30°C)	C14 Laves	(Chen et al. 2022)
TiZrVCrNi	AM	C14 Laves	1.52 (32°C)	C14 Laves	(Kumar et al. 2022)

Table 6.2. *Studies of IM phases in HEAs for hydrogen storage (AM = arc melting, MS = melt spinning, LENS = laser engineered net-shaping, RF = radiofrequency with induction melting)*

Composition	Synthesis	As-cast phase	H_2 capacity (wt. %)	Hydride phase	Ref.
ZrTiNiCrMn	AM	C14 Laves + B2	1.66 (50°C)	–	(Fukagawa et al. 2022)
$(ZrTiVFe)_{0.95}Cu_{0.05}$ $(ZrTiVFe)_{0.90}Cu_{0.10}$ $(ZrTiVFe)_{0.80}Cu_{0.20}$	AM	C14 (major), α-Ti, α-Zr (minor) C14 (major), α-Ti, α-Zr (minor) C14 + $ZrTiCu_2$ + Cu_8Zr_3	1.504 (200°C) 1.357 (200°C) 1.112 (200°C)	C14 + hcp	(Ma et al. 2023)
$Ti_{0.24}V_{0.17}Zr_{0.17}Mn_{0.17}Co_{0.17}Fe_{0.08}$	RF	C14 Laves	0.53 (425°C)	C14 Laves	(Kumar et al. 2023)
TiZrNbCrFeNi	AM	C14 Laves	1.5 (30°C) 1.1 (200°C)	C14 Laves	(Andrade et al. 2023)

Table 6.2. *Continued*

In Table 6.2, different compositions of HEAs with intermetallic phases studied as hydrogen storage materials are listed. These materials are synthesized via AM, LENS or RF with induction melting.

Kao et al. (2010) reported $CoFeMnTi_xV_yZr_z$ alloys with C14 Laves before and after hydrogenation. Zr and Ti, elements with high hydrogen affinity, have been attributed to the increase in capacity to a limit where these elements begin to segregate. In addition, it has been claimed that the kinetics are improved, and the equilibrium plateau pressure is reduced as larger interstitial sites are created. The lattice parameter increases with increasing Ti (x = 0.5–2.5) or Zr (y = 0.4–3.0) content because of the large atomic radii of Zr and Ti, although the high affinity of Zr–H hinders desorption by trapping hydrogen in the structure.

Similar results have been reported by Chen et al. (2018) for $Cr_uFe_vMn_wTi_xV_yZr_z$ alloys, which are C14 Laves type AB_2, where Zr and Ti enhance the maximum hydrogen capacity. The main controlling factor is the enthalpy of hydride formation. Zr addition is responsible for the improved kinetics, hydrogen uptake and reduced equilibrium plateau pressure due to the formation of larger interstitial sites.

Kunce et al. (2013) reported a ZrTiVCrFeNi alloy, and it has been established that the shape, size distribution and purity of the initial powder metals synthesized by LENS are important for improving the chemical homogeneity of the alloy.

Additionally, the desorption equilibrium pressure is very low, and it is not possible to completely desorb the material with typical Sieverts equipment.

TiZrCrMnFeNi studied by Edalati et al. (2020) showed good reversible capacity at room temperature with fast kinetics due to the presence of C14 Laves predicted with VEC = 6.4. High-pressure torsion (HPT) is used before hydrogenation to trap the hydride phase for a longer time (i.e. for at least 20 min), which would allow a better study of the crystal structure of the hydride phase and could improve the kinetics through the introduction of lattice defects. HPT is widely used as a technique for plastic deformation, which creates ultrafine-grained structures. Another study reported that the use of HPT induces poor hydrogenation/dehydrogenation reversibility because of the formation of dislocations and ultrahigh density of grain boundaries. Therefore, the presence of these defects and the resulting lattice distortion block hydrogen transport (Edalati et al. 2018a, 2018b).

Floriano et al. (2021) reported that the TiZrNbCrFe composition achieved good reversible hydrogen capacity with a stable phase structure for hydrogen absorption and desorption; however, such reversible capacity was reached only at high temperature (200°C).

These studies have demonstrated the relevance of the intermetallic phases in HEAs. Specifically, Laves phases improve hydrogenation properties such as kinetics, capacity in some compositions, phase stability and cyclability. In conclusion, the elements used in the composition have a decisive influence on the performance of the alloy; therefore, there is still a wide range of possibilities for the design of new alloys with improved hydrogenation properties.

6.4. Conclusions and perspectives

In conclusion, research in the field of high-entropy alloys capable of forming hydrides is still in its infancy, but some compositions already show promising performance. The main advantage of HEAs/MPEAs, which is also the biggest challenge for this class of materials, is the extremely large composition space that promises to finely tune the chemistry to enhance the final hydrogen storage performance. However, the well-known trial-and-error experimental method is limited for this class of alloys, and novel approaches are needed to rationalize the design of high-performing materials. New approaches are currently being developed, such as data-driven predictions facilitated by machine/statistical learning

(Witman et al. 2020), which stand as an alternative to traditional time-consuming computational methods and aim to rapidly select promising compositions from the practically infinite composition space of HEAs/MPEAs. To complete the research loop, the predicted properties, such as capacity and reaction thermodynamics, need laboratory verification. Nonetheless, if strategies for HEAs/MPEAs hydride destabilization can be quantitatively assessed a priori, the scope of necessarily time-consuming experiments can be quickly narrowed to maximize successful realization of hydrides with desired properties. If machine/statistical learning method in materials science usually predict physicochemical properties from basic information, that is, chemical formula, accurate chemical calculations based on electronic structure theory are sometimes needed for theoretical confirmation and in-depth atomic-level understanding of the basic reaction mechanisms.

Various experiments using large instruments suggest that hydrogen occupation in HEAs/MPEAs is not limited to the interstices usually occupied in conventional alloys. One possible explanation is the significant lattice distortion in these alloys, which implies different interstitial volumes depending on the local environment. Further research is needed to understand the effect of the variation in the local environment at the atomic scale and its implications for the macroscopic properties. To shed light on short- and long-range structures and their changes under the effect of hydrogen absorption/desorption, one possibility will be to apply a combined neutron and X-ray scattering methods to probe both deuterium/hydrogen and metallic "networks", respectively, as the interaction between neutron and X-ray with matter is complementary.

With respect to the hydrogen storage properties of HEAs/MPEAs, a large diversity of trends is encountered. It is obvious that the difficulty of relating these effects to the empirical parameters commonly used in the field of metallurgy (δ, VEC, etc.) is the cause of a still sparse understanding of the interaction of these materials with hydrogen.

In conclusion, the concept of HEAs/MPEAs applied to the solid-state hydrogen storage opens the way for the exploration of many compositions, allowing more flexible adjustment of the desired properties. The countless possibilities of chemical compositions in the multidimensional phase diagrams unlock expectations for discoveries that will clarify the correlations between chemistry, structure, thermodynamics and storage performances.

6.5. References

Amirthan, T. and Perera, M.S.A. (2022). The role of storage systems in hydrogen economy: A review. *J. Nat. Gas Sci. Eng.*, 108, 104843. doi: 10.1016/j.jngse.2022.104843.

Andrade, G., Zepon, G., Edalati, K., Mohammadi, A., Ma, Z., Li, H.-W., Floriano, R. (2023). Crystal structure and hydrogen storage properties of AB-type TiZrNbCrFeNi high-entropy alloy. *Int. J. Hydrog. Energy*, 48, 13555–13565. doi: 10.1016/j.ijhydene.2022.12.134.

Barrett, C.S. (1943). *Structure of Metals*. McGraw-Hill, New York.

Bouzidi, A., Laversenne, L., Zepon, G., Vaughan, G., Nassif, V., Zlotea, C. (2021). Hydrogen sorption properties of a novel refractory Ti-V-Zr-Nb-Mo high entropy alloy. *Hydrogen*, 2, 399–413. doi: 10.3390/hydrogen2040022.

Bouzidi, A., Laversenne, L., Nassif, V., Elkaim, E., Zlotea, C. (2022). Hydrogen storage properties of a new Ti-V-Cr-Zr-Nb high entropy alloy. *Hydrogen*, 3, 270–284. doi: 10.3390/hydrogen3020016.

Cantor, B. (2014). Multicomponent and high entropy alloys. *Entropy*, 16, 4749–4768. doi: 10.3390/e16094749.

Cantor, B. (2021). Multicomponent high-entropy Cantor alloys. *Prog. Mater. Sci.*, 120, 100754. doi: 10.1016/j.pmatsci.2020.100754.

Cantor, B., Chang, I.T.H., Knight, P., Vincent, A.J.B. (2004). Microstructural development in equiatomic multicomponent alloys. *Mater. Sci. Eng. A*, 375(377), 213–218. doi: 10.1016/j.msea.2003.10.257.

Cardoso, K.R., Roche, V., Jorge, A.M., Antiqueira, F.J., Zepon, G., Champion, Y. (2021). Hydrogen storage in MgAlTiFeNi high entropy alloy. *J. Alloys Compd.*, 858, 158357. doi: 10.1016/j.jallcom.2020.158357.

Chanchetti, L.F., Hessel Silva, B., Montero, J., Zlotea, C., Champion, Y., Botta, W.J., Zepon, G. (2023). Structural characterization and hydrogen storage properties of the Ti31V26Nb26Zr12M5 (M = Fe, Co, or Ni) multi-phase multicomponent alloys. *Int. J. Hydrog. Energy*, 48, 2247–2255. doi: 10.1016/j.ijhydene.2022.10.060.

Chen, S.-K., Lee, P.-H., Lee, H., Su, H.-T. (2018). Hydrogen storage of C14-CruFevMnw TixVyZrz alloys. *Mater. Chem. Phys.*, 210, 336–347. doi: 10.1016/j.matchemphys.2017.08.008.

Chen, J., Li, Z., Huang, H., Lv, Y., Liu, B., Li, Y., Wu, Y., Yuan, J., Wang, Y. (2022). Superior cycle life of TiZrFeMnCrV high entropy alloy for hydrogen storage. *Scr. Mater.*, 212, 114548. doi: 10.1016/j.scriptamat.2022.114548.

Cheng, B., Li, Y., Li, X., Ke, H., Wang, L., Cao, T., Wan, D., Wang, B., Xue, Y. (2022). Solid-state hydrogen storage properties of Ti–V–Nb–Cr high-entropy alloys and the associated effects of transitional metals (M = Mn, Fe, Ni). *Acta Metall. Sin. Engl. Lett.*, 36, 1113–1122. doi: 10.1007/s40195-022-01403-9.

Couzinié, J.-P. and Dirras, G. (2019). Body-centered cubic high-entropy alloys: From processing to underlying deformation mechanisms. *Mater. Charact.*, 147, 533–544. doi: 10.1016/j.matchar.2018.07.015.

Couzinié, J.P., Dirras, G., Perrière, L., Chauveau, T., Leroy, E., Champion, Y., Guillot, I. (2014). Microstructure of a near-equimolar refractory high-entropy alloy. *Mater. Lett.*, 126, 285–287. doi: 10.1016/j.matlet.2014.04.062.

Couzinié, J.-P., Senkov, O.N., Miracle, D.B., Dirras, G. (2018). Comprehensive data compilation on the mechanical properties of refractory high-entropy alloys. *Data Brief*, 21, 1622–1641. doi: 10.1016/j.dib.2018.10.071.

D'Ippolito, B. and Rüling, C.-C. (2019). Research collaboration in large scale research infrastructures: Collaboration types and policy implications. *Res. Policy*, 48, 1282–1296. doi: 10.1016/j.respol.2019.01.011.

De Boer, F.R., Boom, R., Mattens, W.C.M., Miedema, A.R., Niessen, A.K. (1988). *Cohesion in Metals: Transition Metal Alloys. Cohésion and Structure.* North-Holland, Amsterdam/Oxford/New York.

Deng, Y., Hu, J., Zhao, S., Wang, W., Xie, L., Sun, G., Shen, H., Zu, X., Xiao, H. (2023). Hydrogen storage properties of $Mg_{0.10}Ti_{0.30}V_{0.25}Zr_{0.10}Nb_{0.25}$ lightweight high entropy alloy: A theoretical study. *Int. J. Hydrog. Energy*, S0360319923034912. doi: 10.1016/j.ijhydene.2023.07.075.

Dove, M.T. and Li, G. (2022). Review: Pair distribution functions from neutron total scattering for the study of local structure in disordered materials. *Nucl. Anal.*, 1, 100037. doi: 10.1016/j.nucana.2022.100037.

Edalati, K., Akiba, E., Horita, Z. (2018a). High-pressure torsion for new hydrogen storage materials. *Sci. Technol. Adv. Mater.*, 19, 185–193. doi: 10.1080/14686996.2018.1435131.

Edalati, K., Novelli, M., Itano, S., Li, H.-W., Akiba, E., Horita, Z., Grosdidier, T. (2018b). Effect of gradient-structure versus uniform nanostructure on hydrogen storage of Ti-V-Cr alloys: Investigation using ultrasonic SMAT and HPT processes. *J. Alloys Compd.*, 737, 337–346. doi: 10.1016/j.jallcom.2017.12.053.

Edalati, P., Floriano, R., Mohammadi, A., Li, Y., Zepon, G., Li, H.-W., Edalati, K. (2020). Reversible room temperature hydrogen storage in high-entropy alloy TiZrCrMnFeNi. *Scr. Mater.*, 178, 387–390. doi: 10.1016/j.scriptamat.2019.12.009.

Ek, G., Nygård, M.M., Pavan, A.F., Montero, J., Henry, P.F., Sørby, M.H., Witman, M., Stavila, V., Zlotea, C., Hauback, B.C. et al. (2021). Elucidating the effects of the composition on hydrogen sorption in TiVZrNbHf-based high-entropy alloys. *Inorg. Chem.*, 60, 1124–1132. doi: 10.1021/acs.inorgchem.0c03270.

El-Hadad, S. (2022). High entropy alloys: The materials of future. *Int. J. Mater. Technol. Innov.*, 2, 67–84. doi: 10.21608/ijmti.2022.118565.1046.

Ferraz, M.D.B., Botta, W.J., Zepon, G. (2022). Synthesis, characterization and first hydrogen absorption/desorption of the $Mg_{35}Al_{15}Ti_{25}V_{10}Zn_{15}$ high entropy alloy. *Int. J. Hydrog. Energy*, 47, 22881–22892. doi: 10.1016/j.ijhydene.2022.05.098.

Ferreira, T., Pineda-Romero, N., Botta, W.J., Zepon, G., Zlotea, C. (2023). Tuning the hydride stability of the TiVNb-based alloys by equimolar Cr/Al addition. *Intermetallics*, 161, 107992. doi: 10.1016/j.intermet.2023.107992.

Fitch, A. (2019). Synchrotron radiation and powder diffraction. *International Tables for Crystallography*, 2(2), 51–65. doi: 10.1107/97809553602060000937.

Floriano, R., Zepon, G., Edalati, K., Fontana, G.L.B.G., Mohammadi, A., Ma, Z., Li, H.-W., Contieri, R.J. (2020). Hydrogen storage in TiZrNbFeNi high entropy alloys, designed by thermodynamic calculations. *Int. J. Hydrog. Energy*, 45, 33759–33770. doi: 10.1016/j.ijhydene.2020.09.047.

Floriano, R., Zepon, G., Edalati, K., Fontana, G.L.B.G., Mohammadi, A., Ma, Z., Li, H.-W., Contieri, R.J. (2021). Hydrogen storage properties of new A3B2-type TiZrNbCrFe high-entropy alloy. *Int. J. Hydrog. Energy*, 46, 23757–23766. doi: 10.1016/j.ijhydene.2021.04.181.

Fukagawa, T., Saito, Y., Matsuyama, A. (2022). Effect of varying Ni content on hydrogen absorption–desorption and electrochemical properties of Zr-Ti-Ni-Cr-Mn high-entropy alloys. *J. Alloys Compd.*, 896, 163118. doi: 10.1016/j.jallcom.2021.163118.

Gao, M. and Qiao, J. (2018). High-entropy alloys (HEAs). *Metals*, 8, 108. doi: 10.3390/met8020108.

Gao, M.C., Yeh, J.-W., Liaw, P.K., Zhang, Y. (eds) (2016). *High-Entropy Alloys*. Springer, Berlin. doi: 10.1007/978-3-319-27013-5.

Gao, M.C., Miracle, D.B., Maurice, D., Yan, X., Zhang, Y., Hawk, J.A. (2018). High-entropy functional materials. *J. Mater. Res.*, 33, 3138–3155. doi: 10.1557/jmr.2018.323.

Garcés, J. (2010). The configurational entropy of mixing of interstitials solid solutions. *Appl. Phys. Lett.*, 96, 161904. doi: 10.1063/1.3400221.

Gross, K.J., Russell Carrington, K., Barcelo, S., Karkamkar, A., Purewal, J., Ma, S., Zhou, H.-C., Dantzer, P., Ott, K., Burrell, T. et al. (2012). Recommended best practices for the characterization of storage properties of hydrogen storage materials. US Department of Energy, Office of Energy Efficiency and Renewable Energy [Online]. Available at: https://www.energy.gov/eere/fuelcells/articles/recommended-best-practices-characterization-storage-properties-hydrogen-0.

Guo, S. and Liu, C.T. (2011). Phase stability in high entropy alloys: Formation of solid-solution phase or amorphous phase. *Prog. Nat. Sci. Mater. Int.*, 21, 433–446. doi: 10.1016/S1002-0071(12)60080-X.

Guo, S., Ng, C., Lu, J., Liu, C.T. (2011). Effect of valence electron concentration on stability of fcc or bcc phase in high entropy alloys. *J. Appl. Phys.*, 109, 103505. doi: 10.1063/1.3587228.

Guo, S., Hu, Q., Ng, C., Liu, C.T. (2013). More than entropy in high-entropy alloys: Forming solid solutions or amorphous phase. *Intermetallics*, 41, 96–103. doi: 10.1016/j.intermet.2013.05.002.

He, Q. and Yang, Y. (2018). On lattice distortion in high entropy alloys. *Front. Mater.*, 5, 42. doi: 10.3389/fmats.2018.00042.

Howard, C.J. and Kisi, E.H. (2019). *Applications of Neutron Powder Diffraction*. Oxford University Press, Oxford. doi: 10.1107/97809553602060000938.

Hu, J., Shen, H., Jiang, M., Gong, H., Xiao, H., Liu, Z., Sun, G., Zu, X. (2019). A DFT study of hydrogen storage in high-entropy alloy TiZrHfScMo. *Nanomaterials*, 9, 461. doi: 10.3390/nano9030461.

Huang, L., Long, M., Liu, W., Li, S. (2021). Effects of Cr on microstructure, mechanical properties and hydrogen desorption behaviors of ZrTiNbMoCr high entropy alloys. *Mater. Lett.*, 293, 129718. doi: 10.1016/j.matlet.2021.129718.

Hume-Rothery, W. and Powell, H.M. (1935). On the theory of super-lattice structures in alloys. *Z. Für Krist. Cryst. Mater.*, 91, 23–47. doi: 10.1524/zkri.1935.91.1.23.

Huot, J. and Černý, R. (2016). Neutron powder diffraction. In *Neutron Scattering and Other Nuclear Techniques for Hydrogen in Materials,* Fritzsche, H., Huot, J., Fruchart, D. (eds). Springer, Berlin. doi: 10.1007/978-3-319-22792-4_3.

JEOL (2020). Scanning electron microscope A To Z: Basic knowledge for using the SEM. Report, JEOL Solut. Innov. [Online]. Available at: https://www.jeolusa.com/RESOURCES/Electron-Optics/Documents-Downloads/scanning-electron-microscope-a-to-z-basic-knowledge-for-using-the-sem [Accessed 24 April 2014].

Kao, Y.-F., Chen, S.-K., Sheu, J.-H., Lin, J.-T., Lin, W.-E., Yeh, J.-W., Lin, S.-J., Liou, T.-H., Wang, C.-W. (2010). Hydrogen storage properties of multi-principal-component CoFeMn TixVyZrz alloys. *Int. J. Hydrog. Energy*, 35, 9046–9059. doi: 10.1016/j.ijhydene.2010.06.012.

Karlsson, D., Ek, G., Cedervall, J., Zlotea, C., Møller, K.T., Hansen, T.C., Bednarčík, J., Paskevicius, M., Sørby, M.H., Jensen, T.R. et al. (2018). Structure and hydrogenation properties of a HfNbTiVZr high-entropy alloy. *Inorg. Chem.*, 57, 2103–2110. doi: 10.1021/acs.inorgchem.7b03004.

Kaushik, N., Meena, A., Mali, H.S. (2022). High entropy alloy synthesis, characterisation, manufacturing & potential applications: A review. *Mater. Manuf. Process.*, 37, 1085–1109. doi: 10.1080/10426914.2021.2006223.

Keith, A., Zlotea, C., Szilágyi, P.Á. (2023). Perspective of interstitial hydrides of high-entropy alloys for vehicular hydrogen storage. *Int. J. Hydrog. Energy*, S0360319923002598. doi: 10.1016/j.ijhydene.2023.01.141.

Kern, A. (2019). Instrumentation for laboratory X-ray scattering techniques. *International Tables for Crystallography*, 2(1), 26–50. doi: 10.1107/97809553602060000936.

Kittel, C. (2005). *Introduction to Solid State Physics*. John Wiley & Sons, Inc., New York.

Kong, L., Cheng, B., Wan, D., Xue, Y. (2023). A review on BCC-structured high-entropy alloys for hydrogen storage. *Front. Mater.*, 10, 1135864. doi: 10.3389/fmats.2023.1135864.

Korchinski, D., Duereth, J., McNeil, R., Au, M., Schmid, V. (2019). A brief introduction to scanning electron microscopy. Report, University of British Columbia [Online]. Available at: https://phas.ubc.ca/~berciu/TEACHING/PHYS502/PROJECTS/SEM.pdf [Accessed 24 April 2014].

Kumar, A., Yadav, T.P., Mukhopadhyay, N.K. (2022). Notable hydrogen storage in Ti–Zr–V–Cr–Ni high entropy alloy. *Int. J. Hydrog. Energy*, 47, 22893–22900. doi: 10.1016/j.ijhydene.2022.05.107.

Kumar, A., Yadav, T.P., Shaz, M.A., Mukhopadhyay, N.K. (2023). Hydrogen storage performance of C14 type Ti0.24V0.17Zr0.17Mn0.17Co0.17Fe0.08 high entropy intermetallics. *Trans. Indian Natl. Acad. Eng.* doi: 10.1007/s41403-023-00390-2.

Kunce, I., Polanski, M., Bystrzycki, J. (2013). Structure and hydrogen storage properties of a high entropy ZrTiVCrFeNi alloy synthesized using Laser Engineered Net Shaping (LENS). *Int. J. Hydrog. Energy*, 38, 12180–12189. doi: 10.1016/j.ijhydene.2013.05.071.

Kunce, I., Polanski, M., Bystrzycki, J. (2014). Microstructure and hydrogen storage properties of a TiZrNbMoV high entropy alloy synthesized using Laser Engineered Net Shaping (LENS). *Int. J. Hydrog. Energy*, 39, 9904–9910. doi: 10.1016/j.ijhydene.2014.02.067.

Kunce, I., Polański, M., Czujko, T. (2017). Microstructures and hydrogen storage properties of La Ni Fe V Mn alloys. *Int. J. Hydrog. Energy*, 42, 27154–27164. doi: 10.1016/j.ijhydene.2017.09.039.

Lai, Q., Sun, Y., Wang, T., Modi, P., Cazorla, C., Demirci, U.B., Ares Fernandez, J.R., Leardini, F., Aguey-Zinsou, K. (2019). How to design hydrogen storage materials? Fundamentals, synthesis, and storage tanks. *Adv. Sustain. Syst.*, 3 1900043. doi: 10.1002/adsu.201900043.

Li, Z., Zhao, S., Ritchie, R.O., Meyers, M.A. (2019). Mechanical properties of high-entropy alloys with emphasis on face-centered cubic alloys. *Prog. Mater. Sci.*, 102, 296–345. doi: 10.1016/j.pmatsci.2018.12.003.

Li, P., Hu, J., Huang, G., Zhang, J., Wang, W., Tian, C., Xiao, H., Zhou, X., Shen, H., Long, X. et al. (2022). Electronic structure regulation toward the improvement of the hydrogenation properties of TiZrHfMoNb high-entropy alloy. *J. Alloys Compd.*, 905, 164150. doi: 10.1016/j.jallcom.2022.164150.

Liu, J., Xu, J., Sleiman, S., Ravalison, F., Zhu, W., Liu, H., Cheng, H., Huot, J. (2022). Hydrogen storage properties of $V_{0.3}Ti_{0.3}Cr_{0.25}Mn_{0.1}Nb_{0.05}$ high entropy alloy. *Int. J. Hydrog. Energy*, 47, 25724–25732. doi: 10.1016/j.ijhydene.2022.06.013.

Luo, L., Chen, L., Li, L., Liu, S., Li, Y., Li, C., Li, L., Cui, J., Li, Y. (2024). High-entropy alloys for solid hydrogen storage: A review. *Int. J. Hydrog. Energy*, 50, 406–430. doi: 10.1016/j.ijhydene.2023.07.146.

Ma, X., Ding, X., Chen, R., Cao, W., Song, Q. (2022a). Study on hydrogen storage property of (ZrTiVFe) Al high-entropy alloys by modifying Al content. *Int. J. Hydrog. Energy*, 47, 8409–8418. doi: 10.1016/j.ijhydene.2021.12.172.

Ma, X., Ding, X., Chen, R., Gao, X., Su, Y., Cui, H. (2022b). Enhanced hydrogen storage properties of ZrTiVAl1-xFex high-entropy alloys by modifying the Fe content. *RSC Adv.*, 12, 11272–11281. doi: 10.1039/D2RA01064J.

Ma, X., Ding, X., Chen, R., Zhang, J., Song, Q., Cui, H. (2023). Microstructural features and improved reversible hydrogen storage properties of ZrTiVFe high-entropy alloy via Cu alloying. *Int. J. Hydrog. Energy*, 48, 2718–2730. doi: 10.1016/j.ijhydene.2022.10.130.

Maréchal, Y. (ed.) (2007). H/D Isotopic substitution in H-bonds. In *The Hydrogen Bond and the Water Molecule*. Elsevier, Amsterdam. doi: 10.1016/B978-044451957-3.50008-1.

Marques, F., Pinto, H.C., Figueroa, S.J.A., Winkelmann, F., Felderhoff, M., Botta, W.J., Zepon, G. (2020). Mg-containing multi-principal element alloys for hydrogen storage: A study of the $MgTiNbCr_{0.5}Mn_{0.5}Ni_{0.5}$ and $Mg_{0.68}TiNbNi_{0.55}$ compositions. *Int. J. Hydrog. Energy*, 45 19539–19552. doi: 10.1016/j.ijhydene.2020.05.069.

Marques, F., Balcerzak, M., Winkelmann, F., Zepon, G., Felderhoff, M. (2021). Review and outlook on high-entropy alloys for hydrogen storage. *Energy Environ. Sci.*, 14, 5191–5227. doi: 10.1039/D1EE01543E.

Miedema, A.R., De Châtel, P.F., De Boer, F.R. (1980). Cohesion in alloys. Fundamentals of a semi-empirical model. *Phys. BC*, 100, 1–28. doi: 10.1016/0378-4363(80)90054-6.

Miracle, D.B. and Senkov, O.N. (2017). A critical review of high entropy alloys and related concepts. *Acta Mater.*, 122, 448–511. doi: 10.1016/j.actamat.2016.08.081.

Miracle, D.B., Miller, J., Senkov, O.N., Woodward, C., Uchic, M., Tiley, J. (2014). Exploration and development of high entropy alloys for structural applications. *Entropy*, 16, 494–525. doi: 10.3390/e16010494.

Mohammadi, A., Ikeda, Y., Edalati, P., Mito, M., Grabowski, B., Li, H.-W., Edalati, K. (2022). High-entropy hydrides for fast and reversible hydrogen storage at room temperature: Binding-energy engineering via first-principles calculations and experiments. *Acta Mater.*, 236, 118117. doi: 10.1016/j.actamat.2022.118117.

Montero, J., Zlotea, C., Ek, G., Crivello, J.-C., Laversenne, L., Sahlberg, M. (2019). TiVZrNb multi-principal-element alloy: Synthesis optimization, structural, and hydrogen sorption properties. *Molecules*, 24, 2799. doi: 10.3390/molecules24152799.

Montero, J., Ek, G., Laversenne, L., Nassif, V., Zepon, G., Sahlberg, M., Zlotea, C. (2020). Hydrogen storage properties of the refractory Ti–V–Zr–Nb–Ta multi-principal element alloy. *J. Alloys Compd.*, 835, 155376. doi: 10.1016/j.jallcom.2020.155376.

Montero, J., Ek, G., Laversenne, L., Nassif, V., Sahlberg, M., Zlotea, C. (2021a). How 10 at% Al addition in the Ti-V-Zr-Nb high-entropy alloy changes hydrogen sorption properties. *Molecules*, 26, 2470. doi: 10.3390/molecules26092470.

Montero, J., Ek, G., Sahlberg, M., Zlotea, C. (2021b). Improving the hydrogen cycling properties by Mg addition in Ti-V-Zr-Nb refractory high entropy alloy. *Scr. Mater.* 194, 113699. doi: 10.1016/j.scriptamat.2020.113699.

Murty, B.S., Yeh, J.-W., Ranganathan, S. (2014). *High-entropy Alloys*. Elsevier, Amsterdam.

Murty, B.S., Yeh, J.-W., Ranganathan, S., Bhattacharjee, P.P. (2019). *High-Entropy Alloys*. Elsevier, Amsterdam.

Nygård, M.M., Ek, G., Karlsson, D., Sahlberg, M., Sørby, M.H., Hauback, B.C. (2019a). Hydrogen storage in high-entropy alloys with varying degree of local lattice strain. *Int. J. Hydrog. Energy*, 44, 29140–29149. doi: 10.1016/j.ijhydene.2019.03.223.

Nygård, M.M., Ek, G., Karlsson, D., Sørby, M.H., Sahlberg, M., Hauback, B.C. (2019b). Counting electrons. A new approach to tailor the hydrogen sorption properties of high-entropy alloys. *Acta Mater.*, 175, 121–129. doi: 10.1016/j.actamat.2019.06.002.

Nygård, M.M., Sławiński, W.A., Ek, G., Sørby, M.H., Sahlberg, M., Keen, D.A., Hauback, B.C. (2020). Local order in high-entropy alloys and associated deuterides. A total scattering and reverse Monte Carlo study. *Acta Mater.* 199, 504–513. doi: 10.1016/j.actamat.2020.08.045.

Nygård, M.M., Fjellvåg, Ø.S., Sørby, M.H., Sakaki, K., Ikeda, K., Armstrong, J., Vajeeston, P., Sławiński, W.A., Kim, H., Machida, A. et al. (2021). The average and local structure of TiVCrNbD x (x = 0, 2.2, 8) from total scattering and neutron spectroscopy. *Acta Mater.* 205, 116496. doi: 10.1016/j.actamat.2020.116496.

Otto, F., Yang, Y., Bei, H., George, E.P. (2013). Relative effects of enthalpy and entropy on the phase stability of equiatomic high-entropy alloys. *Acta Mater.*, 61, 2628–2638. doi: 10.1016/j.actamat.2013.01.042.

Pineda-Romero, N. and Zlotea, C. (2022). Uncovering the effect of Al addition on the hydrogen storage properties of the ternary TiVNb alloy. *Materials*, 15, 7974. doi: 10.3390/ma15227974.

Pineda-Romero, N., Witman, M., Stavila, V., Zlotea, C. (2022). The effect of 10 at.% Al addition on the hydrogen storage properties of the $Ti_{0.33}V_{0.33}Nb_{0.33}$ multi-principal element alloy. *Intermetallics*, 146, 107590. doi: 10.1016/j.intermet.2022.107590.

Pineda-Romero, N., Witman, M., Harvey, K., Stavila, V., Nassif, V., Elkaïm, E., Zlotea, C. (2023). Large destabilization of (TiVNb)-based hydrides via (Al, Mo) addition: Insights from experiments and data-driven models. *ACS Appl. Energy Mater.*, 6, 12560–12572. doi: 10.1021/acsaem.3c02696.

Ranganathan, S. (2003). Alloyed pleasures: Multimetallic cocktails. *Curr. Sci*, 85(5), 1404–1406.

Sahlberg, M., Karlsson, D., Zlotea, C., Jansson, U. (2016). Superior hydrogen storage in high entropy alloys. *Sci. Rep.*, 6, 36770. doi: 10.1038/srep36770.

Sarac, B., Zadorozhnyy, V., Berdonosova, E., Ivanov, Y.P., Klyamkin, S., Gumrukcu, S., Sarac, A.S., Korol, A., Semenov, D., Zadorozhnyy, M. et al. (2020). Hydrogen storage performance of the multi-principal-component CoFeMnTiVZr alloy in electrochemical and gas–solid reactions. *RSC Adv.*, 10, 24613–24623. doi: 10.1039/D0RA04089D.

Schlapbach, L. and Züttel, A. (2001). Hydrogen-storage materials for mobile applications. *Nature*, 414, 353–358. doi: 10.1038/35104634.

Serrano, L., Moussa, M., Yao, J.-Y., Silva, G., Bobet, J.-L., Santos, S.F., Cardoso, K.R. (2023). Development of Ti-V-Nb-Cr-Mn high entropy alloys for hydrogen storage. *J. Alloys Compd.*, 945, 169289. doi: 10.1016/j.jallcom.2023.169289.

Shahi, R.R., Gupta, A.K., Kumari, P. (2022). Perspectives of high entropy alloys as hydrogen storage materials. *Int. J. Hydrog. Energy*, S0360319922007297. doi: 10.1016/j.ijhydene.2022.02.113.

Shen, H., Zhang, J., Hu, J., Mao, Y., Xiao, H., Zhou, X., Zu, X. (2019). A novel TiZrHfMoNb high-entropy alloy for solar thermal energy storage. *Nanomaterials*, 9, 248. doi: 10.3390/nano9020248.

Shen, H., Hu, J., Li, P., Huang, G., Zhang, J., Zhang, J., Mao, Y., Xiao, H., Zhou, X., Zu, X. et al. (2020). Compositional dependence of hydrogenation performance of Ti-Zr-Hf-Mo-Nb high-entropy alloys for hydrogen/tritium storage. *J. Mater. Sci. Technol.*, 55, 116–125. doi: 10.1016/j.jmst.2019.08.060.

Silva, B.H., Zlotea, C., Champion, Y., Botta, W.J., Zepon, G. (2021). Design of TiVNb-(Cr, Ni or Co) multicomponent alloys with the same valence electron concentration for hydrogen storage. *J. Alloys Compd.*, 865, 158767. doi: 10.1016/j.jallcom.2021.158767.

Singh, S., Wanderka, N., Murty, B.S., Glatzel, U., Banhart, J. (2011). Decomposition in multi-component AlCoCrCuFeNi high-entropy alloy. *Acta Mater.*, 59, 182–190. doi: 10.1016/j.actamat.2010.09.023.

Sleiman, S. and Huot, J. (2022). Microstructure and first hydrogenation properties of TiHfZrNb1−xV1+x Alloy for x = 0, 0.1, 0.2, 0.4, 0.6 and 1. *Molecules*, 27, 1054. doi: 10.3390/molecules27031054.

Stein, F. and Leineweber, A. (2021). Laves phases: A review of their functional and structural applications and an improved fundamental understanding of stability and properties. *J. Mater. Sci.*, 56, 5321–5427. doi: 10.1007/s10853-020-05509-2.

Strozi, R.B., Leiva, D.R., Huot, J., Botta, W.J., Zepon, G. (2021a). Synthesis and hydrogen storage behavior of Mg–V–Al–Cr–Ni high entropy alloys. *Int. J. Hydrog. Energy*, 46, 2351–2361. doi: 10.1016/j.ijhydene.2020.10.106.

Strozi, R.B., Leiva, D.R., Zepon, G., Botta, W.J., Huot, J. (2021b). Effects of the chromium content in (TiVNb)100−xCrx body-centered cubic high entropy alloys designed for hydrogen storage applications. *Energies*, 14, 3068. doi: 10.3390/en14113068.

Strozi, R.B., Silva, B.H., Leiva, D.R., Zlotea, C., Botta, W.J., Zepon, G. (2023a). Tuning the hydrogen storage properties of Ti-V-Nb-Cr alloys by controlling the Cr/(TiVNb) ratio. *J. Alloys Compd.*, 932, 167609. doi: 10.1016/j.jallcom.2022.167609.

Strozi, R.B., Witman, M., Stavila, V., Cizek, J., Sakaki, K., Kim, H., Melikhova, O., Perrière, L., Machida, A., Nakahira, Y. et al. (2023b). Elucidating primary degradation mechanisms in high-cycling-capacity, compositionally tunable high-entropy hydrides. *ACS Appl. Mater. Interfaces*, 15, 38412–38422. doi: 10.1021/acsami.3c05206.

Tarhan, C. and Çil, M.A. (2021). A study on hydrogen, the clean energy of the future: Hydrogen storage methods. *J. Energy Storage*, 40, 102676. doi: 10.1016/j.est.2021.102676.

Tsai, M.-H. and Yeh, J.-W. (2014). High-entropy alloys: A critical review. *Mater. Res. Lett.*, 2, 107–123. doi: 10.1080/21663831.2014.912690.

Usman, M.R. (2022). Hydrogen storage methods: Review and current status. *Renew. Sustain. Energy Rev.*, 167, 112743. doi: 10.1016/j.rser.2022.112743.

Wang, W.H. (2014). High-entropy metallic glasses. *JOM*, 66, 2067–2077. doi: 10.1007/s11837-014-1002-3.

Wang, X., Tan, S., Yang, X.-Q., Hu, E. (2020). Pair distribution function analysis: Fundamentals and application to battery materials. *Chin. Phys. B*, 29, 028802. doi: 10.1088/1674-1056/ab6656.

Wei, T.Y., Lim, K.L., Tseng, Y.S., Chan, S.L.I. (2017). A review on the characterization of hydrogen in hydrogen storage materials. *Renew. Sustain. Energy Rev.*, 79, 1122–1133. doi: 10.1016/j.rser.2017.05.132.

Williams, D.E. and Levingston, H.L. (1961). ARC melting in the tungsten electrode furnace. US Department Of Energy, Office of Scientific and Technical Information [Online]. Available at: https://www.osti.gov/servlets/purl/4820818 [Accessed 24 April 2014].

Witman, M., Ling, S., Grant, D.M., Walker, G.S., Agarwal, S., Stavila, V., Allendorf, M.D. (2020). Extracting an empirical intermetallic hydride design principle from limited data via interpretable machine learning. *J. Phys. Chem. Lett.*, 11, 40–47. doi: 10.1021/acs.jpclett.9b02971.

Witman, M., Ek, G., Ling, S., Chames, J., Agarwal, S., Wong, J., Allendorf, M.D., Sahlberg, M., Stavila, V. (2021). Data-driven discovery and synthesis of high entropy alloy hydrides with targeted thermodynamic stability. *Chem. Mater.*, 33, 4067–4076. doi: 10.1021/acs.chemmater.1c00647.

Yadav, T.P., Kumar, A., Verma, S.K., Mukhopadhyay, N.K. (2022). High-entropy alloys for solid hydrogen storage: Potentials and prospects. *Trans. Indian Natl. Acad. Eng.*, 7, 147–156. doi: 10.1007/s41403-021-00316-w.

Yang, X. and Zhang, Y. (2012). Prediction of high-entropy stabilized solid-solution in multi-component alloys. *Mater. Chem. Phys.*, 132, 233–238. doi: 10.1016/j.matchemphys.2011.11.021.

Yang, X., Chen, S.Y., Cotton, J.D., Zhang, Y. (2014). Phase stability of low-density, multiprincipal component alloys containing aluminum, magnesium, and lithium. *JOM*, 66, 2009–2020. doi: 10.1007/s11837-014-1059-z.

Yang, S., Lu, J., Xing, F., Zhang, L., Zhong, Y. (2020). Revisit the VEC rule in high entropy alloys (HEAs) with high-throughput CALPHAD approach and its applications for material design. A case study with Al–Co–Cr–Fe–Ni system. *Acta Mater.* 192, 11–19. doi: 10.1016/j.actamat.2020.03.039.

Yang, F., Wang, J., Zhang, Y., Wu, Z., Zhang, Z., Zhao, F., Huot, J., Grobivć Novaković, J., Novaković, N. (2022). Recent progress on the development of high entropy alloys (HEAs) for solid hydrogen storage: A review. *Int. J. Hydrog. Energy*, 47, 11236–11249. doi: 10.1016/j.ijhydene.2022.01.141.

Ye, Y.F., Wang, Q., Lu, J., Liu, C.T., Yang, Y. (2016). High-entropy alloy: Challenges and prospects. *Mater. Today*, 19, 349–362. doi: 10.1016/j.mattod.2015.11.026.

Yeh, J.-W. (2013). Alloy design strategies and future trends in high-entropy alloys. *JOM*, 65, 1759–1771. doi: 10.1007/s11837-013-0761-6.

Yeh, J.-W., Chen, S.-K., Lin, S.-J., Gan, J.-Y., Chin, T.-S., Shun, T.-T., Tsau, C.-H., Chang, S.-Y. (2004). Nanostructured high-entropy alloys with multiple principal elements: Novel alloy design concepts and outcomes. *Adv. Eng. Mater.*, 6, 299–303. doi: 10.1002/adem.200300567.

Zadorozhnyy, V., Sarac, B., Berdonosova, E., Karazehir, T., Lassnig, A., Gammer, C., Zadorozhnyy, M., Ketov, S., Klyamkin, S., Eckert, J. (2020). Evaluation of hydrogen storage performance of ZrTiVNiCrFe in electrochemical and gas-solid reactions. *Int. J. Hydrog. Energy*, 45, 5347–5355. doi: 10.1016/j.ijhydene.2019.06.157.

Zepon, G., Leiva, D.R., Strozi, R.B., Bedoch, A., Figueroa, S.J.A., Ishikawa, T.T., Botta, W.J. (2018). Hydrogen-induced phase transition of $MgZrTiFe_{0.5}Co_{0.5}Ni_{0.5}$ high entropy alloy. *Int. J. Hydrog. Energy*, 43, 1702–1708. doi: 10.1016/j.ijhydene.2017.11.106.

Zepon, G., Silva, B.H., Zlotea, C., Botta, W.J., Champion, Y. (2021). Thermodynamic modelling of hydrogen-multicomponent alloy systems: Calculating pressure-composition-temperature diagrams. *Acta Mater.*, 215, 117070. doi: 10.1016/j.actamat.2021.117070.

Zhang, Y. and Zhou, Y.J. (2007). Solid solution formation criteria for high entropy alloys. *Mater. Sci. Forum*, 561–565, 1337–1339.

Zhang, Y., Zhou, Y.J., Lin, J.P., Chen, G.L., Liaw, P.K. (2008). Solid-solution phase formation rules for multi-component alloys. *Adv. Eng. Mater.*, 10, 534–538. doi: 10.1002/adem.200700240.

Zhang, Y., Yang, X., Liaw, P.K. (2012). Alloy design and properties optimization of high-entropy alloys. *JOM*, 64, 830–838. doi: 10.1007/s11837-012-0366-5.

Zhang, Y., Zuo, T.T., Tang, Z., Gao, M.C., Dahmen, K.A., Liaw, P.K., Lu, Z.P. (2014). Microstructures and properties of high-entropy alloys. *Prog. Mater. Sci.*, 61, 1–93. doi: 10.1016/j.pmatsci.2013.10.001.

Zhang, C., Wu, Y., You, L., Cao, X., Lu, Z., Song, X. (2019). Investigation on the activation mechanism of hydrogen absorption in TiZrNbTa high entropy alloy. *J. Alloys Compd.*, 781, 613–620. doi: 10.1016/j.jallcom.2018.12.120.

Zlotea, C., Sow, M.A., Ek, G., Couzinié, J.-P., Perrière, L., Guillot, I., Bourgon, J., Møller, K.T., Jensen, T.R., Akiba, E. et al. (2019). Hydrogen sorption in TiZrNbHfTa high entropy alloy. *J. Alloys Compd.*, 775, 667–674. doi: 10.1016/j.jallcom.2018.10.108.

Zlotea, C., Bouzidi, A., Montero, J., Ek, G., Sahlberg, M. (2022). Compositional effects on the hydrogen storage properties in a series of refractory high entropy alloys. *Front. Energy Res.*, 10, 991447. doi: 10.3389/fenrg.2022.991447.

Züttel, A. (2004). Hydrogen storage methods. *Naturwissenschaften*, 91, 157–172. doi: 10.1007/s00114-004-0516-x.

7

Regenerable Hydrides

Carlos A. CASTILLA-MARTINEZ[1], **Jean-Louis BOBET**[2] **and Umit B. DEMIRCI**[1]

[1]*IEM, CNRS, Université de Montpellier, France*

[2]*ICMCB, CNRS, Université de Bordeaux, Pessac, France*

7.1. Introduction

A regenerable hydride is by definition made up of light elements (such as Li, Na, Mg, B and N), which, after dehydrogenation, form one or more very stable by-products that cannot be easily rehydrogenated under molecular hydrogen pressure. A chemical process is required to transform the by-product(s) into the starting hydride. These hydrides are rich in hydrogen, and, compared with other materials developed for hydrogen storage (porous, metallic or complex hydrides), they have the highest densities (gravimetric and volumetric) of hydrogen, particularly at room temperature (Rivard et al. 2019).

One example is ammonia borane (NH_3BH_3), which has a relatively low molar mass of 30.87 g/mol (Demirci 2020). It has a gravimetric density of 196 g(H_2)/kg(NH_3BH_3) and, because it has a lower density than water (0.78 kg/L), has a specific gravity of 152 g(H_2)/L(NH_3BH_3). By comparison, molecular hydrogen gas pressurized to 700 bar has a density of 40 g(H_2)/L(H_2), which is equivalent to storing 5 kg of hydrogen in a 125-L tank (Zheng et al. 2021).

However, with ammonia borane, the same tank could theoretically store 19 kg of hydrogen. When presented in this way, ammonia borane and, more broadly, regenerable hydrides are exceptional for hydrogen storage.

Hydrogen Storage,
coordinated by Patricia DE RANGO and Fermin CUEVAS. © ISTE Ltd 2025.

The strength of regenerable hydrides lies in the nature of their constituent atoms: hydrogen, lithium, boron, sodium, magnesium, aluminum and potassium (Table 7.1).

First, the relative lightness of these atoms and their ability to bond to one to four hydrogen atoms mean that, for regenerable hydrides, hydrogen densities are generally in excess of 100 g(H_2)/kg(hydride) and therefore high (Dematteis et al. 2022).

Second, hydrogen is the most electronegative atom represented here. M-H bonds are then either ionic or iono-covalent, and this significantly impacts the stability of regenerable hydrides (Mohtadi and Orimo 2017). Under an inert, anhydrous atmosphere and at temperatures below 30°C, regenerable hydrides are stable.

On the other hand, they hydrolyze and release molecular hydrogen in the presence of protic hydrogen atoms ($H^{\delta+}$ from H_2O or CH_3OH) (Ouyang et al. 2022), and/or they destabilize under the effect of temperature, releasing molecular hydrogen (Salman et al. 2022). Destabilization, in the context of these hydrides, is actually an advantage. It enables molecular hydrogen to be produced under mild conditions (e.g. 0–40°C for the hydrolysis of the sodium borohydride $NaBH_4$), in any case milder than for high-temperature processes such as steam reforming (Martino et al. 2021).

	H	Li	B	Na	Mg	Al	K
N	1	2	2	3	3	3	4
Z	1	3	5	11	12	13	19
A_r	1.008	6.94	10.811	22.990	24.305	26.982	39.098
χ	2.2	0.98	2.04	0.93	1.31	1.61	0.82

Table 7.1. *Constituent atoms of regenerable hydrides, with information on their period (n) in the periodic table, atomic number (Z), relative atomic masses (A_r in g/mol), and electronegativity according to Pauling (χ)*

There is, however, a counterpart to the advantageous features discussed above. The dehydrogenation of regenerable hydrides is an exothermic process. For example, the enthalpy of dehydrogenation is –277 kJ/mol in the case of the hydrolysis of magnesium hydride MgH_2 (Chen et al. 2020). This implies that energy input (+277 kJ/mol) is required to close the hydrogen cycle with MgH_2. This is technically impossible for an on-board application, such as automobiles, which require hydrogen pressures below 100 bar and temperatures below 100°C (Li et al.

2019c). Reversible storage is therefore out of the question. Thé only possible way to regenerate this type of hydride is by rehydrogenation via a single- or multi-stage chemical process, which can only be carried out at an industrial site (Figure 7.1). These are known as regenerable hydrides.

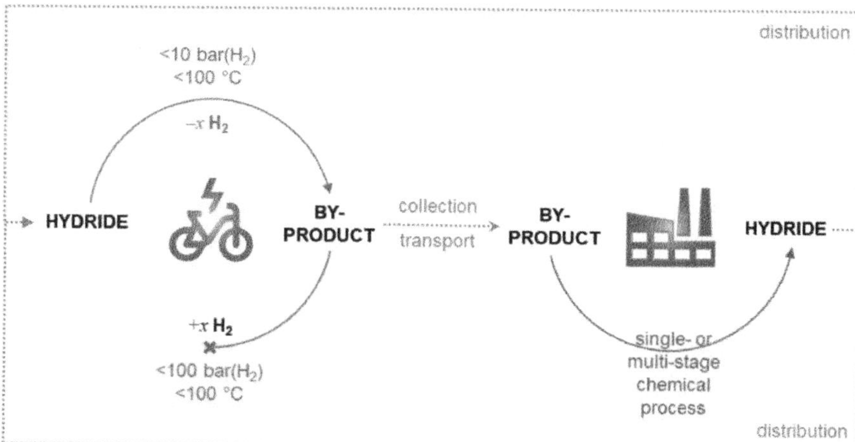

Figure 7.1. *Hydrogen cycle using a regenerable hydride. For a color version of this figure, see www.iste.co.uk/derango-cuevas/hydrogen.zip*

The last 20 years of research have seen, on the one hand, the emergence of *new* regenerable hydrides (such as ammonia borane and its derivatives) and, on the other hand, the rise of a large number of studies into the dehydrogenation of the *older* hydrides discovered in the middle of the last century (MgH_2 and $LiBH_4$, for example). Unfortunately, far less effort has been devoted to developing prototypes, such as those with hydrolysis reactors, and to developing efficient and economically viable processes for hydride regeneration.

In fact, the level of technological maturity remains relatively low. This is particularly true of older hydrides, even though they offer the greatest potential for technological maturation. With this in mind, the aim of this chapter is twofold: (i) to present the advantages and disadvantages of regenerable hydrides with the potential for hydrogen storage, on the basis of the most mature achievements; (ii) to highlight the main obstacles and challenges to overcome to achieve technological maturity.

7.2. Alkaline hydrides

7.2.1. *MH hydrides with M for Li, Na or K*

The simplest hydrides are those consisting of a hydrogen atom $H^{\delta-}$ and one of the most electropositive metals, namely, alkali metals. This leads to the alkaline hydrides MH (with M for Li, Na and K). Rubidium and cesium are not considered here for two reasons: (i) they are too heavy (85.5 and 132.91 g/mol, respectively), which limits their hydrogen storage capacity (e.g. 12 g(H_2)/kg(RbH) and 8 g(H_2)/kg(CsH)), and (ii) the hydrides RbH and CsH are not commercial due to their instability.

The lithium hydride LiH is the lightest hydride (7.95 g/mol). It is a crystalline ionic solid (NaCl-type: face-centered cubic structure, space group $Fm\bar{3}m$). Its density is 0.775 kg/L. It has a gravimetric hydrogen density of 127 g(H_2)/kg(LiH) and a bulk density of 104 g(H_2)/L(LiH). LiH is produced by the reaction of lithium metal with molecular hydrogen (see [7.1]). The reaction is exothermic, with an enthalpy of formation of –91 kJ/mol. LiH is synthesized at near-ambient temperatures, but yields and kinetics are greatly improved by increasing the temperature (Zou et al. 2023). LiH is thermally stable, with a melting temperature of 688°C. It is chemically stable when stored in an inert atmosphere. In the presence of water, it is destabilized and hydrolyzes spontaneously (equation [7.2]), and the enthalpy of the reaction is –132 kJ/mol (Pretzel et al. 1960; Montgomery 1973). In other words, LiH has all the characteristics of a potential hydrogen carrier, specifically a regenerable hydride capable of producing molecular hydrogen by hydrolysis at room temperature:

$$2Li + H_2 \rightarrow 2LiH \qquad [7.1]$$

$$LiH + H_2O \rightarrow LiOH + H_2 \qquad [7.2]$$

Sodium hydride NaH (24 g/mol) is an ionic solid with the same crystalline structure as LiH. Its density is 1.39 kg/L. It has a gravimetric hydrogen density of 42 g(H_2)/kg(NaH) and a bulk density of 58 g(H_2)/L(NaH). NaH is therefore less hydrogen-rich than LiH. NaH is produced by the reaction of metallic sodium with molecular hydrogen (see [7.3]), and the enthalpy of the reaction is –56 kJ/mol. NaH is thermally and chemically stable in an inert, dry atmosphere. However, it reacts rapidly with water by hydrolysis (see [7.4]), and the reaction is as exothermic as with LiH, with an enthalpy of –132 kJ/mol (Messer et al. 1955; Addison and Manning 1964). In fact, the reaction of NaH with water is more violent than for LiH because the reaction is total.

As discussed in the next section, with LiH, complete hydrolysis is prevented by the formation of oxide, which acts as a reaction inhibitor (Kong et al. 2003). With NaH, the risk of spontaneous ignition in the presence of water is significant. The reactivity of NaH is a well-known feature, particularly in organic chemistry, where NaH is used as a strong base and deprotonation agent (Gawley and Hennings 2006; Hesek et al. 2009; Verma 2010; Czerwiński and Furman 2020). With such properties, NaH is not a hydride to be considered for hydrogen storage. For example, in 2021, we experimented with the hydrolysis of NaH and found that it reacts violently with water, as well as igniting; the difficulty of controlling the reaction led us to abandon this study:

$$2Na + H_2 \rightarrow 2NaH \quad [7.3]$$

$$NaH + H_2O \rightarrow NaOH + H_2 \quad [7.4]$$

Potassium hydride KH is very similar to NaH. It is an ionic solid. It is isostructural to LiH and NaH. Its density is 1.43 kg/L, and it is less rich in hydrogen (gravimetric density of 25 $g(H_2)/kg(KH)$ and bulk density of 36 $g(H_2)/L(KH)$). KH is synthesized by the reaction of potassium metal with hydrogen (see [7.5]); the enthalpy of formation is –58 kJ/mol. KH is thermally and chemically stable under anhydrous conditions and reacts violently with water by hydrolysis (see [7.6]). KH decomposes via thermolysis at approximately 350°C (Ares et al. 2016). KH is a superbase that is more basic than NaH (Davy 1808; Brown 1974; Pinnick 1983; Taber and Nelson 2006). To our knowledge, KH has never been published in open literature as to its potential hydrogen storage capabilities:

$$2K + H_2 \rightarrow 2KH \quad [7.5]$$

$$KH + H_2O \rightarrow KOH + H_2 \quad [7.6]$$

To summarize, NaH and KH, because of their high reactivity with water, are not safe hydrides to consider for hydrogen storage. LiH is more interesting, partly because it is less reactive than its counterparts and it has a higher hydrogen density. The following sections focus on LiH hydrolysis and regeneration.

7.2.2. *LiH hydrolysis and hydrogen production*

The hydrolysis of LiH is actually a more complex reaction than that indicated by equation [7.2]. Li_2O is also formed (see [7.7]) on the surface of LiH grains (reaction enthalpy of –170 kJ/mol). This passivation layer reduces the reactivity of Li to water. The water then diffuses through this layer and reacts not only with Li_2O

(enthalpy of reaction −124 kJ/mol) to form lithium hydroxide (LiOH) (see [7.8]) but also with LiH to produce additional Li_2O. The hydrolysis of LiH under these conditions can be described by equation [7.9], where *w*, *x*, *y* and *z* represent the stoichiometric coefficients of each compound normalized to one mole of water (Phillips et al. 1995).

Ultimately, the surface of a LiH grain consists of a layer of Li_2O, a layer of LiOH, and also an outer, water-contacting layer of the lithium hydroxide monohydrate $LiOH{\cdot}H_2O$ (Figure 7.2) (Haertling et al. 2007; Awbery et al. 2008; Garlea et al. 2017; Weber et al. 2018):

$$2LiH + H_2O \rightarrow Li_2O + 2H_2 \quad [7.7]$$

$$Li_2O + H_2O \rightarrow 2LiOH \quad [7.8]$$

$$wLiH + H_2O \rightarrow xLiOH + yLi_2O + zH_2 \quad [7.9]$$

The complete hydrolysis of LiH therefore requires an excess of water, up to 20 times the stoichiometric proportion (Pitcher and Kavarnos 1997). With such an excess of water, the gravimetric hydrogen density of the LiH-$20H_2O$ couple is only 5.5 g(H_2)/kg(LiH-$20H_2O$) or 0.55% wt.

Water can also be recovered from air humidity by diffusion through a Nafion-type membrane. In such a configuration, the weight of the water no longer needs to be taken into account. In other words, the hydrogen density of LiH alone is taken into account. This is the approach that has been considered with a view to developing a hydrogen generator for a low-power fuel cell (250 mW) and for long-term use (3–5 years).

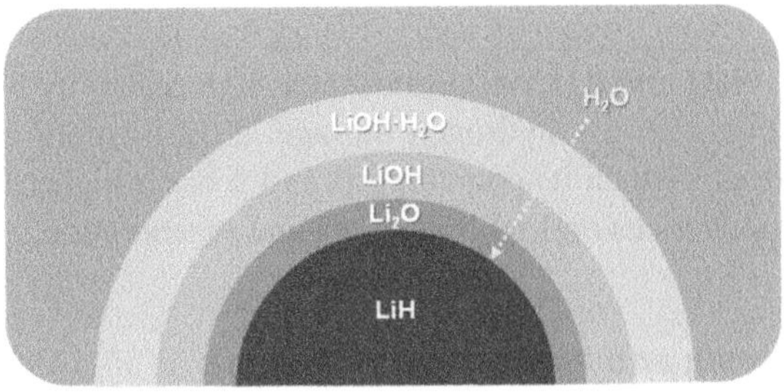

Figure 7.2. *Schematic representation of the product layers formed during hydrolysis of a LiH grain at room temperature and atmospheric pressure. Adapted from Haertling et al. (2007). For a color version of this figure, see www.iste.co.uk/derango-cuevas/hydrogen.zip*

However, hydrogen production failed after 2 months (and after hydrolysis of 48.8 g LiH) because of the formation of a LiOH passivation layer (Strawser et al. 2014). Yields of 90% have been reported for hydrolysis with steam but at temperatures of 150–500°C (Khzouz et al. 2020). Moreover, under a controlled atmosphere, with a relative humidity below 0.04%, LiH is relatively stable at room temperature; very little LiOH is formed, suggesting prospects for storage under these conditions (Guichard et al. 2015).

The hydrolysis of LiH has been the subject of dozens of studies, most of which have sought to understand whether LiOH or Li_2O is formed first (Haertling et al. 2006). The rate of hydrogen production is controlled by the diffusion of water (or unidentified intermediates) through the passivation layers formed on the surface of the LiH grains.

From a technological point of view, the use of water in its liquid state is compromised by the need for excess water. The alternative is to vaporize the water. However, this approach is not yet mature enough and requires further efforts to improve hydrogen production yields and kinetics. The main stumbling block is therefore the limitation induced by the formation of passivation layers due to Li_2O and LiOH.

7.2.3. *LiH regeneration*

Although the hydrolysis of LiH is a rather complex process, its complete hydrolysis implies that all the Li_2O formed (see [7.7]) is converted to LiOH at the end of the process (see [7.8]). Even if traces of Li_2O remain, they are easily converted to LiOH and/or $LiOH \cdot H_2O$ by adding water (Awbery et al. 2008).

LiOH is one of the end products of lithium extraction from ores (Li et al. 2019b). Another end product is lithium carbonate Li_2CO_3 (Talens Peiró et al. 2013), which incidentally forms spontaneously by the reaction of LiOH and carbon dioxide (see [7.10]) in aqueous solution (Swain 2017). Note also that LiOH (as well as $LiOH \cdot H_2O$) and Li_2CO_3 are the two compounds targeted for recovering lithium from end-of-life lithium-ion batteries (Liu et al. 2021a; Maria et al. 2022).

To regenerate LiH, we need to consider only the abovementioned contexts. The process to be considered is then, in the first step, the production of lithium metal from LiOH and, in the second step, that of LiH (see [7.1]). The first step (Sadoway 1998; Laude et al. 2010; Sato and Takeda 2013; Tran and Luong 2015) could be achieved by high-temperature carbon reduction (see [7.11]), high-temperature metal reduction (see [7.12]), or molten-salt electrolysis based on the LiCl-KCl eutectic (see [7.13]):

$$2LiOH + CO_2 \rightarrow Li_2CO_3 + H_2O \qquad [7.10]$$

$$6LiOH + 2C \rightarrow 2Li + 2Li_2CO_3 + 2H_2 \qquad [7.11]$$

$$2LiOH + Al \rightarrow Li + LiAlO_2 + H_2 \qquad [7.12]$$

$$4LiOH \rightarrow 4Li + 2H_2O + O_2 \qquad [7.13]$$

To our knowledge, there are no studies in open literature that have targeted the direct synthesis of LiH from LiOH or from Li_2CO_3. These are avenues that could be explored.

7.3. Magnesium hydrides

7.3.1. *Introduction*

Magnesium from the word *magnesia* (a district of Thessaly in Greece, where the ore magnesia alba – magnesium oxide – was first discovered) is the eighth most abundant element in the Earth's crust (≈2.5% wt.). In its natural state, it occurs as carbonates (such as magnesite $MgCO_3$ or dolomite $CaMg(CO_3)_2$), silicates (such as talc $Mg_3Si_4O_{10}(OH)_2$ or forsterite (Mg_2SiO_4)), hydroxide (brucite $Mg(OH)_2$), chloride (carnallite $KMgCl_3 \cdot 6H_2O$) and sulfate (kieserite $MgSO_4 \cdot H_2O$).

Since its discovery in 1775 by Joseph Black, magnesium and its alloys have been considered to have great application potential in the aerospace, transportation, electronics, biomedical and energy sectors because of their excellent physical and chemical properties (Mordike and Ebert 2001; Gwynne and Lyon 2007; Kulekci 2008; Luo 2013; Kight 2015). Table 7.2 shows the main physicochemical characteristics of magnesium that justify the diversity of its applications (Ghali 2011).

Magnesium is often considered a reversible hydrogen storage material. It has one of the highest gravimetric hydrogen densities at 76 $g(H_2)/kg(MgH_2)$. The hydrogen sorption–desorption reaction is reversible (see [7.14]):

$$Mg + H_2 \leftrightarrow MgH_2 \qquad [7.14]$$

An equilibrium pressure of 1 bar is reached at 287°C, which considerably limits its application. However, even if the very high stability of MgH_2 is a major drawback, it remains a candidate for consideration, particularly in view of its moderate cost ($2–4/kg Mg) and its abundance in the Earth's crust (as mentioned above).

Property	Value
Visual appearance	Silver white
Oxidation state	2
Molar mass (g/mol)	24.304
Density at 20°C (g/cm^3)	1.738
Crystal structure	Compact hexagonal
Metal radius (Å)	1.6
Melting point (°C)	649.5
Boiling point (°C)	1091
Thermal conductivity (W/m·K)	155
Electrical resistivity (nΩ/cm)	44.6

Table 7.2. *Some properties of pure magnesium according to (Ghali 2011)*

7.3.2. MgH_2 hydride

Unlike other hydrides, hydrogen absorption by magnesium results in a structural change. The compact hexagonal structure of magnesium becomes quadratic with a rutile-like structure (Figure 7.3).

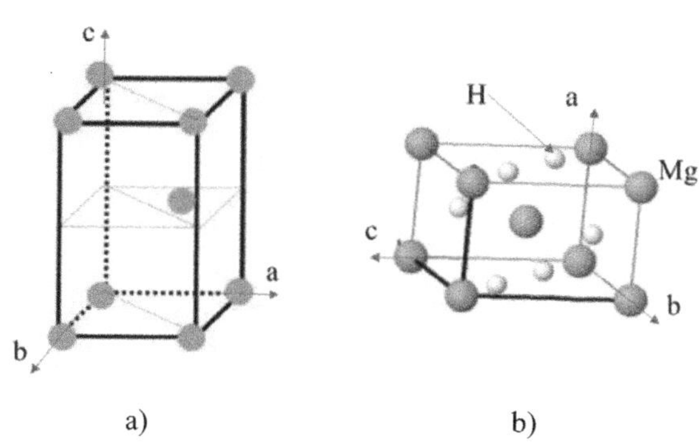

Figure 7.3. *(a) Crystal structure of hexagonal magnesium and (b) quadratic magnesium hydride. For a color version of this figure, see www.iste.co.uk/derango-cuevas/hydrogen.zip*

The hydrolysis reaction of MgH_2 (see [7.15]) is less exothermic and rapid than that of the metal (–277 vs. –354 kJ/mol) but remains spontaneous. Hydrogen is thus extracted from both water and hydride. Considering water, this gives a gravimetric hydrogen density of 64 $g(H_2)/kg(MgH_2)$, but without considering the mass of water, this density increases to 152 $g(H_2)/kg(MgH_2)$:

$$MgH_2 + 2H_2O \rightarrow Mg(OH)_2 + 2H_2 \qquad [7.15]$$

According to the Pourbaix diagram, the hydrolysis reaction can be carried out at neutral pH. However, the formation of a hydroxide layer on the surface of the particles (linked to the increase in pH of the solution by OH^- ion formation) prevents the reaction from proceeding. Controlling pH is therefore a key parameter in the progress of the reaction. An acidic solution increases the amount of Mg that reacts before the pH increases. However, another widely employed strategy does not require the use of acid.

This involves the use of a solution containing Cl^- chloride ions. A 0.6 M sodium chloride (NaCl) solution (simulated seawater) is often used as the electrolyte. This approach is considered completely harmless and available in virtually unlimited quantities. Figure 7.4 illustrates the mechanism that leads to the destruction of the $Mg(OH)_2$ passive layer (Awad et al. 2016; Al Bacha et al. 2019; Pighin et al. 2020) when Cl^- ions are present in solution.

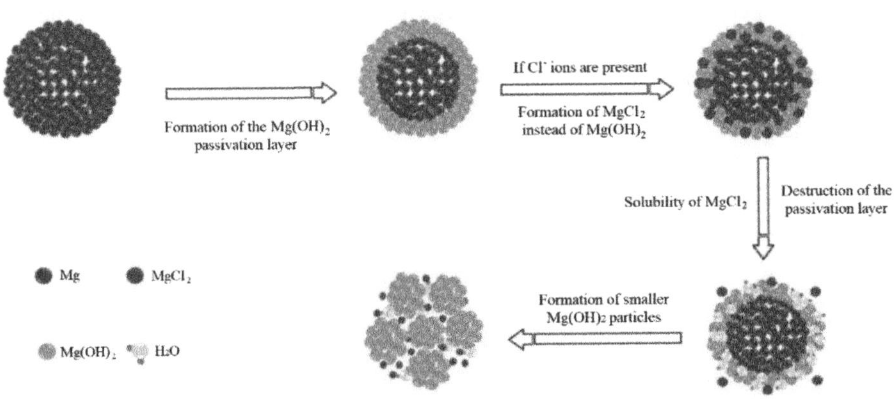

Figure 7.4. *Schematic diagram of the mechanism of destruction of the $Mg(OH)_2$ passive layer by chloride ions. Adapted from Awad et al. (2016). For a color version of this figure, see www.iste.co.uk/derango-cuevas/hydrogen.zip*

They form magnesium chloride ($MgCl_2$), which has a much higher solubility than $Mg(OH)_2$ (543 g/L in pure water at 25°C vs. 0.0096 g/L for $Mg(OH)_2$). In addition, the enthalpy of formation of chloride is greater than that of hydroxide (–924 vs. –641 kJ/mol) (Johnson and Gilliland 1961; Ropp 2013; National Institute of Standards and Technology 2021b, 2021a; American Elements 2022; ChemEurope 2022; Chemical Book 2022; Chemistry Learner 2022).

Thus, the formation of $MgCl_2$ particles within the hydroxide layer will create pathways for water to pass through the hydride, all the way to the core of the material (Tayeh et al. 2014).

However, the reaction kinetics remain low. It is therefore necessary to increase the reaction surface area between the water and hydride. The most commonly used method is mechanical grinding. This is a simple and effective technique involving particle fracture and welding.

In this respect, Huot et al. (2003) were the first to show that nanocrystalline MgH_2 magnesium hydride (obtained by mechanical grinding) exhibited significantly improved kinetics and hydrolysis rate in pure water compared with conventional hydride.

In brief, Kojima et al. (2004a) significantly improved the hydrolysis kinetics of MgH_2 with an alternative strategy of catalyzing the reaction with a $LiCoO_2$ mixture. The work of Grosjean et al. highlighted several key parameters influencing the hydrolysis reaction in the presence of Mg or MgH_2:

– The grinding protocol for powder activation can be optimized (duration, energy used, nature and size of grinding balls, etc.) (Grosjean et al. 2005; Grosjean and Roué 2006).

– The addition of nickel (or another transition metal) as an additive during grinding modifies the morphology of Mg particles and accelerates the hydrolysis reaction by galvanic coupling (Grosjean et al. 2006).

– The addition of salts as an additive during grinding modifies the morphology of Mg particles and accelerates the hydrolysis reaction because of their exothermic dissolution when they are brought into contact with water (Grosjean and Roué 2006).

Based on this work, numerous strategies for optimizing the hydrolysis reaction with Mg and/or MgH_2 involving different grinding protocols and additives of various kinds have been applied (Ouyang et al. 2012, 2013; Liu et al. 2014, 2015; Tayeh et al. 2014; Huang et al. 2017; Alasmar et al. 2018; Tan et al. 2019;

Yang et al. 2019; Hou et al. 2021; Yang et al. 2021; Zhong et al. 2021). In particular, the use of carbon (carbon black, graphite, carbon fibers, etc.) not only improves grinding efficiency but also creates a protective layer on the hydride surface that blocks oxidation. Figure 7.5 shows the effects of grinding with an additive (in this case, KCl) on the powder size and morphology.

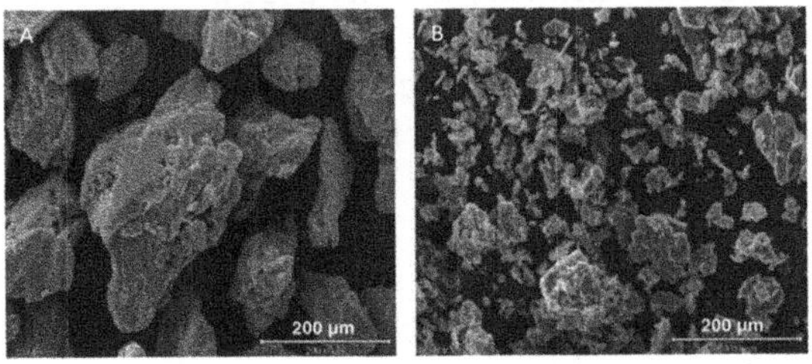

Figure 7.5. *Effect of grinding on the size and morphology of magnesium powders: (A) no grinding and (B) 30 min of grinding with 10 mol% KCl. Adapted from Grosjean et al. (2005)*

7.3.3. *Magnesium alloys*

The use of magnesium for certain applications (notably structural material) has been considerably limited by its poor mechanical properties and corrosion resistance. However, for some applications (e.g. sacrificial anodes), magnesium's corrosion resistance is an advantage (Song and Atrens 1999; Jia et al. 2005). An alternative is the use of magnesium alloys, which have developed widely, mainly with aluminum (Gusieva et al. 2015; Esmaily et al. 2017), to enhance both mechanical properties and corrosion resistance. These magnesium alloys are particularly characterized by an excellent strength-to-mass ratio. They are typically multiphase materials consisting of a dispersion of secondary phase particles (e.g. $Mg_{17}Al_{12}$-type intermetallics) in a magnesium matrix.

For hydrolysis, the presence of a phase whose nobility is different from that of magnesium and/or hydride will create galvanic coupling between the two phases. This should facilitate hydrolysis. Hydrolysis thus appears to be a route offering magnesium alloys that are not recycled (due to the cost of recycling) a "second life" (Uan et al. 2007, 2009a, 2009b; Yu et al. 2012; Kantürk Figen et al. 2015; Kantürk Figen and Coşkuner Filiz 2015). Two environmental issues could thus be resolved: (i) waste recycling and (ii) hydrogen production for clean energy. From an application point of view, it also makes it possible to transform a potentially dangerous product (Mg and/or its alloys are sensitive to humidity and can become flammable in particular) into a perfectly harmless product ($Mg(OH)_2$).

Uan et al. (2007) were the first to replace Mg with Mg-based alloy scrap for the production of H_2 by hydrolysis. To accelerate the hydrolysis of the AZ91D alloy (9% Al, 1% Zn and 90% Mg; D indicates high alloy purity) in a 5% NaCl solution, a catalyst consisting of a titanium (Ti) lattice coated with a layer of platinum (Pt) was used. The production of H_2 by the hydrolysis of AZ91D in the presence of a catalyst is superior to that of $NaBH_4$ (Amendola et al. 2000).

Two catalyst arrays (one Pt-coated Ti and the other AISI 304 stainless steel) were subsequently immersed in a molten Mg-based alloy (low-grade magnesium scraps [LGMS]) (Uan et al. 2009b). The Pt/Ti network produced more H_2 than did the AISI 304 stainless steel (28.2 ± 5.7 L/g in 50 min with Pt/Ti vs. 16.1 ± 7.8 L in 50 min with AISI304). The difference between the electrochemical potential of Pt and that of Mg (ΔE(Mg–Pt) = 2.55 V) is much greater than that between Mg and (steel-like) Fe (ΔE(Mg–Fe) = 1.29 V), ensuring a greater driving force for the LGMS corrosion reaction in aqueous solution. Yu et al. (2012) tested the hydrolysis of another Mg–Al alloy and reported a complete reaction in a 30% wt. solution of citric acid in seawater. The activation energy of the reaction increases with citric acid concentration due to the decreased mobility of H^+ cations.

In addition, the acidic strength of citric acid (the dissociation of the three H atoms) also affects the activation energy for H_2 generation. As a result, the dissociable H^+ ions in the concentrated aqueous citric acid solution cannot be used efficiently for H_2 generation, and the activation energy of the highly concentrated solution increases.

Kantürk Figen et al. (2015) ground a waste Mg alloy without additives. By extending the grinding time from 3 to 30 h, it was possible to switch from chips to small particles. The average grain size decreased from over 20–12 μm, and the specific surface area increased from 0.08 to 2.56 m^2/g. However, morphological analysis of shredded waste showed that prolonged shredding beyond 15 h did not significantly alter the particle size. In pure water, the alloy ground for 15 h (named W/Mg 15) did not produce hydrogen. The use of concentrated 1 M solutions of transition metal chlorides (e.g. $CoCl_2$, $CuCl_2$, $FeCl_3$, $MnCl_2$ and $NiCl_2$) and/or seawater (e.g. water from the Marmara or Aegean Sea with 2.3% and 3.7% wt. NaCl, respectively) improves hydrogen production.

Thus, hydrolysis in a 1 M $NiCl_2$ solution resulted in the best performance, with total hydrolysis occurring within 25 s (Kantürk Figen et al. 2015a). This was attributed to the increased ionic conductivity of the medium and the reduction in pH. The alloy used generated 70% of the theoretical H_2 volume in 260 s in Marmara seawater and 100% in 233 s in Aegean seawater (Kantürk Figen and Coşkuner Filiz

2015). This variation is directly correlated with the level of Cl^- ions present in both waters.

7.3.4. *Regeneration*

The product of the Mg hydrolysis reaction is a solid ($Mg(OH)_2$), which is sparingly soluble in water and has a wide range of applications. This material is particularly noteworthy for its antibacterial activity, non-toxicity and thermal stability. The latter property makes $Mg(OH)_2$ an additive (non-halogenated, non-polluting and non-corrosive) capable of delaying the ignition time of polymeric materials (Song et al. 2010; Ma et al. 2013). It is used in the food industry to regulate the acidity or alkalinity of foods. It is a food additive approved by the US Food and Drug Administration (FDA) and known by the code E528 (National Center for Biotechnology Information 2022).

The key step in upgrading magnesium hydroxide (which will first be transformed into MgO oxide) is to reduce it to the metallic state (Mg^0). The thermodynamics of the reduction of an oxide by a reducing agent is represented by Ellingham diagrams (Jeffes 2001). A given reducing agent (e.g. a metal or carbon) can reduce the oxides of all the other metals whose lines lie above it on the diagram (the free enthalpy of the reduction reaction is then negative). Magnesium oxide is classified as one of the most stable oxides, and there are few agents that can reduce it (Jeffes 2001). However, above 1,800°C (Figure 7.6), the reduction of magnesium oxide by carbon becomes thermodynamically favorable: this is carbothermal reduction (see [7.16]) (Chubukov et al. 2016). To date, this is the only possible route for the reduction of Mg hydroxide (which must first be converted to MgO):

$$Mg(OH)_2 + C \leftrightarrow Mg + CO + H_2O \qquad [7.16]$$

The stoichiometric amount of graphite (C) required is 21% of the mass of $Mg(OH)_2$. However, to shift the reaction to the right (and ensure that the limiting reagent is hydroxide), the addition of excess graphite is preferable. Hu et al. (2013) have shown that MgO is reduced to Mg in the presence of graphite if the experimental conditions (T and P) verify equation [7.17]:

$$T \geq 1199.2\ P^{0.0449} \qquad [7.17]$$

Theoretically, $Mg(OH)_2$ or MgO can therefore be reduced to Mg at temperatures below 1,800°C (Figure 7.6), provided that we work at pressures below atmospheric pressure. Other techniques, such as spark plasma sintering (SPS), are also being investigated. However, the increase in pressure (in SPS, the system can be considered closed) induced by the formation of CO and/or CO_2 can be detrimental (see [7.18]):

$$xMgO + C \rightarrow xMg + CO_x \quad 1 \leq x \leq 2 \qquad [7.18]$$

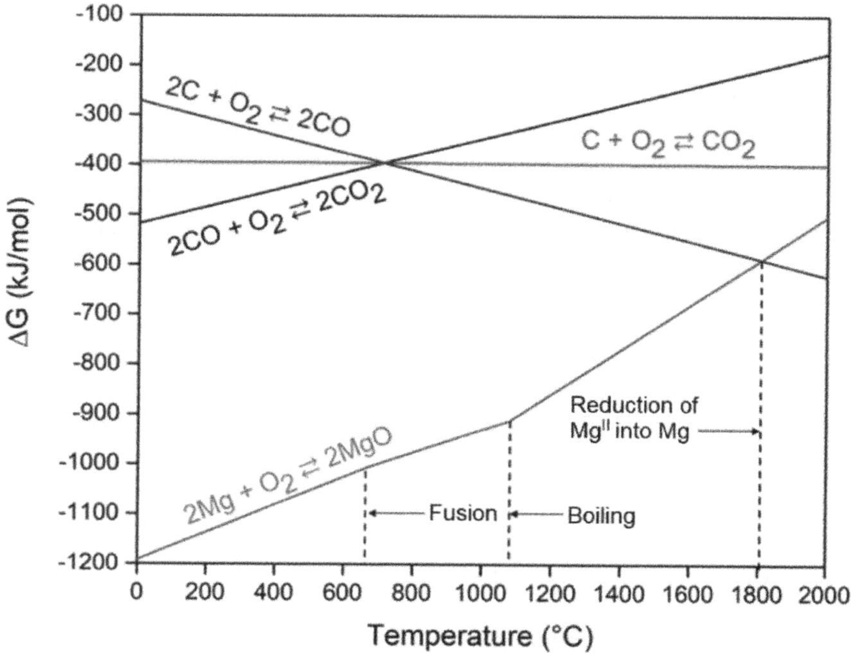

Figure 7.6. *Ellingham diagram for magnesium and carbon. Adapted from Jeffes (2001). For a color version of this figure, see www.iste.co.uk/derango-cuevas/hydrogen.zip*

Finally, to develop the "greenest" solution possible, it is conceivable to use a solar furnace that can easily reach temperatures in excess of 1,800°C (Berro et al. 2021a, 2021b).

7.4. Aluminum hydrides

7.4.1. *Introduction*

Among the materials considered for hydrogen storage, aluminum-based hydrides offer a number of advantages. Aluminum, an abundant element on Earth, is a light element. When combined with other elements, it forms hydrides with gravimetric hydrogen densities between 70 and 105 g(H_2)/kg(hydride). Notably, the dehydrogenation reaction of some aluminum hydrides has been shown to be (partially) reversible, and these reactions have therefore been used to construct prototypes (Na Ranong et al. 2009).

Alane AlH_3 and metal alanates $M[AlH_4]_n$, where M is generally an element from group I or II of the periodic table, are candidates for hydrogen storage. Some of these materials produce hydrogen by hydrolysis or thermolysis, with the latter reaction being reversible under conditions requiring high temperature and pressure. In their solid state, aluminum hydrides present significant kinetic barriers to dehydrogenation and/or rehydrogenation (Jensen et al. 2008), which explains why they were "sidelined" for a time from the field of hydrogen storage.

This study was published in 1997 that once again drew attention to aluminum hydrides. Sodium alanate $NaAlH_4$ had been doped with titanium-based compounds (Bogdanović and Schwickardi 1997), and the binary mixtures thus obtained showed a notable improvement in dehydrogenation kinetics as well as reversibility of the process (see [7.19] and [7.20]):

$$NaAlH_4 \leftrightarrow 1/3\ Na_3AlH_6 + 2/3\ Al + H_2 \qquad [7.19]$$

$$Na_3AlH_6 \leftrightarrow 2NaH + LiH + Al + 3/2H_2 \qquad [7.20]$$

Since then, aluminum hydrides have been extensively studied. Today, there are two categories of aluminum-based hydrides: those ($NaAlH_4$ and $KAlH_4$) that enable reversible hydrogen storage (Zhang et al. 2013) and those (AlH_3 and $LiAlH_4$) for which there is a need for post-dehydrogenation regeneration to return to the original compound. Section 7.4.2 is dedicated to this second category.

7.4.2. *AlH_3 alane*

AlH_3 (30 g/mol) has a gravimetric hydrogen density of 101 g(H_2)/kg(AlH_3) and a bulk density of 149 g(H_2)/L(AlH_3). It is a polymorphic solid with at least seven crystalline structures: α, α', β, γ, δ, ε and ζ (Brower et al. 1976). Polymorph formation depends on the reaction conditions, such as the temperature, reagents used, atmosphere, presence of impurities and reaction time (Liu et al. 2021b). The α phase is the most stable, with a rhombohedral structure (space group $R\bar{3}c$). This is followed by the β (cubic, $Fd\bar{3}m$) and γ (orthorhombic, *Pnnm*) phases. The latter phases transform into the α phase at 80–90°C (Graetz and Reilly 2006).

The synthesis of AlH_3, in the form of $AlH_3 \cdot 2N(CH_3)_3$, was first reported in 1942 (Stecher and Wiberg 1942). Five years later, reaction between LiH or $LiAlH_4$ with $AlCl_3$ in diethyl ether was achieved to form $AlH_3 \cdot nEt_2O$ (Finholt et al. 1947).

However, solvent extraction was not possible without dehydrogenating AlH_3. The procedure for obtaining non-solvated AlH_3 (see [7.21] and [7.22]) was proposed in 1975 (Brower et al. 1976). The key was the use of excess $LiAlH_4$. $LiAlH_4$ and $LiBH_4$ are extracted from the mixture by washing with diethyl ether:

$$4LiAlH_4 + AlCl_3 + nEt_2O \rightarrow 4AlH_3 \cdot nEt_2O + 3LiCl + LiAlH_4 \quad [7.21]$$

$$4AlH_3 \cdot nEt_2O + LiAlH_4 + LiBH_4 \rightarrow 4AlH_3 + nEt_2O + LiAlH_4 + LiBH_4 \quad [7.22]$$

Other synthetic routes have also been proposed. For example, the mechanochemical reaction between $LiAlD_4$ and $AlCl_3$ was the first synthesis to form the deuterated alane AlD_3 (Brinks et al. 2006). This led, at room temperature, to a mixture of the two phases α and α’, aluminum metal and LiCl, indicating decomposition of the AlD_3.

To avoid this decomposition, grinding was carried out at −196 °C/77 K (liquid N_2); from then on, only AlD_3 (α and α’ phases) and LiCl were obtained. The use of other reagents can produce different phases, as is the case with MgH_2 and $AlCl_3$, which generate γ-AlH_3 (Duan et al. 2015). Overall, the α-AlH_3 synthesis route described by equations [7.21] and [7.22] is the most widely used, as the use of other reagents or the application of other elaboration techniques leads to a less stable material (Graetz et al. 2011).

Dehydrogenation of AlH_3 occurs at approximately 100°C in a single endothermic step, with an enthalpy of 7.6 kJ/mol H_2 (see [7.23]) (Orimo et al. 2006). Direct rehydrogenation of the aluminum produced requires heating (to 330°C) under extreme hydrogen pressure (49,000 bar) (Saitoh et al. 2008). In other words, gentler regeneration processes are required:

$$AlH_3 \leftrightarrow Al + 3/2H_2 \quad [7.23]$$

To date, there are two ways of regenerating AlH_3. The first method is electrochemical and takes place at low pressure. The first study of this type dates back to 1961 (Clasen 1961). In an electrochemical cell, aluminum metal is used as the cathode, mercury is used as the anode, and $NaAlH_4$ in tetrahydrofuran (THF) is used as the electrolyte. During the reaction, molecular hydrogen is released at the anode, and AlH_3 is formed in the solution, but this has not been demonstrated by analysis.

In the 2000s, pure AlH_3 was synthesized electrochemically, using platinum at the anode (Figure 7.7) (Zidan et al. 2009). Sodium metal was formed at the cathode, and the AlH_4^- ion combined with the Al^{3+} ion to form $AlH_3 \cdot nTHF$ at the anode (see [7.24]). The latter is then separated, triethylamine (TEA) is added to substitute for THF, and vacuum heat treatment is applied to obtain α-AlH_3:

$$3AlH_4^- + Al^{3+} + nTHF \rightarrow 4AlH_3 \cdot nTHF + 3e^- \quad [7.24]$$

The second regeneration pathway is based on the formation of adducts with amines (Figure 7.8), which takes place in three stages (Graetz et al. 2007; Lacina et al. 2010).

The first step is the hydrogenation of aluminum metal at low pressure, and an NR_3 amine and a titanium-based catalyst are used to produce the $AlH_3 \cdot NR_3$ adduct.

The second step is transamination: the NR_3 group is replaced by another, less stable amine, such as TEA.

In the third step, TEA is separated from AlH_3 by heating to 75°C under nitrogen flow. Amine selection is the key step here. Amines that readily form an adduct with AlH_3 (e.g. triethylenediamine) tend to be more thermally stable, yet separation of the amine and AlH_3 becomes thermodynamically difficult (Graetz et al. 2011).

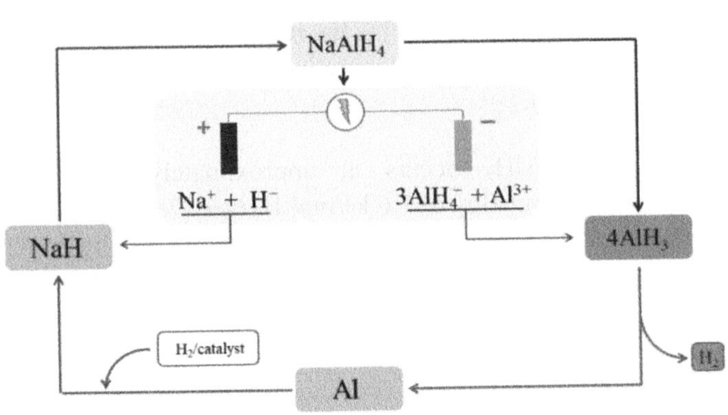

Figure 7.7. *Procedure for the dehydrogenation and electrochemical regeneration of AlH3. Adapted from Zidan et al. (2009). For a color version of this figure, see www.iste.co.uk/derango-cuevas/hydrogen.zip*

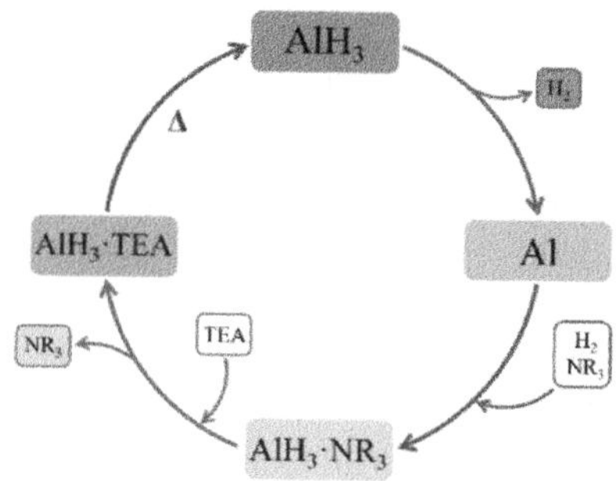

Figure 7.8. *Regeneration of alane by (i) the formation of $AlH_3·NR_3$ adducts, (ii) transamination with triethylamine, and (iii) obtaining AlH_3. For a color version of this figure, see www.iste.co.uk/derango-cuevas/hydrogen.zip*

7.4.3. *Lithium alanate $LiAlH_4$*

Lithium alanate ($LiAlH_4$) (37.95 g/mol) has a gravimetric hydrogen density of 105 g(H_2)/kg($LiAlH_4$). It is an ionic crystalline solid (monoclinic structure, space group $P2_1/c$), where the Al–H bond has a covalent character and the Li–Al bond has an ionic character (Sklar and Post 1967). $LiAlH_4$ can be prepared from LiH and $AlCl_3$ via the Schlesinger method, where the reactants are mixed in diethyl ether under an inert atmosphere (see [7.25]) (Finholt et al. 1947; Zhang et al 2013). Other synthetic routes have been investigated, such as mechanochemistry, but the yields are not satisfactory (Kojima et al. 2007):

$$4LiH + AlCl_3 \rightarrow LiAlH_4 + 3LiCl \quad [7.25]$$

Dehydrogenation of $LiAlH_4$ is a three-step process (Block and Gray 1965). The first (see [7.26]) is exothermic; it takes place between 185 and 220°C, and 5.3% wt. of hydrogen is released. The second (see [7.27]) is an endothermic reaction, releasing 2.6% wt. of hydrogen, and takes place between 225 and 280°C. The third (see [7.28]) is the endothermic decomposition of LiH, which leads to the release of the remaining hydrogen (2.6% wt.) between 370 and 483°C:

$$3LiAlH_4 \rightarrow Li_3AlH_6 + 2Al + 3H_2 \quad [7.26]$$

$$Li_3AlH_6 \rightarrow 3LiH + Al + 3/2H_2 \quad [7.27]$$

$$3LiH \rightarrow 3Li + 3/2H_2 \quad [7.28]$$

There are two issues with the dehydrogenation of $LiAlH_4$. The first is that dehydrogenation temperatures are too high from an application point of view. To overcome this, destabilization of $LiAlH_4$ via various methods, such as nanoconfinement, dopant addition or mechanical grinding, has been proposed (Zhao et al. 2021). The second issue is that the rehydrogenation of $LiAlH_4$ is thermodynamically irreversible under mild conditions (Ke and Chen 2007). $LiAlH_4$ must therefore be regenerated.

The proposed regeneration route for $LiAlH_4$ is based on the reactions among LiH, aluminum metal and H_2 in THF (equation [7.29]). This reaction can be carried out under various conditions, for example, at 35°C and 30 bar or at 120°C and 345 bar, enabling a conversion rate of over 95% and a purity of 99% (Clasen 1961; Ashby et al. 1963):

$$LiH + Al + 3/2H_2 + 4THF \leftrightarrow LiAlH_4 \cdot 4THF \qquad [7.29]$$

Wang et al. (2006) proposed a five-step method in which a titanium-based catalyst, THF and $LiAlH_4$ decomposition products are milled under 97.5 bar of hydrogen. The steps include catalyst dispersion, dehydrogenation, rehydrogenation, vacuum filtration and vacuum drying.

Graetz et al. (2008) presented another, simpler, two-step approach that does not require mechanical grinding. They directly hydrogenated LiH and aluminum metal in the presence of a titanium-based catalyst under 13 bar of hydrogen in THF. However, this route has certain shortcomings. First, the catalyst can destabilize $LiAlH_4$. Second, the solvent cannot be replaced by another, given that diethyl ether only allows a low yield and that solvents such as diglyme, which are strongly bound to alanate, will be more complicated to dissociate.

7.5. Alkali borohydrides

7.5.1. *Introduction*

Alkali borohydrides are among the older regenerable hydrides discovered in the middle of the last century (Demirci 2017). The first to be reported was aluminum borohydride $Al(BH_4)_3$, a gaseous and unstable compound, which limited its development for application (Schlesinger et al. 1939). It was then succeeded by other metal borohydrides, which can be described as derivatives, such as the volatile compounds $Be(BH_4)_2$ and $U(BH_4)_4$, and the crystalline ionic solids $LiBH_4$, $NaBH_4$ and KBH_4 (Burg and Schlesinger 1940; Schlesinger and Brown 1953; Schlesinger et al. 1953a, 1953b). The latter three are stable in anhydrous and dry atmospheres and at temperatures up to at least 240°C for $LiBH_4$.

In the presence of water, however, they hydrolyze more or less rapidly, releasing pure molecular hydrogen, the by-product being an alkali tetrahydroxyborate $MB(OH)_4$ with M for Li, Na or K (see [7.30]–[7.32]). In the 1950s, these alkali borohydrides were referred to as "hydrogen generators" and were studied as such (Schlesinger et al. 1953c) before being forgotten. It was not until the late 1990s that they reemerged for on-board hydrogen storage and production (Demirci et al. 2010):

$$LiBH_4 + 4H_2O \rightarrow LiB(OH)_4 + 4H_2 \quad [7.30]$$

$$NaBH_4 + 4H_2O \rightarrow NaB(OH)_4 + 4H_2 \quad [7.31]$$

$$KBH_4 + 4H_2O \rightarrow KB(OH)_4 + 4H_2 \quad [7.32]$$

Many other borohydrides have been discovered over the last two decades. In addition to monometallic borohydrides (Rude et al. 2011), such as $Mg(BH_4)_2$ (Lv and Wu 2021), the prospects for new compounds have been broadened by the targeting of bimetallic borohydrides (Suárez-Alcántara and García 2021; Comanescu 2022), borohydrides with at least one substituted hydrogen (Paskevicius et al. 2017) and functionalized borohydrides (Grinderslev et al. 2022).

All these borohydrides, however, have been targeted for purely reversible hydrogen storage. They are not considered regenerable hydrides (Singh 2022). Consequently, the following sections will be dedicated exclusively to the regenerable borohydrides $LiBH_4$, $NaBH_4$ and KBH_4.

7.5.2. *Lithium borohydride $LiBH_4$*

$LiBH_4$ is a crystalline ionic solid with orthorhombic symmetry (*Pnma* space group) at room temperature. It has a gravimetric hydrogen density of 185 g(H_2)/kg($LiBH_4$) and a bulk density of 121 g(H_2)/L($LiBH_4$), making it an attractive hydrogen carrier for exploration. Numerous routes to the synthesis of $LiBH_4$ have been developed since the mid-19th century, but today it is produced mainly from its $NaBH_4$ counterpart (see [7.33]) by salt metathesis (Simbara et al. 2020). The consequence is that $LiBH_4$ is seven times more expensive than $NaBH_4$:

$$NaBH_4 + LiBr \rightarrow LiBH_4 + NaBr \quad [7.33]$$

$LiBH_4$ is stable in an inert atmosphere and in dry air. When heated, it undergoes a phase transformation (orthorhombic → hexagonal) at approximately 110°C, with a loss of approximately 0.1% wt. hydrogen (Züttel et al. 2003). The melting point of $LiBH_4$ is 275°C, and this change in state is accompanied by the loss of 0.5–1% wt. hydrogen. Most of the hydrogen, just over 9% wt., is generated between 400 and

680°C. LiH and boron are then formed as by-products (see [7.34]). When heated to 600°C and pressurized to 155 bar of hydrogen, the by-products partially hydrogenate, enabling 75% of the initial $LiBH_4$ to be regenerated (Mauron et al. 2008). As a result, $LiBH_4$ is a complex hydride (Ding et al. 2020) and not a regenerable hydride when hydrogen is produced by thermolysis:

$$LiBH_4 \leftrightarrow LiH + B + 3/2H_2 \qquad [7.34]$$

$LiBH_4$ is a hygroscopic solid, rapidly adsorbing water and then hydrolyzing it. With the hydrolysis reaction (see [7.30]), water theoretically participates in the reaction to the tune of four equivalents for one equivalent of $LiBH_4$ (Kojima et al. 2004b).

In other words, half of the four equivalents of molecular hydrogen produced come from the four water molecules. The $LiBH_4$-$4H_2O$ couple can then be considered to have a gravimetric hydrogen density of 86 g(H_2)/kg($LiBH_4$-$4H_2O$). Moreover, the molecular hydrogen produced is generally pure. Initial work on $LiBH_4$ hydrolysis dates back to the 2000s.

However, the reaction was rather quickly neglected due to several disadvantages, such as excessive hygroscopicity, incomplete hydrolysis, a complex reaction with numerous reaction intermediates and high thermal stability of hydrolysis by-products (Li et al. 2011). Added to this are lithium supply tensions, first because lithium reserves are low (0.01% Li in the Earth's crust), second because of competition from lithium-ion batteries and third because lithium extraction is expensive (Li et al. 2019a).

$LiBH_4$ is also unstable in the presence of an alcohol such as methanol CH_3OH (see [7.35]), producing hydrogen by methanolysis (Schlesinger and Brown 1940). A 2022 study revealed that 93% of $LiBH_4$ can be dehydrogenated by reacting with CH_3OH in the presence of ethylene glycol (Chen et al. 2022). Under these conditions, an effective hydrogen storage capacity of 4.6% wt. has been calculated. Even more interesting is the possibility of closing the hydrogen cycle with $LiBH_4$ under these conditions.

Indeed, the methanolysis by-product $LiB(OCH_3)_4$ is unstable in aqueous media and hydrolyzes, resulting in the regeneration of CH_3OH and the production of $LiB(OH)_4$ (see [7.36]). However, $LiBH_4$ can be regenerated from $LiB(OH)_4$ in the presence of MgH_2 (see [7.37]) via a mechanochemical process under an argon atmosphere at room temperature (Chen et al. 2021). This process generates by-products such as molecular hydrogen, which can be captured for use, but above all magnesium oxide (MgO). Since this process was inspired in its entirety by work on $NaBH_4$, MgO recycling, already discussed in section 7.3.4, will be dealt with again in section 7.5.3 using a different method:

$$LiBH_4 + 4CH_3OH \rightarrow LiB(OCH_3)_4 + 4H_2 \quad [7.35]$$

$$LiB(OCH_3)_4 + 4H_2O \rightarrow LiB(OH)_4 + 4CH_3OH \quad [7.36]$$

$$LiB(OH)_4 + 4MgH_2 \rightarrow LiBH_4 + 4MgO + 4H_2 \quad [7.37]$$

7.5.3. *Sodium borohydride $NaBH_4$*

$NaBH_4$ is an ionic solid. At room temperature and atmospheric pressure, it has a cubic structure that is isomorphous to that of NaCl (Filinchuk and Hagemann 2008). It has a gravimetric hydrogen density of 107 $g(H_2)/kg(NaBH_4)$ and a bulk density of 114 $g(H_2)/L(NaBH_4)$. $NaBH_4$ is a commercial product that is notably used in the chemical (carbonyl function reducer) and paper (bleaching agent) industries. It is produced via the Brown–Schlesinger process (Figure 7.9) based, on the one hand, on the production of NaH from methane CH_4 and NaCl and, on the other hand, on the production of trimethoxyborate $B(OCH_3)_3$ from borax $Na_2B_4O_7$ and CH_3OH; $NaBH_4$ is then produced by reaction of NaH with $B(OCH_3)_3$.

$NaBH_4$ is chemically stable under an inert, anhydrous atmosphere and is thermally more stable than $LiBH_4$. It hydrolyzes spontaneously (see [7.31]), and the enthalpy of the reaction is in the order of –240 to –210 kJ/mol (Zhang et al. 2006; Damjanović et al. 2011). However, hydrolysis kinetics decrease very markedly at approximately 25% $NaBH_4$ conversion because the pH increases to values above 12 (Andrieux et al. 2011). In an alkaline environment, $NaBH_4$ hydrolysis can therefore be (very) reduced, even if it cannot be completely prevented (Alligier et al. 2022).

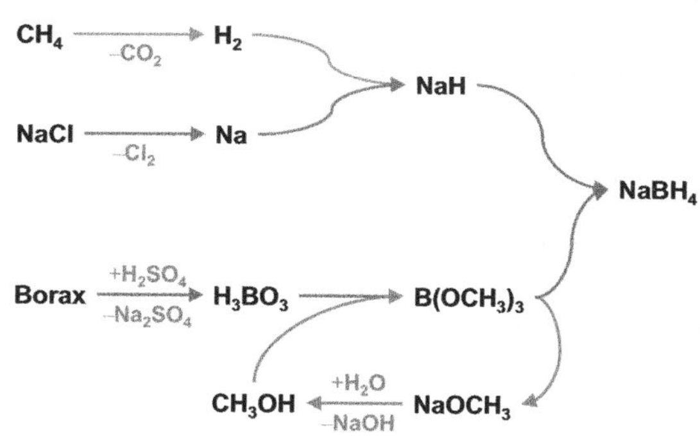

Figure 7.9. *Production of $NaBH_4$ by the industrial Brown–Schlesinger process*

The hydrolysis of $NaBH_4$ has often been shown to be outstanding for three main reasons. First, the $NaBH_4$-$4H_2O$ couple (see [7.31]) has a gravimetric hydrogen density of 73 g(H_2)/kg($NaBH_4$-$4H_2O$). Second, the hydrogen produced is assumed to be pure, with the other reaction product being the ionic solid $NaB(OH)_4$. Finally, the reaction kinetics can be modified by choosing the right catalyst (Abdelhamid 2021). However, the reality is more complex (Demirci 2018).

Under real-life conditions, for example, for a drone equipped with a 200 W fuel cell (Lapeña-Rey et al. 2017), it is necessary to work with excess water, which results in the hydrogen storage capacity being reduced to values well below 35 g(H_2)/kg($NaBH_4$-$4H_2O$).

In addition, water vapors carry hydrated borates (Petit et al. 2016). These borates then condense on the pipe walls and crystallize, which then requires frequent cleaning. Another major problem is that the by-product $NaB(OH)_4$ must be regenerated to close the hydrogen cycle with $NaBH_4$ (Liu and Chen 2015).

However, this challenge has not yet been satisfactorily met (Demirci 2015). Importantly, $NaBH_4$ in the presence of an alcohol (ROH), such as methanol (Davis and Gottbrath 1962), or a glycol (Ould-Amara et al. 2018) is unstable, solvolyzing (equation [7.38]) and producing hydrogen with acceptable kinetics in the absence of a catalyst:

$$NaBH_4 + 4ROH \rightarrow NaB(OR)_4 + 4H_2 \qquad [7.38]$$

The regeneration of $NaBH_4$ from $NaB(OH)_4$ has, despite everything, progressed as a result of one-off studies and the development of reduction processes. This task is difficult because the B–O bond of $NaB(OH)_4$ (536 kJ/mol) has a dissociation energy that lies between that of the C–O bond (358 kJ/mol) and that of the C=O bond (799 kJ/mol); this suggests the thermochemical challenge behind $NaBH_4$ regeneration. Among the regeneration processes reported, we can list four types (Figure 7.10): (i) the three-step Brown–Schlesinger process, which is largely based on the industrial $NaBH_4$ production process (Figure 7.9); (ii) the modified two-stage Bayer process, with the first step being the dehydrogenation of $NaB(OH)_4$ to $NaBO_2$ and the second being the reduction of $NaBO_2$ to $NaBH_4$ by $MgH_2$7; (iii) reduction processes using reductants such as CH_4, H_2 and the carbon-H_2 couple; and (iv) the electroreduction process, which takes place in an aqueous alkaline medium (Nunes et al. 2021). However, all these processes are too costly (Demirci et al. 2009).

Recently, a simplified regeneration process using mechanosynthesis has been developed (Ouyang et al. 2018). It is based on the modified Bayer process, as shown in Figure 7.10. $NaB(OH)_4$ reacts directly with MgH_2; approximately 90% of the initial $NaBH_4$ is thus reformed (see [7.39]). Like with $LiBH_4$, the reaction takes place at room temperature, with the necessary energy supplied by grinding (Chen et al. 2017). This reaction has been successfully extended to other borates, such as hydrated borax $Na_2B_4O_7 \cdot 10H_2O$ (Zhu et al. 2020).

Another variant is to use the dihydrate version of $NaB(OH)_4$ and magnesium metal or an alloy consisting mainly of magnesium under 70 bar of hydrogen (see [7.40]) and at 25–40°C (Le et al. 2019).

For the latter approach, conditions have been optimized to regenerate over 99% of the initial $NaBH_4$. However, the separation of $NaBH_4$ from the MgO by-product and/or other residues (with MgH_2 and/or Mg remains) is a delicate operation. It is complicated by three factors linked to the magnesium by-products mentioned above: (i) these are sensitive to humidity (hydrolysis and exothermic H_2 production); (ii) when they form nanoparticles through the grinding mechanism, they are pyrophoric; and (iii) these, precisely for the smallest nanoparticles, are almost impossible to precipitate over time (even after 2 months) or via centrifugation operations at rotational speeds in excess of 20,000 rpm.

Returning to the MgO by-product, it is obvious that this must be recycled to close the cycle with all the atoms involved in the processes associated with $NaBH_4$ regeneration. One possible sequence is that presented by equations [7.41] to [7.43]: production of $MgCl_2$ at 900°C, followed by production of Mg at 710°C, and hydrogenation to obtain MgH_2 at 25°C (Chen et al. 2021):

$$NaB(OH)_4 + 4MgH_2 \rightarrow NaBH_4 + 4MgO + 4H_2 \quad [7.39]$$

$$NaB(OH)_4 \cdot 2H_2O + 6Mg \rightarrow NaBH_4 + 6MgO + 2H_2 \quad [7.40]$$

$$MgO + Cl_2 \rightarrow MgCl_2 + 1/2O_2 \quad [7.41]$$

$$MgCl_2 \rightarrow Mg + Cl_2 \quad [7.42]$$

$$Mg + H_2 \rightarrow MgH_2 \quad [7.43]$$

Regeneration of $NaBH_4$ necessarily raises the question of overall cost and therefore the cost of the hydrogen stored and produced. Because of the energy stability of the B–O bond, the above-mentioned processes require energy and strong reducing agents, which in turn need to be regenerated. The solution could be the use of renewable energies. The aim is to reduce the overall cost of $NaBH_4$ to approximately €18–19/kg (Monteverde and Magistri 2012). In any case, this will provide a viable and realistic starting point for possible commercial use of $NaBH_4$ technology.

Brown-Schlesinger process

$$\mathbf{NaB(OH)_4} + 1/2\ H_2SO_4 \rightarrow B(OH)_3 + 1/2\ Na_2SO_4 + H_2O$$
$$B(OH)_3 + 3\ CH_3OH \rightarrow B(OCH_3)_3 + 3\ H_2O$$
$$B(OCH_3)_3 + 4\ NaH \rightarrow \mathbf{NaBH_4} + 3\ NaOCH_3$$

Modified Bayer process

$$\mathbf{NaB(OH)_4} \rightarrow NaBO_2 + 2\ H_2O$$
$$NaBO_2 + 2\ MgH_2 \rightarrow \mathbf{NaBH_4} + 2\ MgO$$

Reduction

$$\mathbf{NaB(OH)_4} \rightarrow NaBO_2 + 2\ H_2O$$
$$NaBO_2 + CH_4 \rightarrow \mathbf{NaBH_4} + CO_2$$
ou $$NaBO_2 + 4\ H_2 \rightarrow \mathbf{NaBH_4} + 2\ H_2O$$
ou $$NaBO_2 + C + 2\ H_2 \rightarrow \mathbf{NaBH_4} + CO_2$$

Electroreduction

$$\mathbf{B(OH)_4^-} + 4\ H_2O + 8\ e^- \rightarrow \mathbf{BH_4^-} + 8\ OH^-$$

Figure 7.10. *Examples of $NaBH_4$ regeneration processes*

7.5.4. *Potassium borohydride KBH_4*

KBH_4 is a white ionic solid. It is isostructural to $NaBH_4$ (cubic structure, space group $Fm\bar{3}m$) (Babanova et al. 2010). KBH_4 has a gravimetric hydrogen density of 64 g(H_2)/kg(KBH_4) and a bulk density of 75 g(H_2)/L(KBH_4).

KBH_4 is synthesized from $NaBH_4$ by precipitation (see [7.44]). The reaction can be carried out in cold methanol or in aqueous alkaline medium, bearing in mind that the solubility of KBH_4 is three times lower than that of $NaBH_4$, with 19 g(KBH_4) versus 55 g($NaBH_4$) per 100 g(H_2O) (Hagemann and Černý 2010). Another route to KBH_4 synthesis is based on the use of MgH_2 and KBO_2 and is achieved by mechanical grinding (see [7.45]). In the presence of excess MgH_2 (up to 35%), the yield of KBH_4 is 100% (Li et al. 2003). One variant uses boron oxide B_2O_3, KCl (see [7.46]) and MgH_2 as reducing agent (Bilen et al. 2017):

$$NaBH_4 + KOH \rightarrow KBH_4 + NaOH \quad [7.44]$$

$$KBO_2 + 2MgH_2 \rightarrow KBH_4 + 2MgO \quad [7.45]$$

$$KCl + 1/2B_2O_3 + 2MgH_2 \rightarrow KBH_4 + 3/2MgO + 1/2MgCl_2 \quad [7.46]$$

In terms of thermal and chemical stability, KBH_4 is very comparable to $NaBH_4$. KBH_4 hydrolyzes spontaneously (–220 kJ/mol), and hydrolysis kinetics are slower as the solution pH increases (Brown and Svensson 1957; Sahin et al. 2007; Damjanović et al. 2010). The hydrolysis of KBH_4 (see [7.32]) has been less studied than that of $NaBH_4$, but the general problem for KBH_4 hydrolysis is similar to that of $NaBH_4$.

In recent years, there seems to be renewed interest in KBH_4 and its ability to produce hydrogen at room temperature by hydrolysis. The KBH_4-$4H_2O$ couple is 13% less hydrogen rich than $NaBH_4$-$4H_2O$ but has other advantages. First, KBH_4 is less hygroscopic (Darin et al. 1999), making it easier to handle in air. Second, KBH_4 hydrolysis generates less heat than $NaBH_4$ (Damjanović et al. 2010). Finally, compared with the NaB(OH)4 analog, the $KB(OH)_4$ by-product is less prone to crystallization and precipitation (Churikov et al. 2011).

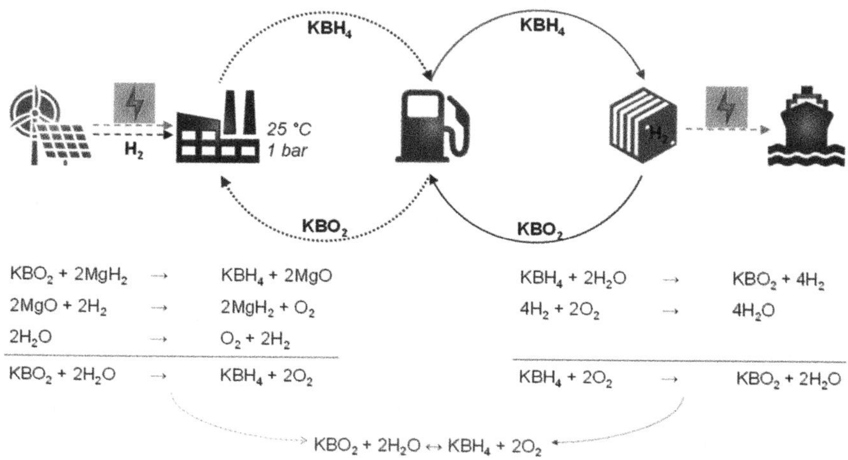

Figure 7.11. *The hydrogen cycle for KBH_4 hydrolysis for use on board marine vessels*

COMMENT ON FIGURE 7.11.– *Hydrogen for KBH_4 production is produced by electrolysis. Renewable energy provides the necessary electricity. KBH_4 is regenerated from KBO_2 in the presence of MgH_2 (grinding at 25°C and 1 bar). KBH_4 produces hydrogen by hydrolysis (optimized at 80°C and 3 bar of hydrogen), and the hydrogen is oxidized in a fuel cell. The cycle is beneficial for each of the atoms involved in the overall process. Adapted from Düll et al. (2022).*

A recent simulation study evaluated the energy efficiency of the hydrogen cycle on the basis of the hydrolysis of KBH_4 (Düll et al. 2022). The cycle (Figure 7.11) considers the use of renewable electricity, the electrolysis of water as a source of H_2, the rehydrogenation of KBO_2 (the anhydrous equivalent of $KB(OH)_4$) in the presence of MgH_2 and a planetary mill at 25°C and 1 bar of hydrogen, the hydrolysis of KBH_4 at 80°C in the presence of a metal catalyst, and the use of a fuel cell for H_2 oxidation and on-board electricity generation. The cycle of each of the atoms involved in the process is closed, making it virtuous. The main challenge remains KBH_4 regeneration.

Two problems have been identified. The first concerns the recovery of MgO and its recycling to produce MgH_2. The second is the final cost of hydrogen. In particular, an energy equivalent of 521 MJ is needed to produce 1 kg of hydrogen, which is comparable to the energy requirements for the production of diesel-type hydrocarbons using the Fisher–Tropsch process.

7.6. BNH compounds

7.6.1. *Introducing BNH compounds*

BNH compounds are boron- and nitrogen-based materials with a high hydrogen content. One of the most widely studied BNH compounds is ammonia borane. It was discovered in the 1950s and "rediscovered" in the 2000s as a hydrogen storage material. Many other BNH compounds, which can be seen as derivatives, have since been synthesized and studied (Kumar et al. 2019).

Compared with other materials, BNH compounds have two unique features that make them good candidates for hydrogen storage. First, boron and nitrogen atoms are light, and each can form multiple bonds with hydrogen atoms, so these compounds have a high hydrogen storage capacity. Second, there are two types of hydrogen in these molecules: protic hydrogens ($H^{\delta+}$) bonded to the nitrogen atom and hydride hydrogens ($H^{\delta-}$) bonded to the boron atom. This leads to the occurrence of dihydrogen bonds, which are electrostatic interactions between these two types of hydrogen ($H^{\delta+}\cdots H^{\delta-}$). These bonds lead to the stabilization of BNH compounds in the solid state (Chen et al. 2013). Compared with other hydrides, hydrogen release occurs at low temperatures because of these interactions. However, homopolar $H^{\delta+}\cdots H^{\delta+}$ and $H^{\delta-}\cdots H^{\delta-}$ interactions also participate, admittedly to a lesser extent, in the stabilization of BNH compounds (Wolstenholme et al. 2011).

To date, various BNH compounds have been considered for hydrogen storage. When heated, these compounds produce hydrogen exothermically. The solids produced after thermolysis are mixtures of boron- and nitrogen-based polymers, and these cannot be easily rehydrogenated. They are therefore considered regenerable hydrides. The following sections describe such BNH compounds.

7.6.2. *Ammonia borane NH_3BH_3*

NH_3BH_3 (30.8 g/mol) is a crystalline solid (centered quadratic structure, space group *I4mm*), stable at room temperature and in an inert atmosphere. NH_3BH_3 is the most studied BNH compound for hydrogen storage, as it has high gravimetric and volumetric hydrogen densities: 196 $g(H_2)/kg(NH_3BH_3)$ and 152 $g(H_2)/L(NH_3BH_3)$, respectively.

NH_3BH_3 was first synthesized in 1955 (Shore and Parry 1955) by metathesis of $LiBH_4$ and ammonium chloride in diethyl ether (see [7.47]). Today, this synthetic route is the main method for producing NH_3BH_3 since both salts can be easily replaced by analogs (see [7.48]) (Ramachandran and Gagare 2007):

$$LiBH_4 + NH_4Cl \rightarrow NH_3BH_3 + LiCl + H_2 \quad [7.47]$$

$$2NaBH_4 + (NH_4)_2SO_4 \rightarrow 2NH_3BH_3 + Na_2SO_4 + 2H_2 \quad [7.48]$$

Dehydrogenation of NH_3BH_3 can be achieved by hydrolysis or thermolysis. Catalytic hydrolysis has been extensively studied, and this reaction is generally described by equation [7.49], where the main product is boric acid (Liu et al. 2012). Nevertheless, the formation of other borates has been described (Chandra and Xu 2006):

$$NH_3BH_3 + 3H_2O \rightarrow B(OH)_3 + NH_3 + 3H_2 \quad [7.49]$$

The hydrolysis reaction has advantages, such as the absence of parallel reactions. However, it also has disadvantages. First, the hydrolysis of NH_3BH_3 is less efficient than that of $NaBH_4$. Second, ammonia gas (NH_3) is released. If this gas is not recovered, the overall efficiency of the system is lower, and the hydrogen released is polluted.

Additionally, only the BH_3 group releases hydrogen. Finally, and most importantly, there are no specific studies on the direct regeneration of $B(OH)_3$ to NH_3BH_3. The only possible route seems to be the synthesis of $NaBH_4$ from $B(OH)_3$ (by the methods mentioned in section 7.5.3), followed by the synthesis of NH_3BH_3 from $NaBH_4$. This represents one step too many, and as a result, NH_3BH_3 will remain more expensive and less cost-effective than $NaBH_4$.

Dehydrogenation of NH_3BH_3 can also be achieved by methanolysis (see [7.50]). The theoretical hydrogen storage capacity of this system is 3.9 wt.% (Lu et al. 2014). The methanolysis reaction has not been studied in depth. It does, however, offer the possibility of regenerating the $NH_4B(OCH_3)_4$ product directly into NH_3BH_3 (Ramachandran and Gagare 2007). To achieve this (Figure 7.12), $B(OCH_3)_3$ and NH_4Cl are mixed in THF at 0°C, in the presence of $LiAlH_4$. NH_3BH_3 of purity >99% is thus obtained (90% yield). Compared with the hydrolysis reaction, methanolysis is therefore more attractive for the dehydrogenation and regeneration of NH_3BH_3:

$$NH_3BH_3 + 4CH_3OH \rightarrow NH_4B(OCH_3)_4 + 3H_2 \quad [7.50]$$

The most efficient approach to NH_3BH_3 dehydrogenation is probably thermolysis. Dehydrogenation can be explained simply (see [7.51]–[7.53]). At approximately 100°C, NH_3BH_3 melts and releases the equivalent of one mole of hydrogen per mole of NH_3BH_3 between 110 and 130°C. A boron-nitrogen-based polymer is then formed, polyaminoborane (PAB) $[-BH_2-NH_2-]_n$ (Hu et al. 1978).

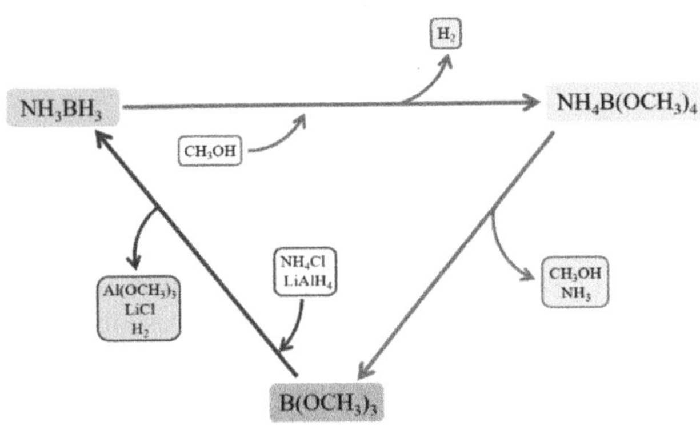

Figure 7.12. *Dehydrogenation cycle of NH_3BH_3 by methanolysis and its regeneration from the by-product $NH_4B(OCH_3)_4$. Adapted from Ramachandran et al. (2012)*

Gaseous by-products such as NH_3 and diborane B_2H_6 are generated. The production of these by-products is a drawback of NH_3BH_3 thermolysis. Between 130 and 200°C, the equivalent of the second mole of hydrogen is released. In terms of gaseous by-products, borazine $B_3N_3H_6$ is also generated. The solid residue that is produced is polyiminoborane (PIB) $[-NH=BH-]_n$, but polyborazylene $[-B_3N_3H_4-]_n$ (PB) is also formed. To release the equivalent of the third mole of hydrogen, heating above 400°C is needed. Crystalline boron nitride (BN) is produced between 1,200 and 1,500°C (Frueh et al. 2011).

However, the formation of BN is not desirable in the field of hydrogen storage because it is a very stable compound, and its rehydrogenation is not thermodynamically favorable. In short, NH_3BH_3 decomposes more than it dehydrogenates:

$$nNH_3BH_3 \rightarrow [-BH_2-NH_2-]_n + nH_2 \quad [7.51]$$

$$[-BH_2-NH_2-]_n \rightarrow [-NH{=}BH-]_n + nH_2 \quad [7.52]$$

$$[-NH{=}BH-]_n \rightarrow [BN]_n + nH_2 \quad [7.53]$$

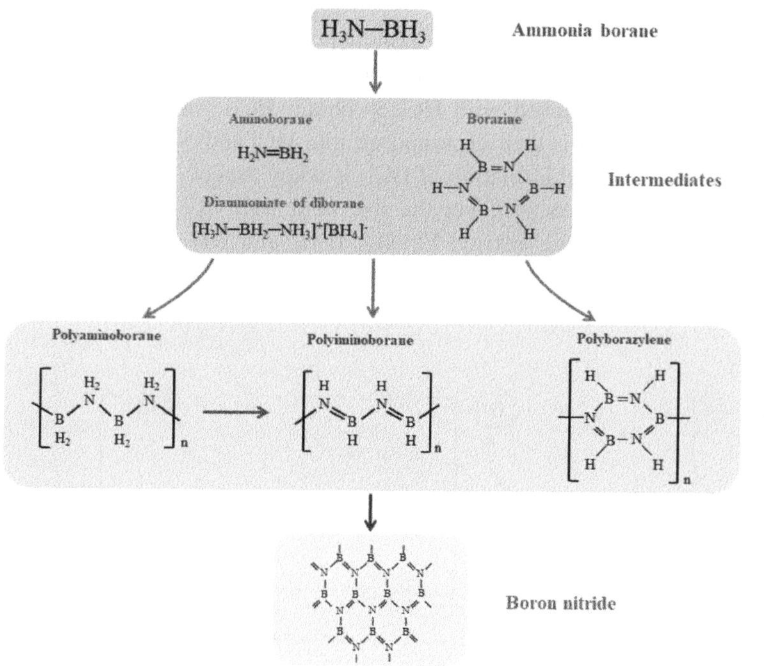

Figure 7.13. *Ammonia borane decomposition products*

In reality, the NH_3BH_3 decomposition reaction is more complicated (Figure 7.13). When NH_3BH_3 melts, highly reactive intermediates are formed. One of these is the ionic dimer under the name diammoniate of diborane (DADB) $[NH_3BH_2NH_3]^+[BH_4]^-$ (Chen et al. 2011). Another compound identified is aminoborane NH_2BH_2, as this is the key reagent for the oligomerization of ammonia borane (Zimmerman et al. 2009). Other gaseous by-products, such as $B_3N_3H_6$, are also involved in polymer formation. Another important point is that NH_3BH_3 decomposition does not follow a single sequence: PAB, PIB and PB polymers are produced almost simultaneously (Miranda and Ceder 2007; Demirci 2021).

BNH polymers are both linear and branched. The final composition of a solid depends on the heat treatment time and temperature (Kobayashi et al. 2014). The identification of these polymers is, and remains, complicated. As a result, the rehydrogenation of BNH polymers has not yet been fully demonstrated, and has remained relatively under-researched. There are, however, several chemical routes for regeneration that have been explored.

The first regeneration route for BNH polymers is based on acid digestion (Hausdorf et al. 2008) (Figure 7.14). A residue obtained after thermolysis of NH_3BH_3, named BNH_x, is attacked with HCl to obtain BCl_3 and NH_4Cl at room temperature after 16 h. NH_4Cl is then decomposed into HCl and NH_3; the latter can be used at a later stage. Hydrochlorination of BCl_3 is a key aspect of this process, as it is an energetically costly reaction, yet the use of a BH_3NR_3 adduct has been proposed to form B_2H_6 and save energy. Finally, B_2H_6 and NH_3 react to produce NH_3BH_3.

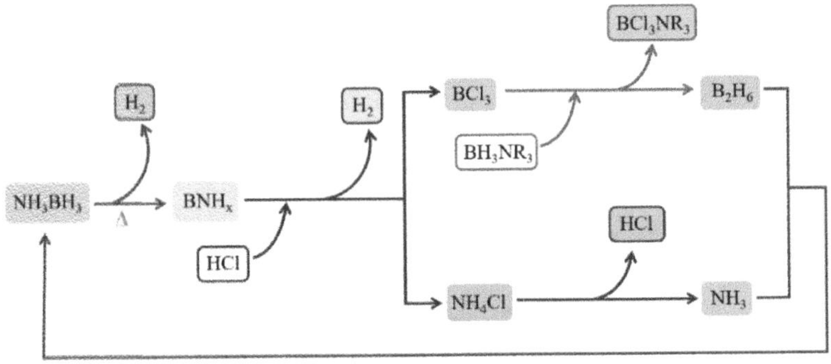

Figure 7.14. *Regeneration of NH_3BH_3 by acid digestion. The polymeric residue BNH_x is converted to BCl_3 and NH_4Cl by acid digestion (HCl). BCl_3 is then converted to B_2H_6, and NH_4Cl is decomposed to NH_3. These two compounds are finally used to produce NH_3BH_3. Adapted from (Hausdorf et al. 2008)*

The second route for regenerating BNH polymers involves the formation of B–S bonds using a thiol (Davis et al. 2009). This route offers two advantages. First, B–S is weak (compared with the B–O bond, for example). Second, the SH functional group is acidic and can react with BNH polymers (Figure 7.15).

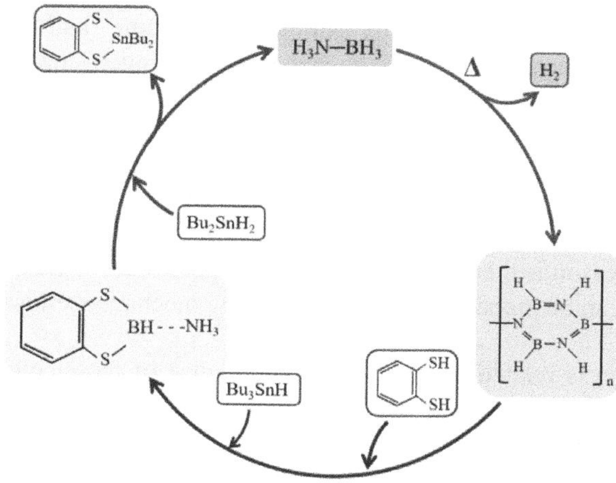

Figure 7.15. *Regeneration of NH_3BH_3 by acid digestion with thiol and reduction with tin hydrides. Adapted from Davis et al. (2009)*

The process was developed for polyborazylene. This is digested by benzenedithiol, leading to a mixture of $(C_6H_4S_2)BH{\cdot}NH_3$ and $[NH_4][B(C_6H_4S_2)_2]$. In 12 h, the second product is reduced by Bu_3SnH to $(C_6H_4S_2)BH{\cdot}NH_3$. Bu_2SnH_2 is then added to the system, and NH_3BH_3 is obtained in 67% yield.

A third route for regenerating BNH polymers involves investigating the reduction of PB (used as a model compound) by N_2H_4 in NH_3 medium (Sutton et al. 2011). The process is carried out in a single step and takes place in a pressurized reactor for 24 h. NH_3BH_3 is produced in 92% yield, the remaining 8% being hydrazine borane $N_2H_4BH_3$.

7.6.3. *Alkaline amidoboranes with M for Li, Na or K*

Metal amidoboranes ($M(NH_2BH_3)_x$) are derivatives of NH_3BH_3 in which one of the latter's protic $H^{\delta+}$ hydrogens is replaced by a metal cation. Amidoboranes offer advantages over NH_3BH_3: hydrogen release occurs at lower temperatures, and little or no undesirable gaseous by-products are formed. Various amidoboranes have been synthesized: alkaline amidoboranes, alkaline-earth equivalents, and those with two cations. Compared with those of NH_3BH_3, the dehydrogenation properties of almost all these compounds were improved.

Of all these compounds, only lithium-, sodium- and potassium-based amidoboranes have the potential to be considered as hydrogen storage materials.

The first reason is their high gravimetric hydrogen storage capacity: $LiNH_2BH_3$ with 136.2 g(H_2)/kg($LiNH_2BH_3$), $NaNH_2BH_3$ with 94.7 g(H_2)/kg($NaNH_2BH_3$) and KNH_2BH_3 with 72.7 g(H_2)/kg(KNH_2BH_3) (Castilla-Martinez et al. 2020).

The second reason is their relatively simple preparation; for example, $LiNH_2BH_3$ and $NaNH_2BH_3$ can be prepared in the solid state by mechanochemistry.

A third reason is related to the relative abundance of the elements sodium and potassium on our planet, which is somewhat less common in the case of lithium.

When heated, $LiNH_2BH_3$ begins to release hydrogen at 70°C and exhibits two dehydrogenation peaks at 90°C and 148°C (Yang et al. 2011). $NaNH_2BH_3$ releases hydrogen at approximately 60°C, and the main dehydrogenation event occurs at 85°C (Shimoda et al. 2011). KNH_2BH_3 begins to dehydrogenate at 55°C, and the process breaks down into two further steps at 74°C and 170°C (Shimoda et al. 2012). Ammonia formation has been noted for $LiNH_2BH_3$ and $NaNH_2BH_3$. The dehydrogenation of amidoboranes occurs via a so-called hydride transfer mechanism assisted by metal cations (Kim et al. 2010).

This mechanism is based on the formation of a *M*H during the dehydrogenation reaction (see [7.54]). Owing to *M*H, amidoborane dehydrogenation occurs at low temperatures. The solids produced after the dehydrogenation of amidoboranes are comparable to the solids produced from NH_3BH_3. For example, an ionic solid similar to the diammoniate of diborane has been proposed for $LiNH_2BH_3$ (e.g. $[(LiNH_2)_2BH_2]^+[BH_4]^-$) (Shevlin et al. 2011). Polymers based on boron, nitrogen and metal $(MNBH)_n$ are also produced (Xiong et al. 2008):

$$MNH_2BH_3 \rightarrow MH\cdots H\text{–}NHBH_2 \rightarrow MNHBH_2 + H_2 \quad [7.54]$$

As the solid residues of amidoborane dehydrogenation are similar to those of NH_3BH_3, the regeneration strategy is also similar. In particular, digestion with methanol followed by reduction, a process similar to that proposed by Ramachandran and Gagare (2007), was developed by Tang et al. (2016). This method was applied to the solid obtained after dehydrogenation of MNH_2BH_3 (M = Li, K) (Figure 7.16). The amidoboranes were regenerated with a purity of 98%.

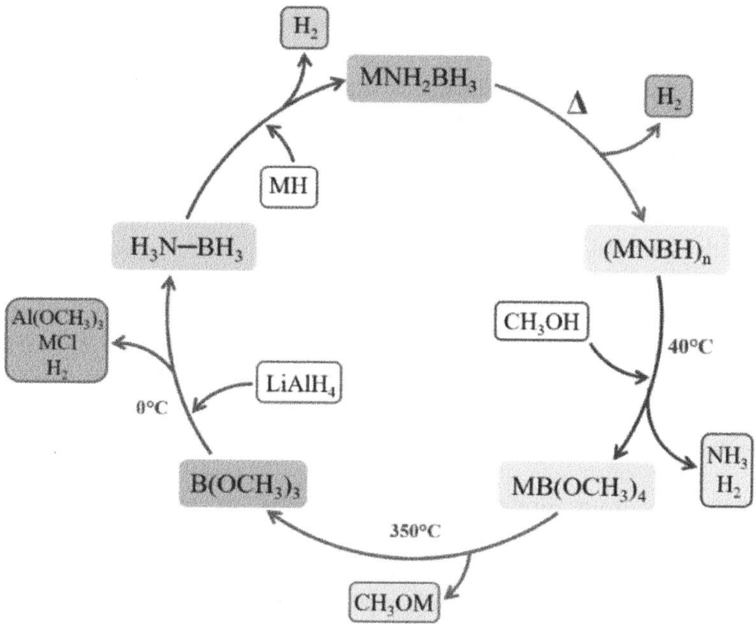

Figure 7.16. *Regeneration of alkaline amidoboranes from the polymeric residue $(MBNH)_n$. Digestion of the solid with methanol is followed by reduction with $LiAlH_4$ to produce NH_3BH_3. Adapted from Tang et al. (2016)*

7.6.4. *Hydrazine borane $N_2H_4BH_3$ and its derivatives*

Hydrazine borane $N_2H_4BH_3$ (45.8 g/mol) is a derivative of NH_3BH_3 where N_2H_4 substitutes the NH_3 group of the molecule. $N_2H_4BH_3$ is a crystalline solid (orthorhombic structure, space group *Pbcn*) with a gravimetric hydrogen density of 154 g(H_2)/kg($N_2H_4BH_3$). It was first reported in 1961 (Goubeau and Ricker 1961), but it was not until 2009 that this compound was considered a material for hydrogen storage (Hügle et al. 2009). The synthesis method is similar to that of NH_3BH_3: it is based on a reaction between a metal borohydride and a hydrazine salt in an organic solvent. A good example of synthesis with 81.6% yield and 99.6% purity is that between $NaBH_4$ and hydrazine hemisulfate $N_2H_4{\cdot}\frac{1}{2}\,H_2SO_4$ (Moury et al. 2012) (see [7.55]):

$$NaBH_4 + N_2H_4{\cdot}\tfrac{1}{2}H_2SO_4 \rightarrow N_2H_4BH_3 + \tfrac{1}{2}\,Na_2SO_4 + H_2 \qquad [7.55]$$

$N_2H_4BH_3$ can be dehydrogenated by hydrolysis or thermolysis. The hydrolysis reaction is more interesting than that of NH_3BH_3 (see [7.56]) because the N_2H_4 group participates in the reaction, producing two more moles of hydrogen per mole of $N_2H_4BH_3$. However, the problem of regenerating the $B(OH)_3$ product is the same as in the case of NH_3BH_3: it involves the initial production of $NaBH_4$, which is then used to produce $N_2H_4BH_3$:

$$N_2H_4BH_3 + 4H_2O \rightarrow B(OH)_3 + N_2 + 5H_2 \quad [7.56]$$

The thermolysis of $N_2H_4BH_3$ offers no more advantages than that of NH_3BH_3. When heated to 200°C, $N_2H_4BH_3$ releases two moles of hydrogen as well as gaseous by-products such as N_2H_4 and NH_3. In addition, if a compound is heated above 300°C, a highly unstable and explosive solid residue is formed. For these reasons, $N_2H_4BH_3$ in its pure state is not a feasible material for hydrogen storage.

To overcome these drawbacks, the synthesis of hydrazinidoboranes was considered. Like amidoboranes, metal hydrazinidoboranes $M(N_2H_3BH_3)_x$ are derivatives of $N_2H_4BH_3$. All alkaline and two alkaline-earth derivatives have been synthesized to date (Castilla-Martinez et al. 2020). Although hydrazinidoboranes have improved dehydrogenation properties compared with $N_2H_4BH_3$, they have notable shortcomings: (i) the synthesis processes are complex (i.e. more complex than those for amidoboranes), (ii) their gravimetric hydrogen density is lower than that of amidoboranes and (iii) the composition of the solid residues produced after their dehydrogenation is still relatively unknown, which explains why, to date, no regeneration process has been proposed for $N_2H_4BH_3$ and its derivatives.

7.6.5. *Toward other BNH compounds*

BNH compounds have high gravimetric hydrogen densities and can be dehydrogenated at moderate temperatures (<200°C). However, their regeneration remains complicated. The idea of circumventing this problem by exploring new compounds naturally arose. This led to the development of cyclic CBNH compounds with C–C, B–N, C–N and C–B bonds. Two typical examples are 1,2-BN cyclohexane and BN-methyl cyclopentane (Figure 7.17) (Luo et al. 2011a, 2011b).

Both compounds are liquid and stable under air and humidity. They are capable of releasing 6 equivalents of hydrogen and their regeneration (by catalytic hydrogenation) seems possible. In fact, these compounds and these studies are inspired by or comparable to what is being developed with liquid organic hydrogen carriers (LOHCs). There is no doubt that in the coming months and years, these compounds will mature and new candidates emerge (Dai et al. 2022).

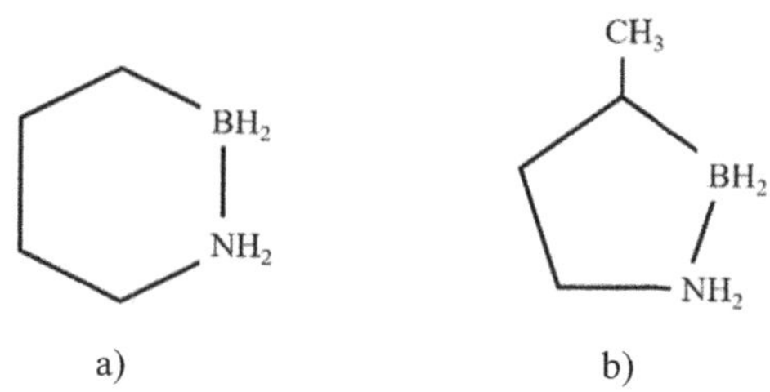

Figure 7.17. *Cyclic structures based on carbon, boron and nitrogen: (a) 1,2-BN cyclohexane and (b) BN-methyl cyclopentane*

7.7. Conclusions and outlook

Regenerable hydrides are compounds made up of light elements, yet rich in hydrogen. They are capable of producing hydrogen on demand under mild conditions, precisely at room temperature and atmospheric pressure (or close to it). These features have attracted a great deal of attention in the field of hydrogen storage. To illustrate this point, we have mentioned a number of much-studied materials based on boron, such as $NaBH_4$, magnesium, including its alloys, and aluminum, such as $LiAlH_4$.

The disadvantage of regenerable hydrides, especially as their dehydrogenation is exothermic, is that they produce stable dehydrogenation products whose hydrogenation requires severe temperature and/or pressure conditions. This is particularly true for $NaB(OH)_4$ and $Mg(OH)_2$, the hydrolysis by-products of $NaBH_4$ and MgH_2. In this context, chemistry has enabled progress to be made through the development of chemical processes for rehydrogenation by reduction. However, efforts devoted to rehydrogenation have remained very limited compared with the resources deployed to optimize the dehydrogenation of regenerable hydrides. This imbalance partly explains why the level of technological maturity of regenerable hydrides has made little progress, despite some twenty years of fairly intensive research.

The other consequence of the lack of an efficient hydrogenation solution is that the cost of hydrogen produced from regenerable hydrides remains high. Even so, there is still room for progress if more time is devoted to optimizing rehydrogenation processes, drawing inspiration from the work and strategies envisaged for CO_2 rehydrogenation and/or water dehydrogenation.

Regrettably, regenerable hydrides have not received much attention and investment in the development of prototypes on a semi-industrial or even industrial scale. This has two immediate consequences.

First, we know very little about the behavior of regenerable hydrides when handled on the scale of tens or hundreds of grams. The heat management of an exothermic reaction based on 100 mg of hydrides is completely different from the management of the same reaction on a 100 g scale. Taking $NaBH_4$ as an example, hydrolysis of 100 g will generate a heat quantity of approximately 630 kJ (175 Wh) that needs to be managed, with consequences for the operation and durability of the system as a whole (storage and fuel cells).

Second, the absence of tried-and-tested prototypes is a limitation in regard to accurately assessing the cost of the technology. Notably, however, solutions are beginning to emerge with cogeneration systems (hydrogen and heat), which could turn this disadvantage into an advantage (in the case of MgH_2, for example, the heat produced during hydrolysis represents 1.5 times the energy of the hydrogen produced).

In addition to regenerable hydrides such as $NaBH_4$ or magnesium-based hydrides, which can be described as *older*, there are *new* ones such as the CBNH materials discussed briefly in section 7.6.5. Like LOHCs, hydrogenation and dehydrogenation cycles under mild conditions seem realistic, provided that active and efficient catalysts are used. In the field of chemical hydrogen storage, the coming years could be those of rapid development.

In conclusion, it is clear that in the energy future, where a one-size-fits-all solution is not realistic, regenerable hydrides have a place to claim, provided that the abovementioned bottlenecks are overcome. Initially, their development seems to be focused on niche applications, even military ones, because of the cost involved, but any major advances toward the establishment of a virtuous hydrogen cycle and a lowering of the cost of stored/produced hydrogen will open wider prospects. The ease with which regenerable hydrides can be used and the ease with which hydrogen can be produced point the way to consumer applications, such as electric bicycles (Legree et al. 2020).

7.8. References

Abdelhamid, H.N. (2021). A review on hydrogen generation from the hydrolysis of sodium borohydride. *Int. J. Hydrog. Energy*, 46(1), 726–765.

Addison, C.C. and Manning, J.A. (1964). The reaction of water vapour with liquid sodium, sodium peroxide, sodium monoxide, and sodium hydride; vapour pressures in the sodium hydroxide–water system. *J. Chem. Soc.*, 4887–4895.

Al Bacha, S., Awad, A.S., El Asmar, E., Tayeh, T., Bobet, J.-L., Nakhl, M., Zakhour, M. (2019). Hydrogen generation via hydrolysis of ball milled WE43 magnesium waste. *Int. J. Hydrog. Energy*, 44(33), 17515–17524.

Alasmar, E., Awad, A.S., Hachem, D., Tayeh, T., Nakhl, M., Zakhour, M., Gaudin, E., Bobet, J.-L. (2018). Hydrogen generation from Nd-Ni-Mg system by hydrolysis reaction. *J. Alloys Compd.*, 740, 52–60.

Alligier, D., Petit, E., Demirci, U.B. (2022). A boron-11 NMR study of the stability of the alkaline aqueous solution of sodium borohydride that is both an indirect fuel and a direct fuel for low-temperature fuel cells. *Int. J. Hydrog. Energy*, 47(55), 23310–23315.

Amendola, S.C., Sharp-Goldman, S.L., Janjua, M.S., Kelly, M.T., Petillo, P.J., Binder, M. (2000). An ultrasafe hydrogen generator: Aqueous, alkaline borohydride solutions and Ru catalyst. *J. Power Sources*, 85(2), 186–189.

American Elements (2022). Berylium chloride [Online]. Available at: https://www.americanelements.com/beryllium-chloride-7787-47-5 [Accessed 16 August 2022].

Andrieux, J., Demirci, U.B., Hannauer, J., Gervais, C., Goutaudier, C., Miele, P. (2011). Spontaneous hydrolysis of sodium borohydride in harsh conditions. *Int. J. Hydrog. Energy*, 36(1), 224–233.

Ares, J.R., Zhang, J., Charpentier, T., Cuevas, F., Latroche, M. (2016). Asymmetric reaction paths and hydrogen sorption mechanism in mechanochemically synthesized potassium alanate ($KAlH_4$). *J. Phys. Chem. C*, 120(38), 21299–21308.

Ashby, E.C., Brendel, G.J., Redman, H.E. (1963) Direct synthesis of complex metal hydrides. *Inorg. Chem.*, 2(3), 499–504.

Awad, A.S., El-Asmar, E., Tayeh, T., Mauvy, F., Nakhl, M., Zakhour, M., Bobet, J.-L. (2016). Effect of carbons (G and CFs), TM (Ni, Fe and Al) and oxides (Nb_2O_5 and V_2O_5) on hydrogen generation from ball milled Mg-based hydrolysis reaction for fuel cell. *Energy*, 95, 175–186.

Awbery, R.P., Broughton, D.A., Tsang, S.C. (2008). In situ observation of lithium hydride hydrolysis by DRIFT spectroscopy. *J. Nucl. Mater.*, 373(1/3), 94–102.

Babanova, O.A., Soloninin, A.V., Stepanov, A.P., Skripov, A.V., Filinchuk, Y. (2010). Structural and dynamical properties of $NaBH_4$ and KBH_4: NMR and synchrotron X-ray diffraction studies. *J. Phys. Chem. C*, 114(8), 3712–3718.

Berro, Y., Kehrli, D., Brilhac, J.-F., Balat-Pichelin, M. (2021a). Metal fuel production through the solar carbothermal reduction of magnesia: Effect of the reducing agent. *Sustain. Energy Fuels*, 5(24), 6315–6327.

Berro, Y., Masse, R., Puig, J., Balat-Pichelin, M. (2021b). Improving the solar carbothermal reduction of magnesia for metallic fuels production through reactor designing, milling and binders. *J. Clean. Prod.*, 315, 128142.

Bilen, M., Gürü, M., Çakanyildirim, Ç. (2017). Conversion of KCl into KBH_4 by mechano-chemical reaction and its catalytic decomposition. *J. Electron. Mater.*, 46(7), 4126–4132.

Block, J. and Gray, A.P. (1965). The thermal decomposition of lithium aluminum hydride. *Inorg. Chem.*, 4(3), 304–305.

Bogdanović, B. and Schwickardi, M. (1997). Ti-doped alkali metal aluminium hydrides as potential novel reversible hydrogen storage materials. *J. Alloys Compd.*, 253(254), 1–9.

Brinks, H.W., Istad-Lem, A., Hauback, B.C. (2006). Mechanochemical synthesis and crystal structure of α'-AlD_3 and α-AlD_3. *J. Phys. Chem. B*, 110(51), 25833–25837.

Brower, F.M., Matzek, N.E., Reigler, P.F., Rinn, H.W., Roberts, C.B., Schmidt, D.L., Snover, J.A., Terada, K. (1976). Preparation and properties of aluminum hydride. *J. Am. Chem. Soc.*, 98(9), 2450–2453.

Brown, C.A. (1974). Saline hydrides and superbases in organic reactions. VII. Potassium hydride, highly active new hydride reagent. Reactivity, applications, and techniques in organic and organometallic reactions. *J. Org. Chem.*, 39(26), 3913–3918.

Brown, J.B. and Svensson, M. (1957). The stability of potassium borohydride in alkaline solutions 1. *J. Am. Chem. Soc.*, 79(15), 4241–4242.

Burg, A.B. and Schlesinger, H.I. (1940) Metallo borohydrides. II. Beryllium borohydride. *J. Am. Chem. Soc.*, 62(12), 3425–3429.

Castilla-Martinez, C.A., Moury, R., Demirci, U.B. (2020). Amidoboranes and hydrazinidoboranes: State of the art, potential for hydrogen storage, and other prospects. *Int. J. Hydrog. Energy*, 45(55), 30731–30755.

Chandra, M. and Xu, Q. (2006). A high-performance hydrogen generation system: Transition metal-catalyzed dissociation and hydrolysis of ammonia–borane. *J. Power Sources*, 156(2), 190–194.

ChemEurope (2022). Magnesium chloride [Online]. Available at: https://www.chemeurope.com/en/encyclopedia/Magnesium_chloride.html [Accessed 13 September 2022].

Chemical Book (2022). Magnesium hydroxide [Online]. Available at: https://www.chemicalbook.com/ChemicalProductProperty_EN_CB0239262.htm [Accessed 13 September 2022].

Chemistry Learner (2022). Berilium hydroxide [Online]. Available at: https://www.chemistrylearner.com/beryllium-hydroxide.html [Accessed 6 August 2022].

Chen, X., Bao, X., Zhao, J.-C.C., Shore, S.G. (2011). Experimental and computational study of the formation mechanism of the diammoniate of diborane: The role of dihydrogen bonds. *J. Am. Chem. Soc.*, 133(36), 14172–14175.

Chen, X., Zhao, J.-C.C., Shore, S.G. (2013). The roles of dihydrogen bonds in amine borane chemistry. *Acc. Chem. Res.*, 46(11), 2666–2675.

Chen, W., Ouyang, L.Z., Liu, J.W., Yao, X.D., Wang, H., Liu, Z.W., Zhu, M. (2017). Hydrolysis and regeneration of sodium borohydride ($NaBH_4$). A combination of hydrogen production and storage. *J. Power Sources*, 359, 400–407.

Chen, Y., Wang, M., Guan, F., Yu, R., Zhang, Y., Qin, H., Chen, X., Fu, Q., Wang, Z. (2020). Study on hydrolysis of magnesium hydride by interface control. *Int. J. Photoenergy*, 2020(1), 1–8.

Chen, K., Zhong, H., Ouyang, L., Liu, F., Wang, H., Liu, J., Shao, H., Zhu, M. (2021). Achieving a novel solvent-free regeneration of $LiBH_4$ combining hydrogen storage and production in a closed material cycle. *J. Magnes. Alloy.*, 11, 1697-1708

Chen, K., Liu, M., Ouyang, L., Wang, H., Liu, J., Zhu, M. (2022). Efficient hydrogen release from $LiBH_4$ alcoholysis in methanol/ethylene glycol-based solutions over a wide temperature range. *J. Alloys Compd.*, 905, 164030.

Chubukov, B.A., Palumbo, A.W., Rowe, S.C., Hischier, I., Groehn, A.J., Weimer, A.W. (2016). Pressure dependent kinetics of magnesium oxide carbothermal reduction. *Thermochim. Acta*, 636, 23–32.

Churikov, A.V., Zapsis, K.V., Khramkov, V.V., Churikov, M.A., Gamayunova, I.M. (2011). Temperature-induced transformation of the phase diagrams of ternary systems $NaBO_2$ + NaOH + H_2O and KBO_2 + KOH + H_2O. *J. Chem. Eng. Data*, 56(3), 383–389.

Clasen, H. (1961). Alanat-synthese aus den elementen und ihre bedeutung. *Angew. Chemie*, 73(10), 322–331.

Comanescu, C. (2022). Complex metal borohydrides: From laboratory oddities to prime candidates in energy storage applications. *Materials*, 15(6), 2286.

Czerwiński, P.J. and Furman, B. (2020). A long-sought reactivity of a sodium hydride. *Trends Chem.*, 2(9), 782–784.

Dai, Y., Zhang, X., Liu, Y., Yu, H., Su, W., Zhou, J., Ye, Q., Huang, Z. (2022). 1,6;2,3-bis-BN Cyclohexane: Synthesis, structure, and hydrogen release. *J. Am. Chem. Soc.*, 144(19), 8434–8438.

Damjanović, L., Bennici, S., Auroux, A. (2010). A direct measurement of the heat evolved during the sodium and potassium borohydride catalytic hydrolysis. *J. Power Sources*, 195(10), 3284–3292.

Damjanović, L., Majchrzak, M., Bennici, S., Auroux, A. (2011). Determination of the heat evolved during sodium borohydride hydrolysis catalyzed by Co_3O_4. *Int. J. Hydrog. Energy*, 36(3), 1991–1997.

Darin, V.A., Neto, A.F., Borges, A.D.L., Miller, J. (1999). Potassium borohydride: Improved use as a reducing agent for organic and organometallic ketones. *Main Gr. Met. Chem.*, 22(8), 505–509.

Davis, R.E. and Gottbrath, J.A. (1962). Boron hydrides. V. Methanolysis of sodium borohydride. *J. Am. Chem. Soc.*, 84(6), 895–898.

Davis, B.L., Dixon, D.A., Garner, E.B., Gordon, J.C., Matus, M.H., Scott, B., Stephens, F.H. (2009). Efficient regeneration of partially spent ammonia borane fuel. *Angew. Chemie*, 121(37), 6944–6948.

Davy, H. (1808). I. The Bakerian Lecture, on some new phenomena of chemical changes produced by electricity, particularly the decomposition of the fixed Alkalies, and the exhibition of the new substances which constitute their bases, and on the general nature of alkaline. *Philos. Trans. R. Soc. London*, 98, 1–44.

Dematteis, E.M., Amdisen, M.B., Autrey, T., Barale, J., Bowden, M.E., Buckley, C.E., Cho, Y.W., Deledda, S., Dornheim, M., de Jongh, P., Grinderslev, J.B., Gizer, G., Gulino, V. et al. (2022). Hydrogen storage in complex hydrides: Past activities and new trends. *Prog. Energy*, 4(3), 032009.

Demirci, U.B. (2015). The hydrogen cycle with the hydrolysis of sodium borohydride: A statistical approach for highlighting the scientific/technical issues to prioritize in the field. *Int. J. Hydrog. Energy*, 40(6), 2673–2691.

Demirci, U.B. (2017). Impact of H.I. Schlesinger's discoveries upon the course of modern chemistry on B–(N–)H hydrogen carriers. *Int. J. Hydrog. Energy*, 42(33), 21048–21062.

Demirci, U.B. (2018). About the technological readiness of the H_2 generation by hydrolysis of B(–N)–H compounds. *Energy Technol.*, 6(3), 470–486.

Demirci, U.B. (2020). Ammonia borane: An extensively studied, though not yet implemented, hydrogen carrier. *Energies*, 13(12), 3071.

Demirci, U.B. (2021). Mechanistic insights into the thermal decomposition of ammonia borane, a material studied for chemical hydrogen storage. *Inorg. Chem. Front.*, 8(7), 1900–1930.

Demirci, U.B., Akdim, O., Miele, P. (2009). Ten-year efforts and a no-go recommendation for sodium borohydride for on-board automotive hydrogen storage. *Int. J. Hydrog. Energy*, 34(6), 2638–2645.

Demirci, U.B., Akdim, O., Andrieux, J., Hannauer, J., Chamoun, R., Miele, P. (2010). Sodium borohydride hydrolysis as hydrogen generator: Issues, state of the art and applicability upstream from a fuel cell. *Fuel Cells*, 10(3), 335–350.

Ding, Z., Li, S., Zhou, Y., Chen, Z., Yang, W., Ma, W., Shaw, L. (2020). $LiBH_4$ for hydrogen storage. New perspectives. *Nano Mater. Sci.*, 2(2), 109–119.

Duan, C.W., Hu, L.X., Xue, D. (2015). Solid state synthesis of nano-sized AlH_3 and its dehydriding behaviour. *Green Chem.*, 17(6), 3466–3474.

Düll, A., Rohlfs, P., Deutschmann, O., Börnhorst, M. (2022). Performance evaluation of KBH_4 as energy carrier for shipping applications. *Chemie Ing. Tech.*, 94(5), 747–759.

Esmaily, M., Svensson, J.E., Fajardo, S., Birbilis, N., Frankel, G.S., Virtanen, S., Arrabal, R., Thomas, S., Johansson, L.G. (2017). Fundamentals and advances in magnesium alloy corrosion. *Prog. Mater. Sci.*, 89, 92–193.

Filinchuk, Y. and Hagemann, H. (2008). Structure and properties of $NaBH_4 \cdot 2H_2O$ and $NaBH_4$. *Eur. J. Inorg. Chem.*, 2008(20), 3127–3133.

Finholt, A.E., Bond, A.C., Schlesinger, H.I. (1947). Lithium aluminum hydride, aluminum hydride and lithium gallium hydride, and some of their applications in organic and inorganic chemistry 1. *J. Am. Chem. Soc.*, 69(5), 1199–1203.

Frueh, S., Kellett, R., Mallery, C., Molter, T., Willis, W.S., King'Ondu, C., Suib, S.L. (2011). Pyrolytic decomposition of ammonia borane to boron nitride. *Inorg. Chem.*, 50(3), 783–792.

Garlea, E., King, M.O., Galloway, E.C., Boyd, T.L., Smyrl, N.R., Bilheux, H.Z., Santodonato, L.J., Morrell, J.S., Leckey, J.H. (2017). Identification of lithium hydride and its hydrolysis products with neutron imaging. *J. Nucl. Mater.*, 485, 147–153.

Gawley, R.E. and Hennings, D.D. (eds) (2006). Sodium hydride. In *Encyclopedia of Reagents for Organic Synthesis*, Crich, D. (ed.). John Wiley & Sons, Chichester.

Ghali, E. (2011). Magnesium and magnesium alloys. In *Uhlig's Corrosion Handbook*, Revie, R.W. (ed.). John Wiley & Sons, Hoboken, NJ.

Goubeau, J. and Ricker, E. (1961). Borinhydrazin und seine pyrolyseprodukte. *Zeitschrift für Anorg. und Allg. Chemie*, 310(3), 123–142.

Graetz, J. and Reilly, J.J. (2006). Thermodynamics of the α, β and γ polymorphs of AlH_3. *J. Alloys Compd.*, 424(1/2), 262–265.

Graetz, J., Chaudhuri, S., Wegrzyn, J., Celebi, Y., Johnson, J.R., Zhou, W., Reilly, J.J. (2007). Direct and reversible synthesis of AlH_3–triethylenediamine from Al and H_2. *J. Phys. Chem. C*, 111(51), 19148–19152.

Graetz, J., Wegrzyn, J., Reilly, J.J. (2008). Regeneration of lithium aluminum hydride. *J. Am. Chem. Soc.*, 130(52), 17790–17794.

Graetz, J., Reilly, J.J., Yartys, V.A., Maehlen, J.P., Bulychev, B.M., Antonov, V.E., Tarasov, B.P., Gabis, I.E. (2011). Aluminum hydride as a hydrogen and energy storage material: Past, present and future. *J. Alloys Compd.*, 509, S517–S528.

Grinderslev, J.B., Amdisen, M.B., Skov, L.N., Møller, K.T., Kristensen, L.G., Polanski, M., Heere, M., Jensen, T.R. (2022). New perspectives of functional metal borohydrides. *J. Alloys Compd.*, 896, 163014.

Grosjean, M. and Roué, L. (2006). Hydrolysis of Mg–salt and MgH_2–salt mixtures prepared by ball milling for hydrogen production. *J. Alloys Compd.*, 416(1/2), 296–302.

Grosjean, M., Zidoune, M., Roué, L. (2005). Hydrogen production from highly corroding Mg-based materials elaborated by ball milling. *J. Alloys Compd.*, 404(406), 712–715.

Grosjean, M., Zidoune, M., Roue, L., Huot, J. (2006). Hydrogen production via hydrolysis reaction from ball-milled Mg-based materials. *Int. J. Hydrog. Energy*, 31(1), 109–119.

Guichard, J., Bouyer, F., Sciora, E., Bernard, F., Lecoq, H., Besnard, R., Bellat, J.P. (2015). Hydrolysis of lithium hydride under low relative humidity. *Int. J. Hydrog. Energy*, 40(37), 12736–12744.

Gusieva, K., Davies, C.H.J., Scully, J.R., Birbilis, N. (2015). Corrosion of magnesium alloys: The role of alloying. *Int. Mater. Rev.*, 60(3), 169–194.

Gwynne, B. and Lyon, P. (2007). Magnesium alloys in aerospace applications, past concerns, current solutions. In *Triennial International Aircraft Fire &. Cabin Safety Research Conference*, 29 October – 1 November, Atlantic City, NJ.

Haertling, C., Hanrahan, R.J., Smith, R. (2006). A literature review of reactions and kinetics of lithium hydride hydrolysis. *J. Nucl. Mater.*, 349(1/2), 195–233.

Haertling, C., Hanrahan, R.J., Tesmer, J.R. (2007). Hydrolysis studies of polycrystalline lithium hydride. *J. Phys. Chem. C*, 111(4), 1716–1724.

Hagemann, H. and Černý, R. (2010). Synthetic approaches to inorganic borohydrides. *Dalt. Trans.*, 39(26), 6006.

Hausdorf, S., Baitalow, F., Wolf, G., Mertens, F.O.R.L. (2008). A procedure for the regeneration of ammonia borane from BNH-waste products. *Int. J. Hydrog. Energy*, 33(2), 608–614.

Hesek, D., Lee, M., Noll, B.C., Fisher, J.F., Mobashery, S. (2009). Complications from dual roles of sodium hydride as a base and as a reducing agent. *J. Org. Chem.*, 74(6), 2567–2570.

Hou, X., Shi, H., Yang, L., Hou, K., Wang, Y., Feng, L., Suo, G., Ye, X., Zhang, L., Yang, Y. (2021). H_2 generation kinetics/thermodynamics and hydrolysis mechanism of high-performance La-doped Mg-Ni alloys in NaCl solution – A large-scale and quick strategy to get hydrogen. *J. Magnes. Alloy.*, 9(3), 1068–1083.

Hu, M.G., Geanangel, R.A., Wendlandt, W.W. (1978). The thermal decomposition of ammonia borane. *Thermochim. Acta*, 23(2), 249–255.

Hu, F., Pan, J., Ma, X., Zhang, X., Chen, J., Xie, W. (2013). Preparation of Mg and Ca metal by carbothermic reduction method--A thermodynamics approach. *J. Magnes. Alloy.*, 1(3), 263–266.

Huang, M., Ouyang, L., Chen, Z., Peng, C., Zhu, X., Zhu, M. (2017). Hydrogen production via hydrolysis of Mg-oxide composites. *Int. J. Hydrog. Energy*, 42(35), 22305–22311.

Hügle, T., Kühnel, M.F., Lentz, D., Hu, T., Ku, M.F., Lentz, D., Hügle, T., Kühnel, M.F., Lentz, D. (2009). Hydrazine borane: A promising hydrogen storage material. *J. Am. Chem. Soc.*, 131(21), 7444–7446.

Huot, J., Liang, G., Schulz, R. (2003). Magnesium-based nanocomposites chemical hydrides. *J. Alloys Compd.*, 353(1/2), L12–L15.

Jeffes, J.H.E. (ed.) (2001). Ellingham diagrams. In *Encyclopedia of Materials: Science and Technology*. Elsevier, Amsterdam.

Jensen, C., Wang, Y., Chou, M.Y. (eds) (2008). Alanates as hydrogen storage materials. In *Solid-State Hydrogen Storage*. Elsevier, Amsterdam.

Jia, J.X., Atrens, A., Song, G., Muster, T.H. (2005). Simulation of galvanic corrosion of magnesium coupled to a steel fastener in NaCl solution. *Mater. Corros.*, 56(7), 468–474.

Johnson, W.H. and Gilliland, A.A. (1961). Heat of formation of beryllium chloride. *J. Res. Natl. Bur. Stand. Sect. A Phys. Chem.*, 65A(1), 59.

Kantürk Figen, A. and Coşkuner Filiz, B. (2015). Hydrogen production by the hydrolysis of milled waste magnesium scraps in nickel chloride solutions and nickel chloride added in Marmara Sea and Aegean Sea Water. *Int. J. Hydrog. Energy*, 40(46), 16169–16177.

Kantürk Figen, A., Coşkuner Filiz, B., Pişkin, S. (2015). Hydrogen generation from waste Mg based material in various saline solutions ($NiCl_2$, $CoCl_2$, $CuCl_2$, $FeCl_3$, $MnCl_2$). *Int. J. Hydrog. Energy*, 40(24), 7483–7489.

Ke, X. and Chen, C. (2007). Thermodynamic functions and pressure-temperature phase diagram of lithium alanates by ab initio calculations. *Phys. Rev. B*, 76(2), 024112.

Khzouz, M., Gkanas, E.I., Girella, A., Statheros, T., Milanese, C. (2020). Sustainable hydrogen production via LiH hydrolysis for unmanned air vehicle (UAV) applications. *Int. J. Hydrog. Energy*, 45(8), 5384–5394.

Kight, B. (2015). Magnesium alloys in the automotive market. In *American Metal Market's North American Automotive Networking Conference,* 12–13 January, Detroit, MI.

Kim, D.Y., Lee, H.M., Seo, J., Shin, S.K., Kim, K.S. (2010). Rules and trends of metal cation driven hydride-transfer mechanisms in metal amidoboranes. *Phys. Chem. Chem. Phys.*, 12(20), 5446.

Kobayashi, T., Gupta, S., Caporini, M.A., Pecharsky, V.K., Pruski, M. (2014). Mechanism of solid-state thermolysis of ammonia borane: A ^{15}N NMR study using fast magic-angle spinning and dynamic nuclear polarization. *J. Phys. Chem. C*, 118(34), 19548–19555.

Kojima, Y., Kawai, Y., Kimbara, M., Nakanishi, H., Matsumoto, S. (2004a). Hydrogen generation by hydrolysis reaction of lithium borohydride. *Int. J. Hydrog. Energy*, 29(12), 1213–1217.

Kojima, Y., Suzuki, K.I., Kawai, Y. (2004b). Hydrogen generation by hydrolysis reaction of magnesium hydride. *J. Mater. Sci.*, 39(6), 2227–2229.

Kojima, Y., Kawai, Y., Haga, T., Matsumoto, M., Koiwai, A. (2007). Direct formation of $LiAlH_4$ by a mechanochemical reaction. *J. Alloys Compd.*, 441(1/2), 189–191.

Kong, V., Kirk, D., Foulkes, F., Hinatsu, J. (2003). Development of hydrogen storage for fuel cell generators II: Utilization of calcium hydride and lithium hydride. *Int. J. Hydrog. Energy*, 28(2), 205–214.

Kulekci, M.K. (2008). Magnesium and its alloys applications in automotive industry. *Int. J. Adv. Manuf. Technol.*, 39(9/10), 851–865.

Kumar, R., Karkamkar, A., Bowden, M., Autrey, T. (2019). Solid-state hydrogen rich boron-nitrogen compounds for energy storage. *Chem. Soc. Rev.*, 48(21), 5350–5380.

Lacina, D., Wegrzyn, J., Reilly, J., Celebi, Y., Graetz, J. (2010). Regeneration of aluminium hydride using dimethylethylamine. *Energy Environ. Sci.*, 3(8), 1099.

Lapeña-Rey, N., Blanco, J.A., Ferreyra, E., Lemus, J.L., Pereira, S., Serrot, E. (2017). A fuel cell powered unmanned aerial vehicle for low altitude surveillance missions. *Int. J. Hydrog. Energy*, 42(10), 6926–6940.

Laude, T., Kobayashi, T., Sato, Y. (2010). Electrolysis of LiOH for hydrogen supply. *Int. J. Hydrog. Energy*, 35(2), 585–588.

Le, T.T., Pistidda, C., Puszkiel, J., Milanese, C., Garroni, S., Emmler, T., Capurso, G., Gizer, G., Klassen, T., Dornheim, M. (2019). Efficient synthesis of alkali borohydrides from mechanochemical reduction of borates using magnesium–aluminum-based waste. *Metals*, 9(10), 1061.

Legree, M., Sabatier, J., Mauvy, F., Awad, A.S., Faessel, M., Bos, F., Bobet, J.-L. (2020). Autonomous hydrogen production for proton exchange membrane fuel cells PEMFC. *J. Energy Power Technol.*, 2(2), 1–18.

Li, Z.P., Liu, B.H., Morigasaki, N., Suda, S. (2003). Preparation of potassium borohydride by a mechano-chemical reaction of saline hydrides with dehydrated borate through ball milling. *J. Alloys Compd.*, 354(1/2), 243–247.

Li, C., Peng, P., Zhou, D.W., Wan, L. (2011). Research progress in $LiBH_4$ for hydrogen storage: A review. *Int. J. Hydrog. Energy*, 36(22), 14512–14526.

Li, F., Wei, Z., Manthiram, A., Feng, Y., Ma, J., Mai, L. (2019a). Sodium-based batteries: From critical materials to battery systems. *J. Mater. Chem. A*, 7(16), 9406–9431.

Li, H., Eksteen, J., Kuang, G. (2019b). Recovery of lithium from mineral resources: State-of-the-art and perspectives – A review. *Hydrometallurgy*, 189, 105129.

Li, M., Bai, Y., Zhang, C., Song, Y., Jiang, S., Grouset, D., Zhang, M. (2019c). Review on the research of hydrogen storage system fast refueling in fuel cell vehicle. *Int. J. Hydrog. Energy*, 44(21), 10677–10693.

Liu, C.H. and Chen, B.H. (2015). The concept about the regeneration of spent borohydrides and used catalysts from green electricity. *Materials*, 8(6), 3456–3466.

Liu, C.H., Wu, Y.C., Chou, C.C., Chen, B.H., Hsueh, C.L., Ku, J.R., Tsau, F. (2012). Hydrogen generated from hydrolysis of ammonia borane using cobalt and ruthenium based catalysts. *Int. J. Hydrog. Energy*, 37(3), 2950–2959.

Liu, Y., Wang, X., Liu, H., Dong, Z., Cao, G., Yan, M. (2014). Exploration of hydrogen generation from an Mg–$LiBH_4$ system improved by $NiCl_2$ addition. *J. Power Sources*, 251, 459–465.

Liu, P., Wu, H., Wu, C., Chen, Y., Xu, Y., Wang, X., Zhang, Y. (2015). Microstructure characteristics and hydrolysis mechanism of Mg–Ca alloy hydrides for hydrogen generation. *Int. J. Hydrog. Energy*, 40(10), 3806–3812.

Liu, F., Peng, C., Ma, Q., Wang, J., Zhou, S., Chen, Z., Wilson, B.P., Lundström, M. (2021a). Selective lithium recovery and integrated preparation of high-purity lithium hydroxide products from spent lithium-ion batteries. *Sep. Purif. Technol.*, 259, 118181.

Liu, H., Zhang, L., Ma, H., Lu, C., Luo, H., Wang, X., Huang, X., Lan, Z., Guo, J. (2021b). Aluminum hydride for solid-state hydrogen storage: Structure, synthesis, thermodynamics, kinetics, and regeneration. *J. Energy Chem.*, 52, 428–440.

Lu, Z.H., Yao, Q., Zhang, Z., Yang, Y., Chen, X. (2014). Nanocatalysts for hydrogen generation from ammonia borane and hydrazine borane. *Journal of Nanomaterials*, 2014(1), 1–11.

Luo, A.A. (2013). Magnesium casting technology for structural applications. *J. Magnes. Alloy.*, 1(1), 2–22.

Luo, W., Campbell, P.G., Zakharov, L.N., Liu, S.-Y. (2011a). A single-component liquid-phase hydrogen storage material. *J. Am. Chem. Soc.*, 133(48), 19326–19329.

Luo, W., Zakharov, L.N., Liu, S.-Y. (2011b). 1,2-BN cyclohexane: Synthesis, structure, dynamics, and reactivity. *J. Am. Chem. Soc.*, 133(33), 13006–13009.

Lv, Y. and Wu, Y. (2021). Current research progress in magnesium borohydride for hydrogen storage: A review. *Prog. Nat. Sci. Mater. Int.*, 31(6), 809–820.

Ma, H., Chen, Z., Mao, Z. (2013). Controlled growth of magnesium hydroxide crystals and its effect on the high-temperature properties of cotton/magnesium hydroxide composites. *Vacuum*, 95, 1–5.

Maria, A., Elghoul, Z., Acker, K. (2022). Environmental assessment of an innovative lithium production process. *Procedia CIRP*, 105, 672–677.

Martino, M., Ruocco, C., Meloni, E., Pullumbi, P., Palma, V. (2021). Main hydrogen production processes: An overview. *Catalysts*, 11(5), 547.

Mauron, P., Buchter, F., Friedrichs, O., Remhof, A., Bielmann, M., Zwicky, C.N., Züttel, A. (2008). Stability and reversibility of $LiBH_4$. *J. Phys. Chem. B*, 112(3), 906–910.

Messer, C.E., Fasolino, L.G., Thalmayer, C.E. (1955). The heats of formation of lithium, sodium and potassium hydrides. *J. Am. Chem. Soc.*, 77(17), 4524–4526.

Miranda, C.R. and Ceder, G. (2007). Ab initio investigation of ammonia-borane complexes for hydrogen storage. *J. Chem. Phys.*, 126(18), 184703.

Mohtadi, R. and Orimo, S. (2017). The renaissance of hydrides as energy materials. *Nat. Rev. Mater.*, 2(3), 16091.

Monteverde, M. and Magistri, L. (2012). Hydrogen from sodium borohydride and fossil source: An energetic and economical comparison. *Int. J. Hydrog. Energy*, 37(6), 5452–5460.

Montgomery, C.D. (1973). Fabrication and properties of lithium hydride. *Nucl. Eng. Des.*, 25(2), 309–314.

Mordike, B. and Ebert, T. (2001). Magnesium: Properties, applications, potential. *Mater. Sci. Eng. A*, 302(1), 37–45.

Moury, R., Moussa, G., Demirci, U.B., Hannauer, J., Bernard, S., Petit, E., van der Lee, A., Miele, P. (2012). Hydrazine borane: Synthesis, characterization, and application prospects in chemical hydrogen storage. *Phys. Chem. Chem. Phys.*, 14(5), 1768–1777.

Na Ranong, C., Höhne, M., Franzen, J., Hapke, J., Fieg, G., Dornheim, M., Eigen, N., Bellosta von Colbe, J.M., Metz, O. (2009). Concept, design and manufacture of a prototype hydrogen storage tank based on sodium alanate. *Chem. Eng. Technol.*, 32(8), 1154–1163.

National Center for Biotechnology Information (2022). PubChem substance record for SID 134979979 [Online]. Available at: https://pubchem.ncbi.nlm.nih.gov/substance/134979979 [Accessed 17 October 2022].

National Institute of Standards and Technology (2021a). Chemistry webbook, magnesium chloride [Online]. Available at: https://webbook.nist.gov/cgi/cbook.cgi?ID=C7786303&Mask=2 [Accessed 13 September 2022].

National Institute of Standards and Technology (2021b). Chemistry webbook, magnesium hydride [Online]. Available at: https://webbook.nist.gov/cgi/cbook.cgi?ID=C1309428&Mask=2 [Accessed 25 August 2022].

Nunes, H.X., Silva, D.L., Rangel, C.M., Pinto, A.M.F.R. (2021). Rehydrogenation of sodium borates to close the $NaBH_4$-H_2 cycle: A review. *Energies*, 14(12), 3567.

Orimo, S., Nakamori, Y., Kato, T., Brown, C., Jensen, C.M. (2006). Intrinsic and mechanically modified thermal stabilities of α-, β- and γ-aluminum trihydrides AlH_3. *Appl. Phys. A*, 83(1), 5–8.

Ould-Amara, H., Alligier, D., Petit, E., Yot, P.G., Demirci, U.B. (2018). Sodium borohydride and propylene glycol, an effective combination for the generation of 2.3 wt % of hydrogen. *Int. J. Hydrog. Energy*, 43(15), 7237–7244.

Ouyang, L., Huang, J.M., Fang, C.J., Zhang, Q.A., Sun, D.L., Zhu, M. (2012). The controllable hydrolysis rate for $LaMg_{12}$ hydride. *Int. J. Hydrog. Energy*, 37(17), 12358–12364.

Ouyang, L., Huang, J.M., Wang, H., Wen, Y.J., Zhang, Q.A., Sun, D.L., Zhu, M. (2013). Excellent hydrolysis performances of Mg_3RE hydrides. *Int. J. Hydrog. Energy*, 38(7), 2973–2978.

Ouyang, L., Zhong, H., Li, H.-W., Zhu, M. (2018). A recycling hydrogen supply system of $NaBH_4$ based on a facile regeneration process: A review. *Inorganics*, 6(1), 10.

Ouyang, L., Liu, M., Chen, K., Liu, J., Wang, H., Zhu, M., Yartys, V. (2022). Recent progress on hydrogen generation from the hydrolysis of light metals and hydrides. *J. Alloys Compd.*, 910, 164831.

Paskevicius, M., Jepsen, L.H., Schouwink, P., Černý, R., Ravnsbæk, D.B., Filinchuk, Y., Dornheim, M., Besenbacher, F., Jensen, T.R. (2017). Metal borohydrides and derivatives – Synthesis, structure and properties. *Chem. Soc. Rev.*, 46(5), 1565–1634.

Petit, E., Miele, P., Demirci, U.B. (2016). By-product carrying humidified hydrogen: An underestimated issue in the hydrolysis of sodium borohydride. *ChemSusChem*, 9(14), 1777–1780.

Phillips, J., Bradford, M.C., Klanchar, M. (1995). A calorimetric study of the mechanism and thermodynamics of the lithium hydride-water reaction at elevated temperatures. *Energy & Fuels*, 9(4), 569–573.

Pighin, S.A., Urretavizcaya, G., Bobet, J.-L., Castro, F.J. (2020). Nanostructured Mg for hydrogen production by hydrolysis obtained by MgH_2 milling and dehydriding. *J. Alloys Compd.*, 827, 154000.

Pinnick, H.W. (1983). Potassium hydride in organic synthesis. A review. *Org. Prep. Proced. Int.*, 15(3), 199–223.

Pitcher, G. and Kavarnos, G.J. (1997). A test assembly for hydrogen production by the hydrolysis of solid lithium hydride. *Int. J. Hydrog. Energy*, 22(6), 575–579.

Pretzel, F.E., Rupert, G.N., Mader, C.L., Storms, E.K., Gritton, G.V., Rushing, C.C. (1960). Properties of lithium hydride I. Single crystals. *J. Phys. Chem. Solids*, 16(1/2), 10–20.

Ramachandran, P. and Gagare, P.D. (2007). Preparation of ammonia borane in high yield and purity, methanolysis, and regeneration. *Inorg. Chem.*, 46(19), 7810–7817.

Ramachandran, P., Raju, B.C., Gagare, P.D. (2012). One-pot synthesis of ammonia–borane and trialkylamine–boranes from trimethyl borate. *Org. Lett.*, 14(24), 6119–6121.

Rivard, E., Trudeau, M., Zaghib, K. (2019). Hydrogen storage for mobility: A review. *Materials*, 12(12), 1973.

Ropp, R.C. (ed.) (2013). Group 16 (O, S, Se, Te) alkaline earth compounds. In *Encyclopedia of the Alkaline Earth Compounds*. Elsevier, Amsterdam.

Rude, L.H., Nielsen, T.K., Ravnsbaek, D.B., Bösenberg, U., Ley, M.B., Richter, B., Arnbjerg, L.M., Dornheim, M., Filinchuk, Y., Besenbacher, F. et al. (2011). Tailoring properties of borohydrides for hydrogen storage: A review. *Phys. Status Solidi*, 208(8), 1754–1773.

Sadoway, D.R. (1998). Toward new technologies for the production of lithium. *JOM*, 50(5), 24–26.

Sahin, O., Dolas, H., Ozdemir, M. (2007). The effect of various factors on the hydrogen generation by hydrolysis reaction of potassium borohydride. *Int. J. Hydrog. Energy*, 32(13), 2330–2336.

Saitoh, H., Machida, A., Katayama, Y., Aoki, K. (2008). Formation and decomposition of AlH_3 in the aluminum-hydrogen system. *Appl. Phys. Lett.*, 93(15), 151918.

Salman, M.S., Lai, Q., Luo, X., Pratthana, C., Rambhujun, N., Costalin, M., Wang, T., Sapkota, P., Liu, W., Grahame, A. et al. (2022). The power of multifunctional metal hydrides: A key enabler beyond hydrogen storage. *J. Alloys Compd.*, 920, 165936.

Sato, Y. and Takeda, O. (eds) (2013). Hydrogen storage and transportation system through lithium hydride using molten salt technology. In *Molten Salts Chemistry*. Elsevier, Amsterdam.

Schlesinger, H.I. and Brown, H.C. (1940). Metallo borohydrides. III. Lithium borohydride. *J. Am. Chem. Soc.*, 62(12), 3429–3435.

Schlesinger, H.I. and Brown, H.C. (1953). Uranium(IV) borohydride. *J. Am. Chem. Soc.*, 75(1), 219–221.

Schlesinger, H.I., Sanderson, R.T., Burg, A.B. (1939). A volatile compound of aluminum, boron and hydrogen. *J. Am. Chem. Soc.*, 61(2), 536–536.

Schlesinger, H.I., Brown, H.C., Abraham, B., Bond, A.C., Davidson, N., Finholt, A.E., Gilbreath, J.R., Hoekstra, H., Horvitz, L., Hyde, E.K. et al. (1953a). New developments in the chemistry of diborane and the borohydrides. I. General summary 1. *J. Am. Chem. Soc.*, 75(1), 186–190.

Schlesinger, H.I., Brown, H.C., Finholt, A.E., Gilbreath, J.R., Hoekstra, H.R., Hyde, E.K. (1953b). Sodium borohydride, its hydrolysis and its use as a reducing agent and in the generation of hydrogen. *J. Am. Chem. Soc.*, 75(1), 215–219.

Schlesinger, H.I., Brown, H.C., Hoekstra, H.R., Rapp, L.R. (1953c). Reactions of diborane with alkali metal Hydrides and their addition compounds. New syntheses of borohydrides. Sodium and potassium borohydrides 1. *J. Am. Chem. Soc.*, 75(1), 199–204.

Shevlin, S.A., Kerkeni, B., Guo, Z.X. (2011). Dehydrogenation mechanisms and thermodynamics of MNH_2BH_3 (M = Li, Na) metal amidoboranes as predicted from first principles. *Phys. Chem. Chem. Phys.*, 13(17), 7649.

Shimoda, K., Zhang, Y., Ichikawa, T., Miyaoka, H., Kojima, Y. (2011). Solid state NMR study on the thermal decomposition pathway of sodium amidoborane $NaNH_2BH_3$. *J. Mater. Chem.*, 21(8), 2609.

Shimoda, K., Doi, K., Nakagawa, T., Zhang, Y., Miyaoka, H., Ichikawa, T., Tansho, M., Shimizu, T., Burrell, A.K., Kojima, Y. (2012). Comparative study of structural changes in NH_3BH_3, $LiNH_2BH_3$, and KNH_2BH_3 during dehydrogenation process. *J. Phys. Chem. C*, 116(9), 5957–5964.

Shore, S.G. and Parry, R.W. (1955). The crystalline compound ammonia-borane[1] H_3NBH_3. *J. Am. Chem. Soc.*, 77(22), 6084–6085.

Simbara, A.T., Muharam, A.L., Hanifa, I., Rizky, K.M., Fadhilla, R.F., Solehudin, S.H., Nandiyanto, A.B.D. (2020). Comparison of the $LiBH_4$ material synthesis method and its application as hydrogen energy storage. *J. Appl. Sci. Environ. Stud.*, 3(4), 232–253.

Singh, R. (2022). Reversible chemical hydrogen storage in borohydrides via thermolysis and hydrolysis: Recent advances, challenges, and perspectives. *Int. J. Hydrog. Energy*, 47(62), 26549–26573.

Sklar, N. and Post, B. (1967). Crystal structure of lithium aluminum hydride. *Inorg. Chem.*, 6(4), 669–671.

Song, G. and Atrens, A. (1999). Corrosion mechanisms of magnesium alloys. *Adv. Eng. Mater.*, 1(1), 11–33.

Song, G., Ma, S., Tang, G., Yin, Z., Wang, X. (2010). Preparation and characterization of flame-retardant form-stable phase change materials composed by EPDM, paraffin and nano magnesium hydroxide. *Energy*, 35(5), 2179–2183.

Stecher, O. and Wiberg, E. (1942). Über einen nichtflüchtigen, polymeren Aluminiumwasserstoff $(AlH_3)_x$ und einige flüchtige Verbindungen des monomeren AlH_3. *Berichte der Dtsch. Chem. Gesellschaft A B Ser.*, 75(12), 2003–2012.

Strawser, D., Thangavelautham, J., Dubowsky, S. (2014). A passive lithium hydride-based hydrogen generator for low power fuel cells for long-duration sensor networks. *Int. J. Hydrog. Energy*, 39(19), 10216–10229.

Suárez-Alcántara, K. and Tena García, J.R. (2021). Metal borohydrides beyond groups I and II: A review. *Materials*, 14(10), 2561.

Sutton, A.D., Burrell, A.K., Dixon, D.A., Garner, E.B., Gordon, J.C., Nakagawa, T., Ott, K.C., Robinson, J.P., Vasiliu, M. (2011). Regeneration of ammonia borane spent fuel by direct reaction with hydrazine and liquid ammonia. *Science*, 331(6023), 1426–1429.

Swain, B. (2017). Recovery and recycling of lithium: A review. *Sep. Purif. Technol.*, 172, 388–403.

Taber, D.F. and Nelson, C.G. (2006). Potassium hydride in paraffin: A useful base for organic synthesis. *J. Org. Chem.*, 71(23), 8973–8974.

Talens Peiró, L., Villalba Méndez, G., Ayres, R.U. (2013). Lithium: Sources, production, uses, and recovery outlook. *JOM*, 65(8), 986–996.

Tan, Z.H., Ouyang, L.Z., Huang, J.M., Liu, J.W., Wang, H., Shao, H.Y., Zhu, M. (2019). Hydrogen generation via hydrolysis of Mg_2Si. *J. Alloys Compd.*, 770, 108–115.

Tang, Z., Zhang, L., Wan, L., Huang, Z., Liu, H., Guo, Z., Yu, X. (2016). Regeneration of alkaline metal amidoboranes with high purity. *Int. J. Hydrog. Energy*, 41(1), 407–412.

Tayeh, T., Awad, A.S., Nakhl, M., Zakhour, M., Silvain, J.-F., Bobet, J.-L. (2014). Production of hydrogen from magnesium hydrides hydrolysis. *Int. J. Hydrog. Energy*, 39(7), 3109–3117.

Tran, T. and Luong, V.T. (eds) (2015). Lithium production processes. In *Lithium Process Chemistry*. Elsevier, Amsterdam.

Uan, J.Y., Cho, C.Y., Liu, K.T. (2007). Generation of hydrogen from magnesium alloy scraps catalyzed by platinum-coated titanium net in NaCl aqueous solution. *Int. J. Hydrog. Energy*, 32(13), 2337–2343.

Uan, J.Y., Lin, M.C., Cho, C.Y., Liu, K.T., Lin, H.I. (2009a). Producing hydrogen in an aqueous NaCl solution by the hydrolysis of metallic couples of low-grade magnesium scrap and noble metal net. *Int. J. Hydrog. Energy*, 34(4), 1677–1687.

Uan, J.Y., Yu, S.H., Lin, M.C., Chen, L.F., Lin, H.I. (2009b). Evolution of hydrogen from magnesium alloy scraps in citric acid-added seawater without catalyst. *Int. J. Hydrog. Energy*, 34(15), 6137–6142.

Verma, A. (2010). Sodium hydride. *Synlett*, 2010(15), 2361–2362.

Wang, J., Ebner, A.D., Ritter, J.A. (2006). Physiochemical pathway for cyclic dehydrogenation and rehydrogenation of $LiAlH_4$. *J. Am. Chem. Soc.*, 128(17), 5949–5954.

Weber, G., Sciora, E., Guichard, J., Bouyer, F., Bezverkhyy, I., Bernard, F., Lecoq, H., Besnard, R., Bellat, J.P. (2018). New insight on the lithium hydride–water vapor reaction system. *Int. J. Hydrog. Energy*, 43(50), 22557–22567.

Wolstenholme, D.J., Titah, J.T., Che, F.N., Traboulsee, K.T., Flogeras, J., McGrady, G.S. (2011). Homopolar dihydrogen bonding in alkali-metal amidoboranes and its implications for hydrogen storage. *J. Am. Chem. Soc.*, 133(41), 16598–16604.

Xiong, Z., Yong, C.K., Wu, G., Chen, P., Shaw, W., Karkamkar, A., Autrey, T., Jones, M.O., Johnson, S.R., Edwards, P.P., David, W.I.F. (2008). High-capacity hydrogen storage in lithium and sodium amidoboranes. *Nat. Mater.*, 7(2), 138–141.

Yang, Z., Wang, Y., Liang, J., Chen, J. (2011). Hydrogen releasing of lithium amidoborane-$LiNH_2BH_3$. *Mater. Trans.*, 52(4), 651–653.

Yang, B., Zou, J., Huang, T., Mao, J., Zeng, X., Ding, W. (2019). Enhanced hydrogenation and hydrolysis properties of core-shell structured Mg-MO_x (M = Al, Ti and Fe) nanocomposites prepared by arc plasma method. *Chem. Eng. J.*, 371, 233–243.

Yang, L., Hou, X., Hou, K., Shi, H., Feng, L., Suo, G., Ye, X., Zhang, L., Yang, Y. (2021). Hydrolysis H_2 generation of Mg–Ni alloy catalyzed by expandable graphite/stannic oxide. *Int. J. Hydrog. Energy*, 46(27), 14024–14035.

Yu, S.H., Uan, J.Y., Hsu, T.L. (2012). Effects of concentrations of NaCl and organic acid on generation of hydrogen from magnesium metal scrap. *Int. J. Hydrog. Energy*, 37(4), 3033–3040.

Zhang, J., Fisher, T., Gore, J., Hazra, D., Ramachandran, P. (2006). Heat of reaction measurements of sodium borohydride alcoholysis and hydrolysis. *Int. J. Hydrog. Energy*, 31(15), 2292–2298.

Zhang, J., Cuevas, F., Percherón-Guégan, A. (2013). Stockage réversible de l'hydrogène dans les alanates. *Tech. l'Ingénieur*, RE111v2, 1–11.

Zhao, L., Xu, F., Zhang, C., Wang, Z., Ju, H., Gao, X., Zhang, X., Sun, L., Liu, Z. (2021). Enhanced hydrogen storage of alanates: Recent progress and future perspectives. *Prog. Nat. Sci. Mater. Int.*, 31(2), 165–179.

Zheng, J., Wang, C.G., Zhou, H., Ye, E., Xu, J., Li, Z., Loh, X.J. (2021). Current research trends and perspectives on solid-state nanomaterials in hydrogen storage. *Research*, 2021, 1–39.

Zhong, S., Wu, C., Li, J., Chen, Y., Wang, Y., Yan, Y. (2021). Effect of cations in chloride solutions on low-temperature hydrolysis performances and mechanisms of Mg–Ca-based hydride. *J. Alloys Compd.*, 851, 156762.

Zhu, Y., Ouyang, L., Zhong, H., Liu, J., Wang, H., Shao, H., Huang, Z., Zhu, M. (2020). Efficient synthesis of sodium borohydride: Balancing reducing agents with intrinsic hydrogen source in hydrated borax. *ACS Sustain. Chem. Eng.*, 8(35), 13449–13458.

Zidan, R., Garcia-Diaz, B.L., Fewox, C.S., Stowe, A.C., Gray, J.R., Harter, A.G. (2009). Aluminium hydride: A reversible material for hydrogen storage. *Chem. Commun.*, 25, 3717.

Zimmerman, P.M., Paul, A., Zhang, Z., Musgrave, C.B. (2009). Oligomerization and autocatalysis of NH_2BH_2 with ammonia–borane. *Inorg. Chem.*, 48(3), 1069–1081.

Zou, R., Li, J., Cui, J., Zhang, W., Lei, G., Cao, H., Chen, P. (2023). Catalyst-free synthesis of lithium hydride at room temperature. *Chem. Commun.*, 59(18), 2660–2663.

Züttel, A., Wenger, P., Rentsch, S., Sudan, P., Mauron, P., Emmenegger, C. (2003). LiBH4 a new hydrogen storage material. *J. Power Sources*, 118(1/2), 1–7.

8

Hydrogen Adsorption in High-Surface Area Porous Materials

Rafael MORALES-OSPINO, Alain CELZARD and Vanessa FIERRO
Institut Jean Lamour, CNRS, Université de Lorraine, Épinal, France

8.1. Introduction

Hydrogen (H_2) adsorption in high-surface area porous materials by adsorption at cryogenic temperatures enables H_2 storage and is an active and growing area of interdisciplinary research between materials science and energy storage. Among the wide variety of available porous adsorbents, carbon-based materials and metal organic frameworks (MOFs) have been the most intensively studied for H_2 storage applications (Liu et al. 2016). More recently, hybrid materials have become increasingly attractive in light of the need to improve the current H_2 storage capacities of traditional adsorbents. Among carbonaceous materials, activated carbons (ACs) have the largest surface areas and porosities, allowing for the adsorption of H_2 molecules onto their surfaces (Yang et al. 2012). MOFs are a type of porous crystalline material consisting of metal ions or clusters linked by organic ligands. MOFs also have a high surface area and tunable pore size, making them versatile materials for H_2 adsorption (Liu et al. 2016).

The ability of a porous material to adsorb H_2 is influenced by various factors, such as surface area, pore size, pressure and temperature. Researchers are working to develop new porous materials with enhanced H_2 adsorption properties and to understand the fundamental mechanisms underlying H_2 adsorption. This knowledge

Hydrogen Storage,
coordinated by Patricia DE RANGO and Fermin CUEVAS.

might lead to the development of more efficient H_2 storage technologies, which could facilitate the use of H_2 as a clean energy carrier.

This chapter addresses several topics related to high-pressure adsorption, specifically H_2 storage in adsorbents, including fundamental aspects of the Gibbs adsorption model and the importance of defining the Gibbs exclusion volume. The chapter also discusses the differences between the adsorbed amount and the excess amount, which are important parameters used to quantify the amount of H_2 stored in adsorbents. Similarly, concepts such as the total amount stored in a tank filled with a given adsorbent and the release capacity upon final application of H_2 are presented as ways that are more practical to account for H_2 storage capabilities. In addition, experimental considerations for high-pressure adsorption isotherms are presented, as are gas adsorption techniques used to characterize adsorbents for hydrogen storage applications. Textural properties commonly used to characterize the adsorbent structure, such as surface area and pore size distribution (PSD), are discussed, as these properties are crucial to understanding H_2 adsorption.

Finally, the H_2 storage capabilities of carbon-based materials and MOFs are presented, as well as recent developments in this area and the challenges that still need to be addressed to make H_2 adsorption, in porous materials, a viable option for energy storage and transportation.

8.2. Adsorption definition and the Gibbs adsorption model

Adsorption is the phenomenon of the accumulation of molecules in a given fluid at an interface, leading to an increase in the density of that fluid near that interface (Rouquerol et al. 2014; Thommes et al. 2015). The solid is called the adsorbent, and the fluid capable of being adsorbed is called the adsorptive fluid. When this fluid changes to an adsorbed state, that is, when its molecules contact the surface of the solid material, it is called an adsorbate (Lowell et al. 2006). If the adsorptive phase is a gas, then its adsorbed phase has a higher density than the bulk gas phase. This leads to the use of adsorption as a gas storage technology. On the other hand, when the fluid is a mixture of different components, the concentration of each component in the adsorbed phase depends on the affinity of the material surface to attract particular molecules of the mixture to the surface. This affinity, often called selectivity, depends on the strength of the gas–solid interactions. The result is an adsorbed phase that, in addition to being denser than the gas phase, may have a different composition. This principle is the basis of adsorption as a separation technology (Shivaji Sircar 2019). In addition, molecules of a given gas can be used as probes to investigate adsorbents. If the area occupied by one of these molecules at the completion of the adsorbed monolayer is known, the surface area of the adsorbent can be determined under specific temperature and pressure conditions. This principle is the basis of adsorption as a characterization technique for porous or divided materials.

H_2 molecules can interact with the surface of an adsorbent primarily through physical interactions of the van der Waals type, which are weak bonds between H_2 and the solid surface. The greater the surface area available for H_2 is, the greater the storage capacity. However, the type of PSD also plays an important role. Energistic solid–gas interactions for H_2 typically have an enthalpy of adsorption between – 3 and – 10 kJ/mol H_2. Therefore, porous materials can be optimized for hydrogen storage by selecting materials with ideal adsorption enthalpy and large hydrogen adsorption area (Schlichtenmayer and Hirscher 2016).

H_2 storage capacities in porous materials are generally reported as *excess* or *adsorbed* amounts, although concepts such as "total H_2 storage" or "H_2 release capacity" are becoming more common due to their practical importance. The H_2 storage capacity per amount of adsorbent can be expressed in various units, such as the weight percentage of H_2 to adsorbent [wt.], grams of H_2 per gram of adsorbent [g H_2/g_{ads}], millimoles of H_2 per gram of adsorbent [mmol H_2/g_{ads}], or kilograms of H_2 per cubic meter of adsorbent [kg $H_2/m^3{}_{ads}$], among others.

Below is a brief theoretical discussion of the meaning, estimation, and implications of the terms *excess* and *adsorbed* amount. Figure 8.1 shows a schematic diagram illustrating the above concepts. The well-known Gibbs adsorption model gave rise to the concept of *excess amount*, also called *surface excess amount*. The validity of the Gibbs adsorption approach relies on the fact that so-called excess amounts can be measured experimentally for a given fluid–solid system and can provide a numerical value representing the extent of adsorption (Shivaji Sircar 2019). The model considers a well-defined geometric surface very close to the adsorbent surface (i.e. at a distance from the surface $z = 0$) called the Gibbs dividing surface (GDS), where adsorption occurs. The amount of adsorbate located on the GDS is known as N_{ex} [g] (region A in Figure 8.1(a)) (Rouquerol et al. 2016). N_a [g] (region A+B in Figure 8.1(a)) is the quantity of adsorbate in a theoretical space of length L called the "adsorption space" or "adsorbed volume" (V_a). The terms N_{ex} and N_a can be related mathematically, as shown in equation [8.1]:

$$N_a = N_{ex} + \rho_g V_a \qquad [8.1]$$

where the second term on the right-hand side represents the amount of adsorptive that would occupy V_a [cm^3] under given conditions of temperature (T) and pressure (P). The term ρ_g [g/cm^3] denotes the density of the gas phase at the same T and P. The value of N_{ex} can be obtained experimentally by manometric or gravimetric methods, whereas N_a, often called the "absolute amount" in the literature, cannot be measured experimentally owing to the difficulty of establishing the boundary of the adsorption space and, therefore, V_a. Region C represents the gas in the dead space, that is, the gas outside the adsorption space boundary. The sum of regions A, B and C represents the total amount of fluid, gas and adsorbed phase in the system.

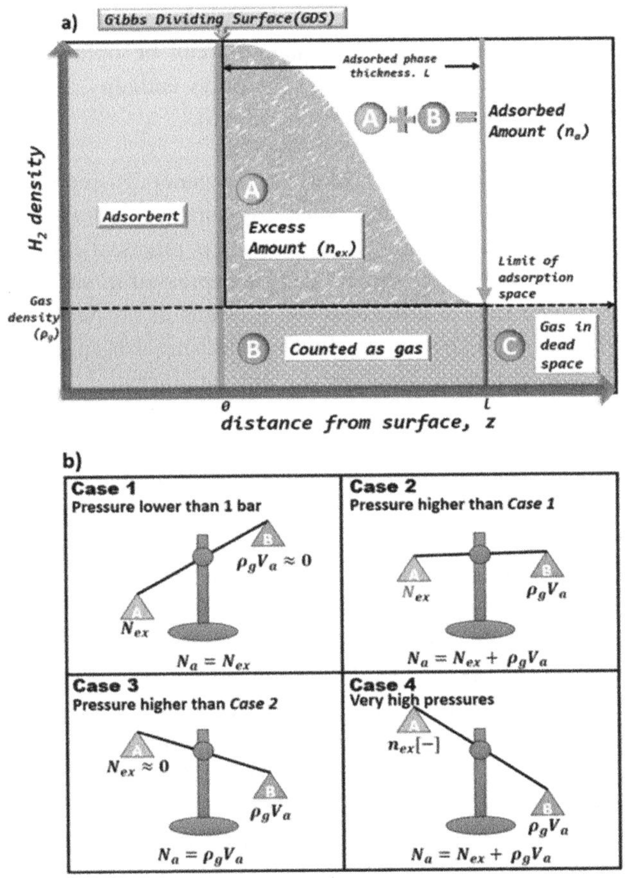

Figure 8.1. *(a) Graphical illustration of the Gibbs adsorption model, which plots the density of the adsorbate as a function of distance z from the adsorbent surface; (b) illustrative cases in the estimation of* N_a*. For a color version of this figure, see www.iste.co.uk/derango-cuevas/hydrogen.zip*

For most gases, the density is very low at pressures typically below 1 bar, that is, region A is much greater than region B (A>>B). Consequently, N_{ex} is almost the same as N_a, as shown in Case 1 of Figure 8.1(b). At higher pressures, the gas density increases, and eventually, region B is no longer negligible compared with region A, making the term $\rho_g V_a$ non-negligible too (Case 2 in Figure 8.1(b)). Notably, although regions A and B appear to have the same value in Case 2, this is merely an illustrative representation indicating that both terms should be considered. If the system reaches even higher pressures, region A becomes almost negligible, so that N_a is represented only by the term $\rho_g V_a$, as in Case 3 of Figure 8.1(b). Finally, at very high pressures (Case 4 in Figure 8.1(b)), Region B is much larger than Region A, to the point that N_{ex} may reach negative values, which has no physical meaning. Therefore, N_a must be systematically calculated since it actually describes the actual amount adsorbed.

The approach to evaluate N_a from N_{ex} in high-pressure adsorption experiments is an open question of notable importance in the field of adsorption, since there is no universal standard on how to estimate and report the adsorbed amount. Some methodologies reported in the open literature to obtain N_a from N_{ex} measurements are discussed in section 8.4.

8.3. How to report excess amount

In most of the literature, the Gibbs exclusion volume V_{GDS} [cm^3] (i.e. the volume that is not accessible to the adsorptive material in a given adsorption experiment) has always been considered to be the solid or skeletal volume of the adsorbent material (V_s [cm^3]), as described in the aforementioned experimental procedures. However, the value of V_s per gram of adsorbent or specific solid volume (v_s [cm^3/g]) is not always reported. In addition to the difficulties in comparing reported experimental data, estimating v_s in microporous adsorbents is tricky because there is a significant source of uncertainty (Rouquerol et al. 2016). In an attempt to standardize the way n_{ex} (the same meaning as N_{ex} [g] but expressed in [$mmol/g_{ads}$]) is reported in high-pressure adsorption experiments, Rouquerol et al. (2016) proposed changing the original position of the GDS (i.e. very close to the adsorbent surface) to a more convenient well-defined position, which would avoid the determination of v_s and thus reduce the uncertainty of the experimental adsorption isotherm data. Although the choice of the specific Gibbs exclusion volume, v_{GDS} [cm^3/g], is a choice of the experimenter, the practical decision has limited practical options as follows:

Choice 1: $v_{GDS} = v_s$

Choice 2: $v_{GDS} = 0.5\ cm^3/g$

Choice 3: $v_{GDS} = 0$

Choice 1 is appropriate for low-pressure experiments, typically below 1 bar, where the value of v_s has no significant influence on the final value of n_{ex} and thus n_a [mmol/g_{ads}]. *Choice 2* corresponds to the assumption of an adsorbent with a skeletal density of 2.0 g/cm^3, which is a typical value for many microporous carbon adsorbents. Although this is an arbitrary value, this choice yields excess amounts that are not far from those obtained when v_s is used, but with the advantage of using a single value, which avoids the risk of introducing the uncertainty of the v_s estimate into the n_{ex} calculation. *Choice 3*, originally proposed by Gumma and Talu (2010), is also known as *net adsorption*. This is a way to represent the experimental adsorption isotherm as raw data, that is, without any type of "correction". Net adsorption is very useful for evaluating the contribution of a certain adsorbent in a gas storage application, as it represents the storage enhancement due to the use of this adsorbent. Therefore, a positive net adsorbed amount indicates the additional amount stored in a tank filled with adsorbent compared with an empty tank, whereas a negative net adsorption value means that the adsorption no longer contributes to gas storage because the density of the gas phase is very high. The pressure at which the net adsorption is zero represents the maximum pressure at which an adsorbent can contribute to gas storage (Gumma and Talu 2010).

After evaluating the three choices above, Rouquerol et al. (2016) considered it more practical for high-pressure adsorption experiments to apply Choices 2 and 3 because they can be used without the risk of confusing v_s and v_{GDS} or falling into conceptual discussions about solid volume calculations. The fact that v_{GDS} is a choice of the experimenter implies that the n_{ex} value depends on the chosen v_{GDS}. In other words, we may have as many different n_{ex} as there are v_{GDS} selected. Therefore, we cannot speak about a "correct or right n_{ex}" but rather a standardized n_{ex}, provided that the chosen v_{GDS} value is indicated. The idea of the Rouquerol et al. (2016) proposal is ultimately to choose a Gibbs exclusion volume that does not necessarily require the estimation of v_s, but rather a simple and standardized way of reporting n_{ex} to facilitate the comparison of experimental data of high-pressure adsorption isotherms.

8.4. Excess amount versus adsorbed amount

For low-pressure gas adsorption or vapor adsorption experiments, the terms n_{ex} and n_a are often used interchangeably. While this is true for isotherms typically up to 1 bar or slightly higher depending on the adsorptive, where n_{ex} can be approximated by n_a, this is not the case for high-pressure adsorption isotherms. From a thermodynamic point of view, n_a is the quantity of interest since it is the one used for adsorbent characterization by gas adsorption (e.g. specific surface area and PSD) and as a reference for comparison with data derived from molecular simulations (Murata and Kaneko 2000; Zhou and Zhou 2009; Rouquerol et al. 2016). To determine the amount adsorbed, the adsorption space must be defined by assigning a value to the adsorbed phase volume V_a or to the adsorbed phase density ρ_a. Neither of these values is likely to be constant in reality, in addition to their inherent difficulty in being calculated with certainty by an experimental technique. Therefore, assuming too large a volume or too low a density will lead to an overestimation of n_a (Broom and Webb 2017). In general, for porous adsorbents, the pore volume has been suggested as the adsorption space (Myers and Monson 2014). The pore volume can be estimated by adsorbing gases with probe molecules such as N_2 or Ar at their boiling point using Gurvich's rule (Lowell et al. 2006; Rouquerol et al. 2014). For highly microporous materials, Myers and Monson (2014) stated that n_{ex} can best be converted to n_a by considering the adsorption space as the micropore volume. However, for macroporous or non-porous solids, n_a is hard to estimate due to the difficulty in drawing the boundary that separates the adsorbed molecules from the bulk gas phase (Myers and Monson 2014). Other studies have estimated the volume of the adsorbed phase by fitting the adsorption isotherm using a rearranged form of equation [8.1] per gram of adsorbent (Sharpe et al. 2013; Sdanghi et al. 2020a; Ramirez-Vidal et al. 2021):

$$n_{ex} = n_a - \rho_g v_a \qquad [8.2]$$

where n_a is determined by a selected adsorption model (e.g. the modified Dubinin–Astakhov model), and the value of v_a is considered an additional parameter that, in most cases, is constant for the set of experimental isotherms for which the fit is applied.

On the other hand, when ρ_a is defined, it is approximated as the liquid density of hydrogen, ρ_{LH2} (70.8 kg/m^3), at its normal boiling point (~ 20 K). Although most high-pressure isotherms for hydrogen storage are performed at 77 K, that is, at a temperature above the critical temperature of hydrogen (~ 33 K), the use of this approximation is widespread. Another way to define ρ_a is to plot n_{ex} against ρ_g to reveal the so-called "saturation regime", as shown in Figure 8.2. Figure 8.2 shows an example of how ρ_a can be estimated graphically by drawing a straight line with the points in the saturation regime to the intersection with the x-axis. The saturation regime begins when n_{ex} decreases linearly after reaching a maximum. At the intersection of the straight line with the x-axis (n_{ex} = 0), the gas phase and the adsorbed phase have the same density (ρ_g = ρ_a) according to equation [8.1], and thus, it is possible to estimate the density of the adsorbed phase at a given temperature (Poirier and Dailly 2008, 2009a, 2009b; Ramirez-Vidal et al. 2022a). Notably, according to the work of Poirier and Dailly (2008, 2009a, 2009b), the estimation of ρ_a or V_a with the saturation regime plot (n_{ex} vs. ρ_a) is best achieved with isotherms closer to the critical temperature of hydrogen with values of ρ_a below ρ_{LH2} for different MOF materials in isotherms experiments between 50 and 55 K.

As shown in Figure 8.2, some values of ρ_a estimated at 77 K exceed the value of ρ_{LH2}. Although this apparently ambiguous outcome may seem unrealistic, as stated by Bastos-Neto et al. (2012), other studies in the open literature have claimed that hydrogen may exhibit even densities closer to the solid-state in the adsorbed phase under cryogenic conditions (Silvera 1980; Jagiello et al. 2006; Romanos et al. 2018). For example, neutron scattering is one of the few experimental techniques allowing direct access to information about the state of H_2 adsorbed inside a porous material. Ting et al. (2015) reported experimental evidence that H_2 can be confined in solid form at temperatures up to 100 K in optimized carbons because of the densification of H_2 molecules in micropores at pressures as low as 0.02 MPa. This finding contradicts the theory that ρ_{LH2} is the theoretical upper limit of nanoconfined H_2 under supercritical conditions. Other studies also support the finding of solid-state H_2 or even a phase transition from solid to liquid with increasing pressure in nanoporous materials at cryogenic temperatures up to 110 K (Bahadur et al. 2017; Olsen et al. 2017).

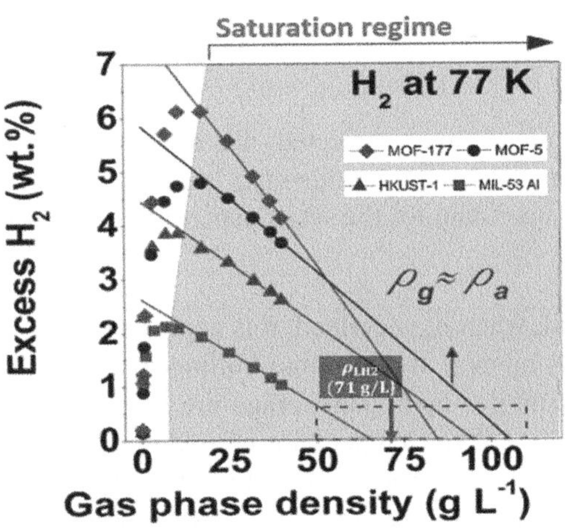

Figure 8.2. *Example of a plot of excess amount versus gas phase density at 77 K to estimate the adsorbed gas phase density for different MOFs (adapted with permission from Ramirez-Vidal et al. (2022a)). Copyright 2022 American Chemical Society). For a color version of this figure, see www.iste.co.uk/derango-cuevas/hydrogen.zip*

8.5. Total stored amount and release capacity

For H_2 storage applications, the total amount of stored hydrogen (n_{total}) is probably the benchmark parameter for comparing adsorption to other storage modes. When a gas is stored in a tank filled with a certain adsorbent, the fluid is stored in both the adsorbed and gas phases. The adsorbed phase is the portion of the fluid near the surface of the adsorbent, whereas the gas phase is stored in the dead space (i.e. in the larger pores of the adsorbent and in the interparticle free space). The higher the storage pressure is, the greater the amount of fluid stored in the gas phase due to the increased bulk gas density. As discussed below, estimating the total amount of H_2 stored has the advantage that it does not require the calculation of n_a. This requires excess adsorption data and some intrinsic physical parameters of the adsorbent, such as the packing density and skeletal density.

The total amount of H_2 stored in a tank in grams (m_{total}) is the sum of the masses of H_2 in both the gas phase (m_g) and the adsorbed phase (m_a):

$$m_{total} = m_g + m_a \quad [8.3]$$

$$m_{total} = \rho_g V_g + m_a \quad [8.4]$$

The volume occupied by the gas phase, V_g, in the tank is the difference between the volume of the tank, V_{tank}, and the sum of the volumes of the adsorbed phase and the solid phase ($V_a + V_s$), while m_a can be defined as the product of the adsorbent mass m_{ads} and n_a in gH_2/g_{ads}.

$$m_{total} = \rho_g (V_{tank} - V_a - V_s) + m_{ads} n_a \quad [8.5]$$

V_a can be expressed as the product of m_a and v_a, and n_a can be defined in terms of n_{ex} plus the amount counted as gas ($\rho_g v_a$), as follows:

$$m_{total} = \rho_g (V_{tank} - m_{ads} v_a - V_s) + m_{ads} (n_{ex} + \rho_g v_a) \quad [8.6]$$

$$m_{total} = \rho_g (V_{tank} - V_s) + m_{ads} n_{ex} \quad [8.7]$$

Dividing the previous equation by m_{ads}, we obtain n_{total} in the tank per gram of adsorbent:

$$n_{total} = \rho_g \left(\frac{1}{\rho_{packing}} - \frac{1}{\rho_s} \right) + n_{ex} \quad [8.8]$$

where $\rho_{packing}$ is the packing density of the adsorbent and ρ_s is the skeletal density of the solid, which is generally obtained by helium pycnometry.

As seen in equation [8.8], n_{total} depends on n_{ex}, which also depends on v_{GDS}. The question, then, is which v_{GDS} can provide a value of n_{total} that can be used to compare adsorption-based systems with other types of gas storage technologies. Although Rouquerol's main idea was to establish a standard method for reporting n_{ex} rather than estimating the "actual or correct n_{ex}", it now seems difficult to know how to proceed to make a fair comparison of adsorption as a gas storage technology. Even if the v_{GDS} value is reported, it is difficult for scientists outside of the adsorption field to understand how adsorption systems can store different amounts of fluid because of the application of a theoretical approach for the same fluid–solid system. For example, choosing net adsorption (*Choice 3*) will lead to lower total stored amounts since n_{ex} is minimal when v_{GDS} is zero.

The present authors' opinion on this issue is that it is probably more accurate to use *Choice 1* ($v_{GDS} = v_s$) for comparing adsorption with other types of gas storage technologies (i.e. for the total amount stored), since ρ_s is already in equation [8.8]. However, n_{ex} could be reported as suggested by Rouquerol by specifying the selected v_{GDS} and the equation of state used to account for the actual gas behavior in high-pressure adsorption experiments.

Furthermore, to evaluate H_2 adsorption storage systems in a real-world scenario, in addition to the storage conditions, the pressure and temperature conditions under which previously stored hydrogen is released must be defined (Ramirez-Vidal et al. 2022b).

The release conditions may change depending on the H_2 application. Once defined, the release capacity (Δn_{H2}) can be obtained by simply subtracting the total amount stored under release conditions $(n_{total})_{release}$ from the total amount stored under the storage conditions $(n_{total})_{storage}$, as shown in equation [8.9] (Dailly and Poirier 2011; Schlichtenmayer and Hirscher 2016; Gómez-Gualdrón et al. 2017; García-Holley et al. 2018; Chen et al. 2020). In general, this release process is achieved by charge–discharge cycles under the defined operational conditions.

$$\Delta n_{H2} = (n_{total})_{storage} - (n_{total})_{release} \qquad [8.9]$$

8.6. Experimental procedure to obtain high-pressure isotherms

Experimental adsorption isotherms for H_2 or any other gas of interest are mainly performed by the gravimetric method or the manometric method; therefore, a brief description of both methods is given in the following subsections.

8.6.1. *Gravimetric method*

High-pressure adsorption experiments performed with commercial gravimetric analyzers are typically performed with magnetic suspension balances (MSBs). MSB has the advantage of separating the microbalance from the sample or measurement space, thus protecting the microbalance from possible damage due to contact with the bulk gas under extreme pressure or temperature conditions (Dreisbach and Lösch 2000; Dreisbach et al. 2003). Therefore, the microbalance is placed in a separate chamber under ambient conditions. Some MSBs allow simultaneous measurement of mass change and density in the gas phase, as shown in Figure 8.3. The operating principle and major parts of a magnetic suspension balance are also shown in this figure. The microbalance can be weighed because of the communication between the electromagnet in the microbalance space and the permanent magnet in the sample space. MSBs with in situ density measurement operate at three operating positions: *zero-point*, *measurement point 1* and *measurement point 2*. At the *zero-point* position, the permanent magnet is freely suspended, and the balance can be tared. At *measurement point 1*, the permanent magnet changes position slightly to allow the microbalance to detect the change in mass. Finally, at *measurement point 2*, the permanent magnet moves again, carrying a titanium sinker with a calibrated volume. The fact that the microbalance weighs the sinker in a gas atmosphere makes it possible to measure the buoyancy effect and thus the density of the gas phase by applying Archimedes' principle.

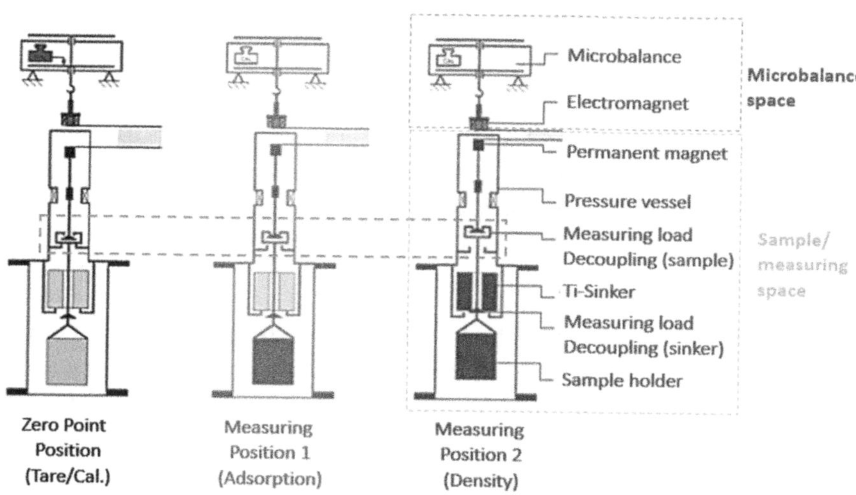

Figure 8.3. *Schematic of a magnetic suspension balance with simultaneous measurement of sample mass and gas phase density (adapted from Dreisbach et al. (2003) with permission from Springer Nature). For a color version of this figure, see www.iste.co.uk/derango-cuevas/hydrogen.zip*

The excess amount, n_{ex}, at a given pressure and temperature is not directly measured by the balance; it is the balance reading or "apparent mass" (Δm) that is obtained. To estimate n_{ex}, Δm must be corrected for the amount of gas displaced by the volume of the adsorbent and balance components due to the buoyancy effect (B_o):

$$n_{ex}(P,T) = \Delta m(P,T) + B_o\,(P,T) \qquad [8.10]$$

where B_o can be defined as follows:

$$B_o = (V_b + V_s)\rho_g(P,T) \qquad [8.11]$$

where V_b is the volume of the balance components and V_s is the solid or skeletal volume of the adsorbent. As shown in equation [8.11], B_o is greater when the density of the gas phase is greater, that is, when the pressure is greater. By substituting B_o, as defined in equation [8.11], into equation [8.10], we obtain equation [8.12]:

$$n_{ex}(P,T) = \Delta m(P,T) + (V_b + V_s)\rho_g(P,T) \qquad [8.12]$$

Equation [8.12] is the standard equation used to determine m_{ex} with a magnetic suspension balance. Dividing equation [8.12] by the mass of the degassed sample (m_{ads}) gives the value of n_{ex} per gram of adsorbent:

$$n_{ex}(P,T)[g/g_{ads}] = \frac{[\Delta m(P,T)+(V_b+V_s)\rho_g(P,T)]}{m_{ads}} \qquad [8.13]$$

To construct the corresponding isotherm, n_{ex} [g/g_{ads}], more commonly expressed in [mmol/g_{ads}], must be calculated and then plotted as a function of relative pressure (P/P_o), where P_o is the saturation pressure of the adsorptive. The experimental procedure for measuring excess adsorption isotherms in an MSB requires prior estimation of the volume of the balance component volume V_b, known as the blank test, and estimation of V_s:

– The *blank test* (without sample, i.e. $n_{ex} = 0$; $V_s = 0$) is performed to estimate V_b. A test without an adsorbent is performed at different pressures by dosing aliquots of gas. Helium is often used for this purpose. A plot of –(Δm) versus ρ_g has a slope equal to V_b according to equation [8.12].

– *Estimation of* V_s: this step occurs after the degassing process, and the experimental procedure is similar to that of the blank test but includes the adsorbent. Helium is used since it is considered an inert gas (i.e. it is not adsorbed in the sample and thus $n_{ex} = 0$). A plot of –(Δm) versus ρ_g has a slope equal to $V_b + V_s$ according to equation [8.12], with V_b being a known value.

Once V_b and V_s are estimated, H_2 adsorption begins with degassing the adsorbent, referred to as the sample regeneration step. Degassing is typically performed by heating under vacuum to remove moisture and/or gases previously adsorbed. Then, a series of H_2 adsorption determinations are performed at several pressure levels, each corresponding to a point on the isotherm taken when Δm remains constant at stabilized temperature and pressure. Temperature control in the sample space for H_2 isotherms is typically achieved using a liquid nitrogen cryocooler for isotherms at 77 K or a gas cooling system for temperatures between 123 and 293 K.

8.6.2. *Manometric method*

Figure 8.4 shows a basic scheme explaining the operation of an automatic manometric system for obtaining high-pressure isotherms at temperatures up to 77 K. The equipment essentially consists of a gas inlet, a manifold, pressure transducers, a depressurization system with a vacuum/atmospheric outlet, and a sample cell. The manifold is used to dose aliquots of gas to the sample cell, where the adsorbent is placed, during the adsorption process, and to gradually remove gas from the sample cell during desorption. The volumes of the manifold and sample cell are known and calibrated at the factory. The volume not occupied by the sample in the sample cell can be measured by expanding helium from the manifold at the beginning or end of the analysis. The latter method is preferable when ultramicroporous adsorbents are used because He can be trapped and cannot be removed by applying a secondary vacuum without heating (García-Díez et al. 2019; Schaefer et al. 2020). When a calibrated sample cell is used, the sample volume is subtracted using v_s.

The operating principle of the equipment is to measure the pressure inside the manifold before and after the gas is dosed to the cell. n_{ex} is then calculated on the basis of a mass balance taking into account the real behavior of the gas:

$$PV = n_{ex}ZRT \tag{8.14}$$

$$n_{ex} = n_{dosed} - n_{void} \tag{8.15}$$

$$n_{ex} = (\Delta PV/ZRT)_{manifold.} - (PV/ZRT)_{cell} \tag{8.16}$$

where Z is the H_2 compressibility factor, R (8.314 Pa m^3/mol/K) is the universal gas law constant, P (Pa) is the pressure and T (K) is the temperature. The difference between the amount of gas dosed at a certain pressure level (n_{dosed}) and the amount of gas remaining unadsorbed in the empty volume of the sample cell, n_{void}, is n_{ex}, as shown in equation [8.15].

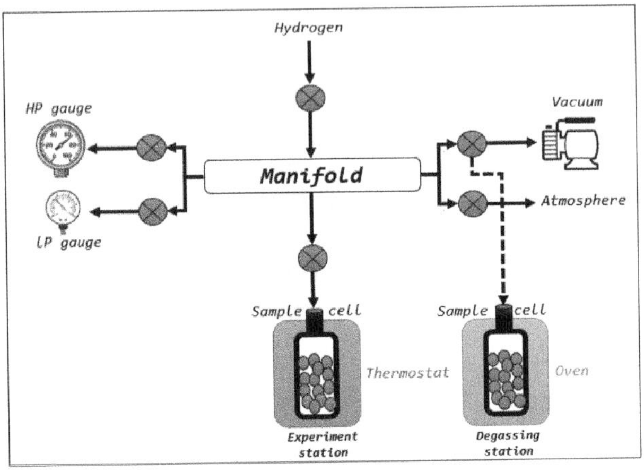

Figure 8.4. *Basic scheme of volumetric equipment for high-pressure hydrogen isotherms at cryogenic temperatures*

The experimental procedure begins with degassing the adsorbent material by heating under vacuum overnight, usually in a different degassing port than the one used to measure isothermal gas adsorption. The heat source often comes from a temperature-controlled oven that uses an electrical resistor, whereas the vacuum can be produced by diffusion or turbomolecular pumps, with the vacuum ranging from 10^{-4} to 10^{-9} mbar. Once the sample degassing is complete, the sample cell is connected to the experimental station, and the entire system, including the manifold and transducers lines, is evacuated. The sample cell is then cooled to the desired temperature, which is kept constant during the experiment using liquid nitrogen, if working at 77 K, or a cryocooler that brings nitrogen to the desired temperature between 77 and 323 K. For temperatures below 77 K, a He compressor allows temperatures as low as 25 K. For temperatures close to room temperature, a circulating bath can be used.

The adsorption experiment begins by admitting a fraction of the adsorptive to reach a target pressure in the manifold, as shown in Figure 8.5(a).

Once the pressure in the manifold has stabilized (see Figure 8.5(b)), the gas is immediately expanded from the manifold to the sample cell to begin the adsorption process, as shown in Figure 8.5(c). The final pressure is recorded when equilibrium is reached, that is, when there is no significant change in pressure or temperature in the sample cell. The difference between the initial manifold pressure and the equilibrium pressure during the adsorption process ($\Delta P_{manifold}$) is used to determine the amount of gas dosed, n_{dosed}. The equipment automatically repeats this process at higher pressure levels until the maximum pressure is set by the user, where the gas dosing is interrupted and the desorption process starts accordingly. During desorption, unlike adsorption, the gas flows from the sample cell to the manifold (see Figure 8.5(d)), while the pressure increase in the manifold is recorded for calculation purposes. As explained above, manometric measurements are indirect since the amount adsorbed/desorbed is obtained from pressure changes and are therefore based on the assumption that no other effects contribute to this pressure difference. Therefore, any gas leakage in the system will result in measurement errors (Broom and Webb 2017).

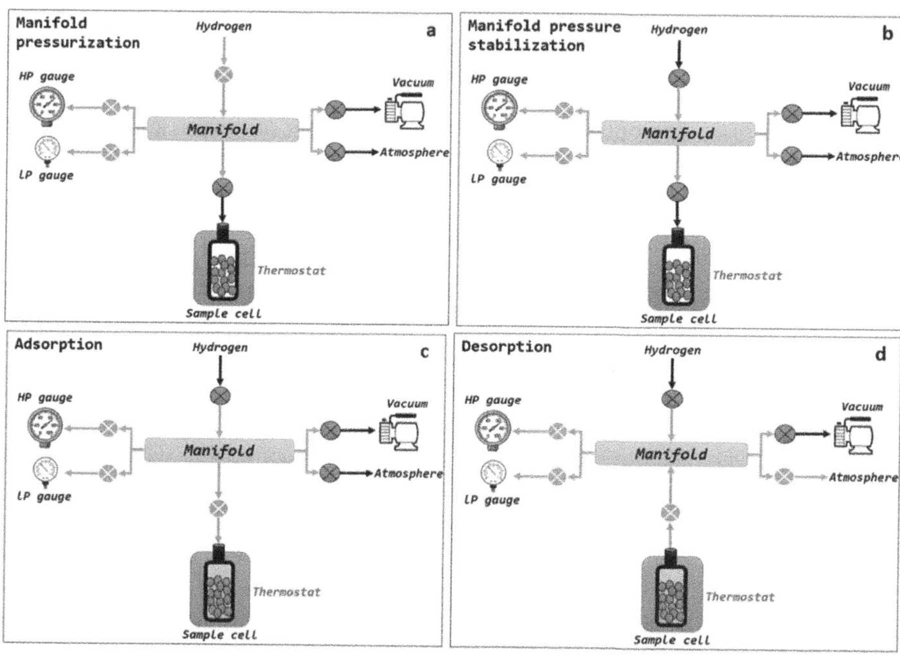

Figure 8.5. *Experimental steps for hydrogen high-pressure isotherms at cryogenic temperatures by the manometric method: (a) manifold pressurization, (b) manifold pressure stabilization, (c) adsorption and (d) desorption*

8.7. Characterization of adsorbents for hydrogen storage

In general, physical adsorption is used to characterize adsorbents and determine their textural properties, such as pore volume, PSD and specific surface area. These textural characteristics can be obtained after a series of adsorption–desorption experiments, carried out under strictly isothermal conditions, using probe molecules at a standardized temperature (e.g. N_2 at 77 K, Ar at 87 K, H_2 at 77 K, and CO_2 at

273 K), and after appropriate treatment of the adsorption data and application of some theoretical approach. The type of isotherm obtained can eventually reveal whether the adsorbent is microporous and/or contains mesopores.

The most common and recognized isotherm types in gas physisorption are classified by the International Union of Pure and Applied Chemistry (IUPAC) (Thommes et al. 2015). IUPAC also defined the type of adsorbent pore based on their pore diameter (width): micropores (< 2 nm), mesopores (between 2 and 50 nm) and macropores (> 50 nm). Owing to its relatively high availability and ease of comparison with literature data, N_2 at 77 K remains the preferred probe molecule for characterizing adsorbent physisorption. However, inaccurate estimates due to interactions of N_2 with polar materials or certain surface functional groups due to its quadrupole moment are also well known. Therefore, Ar adsorption at 87 K is suggested as a more appropriate option for the evaluation of micro- and mesopores due to the lack of interactions with surface functional groups (Thommes et al. 2015).

The selection of the probe molecule is therefore a critical choice that is not always obvious, as a given probe molecule may be more appropriate for a given adsorbent. For example, in well-defined porous materials with very small pore widths such as zeolites, LTA or 4A, N_2 at 77 K or Ar at 87 K are not used because of diffusion problems observed at such low temperatures and in too small pores. In this case, CO_2 at 273 K, with a smaller kinetic diameter and higher adsorption temperature, is a more practical choice. The evaluation of the micropores of zeolite 4A from the CO_2 adsorption data was found to be consistent with the X-ray diffraction results (Thommes et al. 2012). Despite the enhanced CO_2 diffusion at 273 K, CO_2 has a higher quadrupole moment than N_2 does, again raising the question of interactions with surface functional groups (Sdanghi et al. 2020b). For this reason, the use of H_2 at 77 K as a probe molecule for pore size analysis is gaining acceptance, although it is limited to a pore width range below 0.5 nm because of its supercritical state at 77 K (Jagiello and Thommes 2004). For a more detailed discussion of gas physisorption and probe molecule selection, the reader is referred to the latest IUPAC technical report (Thommes et al. 2015), which provides specific guidance on this topic.

8.7.1. *Pore volume and pore size distribution*

The most widely used method to estimate the volume of micropores in carbon materials is the well-known Dubinin–Radushkevich (DR) equation (Dubinin and Radushkevich 1947). Carbon materials were found to provide linear DR plots over a wide range of relative pressures. However, it is not strange that for highly microporous materials, DR plots are linear over a very narrow range of relative pressures (or even nonlinear). For those cases, the Dubinin–Astakhov (DA) approach is recommended. The DA equation (Dubinin and Astakhov 1971) is identical to the DR equation with the difference that the exponent of the former is not a constant with a value of 2. Instead, this exponent is a variable parameter that is adjusted using the gas adsorption data used for the characterization of the adsorbent. Therefore, it may depend on the selected relative pressure range and isotherm temperature (Rouquerol et al. 2014). Dubinin (1975) reported exponent values between two and six. However, owing to the empirical nature of the exponent, it is common to find non-integer values as the "optimal" fitting values. Horvath and Kawazoe (HK) have developed a more advanced methodology that can be used for the micropore size distribution of carbon materials from N_2 adsorption isotherms assuming slit-shaped pores (Horvath and Kawazoe 1983). The HK method has an advantage over the Dubinin-based methods, as it takes into account the specific pore shape and the interaction potential between the adsorbate and the adsorbent. Nevertheless, it fails to provide an accurate representation of micropore filling because it neglects the non-uniformity of the adsorbed molecules in the micropores by assuming that the confined fluid behaves like a bulk fluid (Lowell et al. 2006).

To evaluate the PSD of porous materials, the Barrett–Joyner–Halenda (BJH) method (Lowell et al. 2006; Rouquerol et al. 2014) is a very popular methodology, as it is widely available in most commercial software for processing gas adsorption data. The BJH method (Barrett et al. 1951) is a macroscopic approach with thermodynamic assumptions based on the modified Kelvin equation, which differs from the original Kelvin equation in that it considers the existence of an adsorbed multilayer film prior to condensation in pores by introducing a fluid–wall interaction parameter (Lowell et al. 2006). The main drawback of BJH is its questionable applicability to narrow mesopores where macroscopic concepts lack precision. In particular, the underestimation of the BJH method can reach 25% for pore widths smaller than 10 nm (Gregg and Sing 1982; de la Casa-Lillo et al. 2002).

The issues associated with the aforementioned macroscopic thermodynamic, BJH, method have been overcome by the development of microscopic approaches, such as density functional theory (DFT) of inhomogeneous fluids. Seaton et al. (1989) developed the first local DFT approach for the determination of PSDs of porous carbons from N_2 adsorption isotherms. Nevertheless, predictions with local DFT were unable to provide acceptable approximations for the part of the fluid closest to the adsorbent surface (Rouquerol et al. 2014). For this reason, more advanced DFT approaches, namely, non-local DFT (NLDFT) or quenched solid DFT (QSDFT), were developed. Tarazona (1985) previously proposed a free-energy functional model to describe an inhomogeneous system of hard spheres consisting of a smoothed density approximation procedure improving the density profiles obtained in the vicinity of a solid surface. Tarazona's DFT approach was later considered the pioneer of NLDFT, as discussed in the work of Ravikovitch et al. (2000), where Tarazona's version of DFT was used to introduce a unified way to characterize the pore size of carbonaceous materials from N_2, Ar and CO_2 adsorption isotherms. The NLDFT and QSDFT approaches can provide information about the adsorbed phase at the molecular level by reproducing the density profiles of confined fluids in the pore network of the adsorbent (Horvath and Kawazoe 1983; Monson 2012; Landers et al. 2013). In addition, these methods provide insight into the mechanisms of pore condensation and hysteresis in the mesopore region, which is essential for the proper characterization of carbon materials containing an interconnected network of micromesopores (Lastoskie et al. 1993; Neimark 1995; Tarazona 1995; Olivier 1998).

Reports in the literature (Monson 2012; Thommes et al. 2012; Landers et al. 2013; Thommes et al. 2015; Cychosz et al. 2017) have confirmed the suitability and accuracy of DFT-based methods for PSD and pore analysis in the micromesopore range, compared with traditional macroscopic approaches such as the Kelvin, BJH or Horvath–Kawazoe equations. Despite advances in DFT methods, the characterization of highly disordered porous materials still faces some challenges. Advanced DFT or molecular simulation methodologies have been developed on the basis of the combination of different probe molecules at various temperatures, which have allowed the differentiation of different types of interconnected pore networks and the quantification of the accessible pore volume (Thommes et al. 2012; Cychosz et al. 2017). For example, the combination of N_2 and CO_2 adsorption data at 77 and 273 K, respectively, in addition to supermicro- and mesopores, allowed for the simultaneous evaluation of ultramicroporosity in highly heterogeneous chars and ACs using the 2D-NLDFT approach (Jagiello et al. 2015, 2019). Similarly, attempts to obtain a complete PSD analysis of carbon materials have led to the use of other combinations of gas adsorption data, including H_2 and O_2 at 77 K, by performing conventional and quantum-corrected 2D-NLDFT analyses (Jagiello et al. 2020). On the other hand, the use of the QSDFT was proposed to overcome the issues associated with unsatisfactory NLDFT predictions of solid–liquid adsorbate description by incorporating the roughness of the adsorbent surface (Gor et al. 2012). Therefore, calculations with the QSDFT consider the solid as "rigid" with a roughness parameter, allowing optimization of the grand potential since the density of the "quenched" solid is also taken into account (Rouquerol et al. 2014).

With an appropriate polynomial fit, the PSD data were found to be useful in predicting the amount of H_2 adsorbed on ACs by considering a selected range of pore sizes as parameters of an empirical equation of the following form (Zhao et al. 2011a, 2011b):

$$H_2\,(wt.\%) = \sum_{i=1}^{n} a_i v_i \qquad [8.17]$$

where H_2 (wt.%) is the amount of H_2 adsorbed at each pressure, a_i (wt.% cm^3/g) is referred to as the *adsorbed density factor*, v_i (cm^3/g) is the pore volume of each pore size range, and n is the number of selected pore size ranges. Four pore size ranges were selected to evaluate a family of 17 laboratory-prepared ACs through equation [8.17] (Sdanghi et al. 2020b). The first two pore size ranges were chosen to represent the number of H_2 layers likely to be accommodated considering a kinetic H_2 diameter of 0.296 nm (Jagiello et al. 2020): <0.6 nm (2 H_2 layers) and 0.6–0.9 nm (3 H_2 layers). The third pore size range was selected to complete the microporosity range, that is, between 0.9 and 2 nm, and finally, the fourth pore size range covered the mesopore range (2–50 nm). The objectives of this empirical approach were (i) to apply equation [8.17] to a dataset of PSDs and experimental H_2 (wt.%) of 17 laboratory-prepared ACs to estimate a_i for each level of pressure studied; and (ii) to be able to predict H_2 storage capacities at a given pressure and 77 K on the basis of the textural characterization of an AC. The value of a_i associated with each pore size and pressure level was calculated by minimizing the sum of squared errors and optimizing by the downhill simplex method (Nelder and Mead 1965).

The ACs were labeled according to their A_{BET} (e.g. T1000 refers to a sample with 1000 m^2/g A_{BET}). The results presented in Figures 8.6(a) and (b) allow discrimination of the contribution of the selected pore ranges to the amount of H_2 adsorbed at 77 K for two pressure levels (0.1 and 7 MPa), and they suggest that smaller pores are more relevant for hydrogen storage in carbon materials at lower pressures (Fierro et al. 2010a). Although several works have highlighted that the optimal pore width for H_2 adsorption at 77 K is approximately 0.6 nm (Rzepka et al. 1998; Patchkovskii et al. 2005; Georgiev et al. 2006; Cabria et al. 2007; Celzard et al. 2007; Gogotsi et al. 2009), according to Figure 8.6(b), wider pores are also necessary at the relevant operating pressures for H_2 storage. Figure 8.6(c) shows that the predicted H_2 uptake values are in good agreement with the experimental results, indicating a good fit. The accuracy of the empirical equation [8.17] was tested using PSDs of four commercial ACs, which were therefore not included in the set for estimating a_i, whose H_2 storage capacities were obtained experimentally (Zhao et al. 2011a). The determination coefficient between the predicted and experimental H_2 storage capacities was found to be 0.96 (Sdanghi et al. 2020b). Therefore, this empirical approach proved to be a better analytical tool than A_{BET}, as discussed in the following sections, for predicting the evolution of H_2 adsorption with pressure.

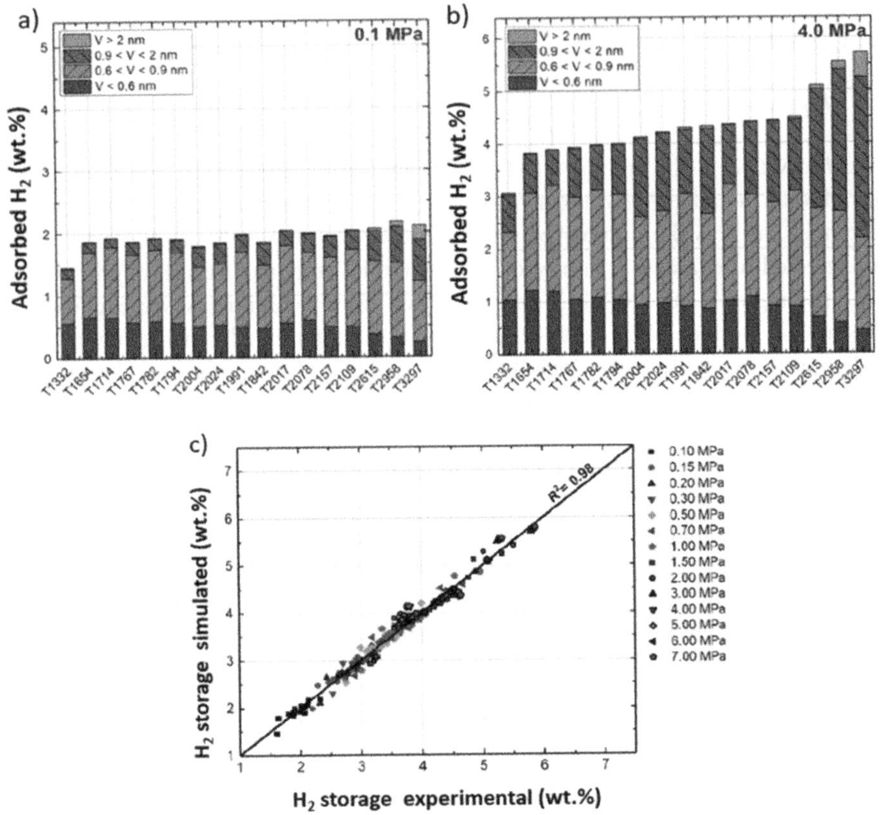

Figure 8.6. *Contribution of each range of pore size to hydrogen adsorption at (a) 0.1 MPa and (b) 4.0 MPa and (c) the correlation between the experimental and predicted H_2 uptakes according to equation [8.17] (taken from Sdanghi et al. (2020b)). For a color version of this figure, see www.iste.co.uk/derango-cuevas/hydrogen.zip*

8.7.2. *Specific surface area*

One of the most important textural characteristics for hydrogen storage in porous materials is the specific surface area. Since hydrogen is primarily adsorbed by physisorption, cryogenic temperatures are required to achieve competitive uptake. It has been shown that, especially at low pressure and cryogenic temperatures, the amount of H_2 adsorbed on the surface of a carbon material is mainly determined by its accessible surface area as well as by its narrow and well-defined micropore size distribution (Fierro et al. 2007; Sdanghi et al. 2020b). The well-known “Chahine’s rule” (Chahine and Bose 1996) considers that, at 77 K, H_2 is adsorbed in a proportion of 1 wt.% for every 500 m^2/g of available surface area. Nevertheless, a reliable assessment of the surface area of nanoporous materials is challenging. The Brunauer–Emmett–Teller (BET) approach is the most widely used standard methodology for surface area evaluation (Rouquerol et al. 2007, 2014; Lowell et al. 2006; Thommes et al. 2015). Although the BET method is suitable for non-porous, macroporous or mesoporous materials with type II or IVa isotherms if applied carefully (Sdanghi et al. 2020b), it is still used for assessing the surface area of microporous adsorbents worldwide. Therefore, special care should be taken when applying the BET method to materials with a significant amount of micropores, as is the case for many ACs or MOFs. To unambiguously determine the linear BET range for microporous materials, Rouquerol et al. (2007) proposed some recommendations that are widely accepted in the gas adsorption community. However, the BET surface area of a microporous adsorbent, which is best calculated from the Ar isotherm at 87 K, should not be considered a true representation of the actual surface area but rather an apparent surface area (A_{BET}), which can, however, serve as a useful "fingerprint" of the adsorbent for comparative purposes.

Although H_2 adsorption on ACs is highly dependent on the specific surface area, the increase in H_2 adsorption with surface area has a limit, as reported by Fierro et al. (2010b). By comparing a set of ACs with increasing BET areas (from 1,200 to more than 3,000 m^2/g) and measuring excess H_2 capacities at 77 K, Fierro et al. (2010b) experimentally demonstrated an upper limit of approximately 6.0 wt.% for an AC with an A_{BET} of 2,500 m^2/g. This experimental observation of the maximum storage capacity is very close to the theoretical calculations developed by Schlapbach and Züttel (2001), in which a H_2 excess of 6.4 wt.% was established for the maximum surface area of a graphene sheet, that is, 2,630 m^2/g, considering both sides of the graphene layer. This theoretical limit was calculated considering that all of the adsorption sites were accessible to hydrogen molecules, so it can be considered an upper limit for H_2 adsorption on ACs.

An alternative to obtain more reliable surface area values is the use of DFT-based methods. DFT methodologies allow the estimation of specific surface area as a function of pore size over the entire micro- and mesopore range (also known as the specific cumulative surface area) and therefore differentiate between micro/mesoporosity and the external surface area. However, accurate results depend on the availability of developed DFT methods applicable to the adsorptive/adsorbent system of interest (Sdanghi et al. 2020b). To evaluate the correlation between surface area and H_2 storage capacity in ACs, Fierro et al. (2010b) synthesized a series of ACs and determined both A_{BET} (using N_2 at 77 K) and S_{NLDFT} (using both N_2 at 77 K and CO_2 at 273 K) for all samples. Figures 8.7(a) and (b) show plots of the experimental n_a of H_2 in wt.% versus A_{BET} and S_{NLDFT} for different pressure levels between 0.1 and 5 MPa.

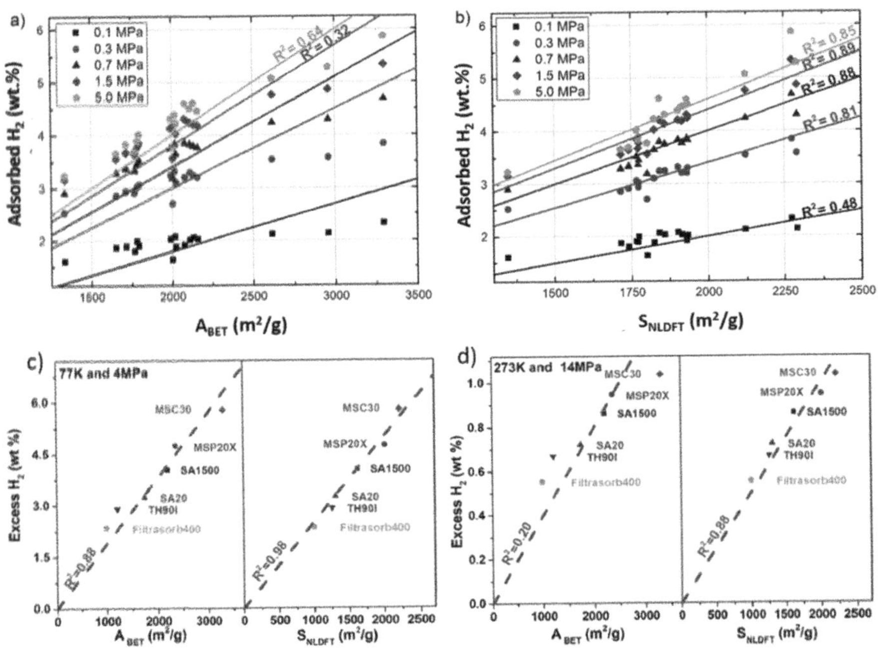

Figure 8.7. *Hydrogen uptake as a function of (a) A_{BET} and (b) S_{NLDFT} for 17 lab-made activated carbons (taken from Sdanghi et al. (2020b)) and excess hydrogen uptake as a function of (c) A_{BET} and (d) S_{NLDFT} for six selected commercial ACs (reprinted with permission from Ramirez-Vidal et al. (2021)). Copyright 2021 American Chemical Society). For a color version of this figure, see www.iste.co.uk/derango-cuevas/hydrogen.zip*

These figures clearly show an improved linear correlation between H_2 adsorption and S_{NLDFT} compared with A_{BET}, confirming the better accuracy of NLDFT calculations to estimate the surfaces' areas. The above result was confirmed by Ramirez-Vidal et al. (2021) by correlating the surface area to the excess H_2 uptake instead of n_a, as shown in Figures 8.7(c) and (d). Furthermore, the excess H_2 uptake at 77 K increased by 1.25 wt.% for every 500 m^2/g of S_{NLDFT}, which could be an alternative to the 1 wt.% suggested by Chahine's rule. Notably, the fact that the S_{NLDFT} calculation was carried out with CO_2 in addition to N_2 allows for the evaluation of pores smaller than 0.6 nm, where hydrogen is preferentially adsorbed (Rzepka et al. 1998; Cabria et al. 2007; Schaefer et al. 2016). Therefore, more accurate surface areas and better linear correlation with hydrogen storage are likely to be obtained, provided that H_2 is used as the probe molecule instead of CO_2 (Jagiello et al. 2020).

8.8. Adsorbents for hydrogen storage

In this section, we focus on two families of high-surface-area adsorbents, namely, carbon-based materials and MOFs, as many types of adsorbents have been considered for H_2 storage. Among the carbon-based materials, some studies showed very promising H_2 storage capacities using carbon nanofibers (CNFs) (Chambers et al. 1998; Kim and Park 2006; Jaybhaye et al. 2007; Kim et al. 2008) and carbon nanotubes (CNTs) (Dillon et al. 1997; Chen et al. 1999; Minami et al. 2010; Lee and Park 2012; Adeniran and Mokaya 2015; Ariharan et al. 2018). However, most of the impressive storage capacities reported could not be replicated by other researchers. Yang (2000), for example, examined the unrealistic H_2 storage values reported for alkali-doped CNTs (Chen et al. 1999) and concluded that a wet H_2 stream was most likely used in temperature-programmed desorption (TPD), leading to a higher n_a. When TPD tests were conducted with a dry H_2 stream, the storage capacities dropped to values not higher than 2 wt.%. Graphene oxides (GOs) (Singh and De 2021), ACs (Wang and Yang 2008; Fierro et al. 2010b; Bader and Ouederni 2016; Blankenship et al. 2017; Romanos et al. 2019; Samantaray et al. 2019; Zhao et al. 2019; Rowlandson et al. 2020) and templated carbons (TCs) (Jiang et al. 2011; Pachfule et al. 2012; Yang et al. 2012; Attia et al. 2013; Masika and Mokaya 2013; Musyoka et al. 2015; Juárez et al. 2018; Carraro et al. 2019) have also been studied. Figure 8.8(a) shows n_{ex} of H_2 (wt.%) at 77 K for selected carbon-based materials as a function of A_{BET} (m^2/g).

Note that the colors in Figure 8.8 represent the different types of carbon materials mentioned above, and the symbols represent the pressure range at which adsorption is performed. Although H_2 storage experiments are typically conducted at low temperatures, as H_2 is mainly adsorbed through physical interactions with low adsorption energy (up to – 10 kJ/mol H_2), some studies have explored H_2 storage at room temperature (298 K), as shown in Figure 8.8(b). As expected, the reported n_{ex} at 298 K are lower than those at 77 K. Notably, in addition to the poorly measured carbon-based materials discussed above, the remaining storage capacities shown in Figure 8.8 do not exceed 7 wt. %, which is consistent with the conclusion of Fierro et al. (2010b) that carbon-based adsorbents have an adsorption cutoff value of approximately 6 wt.%. The adsorbents that have excess quantities below those predicted by the Chahine rule in Figure 8.8(a) are primarily those with the lowest storage pressure (0.1 MPa). This implies that the maximum storage capacity may not yet have been reached at this pressure.

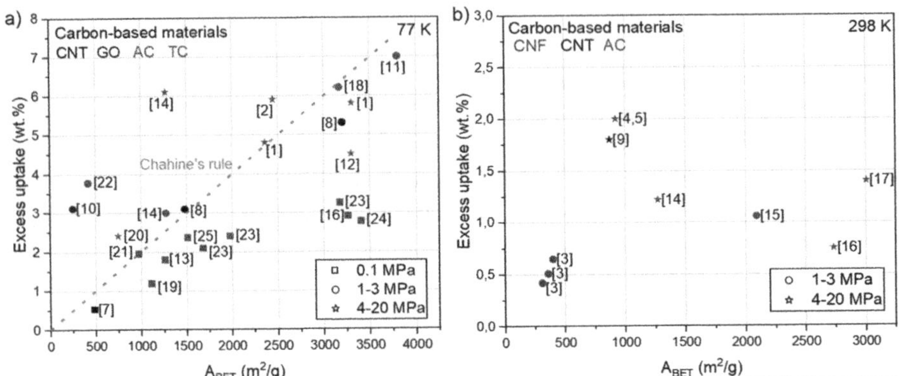

[1]	Sdanghi et al. 2020a	[10]	Singh et al. 2021	[18]	Masika et al. 2013
[2]	Fierro et al. 2010b	[11]	Blankenship et al. 2017	[19]	Musyoka et al. 2015
[3]	Jaybhaye et al. 2007	[12]	Romanos et al. 2019	[20]	Attia et al. 2013
[4,5]	Kim et al. 2006, 2008	[13]	Rowlandson et al. 2020	[21]	Carraro et al. 2018
[6]	Minami et al. 2010	[14]	Bader et al. 2016	[22]	Juarez et al. 2018
[7]	Lee et al. 2012	[15]	Samantara et al. 2019	[23]	Yang et al. 2012
[8]	Adeniran et al. 2015	[16]	Zhao et al. 2019	[24]	Jiang et al. 2011
[8]	Adeniran et al. 2015	[17]	Wang et al. 2008	[25]	Pachfule et al. 2012
[9]	Ariharan et al. 2018				

Figure 8.8. *Excess H_2 uptakes in wt.% at (a) 77 K and (b) 298 K at different pressure levels as a function of A_{BET} for carbon-based materials selected from the literature: carbon nanofibers (CNFs), carbon nanotubes (CNTs), graphene oxides (GOs), activated carbons (ACs) and templated carbons (TCs). For a color version of this figure, see www.iste.co.uk/derango-cuevas/hydrogen.zip*

Owing to their versatility and impressive textural properties, MOFs have very quickly attracted attention for their application in various fields, including gas separation and storage (Gómez-Gualdrón et al. 2017; García-Holley et al. 2018; Ahmed et al. 2019; Chen et al. 2020; Ramirez-Vidal et al. 2022a). Since the report on the synthesis of MOF-5, which has a specific surface area of approximately 2,800 m²/g (Li et al. 1999), MOFs have gained visibility as promising adsorbents for H_2 storage. Rosi et al. (2003) first evaluated the H_2 storage capacity of MOF-5 and reported a H_2 uptake of 4.8 wt.% at 78 K and 2 MPa. Since then, different types of MOFs have been synthesized and evaluated in terms of their H_2 storage capacity.

Figure 8.9 shows the excess uptakes of various types of MOFs (Schlichtenmayer and Hirscher 2012; Gómez-Gualdrón et al. 2017; García-Holley et al. 2018; Ahmed et al. 2019; Chen et al. 2020; Ramirez-Vidal et al. 2022a) at 77 K and 10 MPa as a function of their A_{BET}. MOFs with A_{BET} values below 3,000 m^2/g exhibit behavior consistent with Chahine's rule, as evidenced by their uptake values in Figure 8.9. However, MOFs with A_{BET} values above 2,500 m^2/g have excess uptakes significantly lower than those expected by Chahine's rule. This discrepancy can be attributed to the overestimation of surface area by the BET method for highly supermicroporous materials, that is, those with a large volume of pores with diameters between 0.2 and 2 nm. Using DFT methods to estimate the surface area would likely yield lower values, bringing the excess adsorption values closer to those predicted by Chahine's rule.

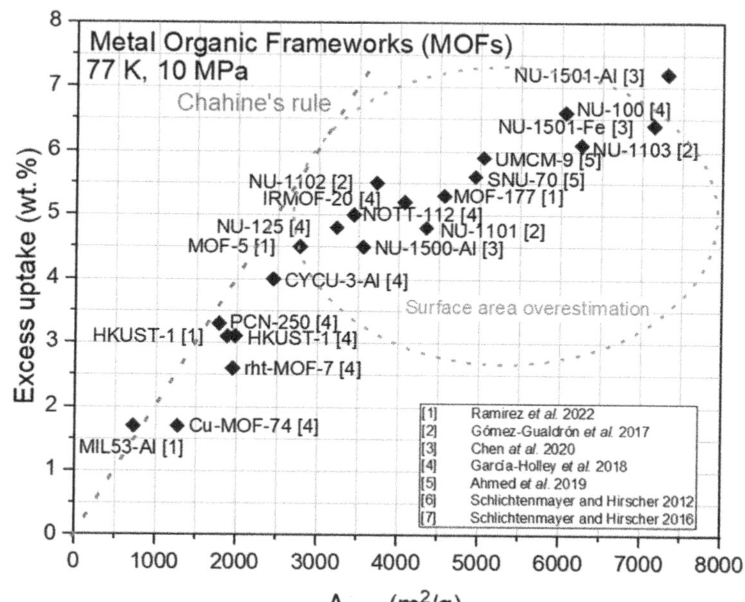

Figure 8.9. *Excess H_2 uptakes in wt.% at 77 K and 10 MPa as a function of the A_{BET} of metal organic frameworks (MOFs) selected from the literature. For a color version of this figure, see www.iste.co.uk/derango-cuevas/hydrogen.zip*

More recently, hybrid materials combining carbon adsorbents and MOFs have been synthesized with the aim of increasing H_2 storage capacity. Yang et al. (2010) reported improved H_2 storage capacities of Pt-loaded CNT/MOF-5 composites of 1.25 wt.% at 298 K and 100 bar and 1.89 wt.% at 77 K and 1 bar, demonstrating better performance than that of pristine materials. Liu et al. (2013) prepared composites based on Cu-BTC and graphene layers with different percentages of GO. The results indicated that H_2 uptake increased from 2.81 wt.% (only Cu-BTC) to 3.58 wt.% at 77 K and 42 bar. Karikkethu Prabhakaran and Deschamps (2015) developed a hybrid composite of lithium-doped MIL-101 and commercial AC NORIT-RB3. The hybrid material with a low concentration of lithium (727 ppm) almost doubled the H_2 uptake capacity of undoped MIL-101, both at 77 K (14.4 wt.%) and 298 K (1.1 wt.%) at 100 bar. In subsequent years, other authors reported enhanced H_2 storage capacities with composite materials (Li et al. 2015; Yu et al. 2017; Policicchio et al. 2020; Rojas-Garcia et al. 2022).

The storage capacities of adsorbents can be expressed either by mass or by volume, the latter taking into account the packing density of the material. Volumetric capacities are of great practical importance since the solid material will be packed in a vessel of a defined volume. However, high volumetric capacities cannot be accompanied by high gravimetric capacities if the adsorbent has a low packing density. In a study by Ramirez-Vidal et al. (2022b), the volumetric and release capacities of two commercial ACs (MSC-30 and MSP-20X) and commercial MOF-5 were compared to compressed H_2 in an empty tank at 298 K and 70 MPa, as shown in Figure 8.10.

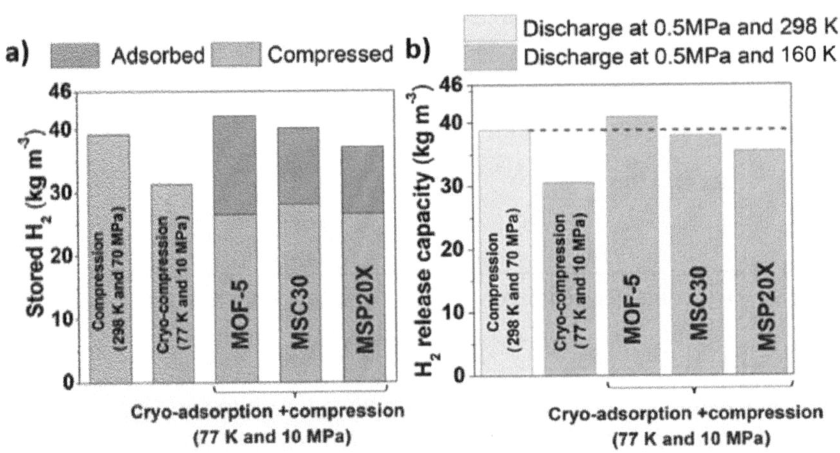

Figure 8.10. *(a) Volumetric and (b) release capacities of a compressed H_2 tank under different conditions and cryoadsorption at 77 K and 10 MPa for selected adsorbents: MOF-5 and commercial ACs MSC30 and MSP20X (reprinted from the publication Ramirez-vidal et al. 2022b, Copyright (2022), with permission from Elsevier). For a color version of this figure, see www.iste.co.uk/derango-cuevas/hydrogen.zip*

The charging and discharging conditions were set to meet the Hydrogen Storage Engineering Center of Excellence (HSECoE) recommendations (Siegel et al. 2015), that is, H_2 is charged at 77 K and 10 MPa and discharged at 160 K and 0.5 MPa. The results showed that both MSC30 and MOF-5 have higher volumetric storage capacities than compressed H_2, whereas MOF-5 has a higher release capacity (41 kg/m^3) than compressed H_2 (39 kg/m^3), MSC-30 (38 kg/m^3) and MSP-20X (36 kg/m^3). However, the amount of H_2 released is limited by the discharge conditions, which means that part of the originally stored H_2 will remain in the tank. Although both MSC-30 and MSP-20X have a slightly lower release capacity, they may be a cost-effective option for H_2 storage compared to MOF-5, given their lower manufacturing price (Ramirez-Vidal et al. 2022b).

The US Department of Energy (DOE) has set a technical target for H_2 storage in light-duty fuel cell vehicles of 6.5 wt.% (or 50 kg/m^3) in the temperature range of 233–333 K (Hydrogen and Fuel Cell Technologies Office of DOE 2020). Nevertheless, the reported storage capacities must consider the weight of the system, that is, for adsorption-based systems, the mass of the tank in addition to the adsorbent material and the cooling system. The current storage capacities achieved using ACs and MOFs are below the target and DOE-recommended temperature conditions, indicating the need for improved use of porous materials for light-duty fuel cell vehicles. However, researchers should also consider the various end-use applications of H_2 rather than focusing solely on the DOE target for automotive applications. Adsorption-based systems, for example, may be more appropriate for stationary H_2 applications.

8.9. Conclusion

High-surface area porous materials, such as carbon-based materials and MOFs, offer high H_2 capacities for energy storage and transportation under cryogenic conditions and moderate pressures. However, several challenges must still be addressed to make H_2 adsorption practical and cost-effective. A key challenge is to achieve and maintain competitive release capacities adapted to the operating conditions. For this purpose, robust and scalable synthesis methods must be developed to produce materials with high H_2 volumetric capacities and suitable and consistent textural properties. In addition, a unified approach for reporting excess amounts and converting excess H_2 to the adsorbed amount is needed. Moreover, to accurately determine the total H_2 storage capacity of an adsorbent, it is essential to specify the amount of H_2 solely stored by compression in the bulk gas phase. To obtain more reliable adsorbent characterization, it is recommended to systematically calculate the surface area using NLDFT with two probe molecules that cover the entire micro- and mesopore range. Finally, to provide a comprehensive H_2 storage capacity of adsorbents, it is suggested to report both gravimetric and volumetric hydrogen capacities by specifying the packing densities.

While researchers are making significant progress in developing new materials and technologies for hydrogen storage, additional efforts are needed to design cryogenic tanks to minimize energy consumption. The development of new porous materials such as hybrid materials may offer several prospects for improving H_2 storage, including increasing the volumetric capacity, moisture resistance, and mechanical properties. To realize these benefits to their fullest extent, synthesis and pelletization scale-up studies are needed while performing economic calculations to determine the viability of large-scale production of solid materials. Finally, the optimization of cryogenic systems with competitive energy consumption offers promising prospects for making H_2 adsorption a practical and sustainable solution for a decarbonized global economy.

This work was funded by the French Research Agency (ANR) under the SOLHYD project (ANR-22-PEHY-0007). This study was partially supported by the TALiSMAN and TALiSMAN2 projects funded by the European Regional Development Fund (ERDF) and the HyPE project (FRCR program funded by Région Grand-Est and FEDER).

8.10. References

Adeniran, B. and Mokaya, R. (2015). Low temperature synthesized carbon nanotube super-structures with superior CO_2 and hydrogen storage capacity. *Journal of Materials Chemistry A*, 3(9), 5148–5161.

Ahmed, A., Seth, S., Purewal, J., Wong-Foy, A.G., Veenstra, M., Matzger, A.J., Siegel, D.J. (2019). Exceptional hydrogen storage achieved by screening nearly half a million metal-organic frameworks. *Nature Communications*, 10(1), 1568.

Ariharan, A., Viswanathan, B., Nandhakumar, V. (2018). Nitrogen-incorporated carbon nanotube derived from polystyrene and polypyrrole as hydrogen storage material. *International Journal of Hydrogen Energy*, 43(10), 5077–5088.

Attia, N.F., Lee, S.M., Kim, H.J., Geckeler, K.E. (2013). Nanoporous carbon-templated silica nanoparticles: Preparation, effect of different carbon precursors, and their hydrogen storage adsorption. *Microporous and Mesoporous Materials*, 173, 139–146.

Bader, N. and Ouederni, A. (2016). Optimization of biomass-based carbon materials for hydrogen storage. *Journal of Energy Storage*, 5, 77–84.

Bahadur, J., Contescu, C.I., Ramirez-Cuesta, A.J., Mamontov, E., Gallego, N.C., Cheng, Y., Daemen, L.L., Melnichenko, Y.B. (2017). Properties of immobile hydrogen confined in microporous carbon. *Carbon*, 117, 383–392.

Barrett, E.P., Joyner, L.G., Halenda, P.P. (1951). The determination of pore volume and area distributions in porous substances. I. Computations from nitrogen isotherms. *Journal of the American Chemical Society*, 73(1), 373–380.

Bastos-Neto, M., Patzschke, C., Lange, M., Möllmer, J., Möller, A., Fichtner, S., Schrage, C., Lässig, D., Lincke, J., Staudt, R. et al. (2012). Assessment of hydrogen storage by physisorption in porous materials. *Energy and Environmental Science*, 5(8), 8294–8303.

Blankenship, L.S., Balahmar, N., Mokaya, R. (2017). Oxygen-rich microporous carbons with exceptional hydrogen storage capacity. *Nature Communications*, 8(1), 1545.

Broom, D.P. and Webb, C.J. (2017). Pitfalls in the characterisation of the hydrogen sorption properties of materials. *International Journal of Hydrogen Energy*, 42(49), 29320–29343.

Cabria, I., López, M.J., Alonso, J.A. (2007). The optimum average nanopore size for hydrogen storage in carbon nanoporous materials. *Carbon*, 45(13), 2649–2658.

Carraro, P.M., García Blanco, A.A., Lener, G., Barrera, D., Amaya-Roncancio, S., Chanquía, C., Troiani, H., Oliva, M.I., Eimer, G.A. (2019). Nanostructured carbons modified with nickel as potential novel reversible hydrogen storage materials: Effects of nickel particle size. *Microporous and Mesoporous Materials*, 273, 50–59.

de la Casa-Lillo, M.A., Lamari-Darkrim, F., Cazorla-Amorós, D., Linares-Solano, A. (2002). Hydrogen storage in activated carbons and activated carbon fibers. *The Journal of Physical Chemistry B*, 106(42), 10930–10934.

Celzard, A., Fierro, V., Marêché, J.F., Furdin, G. (2007). Advanced preparative strategies for activated carbons designed for the adsorptive storage of hydrogen. *Adsorption Science & Technology*, 25(3/4), 129–142.

Chahine, R. and Bose, T.K. (1996). Characterization and optimization of adsorbents for hydrogen storage. *Hydrogen Energy Progress*, 2, 1259–1264.

Chambers, A., Park, C., Baker, R.T.K., Rodriguez, N.M. (1998). Hydrogen storage in graphite nanofibers. *Journal of Physical Chemistry B*, 102(22), 4253–4256.

Chen, P., Wu, X., Lin, J., Tan, K.L. (1999). High H2 uptake by alkali-doped carbon nanotubes under ambient pressure and moderate temperatures. *Science*, 285(5424), 91–93.

Chen, Z., Li, P., Anderson, R., Wang, X., Zhang, X., Robison, L., Redfern, L.R., Moribe, S., Islamoglu, T., Gómez-Gualdrón, D.A. et al. (2020). Balancing volumetric and gravimetric uptake in highly porous materials for clean energy. *Science*, 368(6488), 297–303.

Cychosz, K.A., Guillet-Nicolas, R., García-Martínez, J., Thommes, M. (2017). Recent advances in the textural characterization of hierarchically structured nanoporous materials. *Chemical Society Reviews*, 46(2), 389–414.

Dailly, A. and Poirier, E. (2011). Evaluation of an industrial pilot scale densified MOF-177 adsorbent as an on-board hydrogen storage medium. *Energy & Environmental Science*, 4(9), 3527.

Dillon, A.C., Jones, K.M., Bekkedahl, T.A., Kiang, C.H., Bethune, D.S., Heben, M.J. (1997). Storage of hydrogen in single-walled carbon nanotubes. *Nature*, 386(6623), 377–379.

Dreisbach, F. and Lösch, H.W. (2000). Magnetic suspension balance for simultaneous measurement of a sample and the density of the measuring fluid. *Journal of Thermal Analysis and Calorimetry*, 62, 515–521.

Dreisbach, F., Seif, R., Losch, H.W. (2003). Adsorption equilibria of CO/H_2 with a magnetic suspension balance. Purely gravimetric measurement. *Journal of Thermal Analysis and Calorimetry*, 71(1), 73–82.

Dubinin, M.M. (1975). Physical adsorption of gases and vapors in micropores. *Progress in Surface and Membrane Science*, 9, 1–70.

Dubinin, M.M. and Astakhov, V.A. (1971). Description of adsorption equilibria of vapors on zeolites over wide ranges of temperature and pressure. *Advances in Chemistry*, 102, 69–85.

Dubinin, M.M. and Radushkevich, L.V. (1947). The equation of the characteristic curve of activated charcoal. *Proceedings of the USSR Academy of Sciences*, 55, 327–329.

Fierro, V., Bosch, G., Siperstein, F.R. (2007). Pore size distribution in microporous carbons obtained from molecular modeling and density functional theory. *Studies in Surface Science and Catalysis*, 160, 519–526.

Fierro, V., Szczurek, A., Zlotea, C., Marêché, J.F., Izquierdo, M.T., Albiniak, A., Latroche, M., Furdin, G., Celzard, A. (2010a). Experimental evidence of an upper limit for hydrogen storage at 77 K on activated carbons. *Carbon*, 48(7), 1902–1911.

Fierro, V., Zhao, W., Izquierdo, M.T., Aylon, E., Celzard, A. (2010b). Adsorption and compression contributions to hydrogen storage in activated anthracites. *International Journal of Hydrogen Energy*, 35(17), 9038–9045.

García-Díez, E., Schaefer, S., Sanchez-Sanchez, A., Celzard, A., Fierro, V., Maroto-Valer, M.M., García, S. (2019). Novel porous carbons derived from coal tar rejects: Assessment of the role of pore texture in CO_2 capture under realistic postcombustion operating temperatures. *ACS Applied Materials & Interfaces*, 11(40), 36789–36799.

García-Holley, P., Schweitzer, B., Islamoglu, T., Liu, Y., Lin, L., Rodriguez, S., Weston, M.H., Hupp, J.T., Gómez-Gualdrón, D.A., Yildirim, T. et al. (2018). Benchmark study of hydrogen storage in metal–organic frameworks under temperature and pressure swing conditions. *ACS Energy Letters*, 3(3), 748–754.

Georgiev, P.A., Ross, D.K., Albers, P., Ramirez-Cuesta, A.J. (2006). The rotational and translational dynamics of molecular hydrogen physisorbed in activated carbon: A direct probe of microporosity and hydrogen storage performance. *Carbon*, 44(13), 2724–2738.

Gogotsi, Y., Portet, C., Osswald, S., Simmons, J.M., Yildirim, T., Laudisio, G., Fischer, J.E. (2009). Importance of pore size in high-pressure hydrogen storage by porous carbons. *International Journal of Hydrogen Energy*, 34(15), 6314–6319.

Gómez-Gualdrón, D.A., Wang, T.C., García-Holley, P., Sawelewa, R.M., Argueta, E., Snurr, R.Q., Hupp, J.T., Yildirim, T., Farha, O.K. (2017). Understanding volumetric and gravimetric hydrogen adsorption trade-off in metal–organic frameworks. *ACS Applied Materials & Interfaces*, 9(39), 33419–33428.

Gor, G.Y.U., Thommes, M., Cychosz, K.A., Neimark, A.V. (2012). Quenched solid density functional theory method for characterization of mesoporous carbons by nitrogen adsorption. *Carbon*, 50(4), 1583–1590.

Gregg, S.J. and Sing, K.S.W. (1982). *Adsorption, Surface Area and Porosity*. Academic Press, London.

Gumma, S. and Talu, O. (2010). Net adsorption: A thermodynamic framework for supercritical gas adsorption and storage in porous solids. *Langmuir*, 26(22), 17013–17023.

Horvath, G. and Kawazoe, K. (1983). Method for the calculation of effective pore size distribution in molecular sieve carbon. *Journal of Chemical Engineering of Japan*, 16(6), 470–475.

Hydrogen and Fuel Cell Technologies Office of DOE (2020). DOE technical targets for onboard hydrogen storage for light-duty vehicles [Online]. Available at: https://www.energy.gov/eere/fuelcells/doe-technical-targets-onboard-hydrogen-storage-light-duty-vehicles [Accessed 16 March 2023].

Jagiello, J. and Thommes, M. (2004). Comparison of DFT characterization methods based on N_2, Ar, CO_2, and H_2 adsorption applied to carbons with various pore size distributions. *Carbon*, 42(7), 1227–1232.

Jagiello, J., Ansón, A., Martínez, M.T. (2006). DFT-based prediction of high-pressure H2 adsorption on porous carbons at ambient temperatures from low-pressure adsorption data measured at 77 K. *The Journal of Physical Chemistry B*, 110(10), 4531–4534.

Jagiello, J., Ania, C., Parra, J.B., Cook, C. (2015). Dual gas analysis of microporous carbons using 2D-NLDFT heterogeneous surface model and combined adsorption data of N_2 and CO_2. *Carbon*, 91, 330–337.

Jagiello, J., Kenvin, J., Celzard, A., Fierro, V. (2019). Enhanced resolution of ultra micropore size determination of biochars and activated carbons by dual gas analysis using N_2 and CO_2 with 2D-NLDFT adsorption models. *Carbon*, 144, 206–215.

Jagiello, J., Kenvin, J., Ania, C.O., Parra, J.B., Celzard, A., Fierro, V. (2020). Exploiting the adsorption of simple gases O_2 and H_2 with minimal quadrupole moments for the dual gas characterization of nanoporous carbons using 2D-NLDFT models. *Carbon*, 160, 164–175.

Jaybhaye, S., Sharon, M., Sharon, M., Sathiyamoorthy, D., Dasgupta, K. (2007). Semi-conducting carbon nanofibers and hydrogen storage. *Synthesis and Reactivity in Inorganic, Metal-Organic, and Nano-Metal Chemistry*, 37(6), 473–476.

Jiang, H.-L., Liu, B., Lan, Y.-Q., Kuratani, K., Akita, T., Shioyama, H., Zong, F., Xu, Q. (2011). From metal–organic framework to nanoporous carbon: Toward a very high surface area and hydrogen uptake. *Journal of the American Chemical Society*, 133(31), 11854–11857.

Juárez, J.M., Costa, M.G., Anunziata, O.A. (2018). Direct synthesis of ordered mesoporous carbon applied in hydrogen storage. *Journal of Porous Materials*, 25(5), 1359–1363.

Karikkethu Prabhakaran, P. and Deschamps, J. (2015). Doping activated carbon incorporated composite MIL-101 using lithium: Impact on hydrogen uptake. *Journal of Materials Chemistry A*, 3(13), 7014–7021.

Kim, B.-J. and Park, S.-J. (2006). Preparation of nanoporous carbons from graphite nano-fibres. *Nanotechnology*, 17(17), 4395–4398.

Kim, B.-J., Lee, Y.-S., Park, S.-J. (2008). A study on the hydrogen storage capacity of Ni-plated porous carbon nanofibers. *International Journal of Hydrogen Energy*, 33(15), 4112–4115.

Landers, J., Gor, G.Y., Neimark, A.V. (2013). Density functional theory methods for characterization of porous materials. *Colloids and Surfaces A: Physicochemical and Engineering Aspects*, 437, 3–32.

Lastoskie, C., Gubbins, K.E., Quirke, N. (1993). Pore size distribution analysis of microporous carbons: A density functional theory approach. *The Journal of Physical Chemistry*, 97(18), 4786–4796.

Lee, S.-Y. and Park, S.-J. (2012). Influence of the pore size in multi-walled carbon nanotubes on the hydrogen storage behaviors. *Journal of Solid State Chemistry*, 194, 307–312.

Li, H., Eddaoudi, M., O'Keeffe, M., Yaghi, O.M. (1999). Design and synthesis of an exceptionally stable and highly porous metal-organic framework. *Nature*, 402(6759), 276–279.

Li, J.-S., Tang, Y.-J., Li, S.-L., Zhang, S.-R., Dai, Z.-H., Si, L., Lan, Y.-Q. (2015). Carbon nanodots functional MOFs composites by a stepwise synthetic approach: Enhanced H_2 storage and fluorescent sensing. *CrystEngComm*, 17(5), 1080–1085.

Liu, S., Sun, L., Xu, F., Zhang, J., Jiao, C., Li, F., Li, Z., Wang, S., Wang, Z., Jiang, X. et al. (2013). Nanosized Cu-MOFs induced by graphene oxide and enhanced gas storage capacity. *Energy & Environmental Science*, 6(3), 818.

Liu, X.-W., Sun, T.-J., Hu, J.-L., Wang, S.-D. (2016). Composites of metal–organic frameworks and carbon-based materials: preparations, functionalities and applications. *Journal of Materials Chemistry A*, 4(10), 3584–3616.

Lowell, S., Shields, J.E., Thomas, M.A., Thommes, M. (2006). *Characterization of Porous Solids and Powders: Surface Area, Porosity and Density*. Springer, Berlin.

Masika, E. and Mokaya, R. (2013). Preparation of ultrahigh surface area porous carbons templated using zeolite 13X for enhanced hydrogen storage. *Progress in Natural Science: Materials International*, 23(3), 308–316.

Minami, D., Ohkubo, T., Kuroda, Y., Sakai, K., Sakai, H., Abe, M. (2010). Structural optimization of arranged carbon nanotubes for hydrogen storage by grand canonical Monte Carlo simulation. *International Journal of Hydrogen Energy*, 35(22), 12398–12404.

Monson, P.A. (2012). Understanding adsorption/desorption hysteresis for fluids in mesoporous materials using simple molecular models and classical density functional theory. *Microporous and Mesoporous Materials*, 160, 47–66.

Murata, K. and Kaneko, K. (2000). Nano-range interfacial layer upon high-pressure adsorption of supercritical gases. *Chemical Physics Letters*, 321(5/6), 342–348.

Musyoka, N.M., Ren, J., Langmi, H.W., North, B.C., Mathe, M. (2015). A comparison of hydrogen storage capacity of commercial and fly ash-derived zeolite X together with their respective templated carbon derivatives. *International Journal of Hydrogen Energy*, 40(37), 12705–12712.

Myers, A.L. and Monson, P.A. (2014). Physical adsorption of gases: The case for absolute adsorption as the basis for thermodynamic analysis. *Adsorption*, 20(4), 591–622.

Neimark, A.V. (1995). The method of indeterminate lagrange multipliers in nonlocal density functional theory. *Langmuir*, 11(10), 4183–4184.

Nelder, J.A. and Mead, R. (1965). A simplex method for function minimization. *The Computer Journal*, 7(4), 308–313.

Olivier, J.P. (1998). Improving the models used for calculating the size distribution of micropore volume of activated carbons from adsorption data. *Carbon*, 36(10), 1469–1472.

Olsen, R.J., Gillespie, A.K., Contescu, C.I., Taylor, J.W., Pfeifer, P., Morris, J.R. (2017). Phase transition of H2 in subnanometer pores observed at 75 K. *ACS Nano*, 11(11), 11617–11631.

Pachfule, P., Biswal, B.P., Banerjee, R. (2012). Control of porosity by using isoreticular zeolitic imidazolate frameworks (IRZIFs) as a template for porous carbon synthesis. *Chemistry – A European Journal*, 18(36), 11399–11408.

Patchkovskii, S., Tse, J.S., Yurchenko, S.N., Zhechkov, L., Heine, T., Seifert, G. (2005). Graphene nanostructures as tunable storage media for molecular hydrogen. *Proceedings of the National Academy of Sciences*, 102(30), 10439–10444.

Poirier, E. and Dailly, A. (2008). Investigation of the hydrogen state in IRMOF-1 from measurements and modeling of adsorption isotherms at high gas densities. *The Journal of Physical Chemistry C*, 112(33), 13047–13052.

Poirier, E. and Dailly, A. (2009a). On the nature of the adsorbed hydrogen phase in microporous metal–organic frameworks at supercritical temperatures. *Langmuir*, 25(20), 12169–12176.

Poirier, E. and Dailly, A. (2009b). Thermodynamic study of the adsorbed hydrogen phase in Cu-based metal–organic frameworks at cryogenic temperatures. *Energy & Environmental Science*, 2(4), 420.

Policicchio, A., Florent, M., Celzard, A., Fierro, V., Jagiello, J., Bandosz, T.J. (2020). Enhancing the gas adsorption capacities of UiO-66 by nanographite addition. *Microporous and Mesoporous Materials*, 309, 110571.

Ramirez-Vidal, P., Canevesi, R.L.S., Sdanghi, G., Schaefer, S., Maranzana, G., Celzard, A., Fierro, V. (2021). A step forward in understanding the hydrogen adsorption and compression on activated carbons. *ACS Applied Materials and Interfaces*, 13(10), 12562–12574.

Ramirez-Vidal, P., Canevesi, R.L.S., Celzard, A., Fierro, V. (2022a). Modeling high-pressure hydrogen uptake by nanoporous metal–organic frameworks: Implications for hydrogen storage and delivery. *ACS Applied Nano Materials*, 5(1), 759–773.

Ramirez-Vidal, P., Sdanghi, G., Celzard, A., Fierro, V. (2022b). High hydrogen release by cryo-adsorption and compression on porous materials. *International Journal of Hydrogen Energy*, 47(14), 8892–8915.

Ravikovitch, P.I., Vishnyakov, A., Russo, R., Neimark, A.V. (2000). Unified approach to pore size characterization of microporous carbonaceous materials from N_2, Ar, and CO_2 adsorption isotherms. *Langmuir*, 16(5), 2311–2320.

Rojas-Garcia, E., Castañeda-Ramírez, A.A., Angeles-Beltrán, D., López-Medina, R., Maubert-Franco, A.M. (2022). Enhancing in the hydrogen storage by SWCNT/HKUST-1 composites: Effect of SWCNT amount. *Catalysis Today*, 394(396), 357–364.

Romanos, J., Abou Dargham, S., Roukos, R., Pfeifer, P. (2018). Local pressure of supercritical adsorbed hydrogen in nanopores. *Materials*, 11(11), 2235.

Romanos, J., Beckner, M., Prosniewski, M., Rash, T., Lee, M., Robertson, J.D., Firlej, L., Kuchta, B., Pfeifer, P. (2019). Boron-neutron capture on activated carbon for hydrogen storage. *Scientific Reports*, 9(1), 2971.

Rosi, N.L., Eckert, J., Eddaoudi, M., Vodak, D.T., Kim, J., O'Keeffe, M., Yaghi, O.M. (2003). Hydrogen storage in microporous metal-organic frameworks. *Science*, 300(5622), 1127–1129.

Rouquerol, J., Llewellyn, P., Rouquerol, F. (2007). Is the bet equation applicable to microporous adsorbents? *Studies in Surface Science and Catalysis*, 160, 49–56. [Online]. Available at: https://www.sciencedirect.com/science/article/pii/S0167299107800085.

Rouquerol, F., Rouquerol, J, Sing, K.S.W., Llewellyn, P, Maurin, G., Rouquerol, J., Llewellyn, P., Sing, K. (2014). *Adsorption by Powders and Porous Solids*. Academic Press, London.

Rouquerol, J., Rouquerol, F., Llewellyn, P., Denoyel, R. (2016). Surface excess amounts in high-pressure gas adsorption: Issues and benefits. *Colloids and Surfaces A: Physicochemical and Engineering Aspects*, 496, 3–12.

Rowlandson, J.L., Edler, K.J., Tian, M., Ting, V.P. (2020). Toward process-resilient lignin-derived activated carbons for hydrogen storage applications. *ACS Sustainable Chemistry & Engineering*, 8(5), 2186–2195.

Rzepka, M., Lamp, P., de la Casa-Lillo, M.A. (1998). Physisorption of hydrogen on microporous carbon and carbon nanotubes. *The Journal of Physical Chemistry B*, 102(52), 10894–10898.

Samantaray, S.S., Mangisetti, S.R., Ramaprabhu, S. (2019). Investigation of room temperature hydrogen storage in biomass derived activated carbon. *Journal of Alloys and Compounds*, 789, 800–804.

Schaefer, S., Fierro, V., Izquierdo, M.T., Celzard, A. (2016). Assessment of hydrogen storage in activated carbons produced from hydrothermally treated organic materials. *International Journal of Hydrogen Energy*, 41(28), 12146–12156.

Schaefer, S., Jeder, A., Sdanghi, G., Gadonneix, P., Abdedayem, A., Izquierdo, M.T., Maranzana, G., Ouederni, A., Celzard, A., Fierro, V. (2020). Oxygen-promoted hydrogen adsorption on activated and hybrid carbon materials. *International Journal of Hydrogen Energy*, 45(55), 30767–30782.

Schlapbach, L. and Züttel, A. (2001). Hydrogen-storage materials for mobile applications. *Nature*, 414, 353–358.

Schlichtenmayer, M. and Hirscher, M. (2012). Nanosponges for hydrogen storage. *Journal of Materials Chemistry*, 22(20), 10134.

Schlichtenmayer, M. and Hirscher, M. (2016). The usable capacity of porous materials for hydrogen storage. *Applied Physics A: Materials Science and Processing*, 122(4), 379.

Sdanghi, G., Canevesi, R.L.S., Celzard, A., Thommes, M., Fierro, V. (2020a). Characterization of carbon materials for hydrogen storage and compression. *Journal of Carbon Research*, 6(3), 46.

Sdanghi, G., Schaefer, S., Maranzana, G., Celzard, A., Fierro, V. (2020b). Application of the modified Dubinin–Astakhov equation for a better understanding of high-pressure hydrogen adsorption on activated carbons. *International Journal of Hydrogen Energy*, 45(48), 25912–25926.

Seaton, N.A., Walton, J.P.R.B., Quirke, N. (1989). A new analysis method for the determination of the pore size distribution of porous carbons from nitrogen adsorption measurements. *Carbon*, 27(6), 853–861.

Sharpe, J.E., Bimbo, N., Ting, V.P., Burrows, A.D., Jiang, D., Mays, T.J. (2013). Super-critical hydrogen adsorption in nanostructured solids with hydrogen density variation in pores. *Adsorption*, 19(2/4), 643–652.

Shivaji Sircar (2019). The genius of Gibbsian surface excess (GSE) framework for fluid (gas or liquid). Solid adsorption: A powerful practical tool. *Separation and Purification Technology*, 213, 235–245.

Siegel, D., Hardy, B., Team HSECoE (2015). Engineering an adsorbent-based hydrogen storage system: What have we learned? Report, DOE Materials-Based Hydrogen Storage Summit, NREL.

Silvera, I.F. (1980). The solid molecular hydrogens in the condensed phase: Fundamentals and static properties. *Reviews of Modern Physics*, 52(2), 393–452.

Singh, S.B. and De, M. (2021). Effects of gaseous environments on physicochemical properties of thermally exfoliated graphene oxides for hydrogen storage: A comparative study. *Journal of Porous Materials*, 28(3), 875–888.

Tarazona, P. (1985). Free-energy density functional for hard spheres. *Physical Review A*, 31(4), 2672–2679.

Tarazona, P. (1995). Solid-fluid transition and interfaces with density functional approaches. *Surface Science*, 331(333), 989–994.

Thommes, M., Cychosz, K.A., Neimark, A.V. (eds) (2012). Advanced physical adsorption characterization of nanoporous carbons. In *Novel Carbon Adsorbents*. Elsevier, Amsterdam.

Thommes, M., Kaneko, K., Neimark, A.V., Olivier, J.P., Rodriguez-Reinoso, F., Rouquerol, J., Sing, K.S.W. (2015). Physisorption of gases, with special reference to the evaluation of surface area and pore size distribution (IUPAC Technical Report). *Pure and Applied Chemistry*, 87(9/10), 1051–1069.

Ting, V.P., Ramirez-Cuesta, A.J., Bimbo, N., Sharpe, J.E., Noguera-Diaz, A., Presser, V., Rudic, S., Mays, T.J. (2015). Direct evidence for solid-like hydrogen in a nanoporous carbon hydrogen storage material at supercritical temperatures. *ACS Nano*, 9(8), 8249–8254.

Wang, L. and Yang, R.T. (2008). Hydrogen storage properties of carbons doped with ruthenium, platinum, and nickel nanoparticles. *The Journal of Physical Chemistry C*, 112(32), 12486–12494.

Yang, R.T. (2000). Hydrogen storage by alkali-doped carbon nanotubes–revisited. *Carbon*, 38(4), 623–626.

Yang, S.J., Cho, J.H., Nahm, K.S., Park, C.R. (2010). Enhanced hydrogen storage capacity of Pt-loaded CNT@MOF-5 hybrid composites. *International Journal of Hydrogen Energy*, 35(23), 13062–13067.

Yang, S.J., Kim, T., Im, J.H., Kim, Y.S., Lee, K., Jung, H., Park, C.R. (2012). MOF-derived hierarchically porous carbon with exceptional porosity and hydrogen storage capacity. *Chemistry of Materials*, 24(3), 464–470.

Yu, Z., Deschamps, J., Hamon, L., Karikkethu Prabhakaran, P., Pré, P. (2017). Hydrogen adsorption and kinetics in MIL-101(Cr) and hybrid activated carbon-MIL-101(Cr) materials. *International Journal of Hydrogen Energy*, 42(12), 8021–8031.

Zhao, W., Fierro, V., Zlotea, C., Aylon, E., Izquierdo, M.T., Latroche, M., Celzard, A. (2011a). Activated carbons with appropriate micropore size distribution for hydrogen adsorption. *International Journal of Hydrogen Energy*, 36(9), 5431–5434.

Zhao, W., Fierro, V., Zlotea, C., Aylon, E., Izquierdo, M.T., Latroche, M., Celzard, A. (2011b). Optimization of activated carbons for hydrogen storage. *International Journal of Hydrogen Energy*, 36(18), 11746–11751.

Zhao, W., Luo, L., Chen, T., Li, Z., Zhang, Z., Wang, H., Rao, J., Feo, L., Fan, M. (2019). Synthesis and characterization of Pt-N-doped activated biocarbon composites for hydrogen storage. *Composites Part B: Engineering*, 161, 464–472.

Zhou, Y. and Zhou, L. (2009). Fundamentals of high pressure adsorption. *Langmuir*, 25(23), 13461–13466.

List of Authors

Hugo BENET
ICMPE, CNRS
UPEC
Thiais
France

Nicolas BION
IC2MP, CNRS
Université de Poitiers
France

Jean-Louis BOBET
ICMCB, CNRS
Université de Bordeaux
Pessac
France

Fabien CAN
IC2MP, CNRS
Université de Poitiers
France

Carlos A. CASTILLA-MARTINEZ
IEM, CNRS
Université de Montpellier
France

Alain CELZARD
Institut Jean Lamour, CNRS
Université de Lorraine
Épinal
France

David CHAPELLE
Institut FEMTO-ST, CNRS
Université de Franche-Comté
Besançon
France

Xavier COURTOIS
IC2MP, CNRS
Université de Poitiers
France

Charlotte CROISÉ
IC2MP, CNRS
Université de Poitiers
and
ICMCB, CNRS
Université de Bordeaux
Pessac
France

Fermin CUEVAS
ICMPE, CNRS
UPEC
Thiais
France
and
CNRS
Beijing
China

Umit B. DEMIRCI
IEM, CNRS
Université de Montpellier
France

Frédéric-Victor DONZÉ
ISTerre, CNRS
Université Grenoble Alpes
France

Ghofrane FEDLOUK
ICMPE, CNRS
UPEC
Thiais
France

Vanessa FIERRO
Institut Jean Lamour, CNRS
Université de Lorraine
Épinal
France

Damien HALM
Institut Pprime, CNRS
ISAE-ENSMA
Poitiers
France

Xiaolong JI
IRCELYON, CNRS
Université Claude Bernard Lyon 1
Villeurbanne
France

Mohamed El Amine KRIBÉCHE
IEM, CNRS
Université de Montpellier
and
ETE France
Saint-Laurent-de-la-Salanque
France

Kylia MARCUS
Institut Néel, CNRS
Université Grenoble Alpes
France

Valérie MEILLE
IRCELYON, CNRS
Université Claude Bernard Lyon 1
Villeurbanne
France

Judith MONNIER
ICMPE, CNRS
UPEC
Thiais
France

Rafael MORALES-OSPINO
Institut Jean Lamour, CNRS
Université de Lorraine
Épinal
France

Valérie PAUL-BONCOUR
ICMPE, CNRS
UPEC
Thiais
France

Nayely PINEDA ROMERO
ICMPE, CNRS
UPEC
Thiais
France

Catherine PINEL
IRCELYON, CNRS
Université Claude Bernard Lyon 1
Villeurbanne
France

Patricia DE RANGO
Institut Néel, CNRS
Université Grenoble Alpes
France

Laurent TRUCHE
ISTerre, CNRS
Université Grenoble Alpes
France

Stéphane VILLALONGA
CEA Le Ripault
Monts
France

Junxian ZHANG
ICMPE, CNRS
UPEC
Thiais
France

Claudia ZLOTEA
ICMPE, CNRS
UPEC
Thiais
France

Index